EVOLUTIONARY COMPUTATION IN GENE REGULATORY NETWORK RESEARCH

Wiley Series on

Bioinformatics: Computational Techniques and Engineering

A complete list of the titles in this series appears at the end of this volume.

EVOLUTIONARY COMPUTATION IN GENE REGULATORY NETWORK RESEARCH

Edited by

Hitoshi Iba
The University of Tokyo
Bunkyo, Tokyo, Japan

Nasimul Noman
The University of Newcastle
New South Wales, Australia

WILEY

Published by John Wiley & Sons, Inc., Hoboken, New Jersey
Published simultaneously in Canada

Library of Congress Cataloging-in-Publication Data applied for.

ISBN: 9781118911518

Printed in the United States of America

10 9 8 7 6 5 4 3 2 1

CONTENTS

PREFACE

Since the identification of regulatory sequences associated with genes in the 1960s, the research in the field of gene regulatory network (GRN) is ever increasing—not only for understanding the dynamics of these complex systems but also for uncovering how they control the development, behavior, and fate of biological organisms. Dramatic progress is being made in understanding gene networks of organisms, thanks to the recent revival of evolutionary developmental biology (evo-devo). For example, there have been many startling discoveries regarding the Hox genes (master control genes that define segment structures in most metazoa). At the same time, neuroscientists and evolutionary biologists think that the modularity of gene networks (combination of functionally related structures and separation of unrelated structures) is crucial to the development of complex structures.

Gene control network, which is a representative concept in the evo-devo approach, is considered to be the central process that achieves the functionality of a molecular machine (flow of DNA-RNA-protein-metabolite) and models interactions between genes. Therefore, analysis of gene networks may provide insights into the fundamental mechanisms of life phenomena. These include robustness and possibility of evolution—two mechanisms have been observed at various levels of organisms, from gene control to fitness value of an organism. Stuart Kauffman used the random Boolean graph model to experimentally prove that gene networks in a certain critical condition can be simultaneously robust and capable of evolution under genetic changes. Besides, today it is also believed, based on experimental evidence, that the understanding and control of tumor like complex disease is deep-rooted in completing the GRN wiring diagrams.

As we enter the era of synthetic biology, the research interest and emphasis in GRN research have received a new thrust. After establishing the promise and prospect of this field through the construction of synthetic circuits like oscillators and counters, synthetic biologists now aspire to design complex artificial gene networks that are capable of sensing and adjusting metabolite activities in cells and use those circuits for therapeutic purpose. However, with the growth in size and complexity of the circuit, the experimental construction becomes infeasible and assistance from effective and efficient computational approaches becomes essential.

Because of their enormous capability of generating complex behavior, GRNs are now used for modeling different computational and engineering problems beyond biological realm. Very recently, some fascinating applications of GRN have been used in different fields that ranges from agent control to design. These applications harness the power of knowledge encoding in GRN and the ability of creating complex systems through computer simulations.

All of the research activities related to GRN, whether those are focused on understanding the mechanism of evolution, on uncovering the development of a fatal disease, or on forming an adaptive pattern in swarm robots for monitoring purpose, involve computational approaches. Consequently, the latest development in artificial intelligence and machine learning has been widely applied in the research related to GRN over the last decades. Perhaps evolutionary algorithms and other nature-inspired algorithms (commonly called evolutionary computation (EC)) are the most broadly practiced computational approach, next to machine learning, in this research domain. EC is a branch of optimization that is useful when we do not have enough information regarding the system for which the optimum solution is sought. They are also useful when the problem is non-convex, non-linear, and non-smooth, which makes most techniques incapable of finding the global minimum. Furthermore, EC is also handy when the function to optimize is noisy and irregular, which also dampens the performance of most classic optimization schemes. Since all these characteristics apply in case of GRN analysis and inference, EC has become a very useful methodology and a robust and reliable tool in this research paradigm. Consequently, EC has been used extensively for analysis, reverse-engineering, and automatic construction of GRN both for systems and synthetic biology, thus creating an independent research domain of its own.

The purpose of this book is to create a guidebook for this research field that will be useful for the audience of both background—computer science and biology. This title presents a handbook for research on GRN using EC that contains a compilation of introductory materials for the novice researcher, highlights of the recent progress in this research field for the current practitioner, and guidelines to new prospects and future trends of the field for the advanced researcher. Keeping in mind the diverse backgrounds of the researchers in this interdisciplinary field, this book delivers materials in a way equally attractive for a reader with training in computation or biology.

This book delivers a step-by-step guideline for research in gene regulatory networks using evolutionary computation. Keeping in mind the various applications of EC in GRN research and for addressing the needs of readers from diverse research backgrounds, the book is organized into four parts. Each of these sections, authored by well-known researchers and experienced practitioners, delivers the relevant materials for the interested readers.

The first part gives an introductory background to the field. Taking into account that prospective readers come with either of the two major backgrounds, this introductory material is divided into three chapters providing necessary training on EC for biologists, introducing the relevant concepts and notions of gene regulatory networks for computer scientists, and familiarizing the data sources and analysis methods for GRN research, respectively. Nevertheless, the material presented in this section can be used as a reference by the regular practitioners of the field.

The second part of the book presents the EC approaches for analysis and reconstruction of GRN from gene expression data. The first chapter in this section presents EC as an effective method for information extraction from gene expression data techniques using bi-cluster analysis. Inference of GRN from gene expression data is the sub-field that has seen the most number of applications of EC. Researchers have used different types of models, data, and different classes of EC for reverse engineering GRNs. The other four chapters in this part cover the most recent and advanced usages of EC for reconstruction of GRN from expression profiles using different models and algorithms.

The second largest application of EC in GRN research is the automatic construction of gene regulatory and reaction networks. This field has become particularly attractive for the synthetic biologists to relieve them from the painstaking trial-and-error methods of gene circuit construction. The third part of the book comprises three chapters that covers the contemporary advancements in this topic and gives direction and guideline for future research.

Finally, the last part of this book focuses on applications of GRNs with EC in other fields. We have seen some compelling applications of GRN with EC for constructing complex system or behavior in diverse fields such as art, design, and engineering. These applications have shown promising signs for a new research philosophy and methodology worth further investigation and exploration. Carefully chosen such advanced and cutting edge research topics that have attracted much attention have been organized in four chapters in the last part of the book.

It has been more than 15 years since GRN research started using EC as a useful and effective computational approach. Researchers have used various classes of EC that showed promising results under different topics of the broader research field. Today, EC is an established and effective research methodology in GRN research. In order to sustain and promote research in this active field, some handbook that covers the prospects and challenges of the field is necessary. It is the editors' expectation that this edited title that brings together the background, current status, and future developments of this field will serve this purpose.

HITOSHI IBA AND NASIMUL NOMAN
October 31, 2015

ACKNOWLEDGMENTS

We express our deep sense of gratitude toward all the contributors of this book who enabled us to produce a high-quality work that delivers the state of the art in gene regulatory network research utilizing evolutionary computation. We thank Brett Kurzman, Kathleen Pagliaro, and all of the staffs at Wiley Publishers for their work in producing this book. Last but not the least, we thank our families, friends, and colleagues for their constant support, helpful suggestions, and much-needed encouragement during the production of this book.

CONTRIBUTORS

Tatsuya Akutsu, Bioinformatics Center, Institute for Chemical Research, Kyoto University, Uji, Japan

Martyn Amos, School of Computing, Mathematics and Digital Technology, Manchester Metropolitan University, Manchester, United Kingdom

Nathanael Aubert, Department of Information Science, Ochanomizu University, Bunkyo, Tokyo, Japan

Daniela Besozzi, Dipartimento di Informatica, Università degli Studi di Milano, Milano, Italy
SYSBIO Centre for Systems Biology, Milano, Italy

Paolo Cazzaniga, Dipartimento di Scienze Umane e Sociali, Università degli Studi di Bergamo, Bergamo, Italy
SYSBIO Centre for Systems Biology, Milano, Italy

Madhu Chetty, Faculty of Science and Technology, Federation University Australia, Australia

Ahsan Raja Chowdhury, Faculty of Information Technology, Monash University, Australia

Davide Cipolla, Dipartimento di Informatica, Sistemistica e Comunicazione, Università degli Studi di Milano-Bicocca, Milano, Italy

Sylvain Cussat-Blanc, University of Toulouse – IRIT – CNRS UMR5505, Toulouse, France

Quang Huy Dinh, Department of Information and Communication Engineering, Graduate School of Information Science and Technology, The University of Tokyo, Bunkyo, Tokyo, Japan

Jean Disset, University of Toulouse – IRIT – CNRS UMR5505, Toulouse, France

Yves Duthen, University of Toulouse – IRIT – CNRS UMR5505, Toulouse, France

Masami Hagiya, Department of Computer Science, Graduate School of Information Science and Technology, The University of Tokyo, Tokoyo, Japan

David Holloway, Mathematics Department, British Columbia Institute of Technology, Burnaby, B.C., Canada

Hitoshi Iba, Department of Information and Communication Engineering, Graduate School of Information Science and Technology, The University of Tokyo, Bunkyo, Tokyo, Japan

Yaochu Jin, Department of Computing, University of Surrey, Guildford, UK

Ibuki Kawamata, Department of Bioengineering and Robotics, Graduate School of Engineering, Tohoku University, Miyagi, Japan

Shuhei Kimura, Graduate School of Engineering, Tottori University, Tottori, Japan

Emma Laing, Department of Microbial and Cellular Sciences, University of Surrey, Guildford, UK

Alan Wee-Chung Liew, School of Information and Communication Technology, Griffith University, Queensland, Australia

Michael A. Lones, School of Mathematical and Computer Sciences, Heriot-Watt University, Edinburgh, Scotland, UK

Marco S. Nobile, Dipartimento di Informatica, Sistemistica e Comunicazione, Università degli Studi di Milano-Bicocca, Milano, Italy
SYSBIO Centre for Systems Biology, Milano, Italy

Nasimul Noman, School of Electrical Engineering and Computer Science, Faculty of Engineering and Built Environment, The University of Newcastle, New South Wales, Australia
and
The Priority Research Centre for Bioinformatics, Biomarker Discovery and Information-Based Medicine, The University of Newcastle, New South Wales, Australia

Hyondong Oh, Loughborough University, Loughborough, UK

Daniel Richards, School of Computing, Mathematics and Digital Technology, Manchester Metropolitan University, Manchester, United Kingdom

Yannic Rondelez, LIMMS/CNRSIIS, Institute of Industrial Science, The University of Tokyo, Meguro, Tokyo, Japan

Stéphane Sanchez, University of Toulouse – IRIT – CNRS UMR5505, Toulouse, France

Colin Smith, Department of Microbial and Cellular Sciences, University of Surrey, Guildford, UK

Alexander Spirov, Computer Science and CEWIT, SUNY Stony Brook, Stony Brook, NY, USA; and the Sechenov Institute of Evolutionary Physiology and Biochemistry, St.-Petersburg, Russia

Spencer Angus Thomas, Department of Computing, University of Surrey, Guildford, UK

Yuji Zhang, Department of Epidemiology and Public Health, University of Maryland School of Medicine, Baltimore, MD, USA
and
Division of Biostatistics and Bioinformatics, University of Maryland Greenebaum Cancer Center, Baltimore, MD, USA

I

PRELIMINARIES

1

A BRIEF INTRODUCTION TO EVOLUTIONARY AND OTHER NATURE-INSPIRED ALGORITHMS

Nasimul Noman

School of Electrical Engineering and Computer Science, Faculty of Engineering and Built Environment, The University of Newcastle, New South Wales, Australia
and
The Priority Research Centre for Bioinformatics, Biomarker Discovery and Information-Based Medicine, The University of Newcastle, New South Wales, Australia

Hitoshi Iba

Department of Information and Communication Engineering, Graduate School of Information Science and Technology, The University of Tokyo, Bunkyo, Tokyo, Japan

1.1 INTRODUCTION

When we look at nature, everything seems to be working very systematically. All natural phenomena, ranging from molecular level to ecological level, and from individual level to population level, are functioning effectively. The flawless operation of various natural systems becomes possible due to some underlying governing rules.

From the beginning of human history, people have borrowed ideas and mimicked different natural processes in solving their daily-life problems. With the progress of

Evolutionary Computation in Gene Regulatory Network Research, First Edition.
Edited by Hitoshi Iba and Nasimul Noman.

civilization, we started to analyze and understand the basic laws and fundamental mechanisms behind natural phenomena and imitate those in designing artificial systems. With the beginning of information era, researchers started to investigate these natural processes from the perspective of information processing. We started to mimic how information is stored, processed, and transferred in natural systems in developing new techniques for solving complex problems. Today, a broad field of research is involved in the design, development, and study of intelligent computational systems that are inspired by the mechanisms and principles (often highly simplified versions of those) observed in various natural processes.

Perhaps, the largest natural information processing system that we have studied most widely and understand reasonably is evolution. Evolution refers to the scientific theory that explains how biological hierarchy of DNA, cells, individuals, and populations slowly change over time and give rise to the fantastic diversity that we see around us. Through the evolutionary process, the changes taking place in an organism's genotypes give rise to optimized phenotypic behaviors. Therefore, evolution can be considered as a process capable of finding optimized, albeit not optimal, solutions for problems.

Evolutionary computation (EC) is a branch of computer science, dedicated to the study and development of search and optimization techniques which draw inspiration from Darwinian theory of evolution and molecular genetics. The incremental growth of the field resulted in algorithms with different flavors although all of them utilize the *in silico* simulation of natural evolution. Classically, the most prominent types of evolutionary computation are genetic algorithms (GA), genetic programming (GP), Evolutionary Strategy (ES) and Evolutionary Programming (EP). Although, at the beginning, each class of algorithms had their distinct characteristics, lately, because of hybridization and concept borrowing, it is difficult to categorize some new algorithms as a specific class of EC.

After natural evolution, the artificial intelligence community has been heavily influenced by the social behavior emerged, through information processing and sharing, among relatively simpler life forms. Social insects like ants, termites and bees exhibit remarkable intelligence in improving their way of life, for example, retrieval of food, reducing the threat of predator, division of labour, or nest building. They possess impressive problem-solving capabilities through collaboration and cooperation among fellow members which themselves have very limited intelligence. Many computational algorithms and problem-solving techniques, commonly known as swarm intelligence, have been developed by simulating the coordination and teamwork strategies in social insects.

Other than evolutionary computation and swarm intelligence, many other computational algorithms have been proposed which are inspired by different natural phenomenon such as immune systems of vertebrate, biological nervous systems, chemical systems, or the behavior of different animals such as bat, firefly, and cuckoo. There exist a lot of variation and differences among these algorithms in terms of problem representation and solution searching mechanism; however, the common connection among them is that all of these algorithms extract metaphor and inspiration from nature. These classes of algorithms are commonly known as nature-inspired algorithms or bio-inspired algorithms. In this book, we will mostly focus on evolutionary computation and a few other swarm and nature-inspired algorithms; therefore, we will commonly refer to them as evolutionary computation.

Because of their robust and reliable search performance, these algorithms are preferred for solving many complex problems where traditional computational approaches are found to be inadequate. Gene regulatory networks (GRNs) are complex, nonlinear systems with incomplete understanding of their underlying mechanism at molecular level. Consequently, evolutionary and other nature-inspired algorithms are preferred as the computational approach in different research in GRN which is the topic of this book. Therefore, in this first chapter, we present a gentle introduction of evolutionary and other nature-inspired computation so that the readers can have a better understanding of the more advanced versions of these algorithms presented in subsequent chapters. After the generalized introduction, we also discuss relative advantages/disadvantages and application areas of these algorithms.

1.2 CLASSES OF EVOLUTIONARY COMPUTATION

1.2.1 Genetic Algorithms

Genetic algorithms, which are typical examples of evolutionary computation, have the following characteristics:

- Work with a population of solutions in parallel
- Express candidate solutions to a problem as a string of characters
- Use mutation and crossover to generate next-generation solutions

Elements comprising GAs are data representation (genotypes or phenotypes), selection, crossover, mutation, and alternation of generations. How to implement these elements is a significant issue that determines the search performance. Each element is explained below.

1.2.1.1 Data Representation Data structures in GAs are genotypes (GTYPE) and phenotypes (PTYPE). GTYPE corresponds to genes of organisms, and indicates strings expressing candidate solutions (bit strings with fixed length). Genetic operators, such as crossover and mutation which are discussed later, operate on GTYPE. The implementer can determine how to convert candidate solutions to strings. For instance, GTYPE may be a candidate solution converted into an array of concatenated integers.

On the other hand, PTYPE corresponds to individual organisms, and indicates candidate solutions to a problem based on interpretation of GTYPE. The fitness value that indicates the quality of a candidate solution is calculated using PTYPE.

1.2.1.2 Selection In GAs, individuals that adapt better to the environment leave many children and others are eliminated in line with Darwinian evolution theory.

Individuals that adapt to the environment are candidate solutions that score highly regarding the problem, and the fitness function determines the score. Various methods of selecting parent individuals that generate children comprising the next generation have been proposed. Among these, the roulette selection (each individual generates children with a probability proportional to its fitness value) and the tournament selection (a number of individuals are selected at random and the best individual is chosen as the parent, and this procedure is repeated as necessary) are frequently used.

The elite strategy (best individual always remains in the next generation) is often used in addition to these selection methods. This strategy does not reduce the fitness value of the best individual in subsequent generations (as long as the environment to be evaluated does not change). However, using the elite strategy too much in the initial stages of a search may lead to premature convergence, which means convergence to a local solution.

1.2.1.3 Crossover Crossover is an analogy of sexual reproduction, and is an operation that mates two parent individuals to generate new children. There are a number of crossover methods with different granularity when splitting individuals; examples are one-point crossover and uniform crossover.

One-point crossover selects a crossover point at random and switches parts of two parent individuals at this crossover point for generating children. Figure 1.1 is an example of a one-point crossover. The point between bits 3 and 4 is chosen as the crossover point, and two children are generated. Two-point crossover, where two crossover points are chosen and two switches are made, and multiple-point crossover with three or more crossover points are also possible.

Uniform crossover is the most refined crossover method where the parent value to inherit is determined for each bit. Hitchhiking is a problematic phenomenon regarding

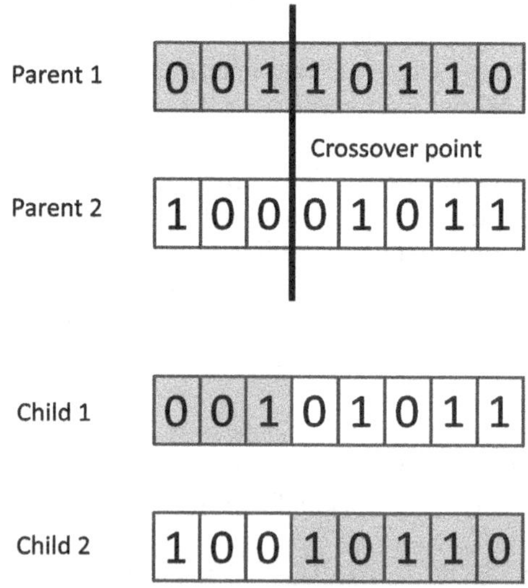

Figure 1.1 One-point crossover in a genetic algorithm.

Figure 1.2 Mutation in a genetic algorithm.

crossover in GAs, in which unnecessary bits existing around a good partial solution spread as parasites to the good partial solution regardless of whether the fitness value is good or not. In general, uniform crossover is considered to suppress hitchhiking.

1.2.1.4 Mutation Mutation corresponds to errors in gene reproduction in nature. In GAs, this operation changes one character in an individual after crossover (in a bit sequence, switches between 0 and 1). Figure 1.2 is one example. Crossover can, in principle, only search for combinations of existing solutions. Therefore, mutation is expected to increase the diversity of the population and broaden the search space by breaking part of a genotype. The reciprocal of the GTYPE length is often used as the mutation rate, which means that on average there is one mutation per genotype. Increasing the mutation rate diversifies the population, but the tradeoff is that there is a higher probability of destroying good partial solutions.

1.2.1.5 Algorithm Flow Summarizing the above, the flow in a GA is as follows.

1. Randomly generate strings (GTYPE) of the initial population.
2. Convert GTYPE to PTYPE and calculate the fitness value for all individuals.
3. Select parents based on the selection method.
4. Generate individuals of the next generation (children) using genetic operators.
5. Check termination conditions; return to 2 if termination conditions are not met.

Generation alternation is a procedure where children generated by operations such as selection, crossover, and mutation replace parent individuals to create the population of the next generation. Typical termination conditions are discovery of an individual with sufficient fitness value or iterating the algorithm for a predetermined number of generations. Instead, one may continue calculations for as long as possible while calculation resources exist, and finish when sufficient convergence is achieved or further improvement of the fitness value is not expected.

1.2.1.6 Extension of GA GTYPE has been explained as a string of fixed length, but improved GAs without this restriction have been proposed. Examples are real coded GA (a vector of real numbers is used as the genotype, see Section 1.2.1.7) and MessyGA where variable length strings are supported by pairing the position in the gene and its value. Genetic programming supporting tree structures, which is explained in the next section, is one example of a variable length GA. Interactive genetic calculation (the user provides the fitness value to simulate breeding, which can be used when applying GAs to fields such as design and art where an objective function cannot be explicitly described) and multi-objective optimization (multiple objective functions are optimized simultaneously; see Section 1.2.6) have also been proposed and are known to be very effective when designing desirable targets.

1.2.1.7 Real Coded GA Function optimization, where a function is optimized in a continuous search space, is one of the important problems that frequently show up in real-world problems. Research on evolutionary computations for function optimization has a long history. Proposed methods are the bit-string GA where gene expressions are binary code or gray code, real coded GA where vectors of real numbers are used as gene expressions, evolution strategy (ES, see Section 1.2.3), differential evolution (DE, see Section 1.2.4), and meta evolutionary programming (meta-EP). This section describes crossover methods and generation alternation models for real coded GA that show good performance among evolutionary computation methods for function optimization.

Function optimization is a problem to find a set of $(x_1, \cdots, x_n)$ that minimizes or maximizes a function $f(x_1, \cdots, x_n)$ consisting of n continuous variables. Intuitively, this is a problem to find the highest or lowest point of the target function. Minimization problems are considered hereafter as these do not lose generality. A unimodal function has only one local solution that is also the global optimum solution in the search space, whereas a multimodal function has many local solutions. Generally speaking, multimodal functions are more difficult to optimize. When considering a function geometrically, there is "dependence between variables" if there are valleys that are not parallel to the coordinate axis, which means that multiple variables must be changed appropriately at the same time to improve the value of the function. Optimization is usually more difficult if the function has dependence between variables.

Design of methods to generate children, such as crossover and mutation, is the key to good performance when applying evolutionary computation methods to optimization problems. Beyer et al. [3] and Kita et al. [18] proposed guidelines for methods to generate children. Beyer et al.'s design guidelines consider dynamic environments where the form of the function changes with time; however, dependencies between variables are not taken into account. On the other hand, Kita et al.'s guidelines assume a static environment and can reflect dependencies between variables. The crossover design guidelines for real coded GAs by Kita et al. are described below.

Design guideline 1 (Inheritance of statistics): The distribution of children generated by crossover should inherit the average vector and the variance-covariance matrix of the parent distribution. In particular, inheritance of covariance is important in optimizing non-separable functions that have strong dependencies between variables. This means that children generated by crossover should have a similar distribution to that of parents.

Design guideline 2 (Generation of diverse solutions): The crossover procedure should be able to generate a population as diverse as possible within the constraint of "inheritance of statistics."

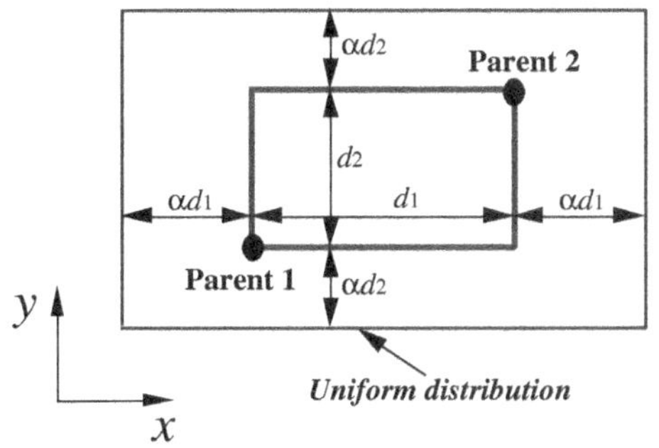

Figure 1.3 Schematic of BLX-α.

Design guideline 3 (Guarantee of robustness): To make the search more robust, the distribution of children should be slightly broader than one that satisfies the design guidelines.

Proposed crossover methods for real coded GA include the blend crossover (BLX-α) by Eshelman et al. [8] and unimodal normal distribution crossover (UNDX) by Ono et al. [21]. BLX-α generates children over a uniform distribution within a hyper-rectangle where each edge determined by parents is parallel to the coordinate axes (Figure 1.3). The algorithm of BLX-α is as follows.

1. Take two parent individuals x^1 and x^2.
2. Each component x_i^c of a child individual x^c is determined independently of each other using a uniform random number within the interval $[X_i^1, X_i^2]$. Here,

$$X_i^1 = \min(x_i^1, x_i^2) - \alpha d_i$$
$$X_i^2 = \max(x_i^1, x_i^2) + \alpha d_i$$
$$d_i = |x_i^1 - x_i^2|$$

   Where x_i^1 and x_i^2 are the i-th component of x^1 and x^2, respectively, and α is a parameter.

On the other hand, UNDX generates children on or near a line connecting two parents using a normal distribution determined by these parents and a third parent. The UNDX algorithm is as follows.

1. Select three parents x^1, x^2, and x^3.
2. Find the center of parents x^1 and x^2, that is, $x^p = (x^1 + x^2)/2$.
3. Define the difference vector of parents x^1 and x^2 as $d = x^1 - x^2$.
4. The primary search line is defined as the line connecting parents x^1 and x^2, and the distance between parent x^3 and the primary search line is denoted as D.

5. Child x^c is generated using the formula

$$\begin{aligned} x^c &= x^p + \xi d + \sum_{i=1}^{n-1} \eta_i D e_i, \\ \xi &\sim N(0, \sigma_\xi^2), \quad \eta_i \sim N(0, \sigma_\eta^2). \end{aligned} \tag{1.1}$$

   Here, n is the dimension of the search space, $N(0, \sigma^2)$ is a normal distribution with average 0 and variance σ^2, and e_i's are orthonormal basis vectors of the subspace normal to the primary search line.

System parameters of each crossover method can be determined to satisfy the above design guideline 1 (inheritance of statistics).

Crossover methods for real coded GAs can be combined with various selection methods. Generation alternation models for a single objective optimization using a single evaluation function include simple GA (SGA) by Goldberg [10], iterated genetic search (IGS) by Ackley [10], steady state (SS) by Syswerda [31] and elitist recombination (ER) by Thierens et al. [32]. Many engineering problems are formulated as multi-objective optimization problems that explicitly handle various evaluation functions in tradeoff relations (see Section 1.2.6). Combination with a generation alternation model that retains a high level of diversity is desirable for maximum crossover performance in real coded GAs for both single objective and multi-objective optimization.

Finally, evolution strategy, which is closely related to real coded GAs in the sense that real number vectors are used as gene expressions, is discussed. ES uses mutation as the main search operator in contrast to real coded GAs that instead use crossover. ES generates children based on a normal distribution around parent individuals, which is similar to some real coded GAs such as UNDX, UNDX-*m*, and extended normal distribution crossover (ENDX). However, ES codes evolution parameters, such as the standard deviation of the normal distribution, into the individual along with the decision variable to be optimized. The region where children are generated is adaptively derived through adaptation of the parameters by mutation. Correlated mutation proposed by Schwefel [27] uses a similar mechanism that considers dependencies between variables and tilts the axis of the normal distribution relative to the coordinate axes.

1.2.2 Genetic Programming

Genetic programming is an evolutionary computation method applicable to many problems and uses tree structures as the genotype. Programming languages such as LISP, relations between concepts, and many knowledge representations including mathematical expressions can be described using tree structures. As a result, GP can be used to apply evolutionary approaches to automatic code generation and problem solving by artificial intelligence. The basic idea of GP was originally proposed by John Koza et al. [19]. The main difference between GP and GAs is the expression of

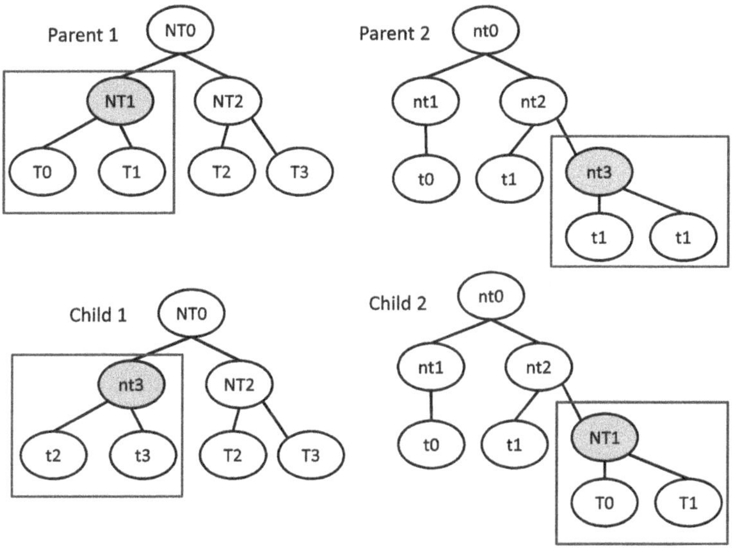

Figure 1.4 Crossover in genetic programming.

GTYPE and operator implementation; selection methods and generation alternation is the same. Data representation and genetic operators unique to GP are described below.

1.2.2.1 Data Representation GP generally expresses GTYPE, which are candidate solutions to a problem, as tree structures. Each node can be categorized into terminal symbols without arguments (corresponding to constants and variables) and nonterminal symbols with arguments (corresponding to functions). Design of GTYPE is carried out by defining usable symbols. As in GAs, the fitness value is obtained by converting each individual into PTYPE (for instance, the results after running code or the evaluated value of a mathematical expression).

1.2.2.2 Crossover Crossover in GP exchanges partial trees between two individuals. The node that would be the crossover point is selected at random in each individual, and partial trees beyond that node are exchanged to generate child individuals. Figure 1.4 is an example of a crossover. NT1 of parent 1 and nt3 of parent 2 are selected as crossover points, and children 1 and 2 are generated by exchanging partial trees beyond these points. However, repeating such simple crossover can lead to unnecessary expansion of tree size as the number of generations increases. This phenomenon is called "bloat" or "fluff," which means to become "structurally complex." The bloat is one factor that inhibits effective search using GP (see Section 1.2.2.4 for details).

1.2.2.3 Mutation Mutation in GP corresponds to replacement of one node by a randomly generated partial tree. Figure 1.5 shows an example of mutation. The

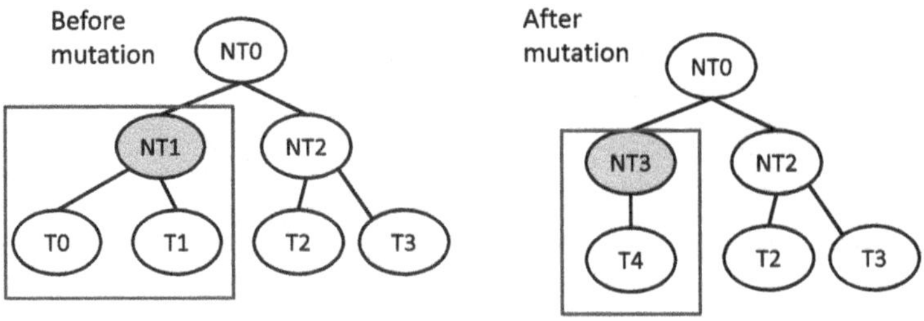

Figure 1.5 Mutation in genetic programming.

effect of mutation in GP is significantly influenced by the node undergoing mutation, thus care is necessary when selecting the node. Examples of mutation are changing a terminal symbol into another terminal symbol, replacing a nonterminal symbol with another nonterminal symbol with the same arguments, changing one nonterminal symbol into a terminal symbol (remove a partial tree), switching nodes in a GTYPE (inversion), and inserting or deleting a terminal symbol.

1.2.2.4 Extension of GP As a method to expand GP, the automatically defined function (ADF) that modularizes and reuses functions to streamline processing has been proposed.

Normal GP can only search combinations of nonterminal and terminal symbols; therefore, the size of GTYPE tends to increase in complex systems (the bloat phenomenon as mentioned above). ADF retains two tree structures per individual, that is, the function definition tree (ADF tree) and the evaluation tree (standard GTYPE). Modularization is achieved by reusing subroutine functions defined in the ADF tree within the evaluation tree. The ADF tree contains dedicated nodes that define functions and arguments, and the evaluation tree takes in functions defined in the ADF tree as nonterminal symbols. Crossover is carried out between ADF trees and between evaluation trees.

Bloat is one of the most persistent issues hindering the efficiency of GP searches. It would cause the following problems:

1. The large programs are difficult for people to understand.
2. The programs require much time and memory space to run.
3. Complicated programs tend to be inflexible and difficult to adapt to general cases, so that they are not very robust.

The following approaches are currently being used to control bloat:

1. Set maxima for tree depth and size. Try to avoid creating tree structures exceeding these upper limits by means of crossover, mutation, etc. This is the easiest way to impose such controls, but the user needs to have a good understanding of the problem at hand, and needs heuristics in order to choose the appropriate settings for maxima.

2. Incorporate program size in the fitness value calculations, that is, penalize large programs for being large. This is called "parsimony." More robust assessment standards using MDL (minimum description length) have been proposed (see Ref. [13] for details).
3. Suppress the tree length by adjustments to genetic operators. For instance, Langdon proposed a homologous crossover or a size-fair crossover to control the tree growth [20]. Other methods to suppress bloat include size-dependent crossover (attempts wherever possible to crossover partial trees of similar size) and depth-dependent crossover (bias crossover such that large partial trees are more likely to be exchanged [15]).

1.2.3 Evolution Strategy

Some research groups in Europe (especially in Germany) have been working on concepts similar to GAs for a long time, that is, "evolution strategy". One leader in ES is Ingo Rechenberg [24]. ES in its early days differed from GAs in the following two ways:

1. Mutation is used as the main operator.
2. Real number expressions are handled.

Individuals in ES are expressed as a pair of real number vectors, $(\vec{x}, \vec{\sigma})$. Here, $\vec{x}$ is a position vector in the search space and $\vec{\sigma}$ is a standard deviation vector. Mutation can be expressed as

$$\vec{x}^{t+1} = \vec{x}^{t} + N(\vec{0}, \vec{\sigma}), \tag{1.2}$$

where $N(\vec{0}, \vec{\sigma})$is a random number from a Gaussian distribution of average $\vec{0}$ and standard deviation $\vec{\sigma}$.

ES in its early days carried out search using a population consisting of one individual. A child ($\vec{x}^{t+1}$ in the above equation) generated by mutation can become a member of the new population (become the parent of the next generation) only when its fitness value is better than that of the parent ($\vec{x}^{t}$).

Quantitative research on ES is more feasible than on GAs because the former is not affected by crossover, and the effect of the mutation rate has been mathematically analyzed. For example, theorems regarding convergence have been proven. In addition, the "$\frac{1}{5}$ rule," that is, "let the probability that a mutation succeeds be $\frac{1}{5}$; if this value is larger (smaller) than $\frac{1}{5}$, increase (reduce) $\vec{\sigma}$." In practice, the probability that a mutation succeeds in the last k generations, $\varphi(k)$, is observed and mutation is controlled such that

$$\vec{\sigma}^{t+1} = \begin{cases} c_d \times \vec{\sigma}^{t}, & \text{if } \varphi(k) < 1/5, \\ c_i \times \vec{\sigma}^{t}, & \text{if } \varphi(k) > 1/5, \\ \vec{\sigma}^{t}, & \text{if } \varphi(k) = 1/5. \end{cases} \tag{1.3}$$

In particular, Schwefel adopted $c_d = 0.82$ and $c_i = 1/0.82$. The intuitive meaning of this rule is: "if successful, continue searching with bigger steps; otherwise, reduce the step size".

ES was later extended to be a search method employing a population of multiple individuals. In addition to the mutation operator mentioned above, the crossover operator and the average operator (an operator that takes the average of two parent vectors) were introduced. Unlike GAs, ES uses the following two selection methods.

1. $(\mu + \lambda) - ES$
   A parent population with μ individuals generates λ children. μ individuals are selected from a total of $(\mu + \lambda)$ individuals to be the parents in the next generation.
2. $(\mu, \lambda) - ES$
   A parent population with μ individuals generates λ children $(\mu < \lambda)$. μ individuals are selected from λ individuals to be the parents in the next generation.

In general, $(\mu, \lambda) - ES$ is considered to perform better in environments that change with time and in problems with noise.

ES has been applied to many optimization problems, and recently is being applied to problems other than real number problems.

1.2.4 Differential Evolution

Differential evolution [30] is one category of evolutionary computation that derives an approximate solution to optimization problems. DE is known to be effective in providing algorithms for various problems such as nonlinear problems, non-differentiable problems, and multimodal problems.

Individuals in DE are real number vectors (points in search space). The flow of this method is outlined below (see Figures 1.6 and 1.7).

Step 1: Input random numbers in each individual (vector) to generate the initial population. Here, the number of elements in the population is N and each individual is denoted as $\vec{x_i}$ $(i = 0, 1, \cdots, N-1)$.

Step 2: Three individuals are randomly chosen from the solution set and are labeled as $\vec{x_{r1}}, \vec{x_{r2}}, \vec{x_{r3}}$ $(r1, r2, r3 \in \{0, 1, \cdots, N-1\}$ and $r1 \neq r2 \neq r3)$. The individual after mutation, $\vec{v_i}$, is generated as

$$\vec{v_i} = \vec{x_{r1}} + F \times (\vec{x_{r2}} - \vec{x_{r3}}) \quad (F \text{ is a constant}). \tag{1.4}$$

This is repeated N times to generate N individuals $\vec{v_0}, \cdots, \vec{v_{N-1}}$.

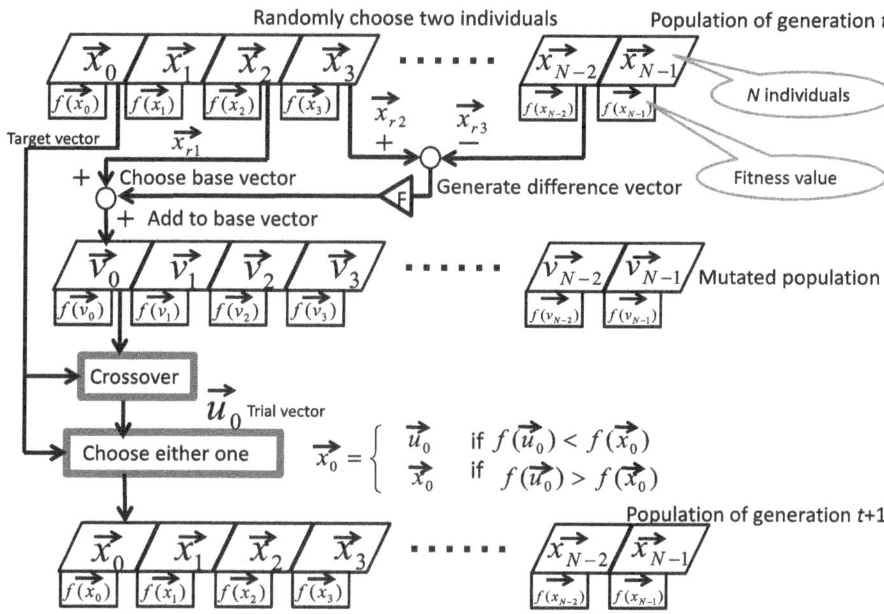

Figure 1.6 Generation alternation in differential evolution (Reprinted with permission from Ref. [23]).

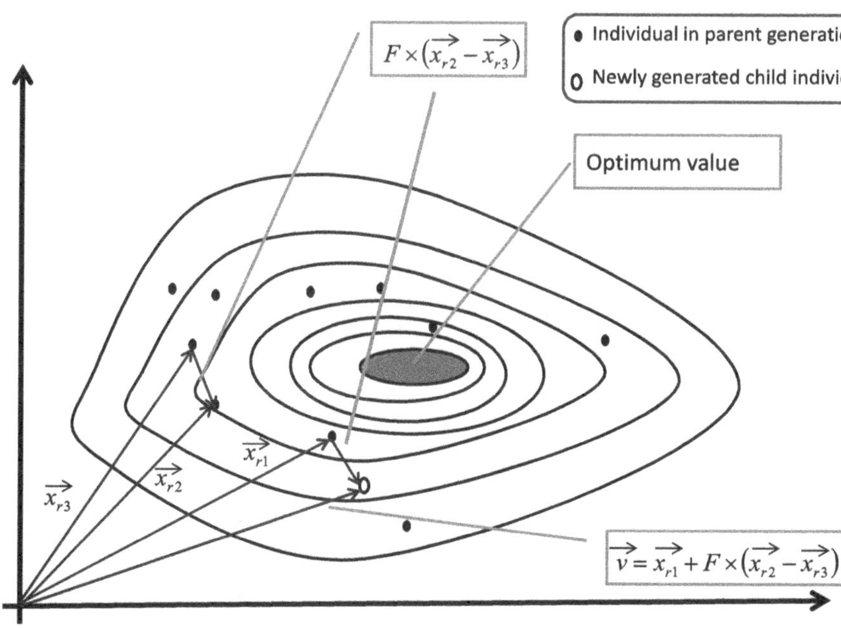

Figure 1.7 Crossover and mutation in differential evolution (Reprinted with permission from Ref. [30]).

Step 3: Generate a child population $\vec{u}_i$ from the parent population $\vec{x}_i$. The elements in $\vec{u}_i$ are selected from elements in $\vec{x}_i$ and $\vec{v}_i$ based on the crossover rate CR.

$$u_{i,j} = \begin{cases} x_{i,j} & \text{if } rand \geq CR, \\ v_{i,j} & \text{if } rand < CR. \end{cases} \tag{1.5}$$

Here, $u_{i,j}, x_{i,j}, v_{i,j}$ are the j-th element of the i-th individual (vector) $\vec{u}_i, \vec{x}_i, \vec{v}_i$, and *rand* is a random number within the interval [0,1]. As a result, the elements in $\vec{u}_i$ contain the elements of both $\vec{x}_i$ and $\vec{v}_i$.

Step 4: Evaluate the child population $\vec{u}_i$ generated in **Step**3 and the parent population $\vec{x}_i$, and decide which solution to adopt.

$$\vec{x}_i = \begin{cases} \vec{x}_i & \text{if } fit(\vec{x}_i) > fit(\vec{u}_i), \\ \vec{u}_i & \text{if } fit(\vec{x}_i) < fit(\vec{u}_i). \end{cases} \tag{1.6}$$

Here, *fit*() is the evaluation function and *fit*(x) is the evaluated value of x.

Step 5: Repeat **Step**2~4 for a fixed number of generations, and output the most valuable individual from the final set of solutions as the optimum solution.

Conventional GAs crossover vectors of two individuals and children obtained by crossover are included in the next generation regardless of their fitness values. Mutation occurs at a fixed parameter value (mutation rate); hence, the amount of mutation does not differ between early generations and later generations near convergence.

In contrast, DE crossovers one individual with (one individual + scaled difference vector of two individuals). Crossover involving a difference vector instead of just position vectors of individuals allows a higher possibility of obtaining children in regions with high fitness value. Faster convergence of the population can be attained because a generated child individual is retained only if it is better than its parent individual. Moreover, mutation in DE is based on the difference vector of individuals; thus the amount of mutation changes depending on the population. As a result, the amount of mutation is large in early generations and becomes smaller in generations near convergence. In other words, evolution progresses effectively and setting of mutation parameters is unnecessary for mutation because the amount of mutation is automatically adjusted.

1.2.5 Swarm Intelligence

Many scientists have tried, using various methods, to reproduce collective behavior in groups of ants, birds, and fish on a computer. Reynolds and Heppner, who have simulated the motion of birds, are well known among such scientists. Reynolds was strongly attracted by the beauty of flocks of birds [25] while Heppner was interested in rules hidden in flocks of birds that instantly gather and scatter. These two researchers had the insight to focus on unpredictable motion of birds. The motion is microscopically very simple, resembling that of cellular automata, but macroscopically is very complex and chaotic. The effect of interactions between individuals has a huge influence in their models as they emphasized the rule that a bird wants to keep an optimum distance between itself and other individuals when considering the overall motion of birds in a flock.

Reynolds' CG animation consists of agents called boids. Each boid determines its motion by combining three vectors, which are (1) the force to move away from the closest neighbor or obstacle, (2) the force to move toward the center of the flock, and (3) the force to move toward the target position. Various patterns of motion can be obtained by adjusting the coefficients used in combining vectors. Complex motion as a whole group emerges when each individual acts based on simple action principles. Technology related to boids is currently widely used for special effects in movies and for animation.

1.2.5.1 Ant Colony Optimization Simple models on the behavior of ants have provided new ideas regarding routing, agents, and distributed control. Applications of ant behavior models have been the focus of many papers and are being established as a research field.

Marching of ants is a cooperative behavior that can be explained by the pheromone trail model. Many cooperative behaviors as a group, such as ant marches, are observed in colonies of ants, and have strongly attracted the interest of entomologists and behavioral scientists. During collecting activities, many types of ants leave a trail of chemical substance when moving from food to their nest, and ants searching for food move along trails that other ants made, if any exists. The chemical substance, which ants generate in their bodies, is called a pheromone.

Ant colony optimization (ACO) is a method that uses the pheromone trail model, for instance, to solve the traveling salesman problem (TSP) [7]. In TSP, there are a number of cities located in different places on a map, and the aim is to look at all of the paths that go through every city exactly once and return to the starting point (called a Hamiltonian cycle or path) and determine the shortest route. There is no efficient algorithm that will solve the traveling salesman problem; in all cases, an exhaustive investigation is required in order to find the optimum solution. Consequently, as the number of cities grows, we see a dramatic leap in the complexity of the problem. This is called a "combinatorial explosion," and is an important issue (an NP-complete problem) in the field of computer science.

ACO optimizes the travel path through the following algorithm:

1. Place ants randomly in each city.
2. Ants move to the next city. The destination is probabilistically determined based on pheromones and given information. Cities already visited are excluded.
3. This procedure is repeated until all cities are visited.
4. Ants completing one loop drop pheromones according to the path length.
5. Return to 1 if a satisfactory solution has not been found.

The length of the path between each city (d_{ij}) and the amount of pheromones on the path are stored in a table, and ants have knowledge about its surroundings. Ants then probabilistically determine the next city to visit. The probability that an ant k at a city i chooses a city j as the next destination, $p_{ij}^k(t)$, is obtained using the reciprocal of the distance $1/d_{ij}$ and the amount of pheromone $\tau_{ij}(t)$ as follows:

$$p_{ij}^k(t) = \frac{[\tau_{ij}(t)] \cdot [d_{ij}]^\alpha}{\sum_{h \in J_i^k} [\tau_{ij}(t)] \cdot [d_{ij}]^\alpha}. \tag{1.7}$$

Here, J_i^k is the set of all cities that ant k can move to from city i. The setting that ants are more likely to select paths with more pheromone reflects positive feedback from past searches and incorporates the heuristic that ants are more likely to select shorter paths. As shown above, information unique to each problem can be adequately reflected in ACO.

The pheromone table is updated using the following two equations. Here, $Q(k)$ is the reciprocal of the length of the loop that ant k found.

$$\Delta\tau_{ij}(t) = \sum_{k \in A_{ij}} Q(k) \tag{1.8}$$

$$\tau_{ij}(t+1) = (1-\rho)\dot{\tau}_{ij}(t) + \Delta\tau_{ij}(t). \tag{1.9}$$

The amount of pheromone to be added to each path is inversely proportional to the length of the loop that an ant found. The score of all ants that passed through a path is reflected in the path. Here, A_{ij} is the set of all ants that passed through the path from city i to city j. Negative feedback to avoid local minima is provided as the pheromone evaporation coefficient. In other words, the pheromone in each path evaporated with a fixed probability (ρ), thereby discarding past information.

The ACO has been applied to, and demonstrated to be effective in combination optimization problems such as the TSP and network routing problems.

1.2.5.2 *Particle Swarm Optimization* Particle swarm optimization (PSO) was introduced by Eberhart and Kennedy in 1995 [17]. The PSO algorithm was inspired by social behavior, and is closely related to code that simulates the collective behavior of birds and fish (for example, of boids by Reynolds). In contrast to GAs that perform genetic operations, PSO decides the next move based on the motion of itself and its neighbors.

The basic PSO proposed by Kennedy et al. consists of many individuals (particles) moving around in a multi-dimensional space and can be applied to real number problems [17]. Each individual remembers its position vector (x_i), velocity vector (v_i), and the position where that individual had its maximum fitness value (p_i). In addition, the position where the group as a whole had its maximum fitness value (p_g) is shared in each individual.

The velocity is updated in each individual based on the best position as a whole and for itself that was found over the generations. The velocity is obtained by

$$v_i = \chi(\omega v_i + \phi_1 \cdot (p_i - x_i) + \phi_2 \cdot (p_g - x_i)).$$

The coefficients used here are the convergence coefficient χ (random number between 0.9 and 1.0) and the decay coefficient ω. In addition, ϕ_1 and ϕ_2 are random numbers equal to or smaller than 2 that are unique to each individual and dimension. The maximum velocity V_{max} is used when the velocity exceeds a given limit. In this way, a search can be performed while keeping individuals in the search space.

The position of each individual is updated in each generation according to the equation

$$x_i = x_i + v_i.$$

Unlike GAs, PSO does not require complex operations such as mutation and crossover, and the structure is very simple. There is theoretical research to derive appropriate values for PSO parameters through mathematical analysis of stability and convergence. PSO is known to give performance comparable to GAs in function optimization properties. Active research is under way for improving the performance of PSO and PSO is being applied to many real-world problems such as power grids and disease diagnosis.

1.2.5.3 Bee Algorithms Bees, together with ants, are well known as social insects. Honey bees can be categorized into three types:

- Employed bees
- Onlooker bees
- Scout bees

Employed bees fly around feeding grounds that they memorize and convey information about food to onlooker bees. Onlooker bees use information from employed bees to selectively find the best food in the feeding ground. When information on a feeding ground becomes too old, employed bees throw away the information, become scout bees, and move to find a new feeding ground. The objective of a beehive is to find the most efficient feeding ground. It is considered that in general about half of the bees in the hive are employed bees, 10–15% are scout bees, and the rest are onlooker bees.

Employed bees waggle-dance (figure-of-eight dance) to convey information to onlooker bees. An employed bee that finds flower nectar or pollen and returns to the nest does a figure-of-eight dance to indicate the direction of the feeding ground to other bees. The direction opposite to gravity corresponds to the direction of the sun and the direction of a straight-line waggle corresponds to the direction of the feeding ground. In other words, bees indicate the angle between the direction of the sun and the direction of the feeding ground to other bees by expressing the angle between the opposite of gravity and the direction of the straight-line waggle. The speed of the waggle represents the distance to the food, and a faster waggle means that the food is nearer. Communication using a similar dance is used to convey the position of a new nest in addition to pollen or the position of water.

Karaboga proposed the artificial bee colony (ABC) optimization algorithm based on the above behavior [16]. The ABC algorithm is a collective search method that mimics food collection by bees. One benefit of the ABC algorithm is the small number of control parameters compared to GAs and the PSO.

The artificial group of bees in the ABC algorithm is separated into employed bees, onlooker bees, and scout bees. N solutions to a problem with d dimensions are generated as feeding grounds. Each employed bee is assigned to a feeding ground $\vec{x}_i$ and finds a new feeding ground $\vec{v}_i$ using the operator

$$v_{ij} = x_{ij} + \text{rand}(-1, 1) \times (x_{ij} - x_{kj}). \tag{1.10}$$

Here, $k \in \{1, 2, \cdots, N\}$, $k \neq i$, and $j \in \{1, 2, \cdots, d\}$ is a randomly chosen index. v_{ij} is the j-th element of vector $\vec{v}_i$. In other words, $\vec{v}_i = (v_{i1}, v_{i2}, v_{i3}, \cdots, v_{id})^{\mathrm{T}}$ and $\vec{x}_i = (x_{i1}, x_{i2}, x_{i3}, \cdots, x_{id})^{\mathrm{T}}$. If the new position is outside the domain, the position is moved to the allowed range. The obtained $\vec{v}_i$ is compared to $\vec{x}_i$, and the better feeding ground is adopted.

In contrast to employed bees, onlooker bees search the feeding ground further using equation (1.10) to select better food. The choosing scheme is based on feedback from employed bees. If a feeding ground cannot be improved for a number of iterations, the feeding ground is abandoned and the bee that was assigned to that feeding ground becomes a scout and reassociates itself with a new feeding ground that is chosen via some principles (in classical ABC it is random initialization).

ABC algorithm is one of the new swarm algorithms that has exhibited very good search performances comparable to many other established algorithms in EC such as DE or PSO.

1.2.5.4 Learning Classifier Systems The classifier system (CS) is a typical example where a GA is applied to machine learning, and has been studied by many researchers such as Holland. Machine learning has two objectives, that is, learning of knowledge in complex systems and generation of appropriate output. A CS uses a GA to enhance and generate rule-based knowledge in achieving these objectives.

Machine learning using a GA is called genetic-based machine learning (GBML). The cognitive system level-1 (CS1) of Holland and Reitman is a famous early example of GBML [11]. Holland and coworkers used this system to learn how to search in maze problems. Smith later developed learning system one (LS1) [28]. LS1 was applied to maze searches and poker game strategy learning, and its effectiveness has been demonstrated.

GBML with its origin in these two systems led to two approaches, that is, the Michigan and Pittsburgh approaches. The difference between these approaches is whether the number of rule sets is just one or more than one. The Michigan approach is based on CS1, where one rule is considered as one individual, and is the main type of CSs.

Machine learning differs from optimization which searches for a solution close to the optimum solution. Instead, machine learning generates new structures and obtains a coordinated set of rules while incorporating given information. Therefore, GBML must consider the following.

1. New rules are continuously generated, and good rules remain while bad rules are discarded.
2. Good rules generated during the learning process are not destroyed in later learning.
3. The number of rules is not limited, and retaining of all necessary rules is possible.
4. Similar rules are sorted out to generate a rule set with little redundancy.

CSs use the following symbol in learning to achieve a creative rule generation mechanism:

```
If <<condition>> then <<action>>
```

This is the same as the production rule often used in expert systems. The rules in CSs consist of a condition that can be expressed with a string of 0, 1, and # (don't care) and an action that can be expressed with a string of 0 and 1. Here, # is a string that matches both 0 and 1. Providing an external message from the environment to a system that learned data in this format results in simultaneous booting of many rules in the system that gives a corresponding output. In other words, the action part of rules where the input message matches the condition part is executed. The following paragraphs describe the key characteristics of Michigan approach and Pittsburgh approach.

(1) Michigan approach

Each rule, called a classifier (CF), corresponds to one individual in the Michigan approach. The system needs a strengthening functionality that provides a "strength" parameter to each CF in addition to functionality to execute learned CFs. Here, strength is a measure of the reliability of CFs in CSs. Moreover, a functionality to generate new classifiers is necessary, and a GA is used in this generation process. The following is an explanation of the functionalities.

- Execution functionality: Searches for a CF that corresponds to input data (state) from the environment and outputs an action resulting from this CF to the environment. The CF that is selected is determined from the strength obtained using past usefulness.
- Strengthening functionality: Observes changes in the environment caused by CF execution and updates the strength of the CF. If the result is good, the CF is determined to be effective and the strength of the executed CF increases. In contrast, the strength decreases if the result is bad. The system is strengthened to be centered on good CFs by repeating this process. Proposed learning methods to update the strength include the bucket brigade algorithm and the profit- sharing plan.

- Generation functionality: The types of CFs in a system are finite and the influence on the environment is limited. Generation of new CFs is necessary to increase the kinds of action to the environment. However, random generation of CFs would require too much time before a useful CF was generated. Therefore, genetic operations are carried out using a GA where CF is the individual and the strength is the fitness function. New individuals generated by the GA replace individuals in the previous population with low strength or high similarity. CFs are generated after every few steps to reflect the strengthening of CFs based on a combination of many actions to the new CFs.

(2) Pittsburgh approach

The Pittsburgh approach considers one individual as a set of rules that comprise one function, and can consider genetic operations for each rule, which are each regarded as one unit. Consequently, GAs can be implemented more directly and therefore more easily. The following are the characteristics of the Pittsburgh approach.

- Rules are considered as genes; thus crossover means recombination of rules and mutation means conversion into a different rule. As a result, the rule strengthening functionality is unnecessary, and the implemented GA will be natural and easy to handle.
- The gene length (number of rules that make up one individual) is variable, not fixed, because many rules comprise one individual. Therefore, some tricks may be necessary for the gene structure in the GA, such as making the apparent length the same for all genes.
- A large number of rules are necessary to avoid premature convergence. As a result, the number of individuals and the number of rules that each individual contains are very large. Furthermore, evaluation of each individual needs to be carried out for the rule set rather than a simple sum of evaluated values of each rule comprising the individual. Consequently, learning takes time.

Reports of methods that improve CF systems include the zeroth level CS system (ZCS) by Wilson [33], the CS system based on accuracy (XCS) by Wilson [34], and the anticipatory CS system by Stolzmann [29].

1.2.5.5 Artificial Immune System The objective of using algorithms mimicking biological systems, such as neural networks and GAs, in engineering applications is mainly to leverage the adaptability and flexibility of natural systems. The same is true for immune system algorithms that are mainly used to achieve diversity. Details of immune systems in organisms are given in Ref. [22]. This section mainly discusses antibody reactions and outlines the mechanisms that are important for engineering applications.

Foreign bodies such as germs and viruses entering from outside the body have antigens, which are "non-self" markers that do not exist in the body. Immune reactions are caused by detection of antigens. T-cells, which are a type of lymphocyte, identify cells that have been changed by antigens and give commands to B-cells by secreting interleukin (IL). B-cells are another type of lymphocyte and secrete antibodies that react only to a specific antigen.

The relation between antigens, lymphocytes, and antibodies is called "relation between keys and keyholes." An antigen invading from outside the body selects the lymphocyte that is the closest match. This lymphocyte becomes active and triggers an immune reaction. This mechanism is called clonal selection, and is used, in combination with a GA, to determine the number of individuals in the next generation that is proportional to its fitness value.

Reactions to antigens encountered in the past are memorized, and swift reaction and repression is possible in subsequent invasions. This is known as immunological memory and is used to memorize good solutions in case-based reasoning (CBR) and evolutionary computation methods. Considering diversity at this time allows searching for diverse solutions and suppression of the number of cases.

There is a limit to the number of antibodies and lymphocytes that can exist in the body while the number of possible antigens is infinite. Moreover, attacking of cells comprising the body must be avoided. The immune system enables both identification of self and non-self as well as retention of diversity.

Somatic mutation in antibody genes is a mechanism to improve the fitness value to an antigen by causing abnormally frequent mutations in a portion of an antibody gene. Affinity maturation is a similar mechanism that is incorporated into evolutionary computations in the form of a step to improve the fitness value (identification capability) of a given individual.

Negative selection is the mechanism in which T-cells generated in bone marrow are sent to the thymus, undergo reaction tests against self-derived cells, and those that did not react are selected. Negative selection is applied to detection of computer viruses and anomalies. Here, normal data (packets and logs) are kept as self-data and a population of detectors is obtained by negative selection that does not react to self-data. This procedure can be used to check and detect viruses.

The immune network is a network that assumes identification between antibodies. This explains why antibodies can stay for a long time in a body as immunological memory long after the corresponding antigen is removed from the body and why a diverse set of antibodies can always be retained. This mechanism has one of the largest number of engineering applications, and can be applied to systems that consist of many elements, including multi-agent systems. Typical examples are coordinated control between agents, detection of abnormal processes, and information visualization through active propagations between keywords.

Artificial immune systems (AIS) are a class of computationally intelligent systems inspired by the principles and processes of the above-mentioned vertebrate immune system. AIS has been successfully applied in a number of areas (see Refs. [4,5] for details).

1.2.6 Multi-Objective EA's

The design of engineering systems must address many needs at the same time, such as enhancement of functionalities and reliability, improvement of user-friendliness, and reduction of manufacturing costs. Multi-objective optimization problems (MOPs) are characterized by the requirement of optimization of multiple objectives simultaneously. In other words, multiple objective functions $\boldsymbol{f} = (f_1, \cdots, f_m)$ are considered which will be minimized simultaneously as

$$\text{(MOP)} \quad \min_{\boldsymbol{x}} \boldsymbol{f}(\boldsymbol{x}) = (f_1(\boldsymbol{x}), \cdots, f_m(\boldsymbol{x})) \quad \boldsymbol{x} \in \mathcal{F}.$$

Here, $\boldsymbol{x}$ is the decision variable, which is a vector, and $\mathcal{F}$ is the feasible region. Objective functions typically have tradeoff relations and a decision variable $\boldsymbol{x}$ that minimizes all objective functions does not necessarily exist.

The concept of "dominance" is introduced in multi-objective optimization. For two solutions $\boldsymbol{x}_1$ and $\boldsymbol{x}_2 \in \mathcal{F}$, $\boldsymbol{x}_1$ dominates $\boldsymbol{x}_2$ if $f_k(\boldsymbol{x}_1) \leq f_k(\boldsymbol{x}_2)$ for all $k = 1, \cdots, m$ and $f_k(\boldsymbol{x}_1) < f_k(\boldsymbol{x}_2)$ for at least one $k = 1, \cdots, m$.

A "Pareto optimal solution" or "non-inferior solution" $\boldsymbol{x}$ is a reasonable solution to a MOP that is not dominant over any other solution. In general, multiple Pareto optimal solutions exist and the entire set of such solutions is called the Pareto optimal set. Therefore, the objective of solving a MOP is to obtain the Pareto optimal set or to appropriately sample solutions in the Pareto optimal set.

The multi-objective GA (MOGA) is a well-researched method to search a number of Pareto optimal solutions to a MOP at the same time by leveraging a GA that searches many points using a solution set. Although standard GAs use a single objective function as the standard for selection and elimination to solve optimization problems, MOGAs need to address a number of requirements in searching the Pareto optimum set:

1. Retain solutions closer to the Pareto optimal set while eliminating distant solutions.
2. The solution set should not be concentrated in a part of the Pareto optimal set but instead be spread out as much as possible.
3. New solutions should be efficiently obtained using crossover and selection from the group of solutions in the Pareto optimal set.

The following measures regarding the above issues are taken when designing the algorithm:

1. To search Pareto optimal solutions, selection and elimination are carried out using "dominance" relations within solutions in the solution population. For instance, Goldberg [10] and Fonseca [9] proposed the Pareto ranking method where solutions in the solution population are ranked based on dominance. The vector evaluated GA (VEGA) by Schaffer [26] and the Pareto tournament strategy by Horn et al. [12] are also demonstrated to be effective.
2. Methods to consider the local density of individuals during selection and elimination are included to disperse solutions in the Pareto optimal set. In other words, the number of other solutions near an individual could be evaluated as the solution density and be reflected in selection and elimination.

3. Generally speaking, good solutions from a GA cannot be obtained from a crossover of very different solutions. This becomes very problematic in MOGAs because the solution population is scattered over the Pareto optimal set. Consequently, one might make efforts such as placing crossover mates close to each other. However, there has been little general discussion regarding this point because the design of the crossover operation in a GA depends heavily on the problem.

MOGAs addressing these problems are rapidly being sorted out. Ref. [6] provides examples of how such algorithms are actually designed.

1.3 ADVANTAGES/DISADVANTAGES OF EVOLUTIONARY COMPUTATION

Evolutionary computations offer some unique advantages over the traditional algorithms for searching and optimization. The classical optimization algorithms such as Quasi-Newtons method, conjugate gradient methods, etc. are often iterative algorithms that can be effective in solving linear, quadratic, convex, or unimodal problems. Often, many of these algorithms have additional requirements such as continuity and/or differentiability of the search space for their working principle. Unfortunately, most of the problems in real life are very complex, nonlinear, non-convex, non-separable, and multi-modal. Often, we do not have a very good understanding of the search spaces or rather knowledge of their continuity or differentiability. Therefore, traditional approaches are often not suitable for searching the optimal solution of these problems.

Generally, evolutionary computation can work in poorly understood search problems with limited or almost no specific knowledge about the search space. Usually, by virtue of their parallel search mechanism, these algorithms show superior performances for multi-modal, nonlinear, non-separable, and non-convex search spaces compared to classical algorithms. One big advantage of these algorithms is their scalability—these algorithms can be readily applied to really large dimensional problems. Another advantage of EC over traditional search algorithms is they are capable of delivering multiple competing solutions which is often desirable in real- world problem solving but not possible in case of most of the traditional algorithms that generally utilize single-point search strategy. EC can also exhibit very good performance in optimizing noisy search spaces and can search with imperfect models as well, which makes them valuable in real-life scenarios because real world is noisy and we often have to work with approximate models for many complex systems such as biological systems. Another advantage of evolutionary algorithms is that they can generate multiple tradeoff solutions by optimizing multiple competitive criteria, which is very useful for practical applications. The parallel nature of EC is an inherent advantage for these algorithms in terms of designing computationally efficient methods.

Although evolutionary algorithms have many benefits when it comes to solving critical problems, they are not free of demerits. The main criticism against EC is that it cannot guarantee finding the optimum solution in a finite amount of time. EC can only guarantee quasi-optimal solution which is particularly useful with very large and complex problems where the optimal solution is unknown. The second shortcoming of these algorithms is that many of these algorithms need to tune various search parameters without proper guidance on how to set them for unknown problems. Today, many adaptive strategies have been incorporated in different algorithms where these parameters can be automatically adjusted online based on the algorithms' search performance. EC is often blamed for utilizing too much exploration; hence it is computationally expensive due to its population-based search approach. However, parallel implementation and other sophisticated approaches like surrogate assistance are used to overcome this limitation. Nevertheless, according to no-free-lunch (NFL) theorem [35], there does not exist any algorithm that has superior performance compared to some other algorithm in solving all optimization problems in general. Therefore, EC cannot be claimed to be superior/inferior to some other algorithm, but they certainly have advantages and limitation over some specific classes of problems.

1.4 APPLICATION AREAS OF EC

Because of their robust and reliable performance in solving complex and odd problems, EC found numerous applications in diverse domains: engineering, science, biology, architecture, arts, music, design, transportation, etc. Almost in every field where we need to solve difficult optimization problems, EC has been successfully used. EC draws researchers' attention through its success in solving different planning problems in the form of routing and scheduling tasks. Different kinds of optimization problems arise in engineering design that ranges from the filter design for digital systems to gearbox or accelerator design for automobiles, to blades, turbine or to engine design for aircrafts. Numerous applications of EC exist in structural engineering, architectural design, environmental engineering, geotechnical and water resource engineering. Today, another broad application area of EC is biological and medical science. In the field of biological sciences, EC is a preferred technique for data analysis, classification, pattern recognition, reverse engineering, and model optimization. In medicine and pharmacology, EC is used for diagnosis, disease data classification, drug design, optimal therapy design for complex disease, etc. In the post-genome era, an increasing surge is observed in analysis and interpretation of the enormous amount of data that is being generated by different studies. EC has also been utilized for solving many problems in finance and economics such as investment planning, market forecasting, etc. Another major application area of EC is control problems where it is applied for fault diagnosis, stability analysis, structure and parameter identification for controllers, etc. Several applications of EC have been observed in robotics which vary from robotic motion planning to automatic learning of cooperation among robots. Besides, EC has found application in other fields as well, such as agriculture, climatology, environmental science and ecology, geo and hydro science, etc.

1.5 CONCLUSION

In this chapter, we have presented a very brief introduction to various algorithms that come under the broad umbrella of evolutionary computation. This introduction is incomplete in every sense —we did not cover many major algorithms in this field such as adaptive evolutionary algorithms, cellular evolutionary algorithms, memetic algorithms (MAs), estimation of distribution algorithms, (EDA) etc.; there are hundreds of variants for each of these algorithms which are not included; parameter setting and relative merits/demerits of each class of algorithms were not discussed. In fact these discussions are not within the purpose and scope of this introduction. The aim of this much generalized introduction is to prepare a background for novice readers to this branch of computation so that they can easily follow the specialized variants of some of these algorithms applied to various GRN research presented in this book.

REFERENCES

1. Ackley, D., "A connectionist machine for genetic hillclimbing," *The Springer International Series in Engineering and Computer Science*, vol. 28, 1987.
2. Aldana, M., Balleza, E., Kauffman, S., and Resendiz, O., "Robustness and evolvability in genetic regulatory networks," *Journal of Theoretical Biology*, vol. 245, no. 3, pp. 433–448, 2006.
3. Beyera, H.-G., Schwefela, H.-P., and Wegener, I., "How to analyse evolutionary algorithms," *Theoretical Computer Science*, vol .287, no. 1, 2002.
4. Castro, de L. N. and Zuben, F. J. V., "Artificial Immune Systems: Part II - A Survey of Applications," Technical Report DCA-RT 02/00 (2000).
5. Dasgupta, D. (Ed.), *Artificial Immune Systems and Their Applications*, Springer-Verlag, 1999.
6. Deb, K., *Multi-Objective Optimization Using Evolutionary Algorithms*, John Wiley & Sons, 2001.
7. Dorigo, M. and Caro, G. Di, "The ant colony optimization meta-heuristic," in *New Ideas in Optimization* (Corne, D., Dorigo, M., and Glover, F. (Eds.)), McGraw-Hill, pp. 11–32, 1999.
8. Eshelman, L. J. and Schaffer, J. D., "Real-coded genetic algorithms and interval-schemata," *Foundations of Genetic Algorithms 2*, pp. 187–202, 1993.
9. Fonseca, C. M. and Fleming, P. J., "Genetic algorithms for multi-objective optimizations," *Proceedings of International First Conference on Genetic Algorithms and Their Applications* C pp. 93–100, 1985D.
10. Goldberg, D. E., *Genetic Algorithms in Search, Optimization, and Machine Learning*, Addison-Wesley, Boston, MA, 1989.
11. Holland, J. H. and Reitman, J. S., "Cognitive systems based on adaptive algorithms," in *Pattern-Directed Inference Systems* (Waterman, D. A. and Hayes-Roth, F. (Eds.)), New York, Academic Press, pp. 313–329, 1978.

12. Horn, J., Nafpliotis, N., and Goldberg, D. E., "A niched Pareto genetic algorithm for multi-objective optimization," *Proceedings of the First IEEE Conference on Evolutionary Computation* C pp. 82–87, 1994D.
13. Iba, H., deGaris, H., and Sato, T., "Genetic programming using a minimum description length principle," in *Advances in Genetic Programming* (Kinnear, Jr. K., (Ed.)), MIT Press, pp. 265–284, 1994.
14. Iba, H., deGaris, H., and Sato, T., "Numerical approach to genetic programming for system identification," *Evolutionary Computation*, vol. 3, no. 4, pp. 417–452, 1996.
15. Ito, T., Iba, H., and Sato, S., "Depth-dependent crossover for genetic programming," *Proceedings of the 1998 IEEE International Conference on Evolutionary Computation* , pp. 775–780, 1998.
16. Karaboga, D., "An idea based on honey bee swarm for numerical optimization," Technical Report TR06, Erciyes University, Engineering Faculty, Computer Engineering Department, 2005.
17. Kennedy, J. and Eberhart, R., "Particle swarm optimization," *Proceedings of IEEE the International Conference on Neural Networks*, 1995.
18. Kita, H., Ono, I., and Kobayashi, S., "Theoretical analysis of the unimodal normal distribution crossover for realcoded genetic algorithms," *Proceedings of ICEC'98*, pp. 529–534, 1998.
19. Koza, J., *Genetic Programming, On the Programming of Computers by Means of Natural Selection*, MIT Press, 1992.
20. Langdon, W. B., "Size fair and homologous tree crossovers for tree genetic programming," *Genetic Programming and Evolvable Machines*, vol. 1, no. 1–2, pp. 95–119, 2000.
21. Ono, I. and Kobayashi, S., "A real-coded genetic algorithm for function optimization using unimodal normal distribution crossover," *Proceedings of 7th ICGA*, pp. 246–253, 1997.
22. Perelson, A. S. and Weisbuch, G., "Immunology for physicists," *Reviews of Modern Physics*, vol. 69, no. 4, 1997.
23. Price, K. V., Storn, R. M., and Lampinen, J. A., *Differential Evolution - A Practical Approach to Global Optimization*, Springer, 2005.
24. Rechenberg, I., *Evolutionsstrategie: Optimierung technischer Systeme nach Prinzipien der biologischen Evolution*, Frommann-Holzboog Verlag, Stuttgart, 1973.
25. Reynolds, C. W., "Flocks, herds and schools: a distributed behavioral model," *Computer Graphics*, vol. 21, no. 4, pp. 25–34, 1987.
26. Schaffer, J. D., "Multiple objective optimization with vector evaluated genetic algorithms," *Proceedings of First International Conference on Genetic Algorithms and Their Application* C pp. 93–100, 1995.
27. Schwefel, H.-P., *Numerical Optimization of Computer Models*, John Wiley & Sons, Inc., New York, NY, 1981.
28. Smith, S. F., "Flexible learning of problem solving heuristics through adaptive search," *Proceedings eighth International Conference on Artificial Intelligence*, pp. 422–425, 1983.
29. Stolzmann, W., *Antizipative Classifier Systems [Anticipatory Classifier Systems]*, Osnabrueck, Germany: Shaker Verlag, Aachen, Germany, 1997.
30. Storn, R. and Price, K., "Differential Evolution – A simple and efficient heuristic for global optimization over continuous spaces," *Journal Global Optimization*, pp. 341–359, 1997.

31. Syswerda, G., "A study of reproduction in generational and steady-state genetic algorithms," *Foundations of Genetic Algorithms*, pp. 94–101, 1991.
32. Thierens, D. and Goldberg, D., "Elitist recombination: an integrated selection recombination GA," *Proceedings of the First IEEE Conference on volutionary Computation*, pp. 508–512, 1994.
33. Wilson, S. W., "A zeroth level classifier system," *Evolutionary Computation*, vol. 2, no. 1, pp. 1–18, 1994.
34. Wilson, S. W., "Classifier fitness based on accuracy," *Evolutionary Computation*, vol. 3, no. 2, pp. 149–175, 1995.
35. Wolpert, D.-H. and Macready, W.-G.,"No free lunch theorems for optimization."*IEEE Transactions on Evolutionary Computation*, vol. 1, no. 1, pp. 67–82, 1997.

2

MATHEMATICAL MODELS AND COMPUTATIONAL METHODS FOR INFERENCE OF GENETIC NETWORKS

Tatsuya Akutsu

Bioinformatics Center, Institute for Chemical Research, Kyoto University, Uji, Japan

2.1 INTRODUCTION

Genes maintain organisms by interacting with each other through messenger ribonucleic acids (mRNAs), proteins, and other types of molecules. Interactions among genes are often represented as networks, which are called *gene regulatory networks*, or *genetic networks* in short. Genetic networks are usually represented as directed graphs, in which nodes correspond to genes and edges correspond to regulatory relationships between two genes.

Deciphering genetic networks is important for understanding complex cellular systems because they play important roles in cells through control of protein production. In order to infer genetic networks, various kinds of data have been used such as gene expression profiles (particularly mRNA expression profiles), CHromatin ImmunoPrecipitation-chip data for transcription binding information, deoxyribonucleic acid (DNA)–protein interaction data, and mRNA seq data generated by using the next-generation DNA sequencing technology. In this chapter, we focus on inference of genetic networks from gene expression time series data (see Figure 2.1) because

Evolutionary Computation in Gene Regulatory Network Research, First Edition.
Edited by Hitoshi Iba and Nasimul Noman.

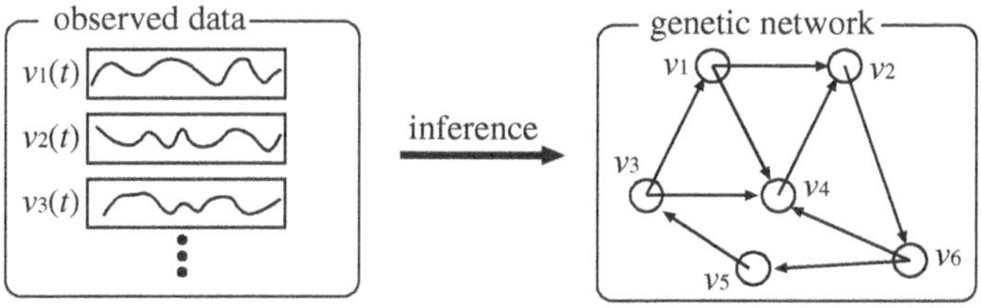

Figure 2.1 Inference of a genetic network from gene expression time series data.

many studies have been done based on this kind of data and many of the developed methods and techniques may be applied to mRNA seq data.

Various mathematical models have been applied and/or developed to infer genetic networks from gene expression profiles, which include Boolean networks, Bayesian networks, dynamic Bayesian networks, linear and nonlinear differential equation models, and graphical Gaussian models (GGMs). Since too many models and methods have been developed and applied, it is impossible to review all or many of them. Therefore, in this chapter, we focus on basic and simple models and methods, and try to explain basic ideas/concepts in them. Readers interested in more advanced models and methods are referred to more comprehensive review articles [6, 18, 23, 48].

2.2 BOOLEAN NETWORKS

A Boolean network is a discrete model of genetic networks [22]. It is one of the simplest models of genetic networks and was proposed in 1960s [21].

A *Boolean network* $G(V, F)$ consists of a set $V = \{v_1, \dots, v_n\}$ of *nodes* and a list $F = (f^1, \dots, f^n)$ of *Boolean functions*, where we use n to denote the number of nodes throughout this chapter. Each node corresponds to a gene and takes either 0 (gene is expressed) and 1 (gene is not expressed) at each discrete time t. The state of node v_i at time t is denoted by $v_i(t)$, and the states of all nodes change synchronously according to given regulation rules.[1] A Boolean function $f^i(v_{i_1}, \dots, v_{i_k})$ with inputs from specified nodes $v_{i_1}, \dots, v_{i_k}$ is assigned to each node, where it represents a regulation rule for v_i and the (ordered) set of input nodes (parent nodes) $\{v_{i_1}, \dots, v_{i_k}\}$ is denoted by pa(v_i). The state of node v_i at time $t + 1$ is determined by

$$v_i(t+1) = f^i(v_{i_1}(t), \dots, v_{i_{k_i}}(t)).$$

The vector consisting of the states of all nodes $\mathbf{v}(t) = (v_1(t), \dots, v_n(t))$ is called a (global) state of a Boolean network at time t. We also write $v_i(t+1) = f^i(\mathbf{v}(t))$ to denote the regulation rule for v_i and $\mathbf{v}(t+1) = \mathbf{f}(\mathbf{v}(t))$ to denote the regulation rule for the whole network.

[1] Although there exist several variants according to state update rules, we focus on synchronous Boolean networks in this chapter.

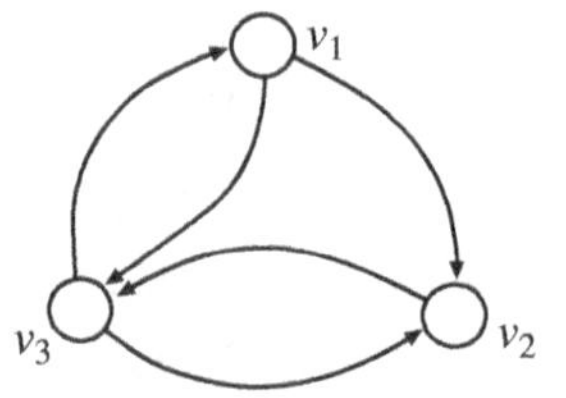

$$v_1(t+1) = v_3(t)$$
$$v_2(t+1) = v_1(t) \wedge \overline{v_3(t)}$$
$$v_3(t+1) = v_1(t) \wedge \overline{v_2(t)}$$

Figure 2.2 Example of a Boolean network.

The structure of a Boolean network can be represented by a directed graph $G(V, E)$, where E is the set of edges defined by $E = \{(v_{i_j}, v_i) \mid v_{i_j} \in \text{pa}(v_i)\}$. An edge from v_{i_j} to v_i means that v_{i_j} directly affects the expression of v_i. The number of input nodes to v_i (i.e., $|\text{pa}(v_i)|$) is called the *indegree* of v_i and we use K to denote the *maximum indegree* of a Boolean network.

Figure 2.2 shows an example of a Boolean network. In this example, the state of node v_1 at time $t + 1$ is determined by the state of node v_3 at time t, the state of node v_2 at time $t + 1$ is determined by the logical AND of the state of v_1 and the negation (i.e., logical NOT) of the state of v_3 at time t, and the state of node v_3 at time $t + 1$ is determined by AND of the state of node v_1 and NOT of the state of node v_2 at time t, where we use $x \wedge y$ and $\overline{x}$ to denote logical AND of x and y, and logical NOT of x, respectively.

The dynamics of a Boolean network can be well described by a *state transition diagram* (see Figure 2.3 for an example). In this diagram, each node corresponds to a global state and each edge corresponds to a transition of global states from time t to $t + 1$. For example, an edge from 110 to 010 in Figure 2.3 means that if the global state is $[1, 1, 0]$ at time t, the global state becomes $[0, 1, 0]$ at time $t + 1$. It is also seen that if $\mathbf{v}(0) = [1, 1, 0]$, the global state changes as

$$[1, 1, 0] \Longrightarrow [0, 1, 0] \Longrightarrow [0, 0, 0] \Longrightarrow [0, 0, 0] \Longrightarrow \cdots,$$

and the same global state $[0, 0, 0]$ is repeated after $t = 1$. It is also seen if $\mathbf{v}(0) = [0, 0, 1]$, $[1, 0, 0]$, and $[0, 1, 1]$ are repeated alternatively after $t = 0$. Sets of repeating states are called *attractors*. Each attractor corresponds to a directed cycle in a state transition table, and the number of elements in an attractor is called the *period* of the attractor. An attractor with period 1 is called a *singleton* attractor,

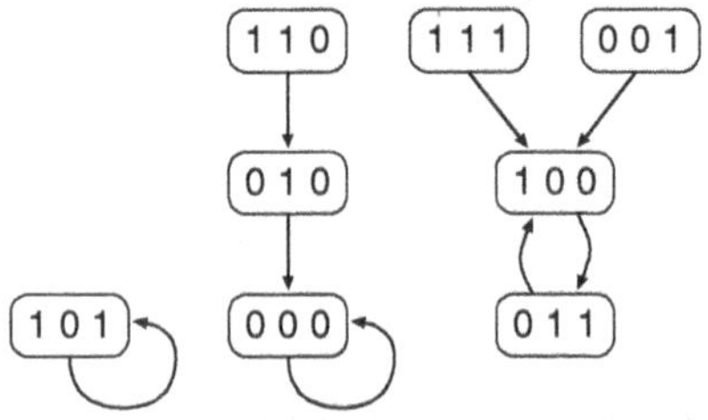

Figure 2.3 State transition diagram for the Boolean network in Figure 2.2.

whereas an attractor with period greater than 1 is called a *cyclic* attractor. In Figure 2.3, we can see two singleton attractors $\{[1,0,1]\}$ and $\{[0,0,0]\}$, and one cyclic attractor $\{[1,0,0],[0,1,1]\}$ with period 2. Because of an interpretation that attractors correspond to types of cells, extensive studies have been done on distribution of attractors [10,43] although no conclusive result has yet obtained. Many studies have also been done to efficiently detect attractors [11,20,32]. However, it is known that detection of an attractor is non-deterministic polynomial-time hard (NP-hard) [1].

The inference problem of a Boolean network from time series gene expression data is defined as follows. Let (I^j, O^j) $(j = 1, \dots, m)$ be a pair of expression profiles (i.e., 0–1 vectors) of $v_1, \dots, v_n$, where I^j corresponds to a global state at time t and O^j corresponds to a global state at time $t+1$. I_i^j (resp. O_i^j) denotes the expression (0 or 1) of gene v_i in I^j (resp. O^j). Each pair (I^j, O^j) is called a *sample*. We say that a Boolean function f^i assigned to node v_i is *consistent* with a sample (I^j, O^j) if $O_i^j = f^i(I_{i_1}^j, \dots, I_{i_k}^j)$ holds, and $G(V, F)$ is consistent with (I^j, O^j) if all Boolean functions in F are consistent with (I^j, O^j). For a set of samples $EX = \{(I^1, O^1), (I^2, O^2), \dots, (I^m, O^m)\}$, we say that $G(V, F)$ (resp., node v_i) is consistent with EX if $G(V, F)$ (resp., node v_i) is consistent with all (I^j, O^j) for $1 \le j \le m$. Then, the inference problem is defined as follows: given the number of nodes n and a set of samples $EX = \{(I^j, O^j) \mid j = 1, \dots, m\}$, find a Boolean network with n nodes that is consistent with EX, where "none" should be reported if it does not exist. For example, suppose that the following set of samples are given:

$$EX = \{([1,0,1],[1,0,1]), ([1,1,0],[0,1,0]), ([0,0,1],[1,0,0])\}.$$

Then, the following functions are consistent with EX:

$$v_1(t+1) = v_3(t), \quad v_1(t+1) = \overline{v_2}(t) \wedge v_3(t),$$

whereas the following functions are inconsistent with EX:

$$v_1(t+1) = v_1(t), \quad v_1(t+1) = v_2(t) \wedge \overline{v_3}(t).$$

In general, this inference problem is NP-hard [15]. However, if the maximum indegree K is a constant (e.g., 2 or 3), this problem can be solved in polynomial time by exhaustive search [2]. Liang et al. developed a faster algorithm by making use of mutual information [29]. Fukagawa and Akutsu showed that a simple greedy-type algorithm can correctly identify a Boolean network for most cases with high probability if samples are given uniformly at random [15].

Akutsu et al. studied the *sample complexity* in order to uniquely identify a Boolean network (i.e., how many samples are needed so that there exists at most only one Boolean network consistent with the samples) and showed that $O(K2^{2K+1} \log n)$ samples are enough on the average under the assumption that each sample (i.e., I^j) is generated uniformly at random [2], where the constant factor was improved by Perkins and Hallett [39]. Since 2^n samples are required (i.e., the whole state transition diagram should be given) if there is no restriction on K, this result suggests that only a small part of the state transition diagram is enough for identification of a Boolean network if K is a constant. It should also be noted that the assumption of uniform randomness does not hold for real-time series data and thus more samples are required in practice.

Recently, a new approach based on *semi-tensor product* has been proposed for analysis of Boolean networks, and has been applied to various problems, which include stability analysis, identification, and control [4].

2.3 PROBABILISTIC BOOLEAN NETWORK

In a Boolean network, the global state at time $t+1$ is determined uniquely from the global state at time t. However, real genetic networks may not necessarily work deterministically. Therefore, it is more realistic to assume that genetic networks work stochastically by means of the effects of noise and elements other than genes. In order to introduce such effects, Shmulevich et al. proposed the *probabilistic Boolean network* (PBN) [46,47].[2]

A PBN is an extension of a Boolean network. The difference between two models lies only on that multiple Boolean functions $f_j^i (j = 1, 2, \ldots, l(i))$ are assigned to each node v_i and one of them is randomly chosen at each time step according to the probability assigned to each function. At each time step, f_j^i is chosen with pre-defined probability $c_j^i \geq 0$, where c_j^i should satisfy $\sum_{j=1}^{l(i)} c_j^i = 1$ for each $i = 1, 2, \ldots, n$. Let $\mathbf{f}_j$ be the jth possible realization of a Boolean network:

$$\mathbf{f}_j = (f_{j_1}^1, f_{j_2}^2, \ldots, f_{j_n}^n), \; 1 \leq j_i \leq l(i), \; i = 1, 2, \ldots, n,$$

where $N = \prod_{i=1}^n l(i)$ is the maximum possible number of different realizations of Boolean networks. The probability of choosing such a realization is given by

$$p_j = \prod_{i=1}^{n} c_{j_i}^i, \quad j = 1, 2, \ldots, N,$$

since the selection of the Boolean function for each node is independent.

Figure 2.4 shows an example of PBN. Suppose that the global state of PBN at time t is $[0, 0, 0]$. If (f_1^1, f_1^2, f_1^3) is selected with probability $0.8 \times 0.7 = 0.56$, the global state at time $t+1$ remains $[0, 0, 0]$. Similarly, if (f_1^1, f_1^2, f_2^3) is selected with probability $0.8 \times 0.3 = 0.24$, the global state at time $t+1$ remains $[0, 0, 0]$. On the other hand, if (f_2^1, f_1^2, f_1^3) is selected with probability $0.2 \times 0.7 = 0.14$ or (f_2^1, f_1^2, f_2^3) is selected with probability $0.2 \times 0.3 = 0.06$, the global state at time $t+1$ becomes

[2]Although there exist several variants, we consider the original one proposed in Ref. [47].

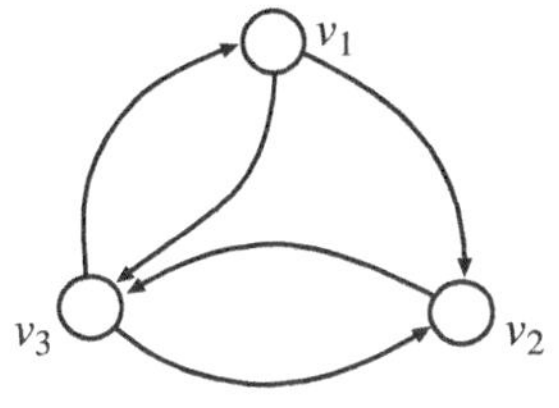

	Boolean Function	Prob.
f_1^1	$v_1(t+1) = v_3(t)$	0.8
f_2^1	$v_1(t+1) = \overline{v_3(t)}$	0.2
f_1^2	$v_2(t+1) = v_1(t) \wedge \overline{v_3(t)}$	1.0
f_1^3	$v_3(t+1) = v_1(t) \wedge \overline{v_2(t)}$	0.7
f_2^3	$v_3(t+1) = v_2(t)$	0.3

Figure 2.4 Example of a probabilistic Boolean network.

$[1, 0, 0]$. Therefore, we have the following transition probabilities:

$$P(\mathbf{v}(t+1) = [0,0,0] \mid \mathbf{v}(t) = [0,0,0]) = 0.8,$$
$$P(\mathbf{v}(t+1) = [1,0,0] \mid \mathbf{v}(t) = [0,0,0]) = 0.2,$$

where the probabilities of the transitions from $[0, 0, 0]$ to the other states are 0. For another example, the (nonzero) transition probabilities from $[0, 1, 0]$ are as follows:

$$P(\mathbf{v}(t+1) = [0,0,0] \mid \mathbf{v}(t) = [0,1,0]) = 0.56,$$
$$P(\mathbf{v}(t+1) = [0,0,1] \mid \mathbf{v}(t) = [0,1,0]) = 0.24,$$
$$P(\mathbf{v}(t+1) = [1,0,0] \mid \mathbf{v}(t) = [0,1,0]) = 0.14,$$
$$P(\mathbf{v}(t+1) = [1,0,1] \mid \mathbf{v}(t) = [0,1,0]) = 0.06.$$

Transitions of global states in a Boolean network were represented by a state transition diagram. In a PBN, transitions are well represented by using $2^n \times 2^n$ matrix A. For each 0–1 vector $\mathbf{a} = [a_1, \ldots, a_n]$ of size n, we assign an integer number $id(\mathbf{a})$ between 1 and 2^n by $id(\mathbf{a}) = 2^{n-1}a_1 + 2^{n-2}a_2 + \cdots + 2a_{n-1} + a_n + 1$. Then, A is defined by

$$A_{ij} = P(\mathbf{v}(t+1) = \mathbf{b} \mid \mathbf{v}(t) = \mathbf{a}),$$

where $i = id(\mathbf{b})$ and $j = id(\mathbf{a})$. For example, the transition matrix of the PBN of Figure 2.4 is as follows:

$$\begin{pmatrix} 0.8 & 0.2 & 0.56 & 0.14 & 0 & 0.06 & 0 & 0.14 \\ 0 & 0 & 0.24 & 0.06 & 0 & 0.14 & 0 & 0.06 \\ 0 & 0 & 0 & 0 & 0.24 & 0 & 0.56 & 0 \\ 0 & 0 & 0 & 0 & 0.56 & 0 & 0.24 & 0 \\ 0.2 & 0.8 & 0.14 & 0.56 & 0 & 0.24 & 0 & 0.56 \\ 0 & 0 & 0.06 & 0.24 & 0 & 0.56 & 0 & 0.24 \\ 0 & 0 & 0 & 0 & 0.06 & 0 & 0.14 & 0 \\ 0 & 0 & 0 & 0 & 0.14 & 0 & 0.06 & 0 \end{pmatrix}.$$

The first column of this matrix represents the transition probabilities from $[0,0,0]$; the second column represents the transition probabilities from $[0,0,1]$; and so on. Since this matrix corresponds to a state transition matrix in a standard Markov chain (with 2^n states), the techniques developed in the field of Markov chain can be applied to PBNs.

In order to infer a PBN from gene expression time series data, we need to determine Boolean functions assigned to each node and the corresponding probabilities. For inference of Boolean functions, exhaustive search or heuristic methods are usually used. For determination of probabilities, the coefficient of determination is widely used, which is a measure of the relative decrease in error from estimating the state of v_i via f_j^i rather than by just the best constant estimate. Many studies have been done on simulation of biological systems using PBNs and on finding control strategies for biological systems. For details of inference, simulation, and control studies on PBNs, see a recent review in Ref. [53].

2.4 BAYESIAN NETWORK

The *Bayesian network* is a statistical model, and has been developed and studied in the fields of statistics and artificial intelligence. A Bayesian network is usually defined by a *directed acyclic graph* and *conditional probabilities*.

Let $G(V,E)$ be a directed acyclic graph with a set of nodes V and a set of directed edges E, where acyclic means that there is no directed cycle (i.e., there is no path beginning and ending with the same node). As in Section 2.2, $\text{pa}(v_i)$ denotes the ordered set of input nodes to v_i. In the rest of this section, we use X_i to denote both the ith node (i.e., v_i) and random variable assigned to this node. In Figure 2.5(a), $\text{pa}(X_1) = \text{pa}(X_2) = \emptyset$ and $\text{pa}(X_3) = (X_1, X_2)$. In the basic version, each X_i takes discrete values. For example, we consider a case in which each X_i takes either 0 or 1. Then, logical AND (i.e., $X_3 = X_1 \wedge X_2$) can be represented by the following conditional probabilities:

$$
\begin{aligned}
P(X_3 = 1 \mid X_1 = 1, X_2 = 1) &= 1, & P(X_3 = 0 \mid X_1 = 1, X_2 = 1) &= 0,\\
P(X_3 = 1 \mid X_1 = 1, X_2 = 0) &= 0, & P(X_3 = 0 \mid X_1 = 1, X_2 = 0) &= 1,\\
P(X_3 = 1 \mid X_1 = 0, X_2 = 1) &= 0, & P(X_3 = 0 \mid X_1 = 0, X_2 = 1) &= 1,\\
P(X_3 = 1 \mid X_1 = 0, X_2 = 0) &= 0, & P(X_3 = 0 \mid X_1 = 0, X_2 = 0) &= 1.
\end{aligned}
$$

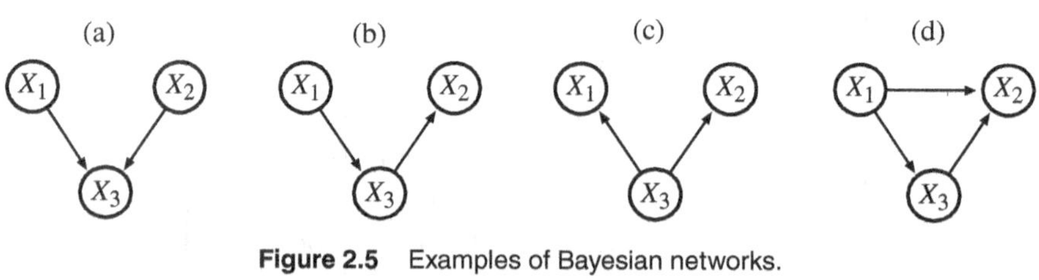

Figure 2.5 Examples of Bayesian networks.

We also denote by $P(X_i \mid \mathrm{pa}(X_i))$ the conditional probability of X_i given $\mathrm{pa}(X_i)$. Using conditional probabilities, the probability of the whole network is given by

$$P(X_1, \dots, X_n) = \prod_{i=1}^{n} P(X_i \mid \mathrm{pa}(X_i)).$$

For each of the networks given in Figure 2.5, this probability is given as follows:

(a) $P(X_1, X_2, X_3) = P(X_3 \mid X_1, X_2)P(X_1)P(X_2)$,
(b) $P(X_1, X_2, X_3) = P(X_3 \mid X_2)P(X_2 \mid X_1)P(X_1)$,
(c) $P(X_1, X_2, X_3) = P(X_1 \mid X_3)P(X_2 \mid X_3)P(X_3)$,
(d) $P(X_1, X_2, X_3) = P(X_2 \mid X_1, X_3)P(X_3 \mid X_1)P(X_1)$.

For inference of Bayesian network structure $G = G(V, E)$ from observed data D, we use the following relation:

$$P(G \mid D) \propto P(D \mid G)P(G).$$

In order to calculate $P(D \mid G)$, a family of probability distributions characterized by a set of parameters θ is employed. Then, $P(D \mid G)$ is given by

$$P(D \mid G) = \int P(D \mid \theta, G)P(\theta \mid G)d\theta.$$

Although computation of this score is very difficult in general, it can be given by a simple formula in some special but useful cases [7].

Then, the inference problem is to find G maximizing $P(D \mid G)$ (under the assumption that the prior probability $P(G)$ is the same for all G). However, this problem is proved to be NP-hard [5]. Therefore, such heuristic algorithms as greedy algorithms and simulated annealing have been utilized. However, it is reported that an optimal solution can be obtained if n is small (e.g., $n \leq 30$) [38]. In the case of application of Bayesian networks to inference of genetic networks, the domain consisting of three states, $\{-1, 0, 1\}$, has been used [13], where 0, -1, and 1 mean the neutral, down-regulated, and up-regulated levels, respectively.

Various extensions and modifications of Bayesian networks have been developed and applied for inference of genetic networks. Although we have considered discrete values for variables, continuous values can also be used. For example, Imoto et al. developed a method to infer Bayesian networks with continuous variables by combining nonparametric regression and information criteria [19].

Another important variant of Bayesian networks is the *dynamic Bayesian network* [40]. In dynamic Bayesian networks, random variable X_i depends on time and is represented as $X_i(t)$. Furthermore, the state of $X_i(t+1)$ depends on states of $\mathrm{pa}(X_i)(t)$

(states of $\mathrm{pa}(X_i)$ at time t). Therefore, dynamic Bayesian networks are defined via the conditional probability,

$$P(X_i(t+1) \mid \mathrm{pa}(X_i)(t)).$$

Different from usual Bayesian networks, loops are allowed in dynamic Bayesian networks. However, time series data are necessary to infer dynamic Bayesian networks although usual Bayesian networks can be inferred from static data (e.g., data obtained by gene knockout and overexpression experiments).

If we consider the binary domain (i.e., each variable takes 0 or 1), dynamic Bayesian networks are almost the same as PBNs. This relationship is discussed in Ref. [26]. Dynamic Bayesian networks have been further extended to time-delayed Bayesian networks in which $X_i(t+1)$ depends not only on $\mathrm{pa}(X_i)(t)$ but also on $\mathrm{pa}(X_i)(t-1), \ldots, \mathrm{pa}(X_i)(t-h)$, where h is some positive integer, and have been applied to inference of genetic networks [30].

2.5 GRAPHICAL GAUSSIAN MODELING

A *graphical Gaussian model* is a statistical model based on *conditional probabilities* and *Gaussian distributions*. A GGM is represented with an undirected graph in which each edge represents the pairwise correlation between two nodes conditioned against the correlations with all other nodes. In the GGM approach, indirect association effects are eliminated by evaluating conditional dependencies in multivariate Gaussian distributions (see also Figure 2.6). It is to be noted that undirected networks are inferred using the GGM approach, different from Boolean and Bayesian network approaches.

As discussed in Sections 2.2 and 2.3, we assume that gene expression time series data are given as $v_i(t)$ $(i = 1, \ldots, n,\ t = 1, \ldots, m)$. However, different from Boolean and Bayesian models, we assume that $v_i(t)$ takes real values. The standard *Pearson correlation coefficient* r_{ij} between v_i and v_j is defined as

$$r_{ij} = \frac{\sum_{t=1}^{m}(v_i(t) - \overline{v}_i)(v_j(t) - \overline{v}_j)}{\sqrt{\sum_{t=1}^{m}(v_i(t) - \overline{v}_i)^2} \cdot \sqrt{\sum_{t=1}^{m}(v_j(t) - \overline{v}_j)^2}},$$

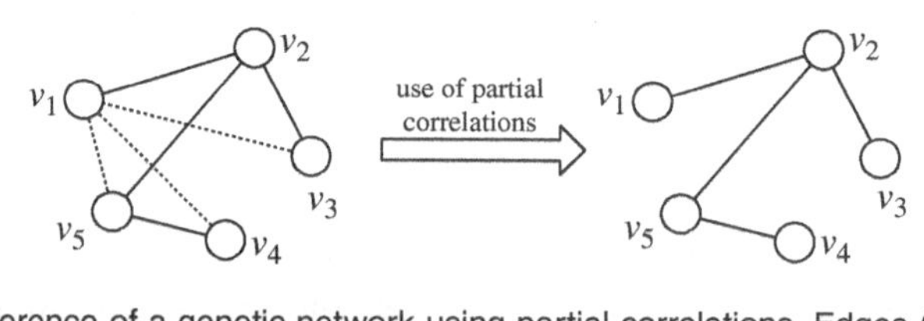

Figure 2.6 Inference of a genetic network using partial correlations. Edges (shown by dotted lines) with small partial correlations are (repeatedly) eliminated in the graphical Gaussian model.

where $\overline{v}_i$ represents the mean value of $v_i(t)$ (i.e., $\overline{v}_i = \frac{1}{m}\sum_{t=1}^{m} v_i(t)$). It is to be noted that $\sqrt{\frac{1}{m}\sum_{t=1}^{m}(v_i(t) - \overline{v}_i)^2}$ is the standard deviation of $v_i(t)$. Let R be the correlation matrix consisting of r_{ij}s (i.e., $R = (r_{ij})$) and let $R^{-1} = (r^{ij})$ be the inverse matrix of R. Then, the *partial correlation* coefficient p_{ij} is defined as

$$p_{ij} = \begin{cases} \dfrac{-r^{ij}}{\sqrt{r^{ii}} \cdot \sqrt{r^{jj}}} & \text{for } i \neq j, \\ 1 & \text{for } i = j. \end{cases}$$

Let P be the partial correlation matrix consisting of p_{ij}s (i.e., $P = (p_{ij})$), where the partial correlation means the pairwise correlation between two nodes conditioned against the correlations with all other nodes. Therefore, if there is no direct correlation between nodes v_i and v_j, $p_{ij} = 0$ holds (under a linear Gaussian model). However, in practice, we cannot expect that $p_{ij} = 0$ holds exactly even if there is no direct correlation. Therefore, we delete edge(s) whose partial correlation is minimum or less than some threshold. Furthermore, we can apply computation of partial correlation and deletion of edges repeatedly, where R for the next step is computed from the modified P in which 0 is assigned to p_{ij}s corresponding to deleted edges.

GGMs have been applied to inference of genetic networks by many groups [8, 25, 45, 52, 55]. Other methods for eliminating indirect edges have also been developed and applied. For example, mutual information [31] and entropy maximization [28] have been applied for that purpose.

Recently, new approaches for eliminating indirect correlations have been proposed: the *silencing* method [3] and the *network deconvolution* method [12]. The silencing method is based on a dynamical model. In this method, the global response matrix,

$$G_{ij} = \frac{dv_i}{dv_j},$$

and the local response matrix,

$$S_{ij} = \frac{\partial v_i}{\partial v_j},$$

are used, where we use v_i to denote both a node and its expression value. G captures the change of v_i's activity in response to changes of v_j's activity, and can be obtained from gene knockout and overexpression experiments. It is to be noted that G includes the effects of indirect correlations whereas S consists of only direct correlations. Therefore, G and S correspond to observed correlations and direct correlations, respectively. These two matrices are related by

$$\begin{cases} \dfrac{dv_i}{dv_i} = 1, \\ \dfrac{dv_i}{dv_j} = \sum_{k=1}^{n} \dfrac{\partial v_i}{\partial v_k} \dfrac{\partial v_k}{\partial v_j}, \qquad i \neq j, \end{cases}$$

which leads to a matrix-based method to compute S from G.

In the network deconvolution method, it is assumed that the true network is given in the form of $n \times n$ matrix G_{dir}, and the observed network G_{obs} has the form of

$$G_{\text{obs}} = G_{\text{dir}} + G_{\text{dir}}^2 + G_{\text{dir}}^3 + \cdots = G_{\text{dir}}(I - G_{\text{dir}})^{-1},$$

where I is the identity matrix. Then, G_{dir} can be computed from G_{obs} by

$$G_{\text{dir}} = G_{\text{obs}}(I + G_{\text{obs}})^{-1}.$$

It is to be noted that

$$G_{\text{dir}}^2 + G_{\text{dir}}^3 + \cdots$$

corresponds to indirect correlations.

In the above-mentioned methods, only elimination of spurious edges/correlations is considered. However, it may also be useful to modify a known network by addition and deletion of edges so that the resulting network is more consistent with observed data. This approach is called *network completion* (see Figure 2.7) and various methods have been proposed although the target networks do not necessarily include genetic networks. Guimerà and Sales-Pardo developed a method that can identify both missing and spurious interactions by using the stochastic block models [16]. Hanneke and Xing [17] developed a method based on random sampling of subnetworks and analyzed confidence intervals from samples [8]. Kim and Leskovec developed the

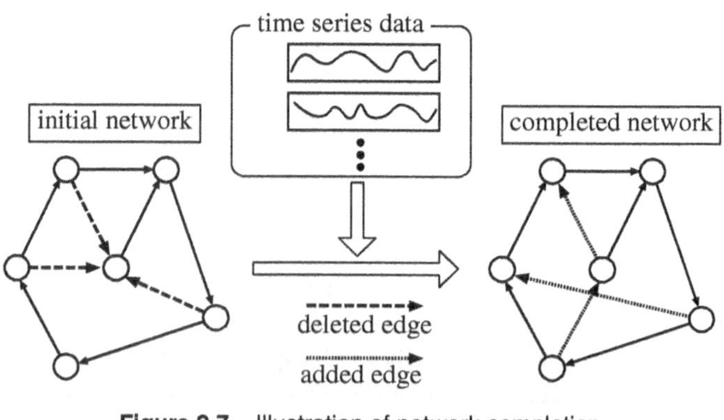

Figure 2.7 Illustration of network completion.

KronEM algorithm by combining the expectation–maximization (EM) method with the Kronecker graphs [24]. Nakajima et al. developed a network completion method by combining dynamic programming and least-squared fitting [36].

2.6 DIFFERENTIAL EQUATIONS

Differential equations have been widely used in many areas in science and engineering. They have also been used for describing dynamics of various biological systems. Therefore, it is reasonable to use differential equations for inference of genetic networks, and many such studies have been done.

First, we consider inference methods based on *linear differential equations*. Let $V = \{v_1, \dots, v_n\}$ be a set of nodes (i.e., a set of genes). Let $v_i(t)$ denote the expression level of gene v_i at time t. Then, we assume that $v_i(t)$s ($i = 1, \dots, n$) are determined by the following differential equations:

$$\frac{dv_i(t)}{dt} = a_{i,0} + a_{i,1}v_1(t) + \cdots + a_{i,n}v_n(t),$$

where $a_{i,j}$s are parameters. We relate these parameters to the network structure by the correspondence that there exists a directed edge (v_j, v_i) if and only if $a_{i,j} \neq 0$. In practice, "$a_{i,j} \neq 0$" should be replaced by "$|a_{i,j}| > \delta$" for some positive constant δ. Furthermore, since only gene expression data for discrete time steps are available, we approximate the differential equations with difference equations:

$$\frac{v_i(t+\Delta) - v_i(t)}{\Delta} = a_{i,0} + a_{i,1}v_1(t) + \cdots + a_{i,n}v_n(t),$$

where Δ is the unit time (i.e., period between two consecutive time steps). The inference problem based on these difference equations is defined as follows: given gene expression time series data $v_i(t)$ for $i = 1, \dots, n$ and $t = 0, \Delta, 2\Delta, \dots, (m-1)\Delta$, infer parameters $a_{i,j}$ for $i = 1, \dots, n, j = 0, \dots, n$ that satisfy the above difference equations.

Then, the difference equations give a system of linear equations in which $v_i(t)$s are constants (since they are given from gene expression data) and $a_{i,j}$s are variables to be determined. However, we cannot expect that this system has a unique solution because the number of equations $n(m-1)$ is not necessarily equal to the number of variables $n(n+1)$. In practice, the number of time points m is much smaller than the number of genes n. Furthermore, we need to consider the effect of observation and system noises. In order to cope with these problems, *linear regression*, precisely, linear regression using *least-squares fitting*, has been applied [9,49]. In this approach, we find a set of parameters, $a_{i,j}$s, that minimizes the sum of least-squared errors:

$$\sum_{t=0}^{m-2} \left| v_i(t+\Delta) - \{v_i(t) + \Delta(a_{i,0} + a_{i,1}v_1(t) + \cdots + a_{i,n}v_n(t))\} \right|^2.$$

This minimization problem can be solved by using matrix computations.

Although the above-mentioned approach is easy to implement, it may cause overfitting owing to unbalance between the number of data and the number of variables. One way to cope with the overfitting problem is to introduce a *regularization term* in the objective function, where L_1 and L_2 regularization terms have widely been used [18]. For example, in the case of L_1 regularization (least absolute shrinkage and selection operator), the objective function is modified as

$$\sum_{t=0}^{m-2} \left| v_i(t+\Delta) - \{v_i(t) + \Delta(a_{i,0} + a_{i,1}v_1(t) + \cdots + a_{i,n}v_n(t))\} \right|^2$$
$$+ \lambda(|a_{i,0}| + |a_{i,1}| + \cdots + |a_{i,n}|),$$

where λ is a parameter to be adjusted (usually by trial and error). It is also useful to incorporate with information criteria such as Akaikes information criteria and Bayesian information criteria [37].

Next, we consider inference based on nonlinear differential equations. Of course, various types of nonlinear differential equations have been applied and thus it is impossible to introduce all or many of them. Therefore, we only introduce the *S-system* [44] because extensive studies have been done on inference of S-systems. In the S-system modeling, differential equations have the following form:

$$\frac{dv_i(t)}{dt} = \alpha_i \prod_{j=1}^{n} v_j(t)^{g_{i,j}} - \beta_i \prod_{j=1}^{n} v_j(t)^{h_{i,j}},$$

where α_i, β_i, $g_{i,j}$, and $h_{i,j}$ $(i = 1, \ldots, n, j = 1, \ldots, n)$ are parameters to be estimated from time series data $v_i(t)$. Again, we approximate these differential equations by difference equations. Different from the linear case, no simple and efficient method is known for determining parameters for this model (and most of other nonlinear models). Therefore, use of evolutionary computation methods is a good choice. Details of such methods are explained in Chapter 5 (by Kimura) and Chapter 8 (by Chowdhury and Chetty) of this book along with other types of nonlinear differential equations.

In addition to differential equation-based modeling, inference methods based on *mass-action modeling* have also been proposed. In this approach, each biochemical reaction R is represented in the following form:

$$a_1S_1 + \cdots + a_NS_N \rightarrow b_1S_1 + \cdots + b_NS_N,$$

where $\{S_1, \ldots, S_N\}$ denotes a set of different molecular species occurring in R and a_i, b_i are non-negative integer parameters called the *stoichiometric coefficients*. Nobel et al. developed a method for inference of genetic networks based on mass-action modeling by combining Cartesian genetic programming and particle swarm optimization. They also discuss an extension of the method to support stochastic simulation algorithms like the Gillespie algorithm. For details of mass-action modeling based approaches, see Chapter 6 (by Nobile et al.).

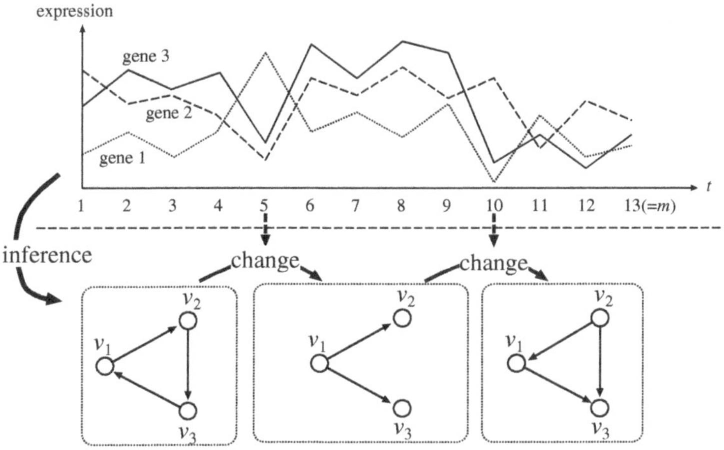

Figure 2.8 Illustration of inference of time-varying networks. In this problem, it is required to identify both change points and network structures.

2.7 TIME-VARYING NETWORK

We have seen various models and methods for inference of genetic networks. All the reviewed models and methods assume that the topology of the network does not change through time. However, the real gene regulatory network in a cell might dynamically change its structure depending on the effects of certain shocks, cell division, and so on. Therefore, it is also important to develop mathematical models and methods for inferring *time-varying genetic networks* based on time series gene expression data (see Figure 2.8). In this problem, it is required to infer both change time points and network structures between two consecutive change time points (including the starting and end time points).

For this problem, various models and methods have been proposed. For example, Yoshida et al. developed a dynamic linear model with Markov switching that represents change points in regimes, which evolve according to a first-order Markov process [54]. Robinson and Hartemink proposed a nonstationary dynamic Bayesian network, based on dynamic Bayesian networks [42]. Fujita et al. proposed a method based on the dynamic autoregressive model by extending the vector autoregression model [14]. Lèbre et al. developed another autoregressive time-varying algorithm by adopting a combination of reversible jump Markov chain Monte Carlo and dynamic Bayesian networks [27]. Thorne and Stumpf developed a method based on modeling of the regulatory network structure between distinct segments with a set of hidden states by applying the hierarchical Dirichlet process and hidden Markov model [51]. Rassol and Bouaynaya developed a method based on constrained and smoothed Kalman filtering [41]. Nakajima and Akutsu extended a network completion method based on dynamic programming and least-squares fitting [36] for time-varying networks [35].

2.8 CONCLUSION

In this chapter, we have briefly reviewed mathematical models and computational methods for inference of genetic networks from gene expression time series data. Since too many methods have been proposed, we could only review basic and simple models and methods. Although we have focused on inference from gene expression data only, integration of various types of data can also be useful for inference of genetic networks, and various methods have been proposed [18] (see also Chapter 3 (by Zhang)).

It is quite difficult to assess which inference method is the best (or better). In order to assess inference methods, we need to know the true genetic networks or gold standard networks. However, it is quite difficult to determine true genetic networks by biological experiments because genes are not directly regulated; they are regulated in a complex way via mRNA, proteins, and other factors (e.g., epigenetic factors). This situation is very different from metabolic networks and protein–protein interaction networks although it is still difficult to determine true protein–protein interactions by biological experiments.

For protein structure prediction, Critical Assessment of protein Structure Prediction (CASP) has been playing an important role in assessment and development of prediction methods [34]. The success of CASP may be due to that accurate structures can be determined by X-ray and NMR experiments. In order to assess inference and analysis methods for gene expression data, DREAM (Dialogue for Reverse Engineering Assessments and Methods) [33] and CAMDA (Critical Assessment of Techniques for Microarray Data Analysis) [50] have been organized, where CAMDA changed its name and scope to Critical Assessment of Massive Data Analysis. Different from CASP, it is difficult to provide the benchmark data and assess the results. Therefore, these projects provide various types of data, which include real gene expression data and simulated gene expression data. It is to be noted that the gold standards are available to assess the prediction results when simulated gene expression data are used. For example, in a recent challenge [34], two subchallenges were proposed: one for parameter estimation for given topology and biochemical structure of a 9-gene regulatory network, and the other for prediction of three missing links for an incomplete topology with 11 genes. An interesting point of this recent challenge is that participants are allowed to (virtually) buy several kinds but limited amount of experimental data that are generated *in silico*. The benchmark data provided in these projects are becoming useful to develop and evaluate new inference and analysis methods.

REFERENCES

1. T. Akutsu, S. Kuhara, O. Maruyama, and S. Miyano. A system for identifying genetic networks from gene expression patterns produced by gene disruptions and overexpressions. *Genome Informatics*, 9:151–160, 1998.

2. T. Akutsu, S. Miyano, and S. Kuhara. Identification of genetic networks from a small number of gene expression patterns under the Boolean network model. In: R. B. Altman, A. K. Dunker, L. Hunter, T. E. Klein (eds.). *Proceedings of Pacific Symposium on Biocomputing 1999*, World Scientific, Singapore, pp. 17–28, 1999.
3. B. Barzel and A.-L. Barabási. Network link prediction by global silencing of indirect correlations. *Nature Biotechnology*, 31:720–725, 2013.
4. D. Cheng, H. Qi, and Z. Li. *Analysis and Control of Boolean Networks. A Semi-Tensor Product Approach.* Springer, Heidelberg, 2011.
5. D. M. Chickering. Learning Bayesian networks is NP-complete. In: D. Fisher and H.-J. Lenz (eds). *Learning from Data: Artificial Intelligence and Statistics V*, Springer, Heidelberg, 1996.
6. K.-H. Cho, S.-M. Choo, S. H. Jung, J.-R. Kim, H.-S. Choi, and J. Kim. Reverse engineering of gene regulatory networks. *IET Systems Biology*, 1:149–163, 2007.
7. G. F. Cooper and E. Herskovits. A Bayesian method for the induction of probabilistic networks from data. *Machine Learning*, 9:309–347, 1992.
8. A. de la Fuente, N. Bing, I. Hoeschele, and P. Mendes. Discovery of meaningful associations in genomic data using partial correlation coefficients. *Bioinformatics*, 20:3565–3574, 2004.
9. P. D'haeseleer, X. Wen, S. Fuhrman, and R. Somogyi. Linear modeling of mRNA expression levels during CNS development and injury. In: R. B. Altman, A. K. Dunker, L. Hunter, T. E. Klein (eds.). *Proceedings of Pacific Symposium on Biocomputing 1999*, World Scientific, Singapore, pp. 41–52, 1999.
10. B. Drossel, T. Mihaljev, and F. Greil. Number and length of attractors in a critical Kauffman model with connectivity one. *Physical Review Letters*, 94:088701, 2005.
11. E. Dubrova and M. Teslenko. A SAT-based algorithm for finding attractors in synchronous Boolean networks. *IEEE/ACM Transactions on Computational Biology and Bioinformatics*, 8:1393–1398, 2011.
12. S. Feizi, D. Marbach, M. Médard, and M. Kellis. Network deconvolution as a general method to distinguish direct dependencies in networks. *Nature Biotechnology*, 31:726–733, 2013.
13. N. Friedman, N. Linial, I. Nachman, and D. Pe'er. Using Bayesian networks to analyze expression data. *Journal of Computational Biology*, 7:601–620, 2000.
14. A. Fujita, J. R. Sato, H. M. Garay-Malpartida, P. A. Morettin, M. C. Sogayar, and C. E. Ferreira. Time-varying modeling of gene expression regulatory networks using the wavelet dynamic vector autoregressive method. *Bioinformatics*, 23:1623–1630, 2007.
15. D. Fukagawa and T. Akutsu. Performance analysis of a greedy algorithm for inferring Boolean functions. *Information Processing Letters*, 93:7–12, 2005.
16. R. Guimerà and M. Sales-Pardo. Missing and spurious interactions and the reconstruction of complex networks. *Proceedings of the National Academy of Sciences of the United States of America*, 106:22073–22078, 2009.
17. S. Hanneke and E. P. Xing. Network completion and survey sampling. *Journal of Machine Learning Research*, 5:209–215, 2009.
18. M. Hecker, S. Lambeck, S. Toepfer, E. van Someren, and R. Guthke. Gene regulatory network inference: data integration in dynamic models—a review. *Biosystems*, 96:86–103, 2009.

19. S. Imoto, T. Higuchi, T. Goto, K. Tashiro, S. Kuhara, and S. Miyano. Combining microarrays and biological knowledge for estimating gene networks via Bayesian networks. *Journal of Bioinformatics and Computational Biology*, 2:77–98, 2004.
20. D. J. Irons. Improving the efficiency of attractor cycle identification in Boolean networks. *Physica D*, 217:7–21, 2006.
21. S. A. Kauffman. Metabolic stability and epigenesis in randomly constructed genetic nets. *Journal of Theoretical Biology*, 22:437–467, 1969.
22. S. A. Kauffman. *The Origins of Order: Self-organization and Selection in Evolution*, Oxford University Press, New York, 1993.
23. G. Karlebach and R. Shamir. Modelling and analysis of gene regulatory networks. *Nature Reviews Molecular Cell Biology*, 9:770–780, 2008.
24. M. Kim and J. Leskovec. The network completion problem: inferring missing nodes and edges in networks. In: B. Liu, H. Liu, C. Clifton, T. Washio, C. Kamath (eds.). *Proceedings of the 2011 SIAM International Conference on Data Mining*, SIAM, Philadelphia, pp. 47–58, 2011.
25. H. Kishino and P. J. Waddell. Correspondence analysis of genes and tissue types and finding genetic links from microarray data. *Genome Informatics*, 11:83–95, 2000.
26. H. Lähdesmäki, S. Hautaniemi, I. Shmulevich, and O. Yli-Harja. Relationships between probabilistic Boolean networks and dynamic Bayesian networks as models of gene regulatory networks. *Signal Processing*, 86:814–834, 2006.
27. S. Lèbre, J. Becq, F. Devaux, M. P. H. Stumpf, and G. Lelandais. Statistical inference of the time-varying structure of gene-regulation networks. *BMC Systems Biology*, 4:130, 2010.
28. T. R. Lezon, J. R. Banavar, M. Cieplak, A. Maritan, and N. V. Fedoroff. Using the principle of entropy maximization to infer genetic interaction networks from gene expression patterns. *Proceedings of the National Academy of Sciences of the United States of America*, 103:19033–19038, 2006.
29. S. Liang, S. Fuhrman, and R. Somogyi. REVEAL, a general reverse engineering algorithm for inference of genetic network architectures. In: R. B. Altman, A. K. Dunker, L. Hunter, T. E. Klein (eds.). *Proceedings of Pacific Symposium on Biocomputing 1998*, World Scientific, Singapore, pp. 18–29, 1998.
30. T.-F. Liu, W.-K. Sung, and A. Mittal. Learning gene network using time-delayed Bayesian framework. *International Journal on Artificial Intelligence Tools*, 15:353–370, 2006.
31. A. A. Margolin, K. Wang, W. K. Lim, M. Kustagi, I. Nemenman, and A. Califano. Reverse engineering cellular networks. *Nature Protocols*, 1:662–671, 2006.
32. A. A. Melkman and T. Akutsu. An improved satisfiability algorithm for nested canalyzing functions and its application to determining a singleton attractor of a Boolean network. *Journal of Computational Biology*, 20:958–969, 2013.
33. P. Meyer, et al. Network topology and parameter estimation: from experimental design methods to gene regulatory network kinetics using a community based approach. *BMC Systems Biology*, 8:13, 2014.
34. J. Moult, K. Fidelis, A. Kryshtafovych, T. Schwede, and A.Tramontano. Critical assessment of methods of protein structure prediction (CASP) round x. *Proteins: Structure, Function, and Bioinformatics*, 82(S2):16, 2014.
35. N. Nakajima and T. Akutsu. Exact and heuristic methods for network completion for time-varying genetic networks. *BioMed Research International*, 2014:684014, 2014.

36. N. Nakajima, T. Tamura, Y. Yamanishi, K. Horimoto, and T. Akutsu. Network completion using dynamic programming and least-squares fitting. *The ScientificWorld Journal*, 2012:957620, 2012.
37. N. Noman, L. Palafox, and H. Iba. On model selection criteria in reverse engineering gene networks using RNN model. In: G. Lee, D. Howard, J. J. Kang, D. Ślęzak (eds.). *Proceedings of the International Conference on Convergence and Hybrid Information Technology*, Springer, Berlin, pp. 155–164, 2012.
38. S. Ott, S. Imoto, and S. Miyano. Finding optimal models for small Gene networks. In: R. B. Altman, A. K. Dunker, L. Hunter, T. E. Klein (eds.). *Proceedings of Pacific Symposium on Biocomputing 2004*, World Scientific, Singapore, pp. 557–567, 2004.
39. T. J. Perkins and M. J. Hallett. A trade-off between sample complexity and computational complexity in learning Boolean networks from time-series data. *IEEE/ACM Transactions on Computational Biology and Bioinformatics*, 7:118–125, 2010.
40. B.-E. Perrin, L. Ralaivola, A. Mazurie, S. Bottani, J. Mallet, and F. dAlchéBuc. Gene networks inference using dynamic Bayesian networks. *Bioinformatics*, 19:ii138–ii148, 2013.
41. G. Rassol and N. Bouaynaya. Inference of time-varying gene networks using constrained and smoothed Kalman filtering. In: *Proceedings of International Workshop on Genomic Signal Processing and Statistics*, IEEE, Piscataway, NJ, pp. 172–175, 2012.
42. J. Robinson and A. Hartemink. Non-stationary dynamic Bayesian networks. In: D. Koller, D. Schuurmans, Y. Bengio, L. Bottou (eds.). *Proceedings of the 22nd Annual Conference on Neural Information Processing Systems*, Curran Associates, Inc., Red Hook, NY, pp. 1369–1376, 2008.
43. B. Samuelsson and C. Troein. Superpolynomial growth in the number of attractors in Kauffman networks. *Physical Review Letters*, 90:098701, 2003.
44. M. Savageau. *Biochemical Systems Analysis. A Study of Function and Design in Molecular Biology.* Addison-Wesley, Massachusetts, 1976.
45. J. Schäfer and K. Strimmer. An empirical Bayes approach to inferring large-scale gene association networks. *Bioinformatics*, 21:754–764, 2005.
46. I. Shmulevich and E. R. Dougherty. *Probabilistic Boolean Networks: The Modeling and Control of Gene Regulatory Networks.* SIAM, Philadelphia, 2010.
47. I. Shmulevich, E. R. Dougherty, S. Kim, and W. Zhang. Probabilistic Boolean networks: a rule-based uncertainty model for gene regulatory networks. *Bioinformatics*, 18:261–274, 2002.
48. C. Sima, J. Hua, and S. Jung. Inference of gene regulatory networks using time-series data: a survey. *Current Genomics*, 10:416–429, 2009.
49. E. P. van Someren, L. F. A. Wessels, and M. J. T. Reinders. Linear modeling of genetic networks from experimental data. In: P. E. Bourne, M. Gribskov, R. B. Altman, N. Jensen, D. A. Hope, T. Lengauer, J. C. Mitchell, E. D. Scheeff, C. Smith, S. Strande, H. Weissig (eds.). *Proceeding of the 18th International Conference on Intelligent Systems for Molecular Biology*, AAAI Press, Palo Alto, CA, pp. 355–366, 2000.
50. C. Telstone. Vital statistics. *Nature*, 424:610–512, 2003.
51. T. Thorne and M. P. H. Stumpf. Inference of temporally varying Bayesian Networks. *Bioinformatics*, 28:3298–3305, 2012.
52. H. Toh and K. Horimoto. Inference of a genetic network by a combined approach of cluster analysis and graphical Gaussian modeling. *Bioinformatics*, 18:287–297, 2002.

53. R. Trairatphisan, A. Mizera, J. Pang, A. A. Tantar, H. Schneider, and T. Sauter. Recent development and biomedical applications of probabilistic Boolean networks. *Cell Communication and Signaling*, 11:46, 2013.
54. R. Yoshida, S. Imoto, and T. Higuchi. Estimating time-dependent gene networks from time series microarray data by dynamic linear models with Markov switching. In: V. Markstein (ed). *Proceedings of 4th Computational Systems Bioinformatics*, IEEE, Piscataway, NJ, pp. 289–298. 2005.
55. Y. Yuan, C-T. Li, and O. Windram. Directed partial correlation: inferring large-scale gene regulatory network through induced topology disruptions. *PLoS One*, 6:e16835, 2011.

3

GENE REGULATORY NETWORKS: REAL DATA SOURCES AND THEIR ANALYSIS

Yuji Zhang

Department of Epidemiology and Public Health, University of Maryland School of Medicine, Baltimore, MD, USA

and

Division of Biostatistics and Bioinformatics, University of Maryland Greenebaum Cancer Center, Baltimore, MD, USA

3.1 INTRODUCTION

In all living organisms, deoxyribonucleic acid (DNA), ribonucleic acid (RNA), and protein are three types of biological macromolecules that are indispensable for all biological processes. They are unbranched polymer chains, formed by the string together of monomeric building blocks drawn from a standard repertoire that is the same for all living cells. These molecules interact with each other frequently, and conditionally depend on each other to provide complex biological functions (e.g., functions of a protein are usually provided by its interactions with other proteins and genes). These molecules and their interactions compose complex networks, called gene regulatory networks (GRNs).

Gene regulatory networks are one of the most important biological networks in the bioinformatics and systems biology field. They play a vital role in almost

Evolutionary Computation in Gene Regulatory Network Research, First Edition.
Edited by Hitoshi Iba and Nasimul Noman.

every biological process, including cell differentiation, metabolism, cell cycle, and signal transduction. By understanding the properties and dynamics of these networks, we can shed light on the mechanisms of diseases that occur when these cellular processes are dysregulated. Analysis and inference of GRNs will also guide biologists in downstream biological experimental designs, as such inferences are more time- and cost-effective than wet lab experiment validations. In general, there are two different types of computational approaches for analysis and inference of GRNs:

1. Topological analysis of GRN: based on the regulatory interactions (e.g., protein–DNA interactions and protein–protein interactions) collected in public databases (e.g., Human Protein Reference Database [1], IntAct [2], Biomolecular Interaction Network Database [3], and Search Tool for the Retrieval of Interacting Genes/Proteins [4]) and genome-wide high-throughput experiments [5, 6], several network analysis approaches have been proposed to investigate the topological properties of GRNs in different organisms [7–9].
2. Inference of gene regulatory relationships: based on gene expression data (e.g., time-series gene expression data), a series of computational approaches have been developed for reconstruction and inference of gene regulatory relationships at genome-wide level [10–12].

This chapter is organized as follows: we will first review the biological data sources available for analysis and inference of GRNs, and then we will introduce the topological analysis approaches for GRNs. We will briefly review different types of computational approaches for GRN inference as well as our proposed approach for GRN inference by integrating prior biological knowledge. Finally, we will conclude the chapter with discussions and future works in the GRN analysis and inference field.

3.2 BIOLOGICAL DATA SOURCES

In this section, we describe multiple sources of biological data that have been used for GRN analysis and inference. This will help us better understand how integration of these different types of biological data brings us a more complete picture in GRN inference.

3.2.1 Gene Expression Data

Gene expression data can be divided into two levels: mRNA level and protein level. In this chapter, we focus on the gene expression data on mRNA level, including cDNA microarrays, high-density oligonucleotide chips, reverse transcriptase polymerase chain reaction (RT-PCR), and RNA-seq.

3.2.1.1 cDNA Microarrays Originally developed at Stanford University, cDNA microarrays are glass slides on which cDNA has been deposited by high-speed robotic printing [13]. They are ideally suited for expression analysis of up to 50,000 cDNA clones per array from expressed sequence tag sequencing projects (e.g., private effort at Incyte Pharmaceuticals and the public Washington University project). Measurements are carried out as differential hybridizations to minimize errors originating from cDNA spotting variability: mRNA from two different sources (e.g., control and drug treated), labeled with two different fluorescent dyes, is passed over the array at the same time. The fluorescence signal from each mRNA population is evaluated independently, and then used to calculate the treated/control expression ratio.

3.2.1.2 *High-Density Oligonucleotide Chips* These chips, produced by Affymetrix [14], consist of small glass plates with thousands of short 20-mer oligonucleotide probes attached to their surface. The oligonucleotides are synthesized directly onto the surface using a combination of semiconductor-based photolithography and light-directed chemical synthesis. Due to the combinatorial nature of the process, very large numbers of mRNAs can be probed at the same time. However, manufacturing and reading of the chips requires expensive equipment. Current chips have over 65,000 different probes, with typically several probes for each mRNA.

3.2.1.3 *RT-PCR* To measure gene expression using RT-PCR, the mRNA is first reverse-transcribed into cDNA, and the cDNA is then amplified to measurable levels using PCR [15]. Using built-in calibration techniques, RT-PCR can achieve high accuracy coupled with an exceptional sensitivity of 10 molecules/10 μl and a dynamic range covering 6–8 orders of magnitude. The method requires PCR primers for all the genes of interest, and is not inherently parallel like the previous two, so automation is crucial to scale up.

Roland Somogyi used this method to measure the expression levels of 112 genes at 9 different time points during the development of rat cervical spinal cord [16], and 70 genes during development and following injury of the hippocampus.

3.2.1.4 *RNA-Seq* RNA-seq, also called "whole transcriptome shotgun sequencing," is a recently developed approach to transcriptome profiling that uses deep-sequencing technologies [17]. Compared to hybridization-based approaches, RNA-seq has the advantages to (1) detect novel transcripts, (2) have very low background noise, (3) contain a large dynamic range of expression levels over which transcripts can be detected, (4) have higher level of reproducibility, and (5) require less RNA sample [17]. It has been applied to various organisms, including *Saccharomyces cerevisiae* [18], *Schizosaccharomyces pombe* [19], *Arabidopsis thaliana* [20], zebrafish [21], mouse [22], and human cells [23].

3.2.2 Protein–Protein Interaction Data

Protein–protein interactions are essential for a wide range of cellular processes and form a network of astonishing complexity. Until recently, our knowledge of this complex network was rather limited. The emergence of large-scale protein–protein interaction maps has given us new possibilities to systematically survey and study the underlying biological system. First attempts to collect protein–protein interactions on a large scale were initiated for model organisms such as *S. cerevisiae*, *Drosophila melanogaster*, and *Caenorhabditis elegans* [24–28]. Evidently, the generated interaction maps offered a rich resource for systematic studies of molecular networks.

After these initial efforts, the focus has moved toward deciphering the human protein–protein interactions. Most currently available human interaction maps can be divided into three classes: (1) maps obtained from the literature search [29–31], (2) maps derived from interactions among orthologous proteins in other organisms [32–34], and (3) maps based on large scans using yeast two-hybrid (Y2H) assays [35, 36]. All of these different mapping strategies have their obvious advantages as well as disadvantages. For example, Y2H-based mapping approaches offer rapid screens between thousands of proteins, but might be compromised by large false-positive rates. The extent, however, of how much the resulting interaction maps are influenced by the choice of mapping strategy is less clear. Thus, it is important to critically assess and compare the quality and reliability of produced maps.

Protein–protein interaction networks are commonly represented in a graph format, with vertices corresponding to proteins and edges corresponding to protein–protein interactions [37]. The network consists of many small subnets (groups of proteins that interact with each other but not with any other protein) and one large, connected subnet comprising more than half of all interacting proteins. The volume of experimental data on protein–protein interactions is rapidly increasing, thanks to high-throughput biotechniques that are able to produce a large number of protein–protein interactions. For instance, yeast contains over 5000 proteins, and currently about 18,000 protein–protein interactions have been identified among the yeast proteins, with hundreds of labs around the world constantly adding to this list [38]. The analogous networks for mammals are expected to be much larger. For example, humans are expected to have around 12,000 proteins and about 10^6 interactions.

3.2.3 Protein–DNA Interaction Data

Currently, there are several different sources available for protein–DNA interactions: (1) experimental data from genome-wide location analysis (GWLA) [39], such as ChIP-chip [40] and ChIP-seq [41], (2) curated binding information in public databases, such as TRANSFAC [42], and (3) putative binding sites based on computational prediction algorithms [43]. As an *in vivo* study, GWLA technology is biologically most significant, but provides only the roughest information about possible binding sites. As more advanced sequencing technologies are being developed, it is expected that such approaches will overcome these limitations in the future. Curated information in public databases, on the other hand, presents a compilation of mostly *in vitro* studies and provides more accurate information, but at the expense of only small coverage of all intergenic regions. The third method is based on *in silico* predictions and provides the most detailed information on DNA-binding site locations, but contains the highest rate of false positives. As more interactions accumulate from different resources, interactions that are identified by more than one resource will be considered as high-confidence interactions. This will help us reduce the false positives arising from different resources.

3.2.4 Gene Ontology

The Gene Ontology (GO) Consortium [44] has developed three separate ontologies—molecular function, biological process, and cellular component—to describe the attributes of gene products, where molecular function defines what a gene product does at the biochemical level without specifying where or when the event actually occurs or its broader context; biological process describes the contribution of a gene product to a biological objective, and cellular component refers to where in the cell a gene product functions. Each GO is structured as a directed acyclic graph, wherein each term is a child of one or multiple parents, and child terms are instances or components of parent terms. For example, in Figure 3.1, the term S phase of meiotic cell cycle (GO: 0051332) is an instance of the term S phase (GO: 0051320) as well as an instance of the term interphase of meiotic cell cycle (GO: 0051328). Such information can be incorporated into GRN analysis and inference approaches to increase the inference accuracy [45,46].

3.3 TOPOLOGICAL ANALYSIS OF GENE REGULATORY NETWORKS

A GRN can be defined as a directed graph: the nodes represent the genes, a directed edge from one node to another indicates that the first gene codes for a transcription factor regulates the second genes, and an undirected edge between two nodes represents their interactions at protein level, which further direct these two proteins to regulate some common downstream genes [46]. The architecture of GRN can be described by means of graph features such as node degree, network diameter, and clustering coefficient. We briefly introduce these concepts, followed by a description of GRN analysis approaches at different network levels.

3.3.1 Node Degree

The degree of a node is defined as the number of edges that connect to this node. In directed networks, the number of incoming edges is called the in-degree, and the number of outgoing edges is called the out-degree. If a node has a high degree, it indicates that this node is connected to many other nodes in the network. Biological networks are not randomly organized but have a scale-free architecture with the typical power law degree distribution, that is, only a small number of nodes have a high degree while most nodes have a small degree. The GRN network, protein–protein interaction network, and metabolic network are all scale-free networks [47]. The advantage of this kind of organization is that the loss of one non-hub link is not as disruptive in scale-free networks as in random networks. In other words, scale-free networks are generally more robust. The hubs are extremely important and usually play essential roles in many biological systems [48].

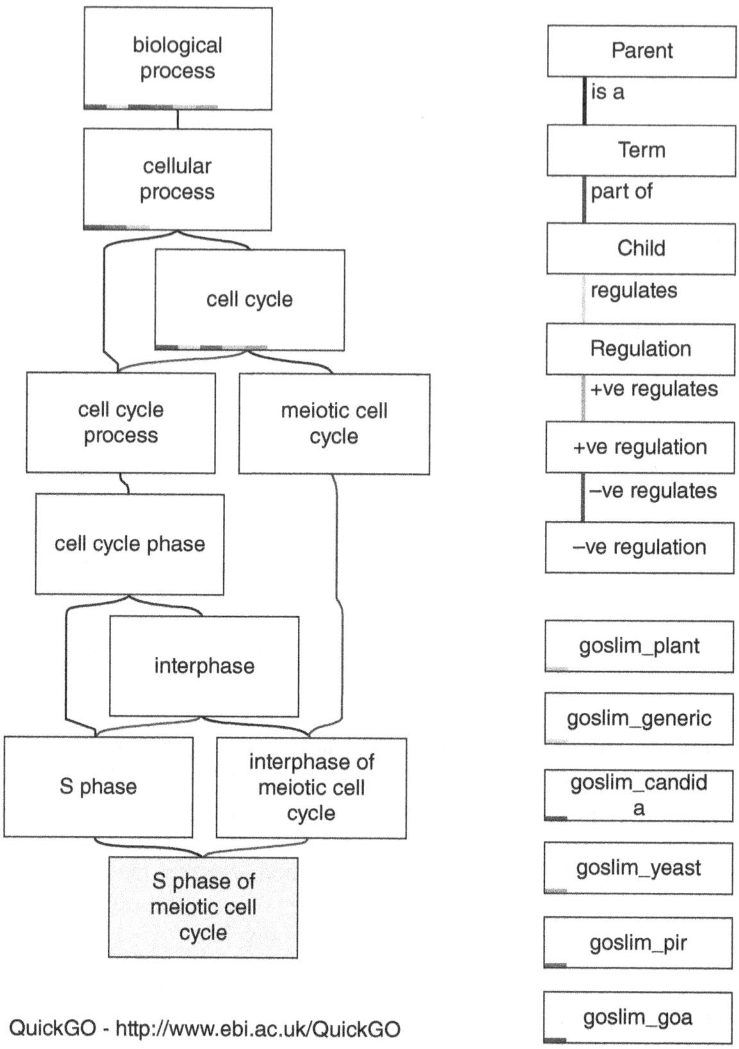

Figure 3.1 The directed acyclic graph induced from the GO term S phase of meiotic cell cycle (GO: 0051332), wherein at the bottom-most level is the GO term of interest itself, and at the upper levels are all its ancestors, adapted from QuickGO GO Browser (http://www.ebi.ac.uk/ego/). GO, Gene ontology.

3.3.2 Neighborhood Connectivity

The connectivity of a node is the number of its neighbors. The neighborhood connectivity of a node is defined as the average connectivity of its all neighbors. In analogy to the in- and out-degree, every node in a directed network has in- and out-connectivity.

3.3.3 Shortest Paths

The shortest path length between two nodes in one network is called the node distance. The network diameter is the maximum length of shortest paths between any two nodes in one network. If a network is disconnected, its diameter is the maximum of all diameters of its connected components. The distribution of shortest path length and network diameter can indicate small-world properties of the analyzed network [49]. Many biological networks, such as GRN and metabolic networks, are known to exhibit this small-world property.

There are also other important network parameters such as clustering coefficient, betweenness centrality, and stress centrality in network topological analysis; please refer to Doncheva et al. [50] for a detailed review.

In addition to the network properties described in Section 3.3, studies have also unveiled that biological networks have modular structure in most organisms [51–54]. Indeed, biological processes consist of pathways that mainly act on their own and cross talk with each other under certain conditions. Therefore, it is expected that the distinct biological processes can be organized in discrete and separable modules. A module in a network can be defined in various ways [55]:

1. One popular definition of a module involves co-expressed genes, with or without environmental context dependence and assigning a regulatory motif or regulator to these genes. This definition of a module will refer to a gene module in the remainder of the present chapter.
2. Another definition of topological module can be defined by means of graph-based approaches. For instance, network motifs (NMs) are one of the smallest modules in GRNs, which will be called as gene regulatory modules in the remainder of the present chapter.

Furthermore, Shen-orr et al. [56] discovered the presence of NMs in the transcriptional network of *Escherichia coli*. Network motifs are the smallest building blocks in networks. They are topologically distinct regulatory interaction patterns that are present more frequently in true biological networks than in random networks. Therefore, these motifs must have a specific biological function: they are postulated to be the basic signal transduction elements, each with its own characteristic properties. Shen-orr et al. were the first to identify these NMs. Examples of NMs are the single input motif and multiple input and feed-forward loop motifs [56].

3.3.4 Reconstruction of Transcriptional Regulatory Network

Besides network properties of GRNs, inference of GRNs is another challenging topic in the bioinformatics field. The purpose of GRN inference is to determine for all transcription factors the regulatory mechanisms they recognize, the conditions in which they are active, the regulators they cooperate with in these conditions, and their target genes in these conditions. The approach can be categorized into two groups: bottom-up and top-down approaches [57, 58]:

- *Bottom-up* approaches start from a comprehensive expert model of known interactions among molecular entities as described in literature and curated databases. Such models can be used to simulate cellular behavior or to predict the outcome of a perturbation experiment. Inconsistencies between observed data and simulations will direct to deficiencies in the current network structure and outline hypotheses of novel interactions that can better explain the observations.
- *Top-down* approaches start from a global view of the behavior of the whole biological system by using high-throughput data. This type of inference methods does not rely on expert knowledge on the relationships among the molecular components. Top-down inference is a data-driven and thus data-demanding approach. Given the current data availability, top-down network inference problems are often underdetermined (i.e., the network that is reconstructed from the data is not unique, and many equally likely solutions can explain the observations). However, the top-down inference can be made increasingly tractable by integrating data from different sources, and holds great promise for future bioinformatics research.

One approach to tackle the above underdetermined problem in top-down network inference is to integrate the multi-complementary high-throughput datasets. Transcriptional regulation is a process that needs to be understood at multiple levels of description [59, 60] (Figure 3.2), including (1) the factor–target gene interaction, in which transcription factors activated under certain conditions interact with their conserved binding site sequences, and (2) transcriptional regulation, which explains how the bindings of transcription factors to their unique recognition sites regulate the expression of specific genes. A single source of information such as gene expression data is aimed at only one level of description (i.e., transcriptional regulation level), and thus it is limited in its ability to obtain a full understanding of the entire regulatory process. Other types of information such as protein–protein interaction [61, 62] and protein–DNA interaction [40] data provide complementary constraints on the models of regulatory processes. By integrating limited but complementary data sources, we can realize a mutually consistent hypothesis bearing stronger similarity to the underlying causal structures [60]. Among the various types of high-throughput biological data available nowadays, time-course gene expression profiles and GWLA data are two complementary sets of information that can be used to infer regulatory components. Time-course gene expression data are advantageous over typical static expression profiles as time can be used to disambiguate causal interactions. GWLA data, on the other hand, provide high-throughput quantitative information about *in vivo* binding of transcription factors to the target regulatory regions of the DNA. Incorporation of prior biological knowledge accumulated in literature will help guide inference from the above datasets, and integration of multiple data sources offers insights into the cellular system at different levels [46].

Another way to reduce the complexity of the GRN inference problem is to decompose it into small units of commonly used network structures, called gene regulatory modules. As we introduced in this section, GRNs are made of repeated occurrences of

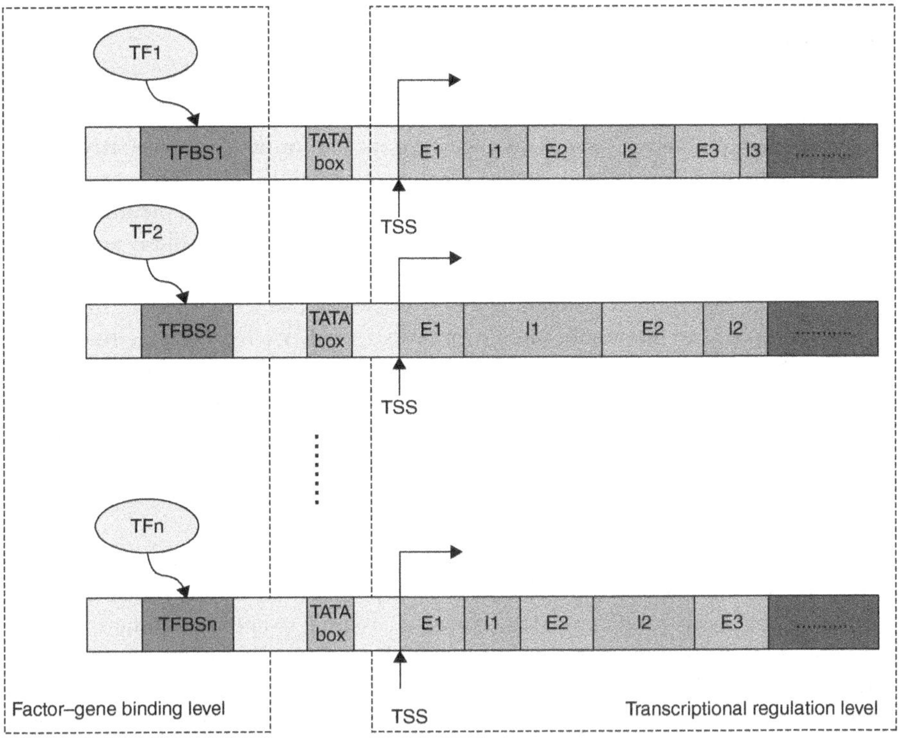

Figure 3.2 The gene transcriptional regulatory program. The gene transcriptional regulatory program can be simplified at two levels. At the factor–gene binding level, the "activated" transcription factors bind to their specific conserved sequence motifs, called transcription factor binding sites. When the binding process is completed, the regulation mechanism instructs the gene transcription from transcriptional start site (DNA to mRNA); first part of the central dogma in molecular biology. This figure was adapted from Zhang et al. [46].

simple patterns–NMs. Since the establishment of the first NM in *E. coli* [56], similar NMs have also been found in eukaryotes including yeast [63], plants, and animals [64–66], suggesting that the general structure of NMs is evolutionarily conserved. One well-known family of NMs is the feed-forward loop [67], which appears in hundreds of gene systems in *E. coli* [56, 68] and yeast [63, 69], as well as in other organisms [64–66, 70–72]. A comprehensive review on NM theory and experimental approaches is presented in Ref. [73]. Knowledge of the NMs to which a given transcription factor belongs facilitates the identification of downstream target gene clusters. In yeast, a GWLA was carried out for 106 transcription factors and 5 NMs were considered significant: autoregulation, feed-forward loop, single input module, multi-input module, and regulator cascade [63]. In Section 3.4, we will review commonly used models for GRN inference, followed by the introduction of our proposed computational approach integrating multi-source biological data for GRN inference.

3.4 GRN INFERENCE BY INTEGRATION OF MULTI-SOURCE BIOLOGICAL DATA

In the last two decades, a variety of continuous or discrete, static or dynamic, and quantitative or qualitative models have been proposed for inference of GRNs. These include biochemically driven methods [74], linear models [75,76], Boolean networks [77], fuzzy logic [78,79], Bayesian networks [80,81], and recurrent neural networks (RNNs) [82–84]. Chapter 2 provides a detailed description of these approaches. However, all these computational approaches described in Section 3.3 still cannot solve the underdetermined problem in GRN inference due to the typical small sample size compared to the number of genes investigated. We hypothesize that we can enhance our understanding of gene interactions in important biological processes and improve the inference accuracy of a GRN by (1) incorporating prior biological knowledge into the inference scheme, (2) integrating multiple biological data sources, and (3) decomposing the inference problem into smaller network modules. In this section, we will introduce our proposed integrative framework to tackle these challenges.

The GRN inference only based on gene expression data is inadequate because of its intrinsic complexity and noise in gene expression data. Integrating data from multiple global assays and curated databases is essential to understand the spatiotemporal interactions within cells. Different experiments measure cellular processes at various widths and depths, while databases contain biological information based on established facts or published data. Integrating these complementary datasets helps infer a mutually consistent transcriptional regulatory network with strong similarity to the structure of the underlying gene regulatory modules. Decomposing the transcriptional regulatory network into a small set of recurring regulatory patterns, called NMs, facilitates the inference. Identifying NMs defined by specific transcription factors establishes the modular framework structure of a transcriptional regulatory network and allows the inference of transcription factor–target gene relationship. This section introduces a computational framework for utilizing data from multiple sources to infer transcription factor–target gene relationships on the basis of NM regulatory modules. The data include time-course gene expression profiles, molecular interaction data, and GO information.

In the proposed framework, we consider two different layers of networks in the GRN. One is the molecular interaction network that includes protein–protein interactions and protein–DNA interactions at the factor–gene binding level. The other is the functional network that incorporates the consequences of these physical interactions, such as the activation or repression of transcription. We used three types of data to reconstruct the GRN, namely protein–protein interactions derived from a collection of public databases, protein–DNA interactions from the TRANSFAC database [42], and time-course gene expression profiles. The first two data sources provided direct network information to constrain the GRN model. The gene expression profiles provided an unambiguous measurement on the causal effects of the GRN model. GO annotation describes the similarities among genes within one network, which facilitates further characterization of the relationships between genes. The goal is to discern

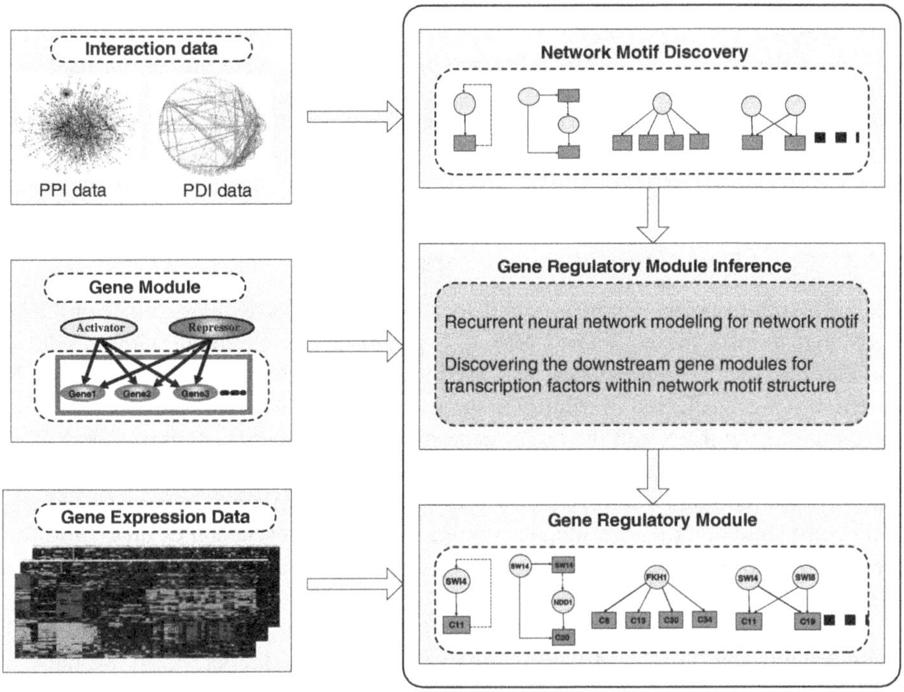

Figure 3.3 Schematic overview of the computational framework used for the gene regulatory module inference. PPI, protein–protein interaction; PDI, protein–DNA interaction.

dependencies between the gene expression patterns and the physical intermolecular interactions revealed by complementary data sources.

The framework model for GRN inference is illustrated in Figure 3.3. Three successive steps were involved in this framework as outlined in the following:

3.4.1 Gene Module Selection

Genes with similar expression profiles were represented by a gene module to address the scalability problem in GRN inference [79]. The assumption is that a subset of genes that are related in terms of expression (co-regulated) can be grouped together by virtue of a unifying *cis*-regulatory element(s) associated with a common transcription factor regulating each and every member of the cluster (co-expressed) [85]. GO information was utilized to define the optimal number of clusters with respect to certain broad functional categories. Since each gene module identified from clustering analysis mainly represents one broad biological or process category as evaluated by FuncAssociate [86], the regulatory network implies that a given transcription factor is likely to be involved in the control of a group of functionally related genes [87].

3.4.2 Network Motif Discovery

To reduce the complexity of the inference problem, NMs were utilized instead of a global GRN inference. The significant NMs in the combined molecular interaction network were first established and assigned to at least one transcription factor. These associations were further used to reconstruct the regulatory modules. This step was implemented using the FANMOD tool.

3.4.3 Gene Regulatory Module Inference

For each transcription factor assigned to an NM, a RNN was trained to model a GRN that mimics the associated NM. GA generated the candidate gene modules, and Particle Swarm Optimization was used to configure the parameters of the RNN. Parameters were selected to minimize the root-mean-square error (RMSE) between the output of the RNN and the target gene module's expression pattern. The RMSE was returned to GA to produce the next generation of candidate gene modules. Optimization continued until either a pre-specified maximum number of iterations were completed or a pre-specified minimum RMSE was reached. The procedure was repeated for all transcription factors. Biological knowledge from databases was used to evaluate the predicted results.

We applied this computational framework to two biological processes: yeast cell cycle progression process [88] and human *Hela* cancer cell cycle [89]. We demonstrate that our method can accurately infer the underlying relationships between transcription factor and the downstream target genes by integrating multi-sources of biological data. The predictive strength of this strategy is based on the combined constraints arising from multiple biological data sources including time-course gene expression data, combined molecular interaction network data, and GO category information.

3.5 CONCLUSIONS AND FUTURE DIRECTIONS

The analysis and inference of GRNs is a major obstacle in bioinformatics majorly due to (1) intrinsic complexity of gene regulation mechanisms, (2) limited sample size compared to gene numbers in one experiment, and (3) noise in gene expression data themselves. Computational approaches integrating multi-source biological data can address this challenge by reverse engineering GRNs at NM level. However, there are still needs for developing models that can integrate more types of biological data.

Biological systems are characterized by many highly interconnected levels. Most approaches analyze the GRNs at transcriptional level. This network may be augmented with additional types of relations, which can then provide further insight into other types of cellular mechanisms. One of the major tasks ahead is therefore the integration of more sources of information. One intriguing dataset to add is that of signal transduction pathways, including directed protein–protein interactions as those between kinases and their substrates, and interactions between signaling molecules (e.g., pheromone) and their targets. Assuming these data were available, it will enable the characterization of signal transduction pathways and their control mechanisms. Attempts to collect genome-wide signaling data are underway, for example, using protein chips designed to test kinase phosphorylation interactions as performed by Snyder et al. [90].

ACKNOWLEDGMENT

This work is supported in part by grant P30 CA 134274-04 from the NCI.

REFERENCES

1. Keshava Prasad, T.S., et al., Human Protein Reference Database—2009 update. *Nucleic Acids Res*, 2009. *37*(Database issue): D767–D772.
2. Kerrien, S., et al., The IntAct molecular interaction database in 2012. *Nucleic Acids Res*, 2012. *40*(Database issue): D841–D846.
3. Willis, R.C. and C.W. Hogue, Searching, viewing, and visualizing data in the Biomolecular Interaction Network Database (BIND). *Curr Protoc Bioinformatics*, 2006. *Chapter 8*: Unit 8.9.
4. Franceschini, A., et al., STRING v9.1: protein–protein interaction networks, with increased coverage and integration. *Nucleic Acids Res*, 2013. *41*(Database issue): D808–D815.
5. Schmidt, D., et al., ChIP-seq: using high-throughput sequencing to discover protein-DNA interactions. *Methods*, 2009. *48*(3): 240–248.
6. Wu, J., et al., ChIP–chip comes of age for genome-wide functional analysis. *Cancer Res*, 2006. *66*(14): 6899–6902.
7. Luscombe, N.M., et al., Genomic analysis of regulatory network dynamics reveals large topological changes. *Nature*, 2004. *431*(7006): 308–312.
8. Peter, I.S. and E.H. Davidson, Modularity and design principles in the sea urchin embryo gene regulatory network. *FEBS Lett*, 2009. *583*(24): 3948–3958.
9. Davidson, E.H., Emerging properties of animal gene regulatory networks. *Nature*, 2010. *468*(7326): 911–920.
10. Marbach, D., et al., Revealing strengths and weaknesses of methods for gene network inference. *Proc Natl Acad Sci USA*, 2010. *107*(14): 6286–6291.
11. Sirbu, A., H.J. Ruskin, and M. Crane, Comparison of evolutionary algorithms in gene regulatory network model inference. *BMC Bioinformatics*, 2010. *11*: 59.
12. De Smet, R. and K. Marchal, Advantages and limitations of current network inference methods. *Nat Rev Microbiol*, 2010. *8*(10): 717–729.
13. Schena, M., et al., Quantitative monitoring of gene expression patterns with a complementary DNA microarray. *Science*, 1995. *270*(5235): 467–470.
14. McGall, G.H. and J.A. Fidanza, Photolithographic synthesis of high-density oligonucleotide arrays. *Methods Mol Biol*, 2001. *170*: 71–101.
15. Joyce, C., Quantitative RT-PCR. A review of current methodologies. *Methods Mol Biol*, 2002. *193*: 83–92.
16. Wen, X., et al., Large-scale temporal gene expression mapping of central nervous system development. *Proc Natl Acad Sci USA*, 1998. *95*(1): 334–339.

17. Wang, Z., M. Gerstein, and M. Snyder, RNA-seq: a revolutionary tool for transcriptomics. *Nat Rev Genet*, 2009. *10*(1): 57–63.
18. Nagalakshmi, U., et al., The transcriptional landscape of the yeast genome defined by RNA sequencing. *Science*, 2008. *320*(5881): 1344–1349.
19. Wilhelm, B.T., et al., Dynamic repertoire of a eukaryotic transcriptome surveyed at single-nucleotide resolution. *Nature*, 2008. *453*(7199): 1239–1243.
20. Lister, R., et al., Highly integrated single-base resolution maps of the epigenome in Arabidopsis. *Cell*, 2008. *133*(3): 523–536.
21. Craig, T.A., et al., Research resource: whole transcriptome RNA sequencing detects multiple 1alpha,25-dihydroxyvitamin D(3)-sensitive metabolic pathways in developing zebrafish. *Mol Endocrinol*, 2012. *26*(9): 1630–1642.
22. Cloonan, N., et al., Stem cell transcriptome profiling via massive-scale mRNA sequencing. *Nat Methods*, 2008. *5*(7): 613–139.
23. Morin, R., et al., Profiling the HeLa S3 transcriptome using randomly primed cDNA and massively parallel short-read sequencing. *Biotechniques*, 2008. *45*(1): 81–94.
24. Gavin, A.C., et al., Functional organization of the yeast proteome by systematic analysis of protein complexes. *Nature*, 2002. *415*(6868): 141–147.
25. Giot, L., et al., A protein interaction map of *Drosophila melanogaster*. *Science*, 2003. *302*(5651): 1727–1736.
26. Ito, T., et al., A comprehensive two-hybrid analysis to explore the yeast protein interactome. *Proc Natl Acad Sci USA*, 2001. *98*(8): 4569–4574.
27. Li, S., et al., A map of the interactome network of the metazoan *C. elegans*. *Science*, 2004. *303*(5657): 540–543.
28. Uetz, P., et al., A comprehensive analysis of protein–protein interactions in *Saccharomyces cerevisiae*. *Nature*, 2000. *403*(6770): 623–627.
29. Bader, G.D., D. Betel, and C.W. Hogue, BIND: the Biomolecular Interaction Network Database. *Nucleic Acids Res*, 2003. *31*(1): 248–250.
30. Peri, S., et al., Development of human protein reference database as an initial platform for approaching systems biology in humans. *Genome Res*, 2003. *13*(10): 2363–2371.
31. Ramani, A.K., et al., Consolidating the set of known human protein–protein interactions in preparation for large-scale mapping of the human interactome. *Genome Biol*, 2005. *6*(5): R40.
32. Lehner, B. and A.G. Fraser, A first-draft human protein-interaction map. *Genome Biol*, 2004. *5*(9): R63.
33. Brown, K.R. and I. Jurisica, Online predicted human interaction database. *Bioinformatics*, 2005. *21*(9): 2076–2082.
34. Persico, M., et al., HomoMINT: an inferred human network based on orthology mapping of protein interactions discovered in model organisms. *BMC Bioinformatics*, 2005. *6*(Suppl 4): S21.
35. Rual, J.F., et al., Towards a proteome-scale map of the human protein–protein interaction network. *Nature*, 2005. *437*(7062): 1173–1178.
36. Stelzl, U., et al., A human protein-protein interaction network: a resource for annotating the proteome. *Cell*, 2005. *122*(6): 957–968.
37. Przulj, N., D.A. Wigle, and I. Jurisica, Functional topology in a network of protein interactions. *Bioinformatics*, 2004. *20*(3): 340–348.

38. Xenarios, I., et al., DIP, the Database of Interacting Proteins: a research tool for studying cellular networks of protein interactions. *Nucleic Acids Res*, 2002. *30*(1): 303–305.
39. Hawkins, R.D. and B. Ren, Genome-wide location analysis: insights on transcriptional regulation. *Hum Mol Genet*, 2006. *15*(Spec No 1): R1–R7.
40. Buck, M.J. and J.D. Lieb, ChIP–chip: considerations for the design, analysis, and application of genome-wide chromatin immunoprecipitation experiments. *Genomics*, 2004. *83*(3): 349–360.
41. Jothi, R., et al., Genome-wide identification of in vivo protein-DNA binding sites from ChIP-seq data. *Nucleic Acids Res*, 2008. *36*(16): 5221–5231.
42. Matys, V., et al., TRANSFAC: transcriptional regulation, from patterns to profiles. *Nucleic Acids Res*, 2003. *31*(1): 374–378.
43. Kel, A.E., et al., MATCH: A tool for searching transcription factor binding sites in DNA sequences. *Nucleic Acids Res*, 2003. *31*(13): 3576–3579.
44. Ashburner, M., et al., Gene ontology: tool for the unification of biology. The Gene Ontology Consortium. *Nat Genet*, 2000. *25*(1): 25–29.
45. Yeung, K.Y., et al., Construction of regulatory networks using expression time-series data of a genotyped population. *Proc Natl Acad Sci USA*, 2011. *108*(48): 19436–19441.
46. Zhang, Y., et al., Network motif-based identification of transcription factor-target gene relationships by integrating multi-source biological data. *BMC Bioinformatics*, 2008. *9*: 203.
47. Barabasi, A.L. and Z.N. Oltvai, Network biology: understanding the cell's functional organization. *Nat Rev Genet*, 2004. *5*(2): 101–113.
48. Yu, H., et al., Genomic analysis of essentiality within protein networks. *Trends Genet*, 2004. *20*(6): 227–231.
49. Vidal, M., M.E. Cusick, and A.L. Barabasi, Interactome networks and human disease. *Cell*, 2011. *144*(6): 986–998.
50. Doncheva, N.T., et al., Topological analysis and interactive visualization of biological networks and protein structures. *Nat Protoc*, 2012. *7*(4): 670–685.
51. Guelzim, N., et al., Topological and causal structure of the yeast transcriptional regulatory network. *Nat Genet*, 2002. *31*(1): 60–63.
52. Hartwell, L.H., et al., From molecular to modular cell biology. *Nature*, 1999. *402*(6761 Suppl): C47–C52.
53. Ravasz, E., et al., Hierarchical organization of modularity in metabolic networks. *Science*, 2002. *297*(5586): 1551–1555.
54. Mitra, K., et al., Integrative approaches for finding modular structure in biological networks. *Nat Rev Genet*, 2013. *14*(10): 719–732.
55. Wolf, D.M. and A.P. Arkin, Motifs, modules and games in bacteria. *Curr Opin Microbiol*, 2003. *6*(2): 125–134.
56. Shen-Orr, S.S., et al., Network motifs in the transcriptional regulation network of *Escherichia coli*. *Nat Genet*, 2002. *31*(1): 64–68.
57. Bruggeman, F.J. and H.V. Westerhoff, The nature of systems biology. *Trends Microbiol*, 2007. *15*(1): 45–50.
58. De Keersmaecker, S.C., et al., Integration of omics data: how well does it work for bacteria? *Mol Microbiol*, 2006. *62*(5): 1239–1250.

59. Walhout, A.J., Unraveling transcription regulatory networks by protein-DNA and protein-protein interaction mapping. *Genome Res*, 2006. *16*(12): 1445–1454.
60. Blais, A. and B.D. Dynlacht, Constructing transcriptional regulatory networks. *Genes Dev*, 2005. *19*(13): 1499–1511.
61. Fields, S. and O. Song, A novel genetic system to detect protein-protein interactions. *Nature*, 1989. *340*(6230): 245–226.
62. Zhu, H., et al., Global analysis of protein activities using proteome chips. *Science*, 2001. *293*(5537): 2101–2105.
63. Lee, T.I., et al., Transcriptional regulatory networks in *Saccharomyces cerevisiae*. *Science*, 2002. *298*(5594): 799–804.
64. Odom, D.T., et al., Control of pancreas and liver gene expression by HNF transcription factors. *Science*, 2004. *303*(5662): 1378–1381.
65. Boyer, L.A., et al., Core transcriptional regulatory circuitry in human embryonic stem cells. *Cell*, 2005. *122*(6): 947–956.
66. Swiers, G., R. Patient, and M. Loose, Genetic regulatory networks programming hematopoietic stem cells and erythroid lineage specification. *Dev Biol*, 2006. *294*(2): 525–540.
67. Mangan, S. and U. Alon, Structure and function of the feed-forward loop network motif. *Proc Natl Acad Sci USA*, 2003. *100*(21): 11980–11985.
68. Mangan, S., A. Zaslaver, and U. Alon, The coherent feedforward loop serves as a sign-sensitive delay element in transcription networks. *J Mol Biol*, 2003. *334*(2): 197–204.
69. Milo, R., et al., Network motifs: simple building blocks of complex networks. *Science*, 2002. *298*(5594): 824–827.
70. Saddic, L.A., et al., The LEAFY target LMI1 is a meristem identity regulator and acts together with LEAFY to regulate expression of CAULIFLOWER. *Development*, 2006. *133*(9): 1673–1682.
71. Iranfar, N., D. Fuller, and W.F. Loomis, Transcriptional regulation of post-aggregation genes in Dictyostelium by a feed-forward loop involving GBF and LagC. *Dev Biol*, 2006. *290*(2): 460–469.
72. Milo, R., et al., Superfamilies of evolved and designed networks. *Science*, 2004. *303*(5663): 1538–1542.
73. Alon, U., Network motifs: theory and experimental approaches. *Nat Rev Genet*, 2007. *8*(6): 450–461.
74. Naraghi, M. and E. Neher, Linearized buffered Ca2+ diffusion in microdomains and its implications for calculation of [Ca2+] at the mouth of a calcium channel. *J Neurosci*, 1997. *17*(18): 6961–6973.
75. Chen, T., H.L. He, and G.M. Church, Modeling gene expression with differential equations. *Pac Symp Biocomput*, 1999. *4*: 29–40.
76. D'Haeseleer, P., et al., Linear modeling of mRNA expression levels during CNS development and injury. *Pac Symp Biocomput*, 1999. *4*:41–52.
77. Shmulevich, I., et al., Probabilistic Boolean Networks: a rule-based uncertainty model for gene regulatory networks. *Bioinformatics*, 2002. *18*(2): 261–274.
78. Woolf, P.J. and Y. Wang, A fuzzy logic approach to analyzing gene expression data. *Physiol Genomics*, 2000. *3*(1): 9–15.

79. Ressom, H., R. Reynolds, and R.S. Varghese, Increasing the efficiency of fuzzy logic-based gene expression data analysis. *Physiol Genomics*, 2003. *13*(2): 107–117.

80. Friedman, N., et al., Using Bayesian networks to analyze expression data. *J Comput Biol*, 2000. *7*: 601–620.

81. Vignes, M., et al., Gene regulatory network reconstruction using Bayesian networks, the Dantzig Selector, the Lasso and their meta-analysis. *PLoS One*, 2011. *6*(12): e29165.

82. Ressom, H.W., Y. Zhang, J. Xuan, Y. Wang, and R. Clarke, Inferring network interactions using recurrent neural networks and swarm intelligence. *The 28th Annual International Conference of the IEEE Engineering in Medicine and Biology Society (EMBS 2006)*, IEEE, New York City, NY. 2006. 4241–4244.

83. Maraziotis, I., A. Dragomir, and A. Bezerianos, Gene networks inference from expression data using a recurrent neuro-fuzzy approach. *Conf Proc IEEE Eng Med Biol Soc*, 2005. *5*: 4834–4837.

84. Chiang, J.H. and S.Y. Chao, Modeling human cancer-related regulatory modules by GA-RNN hybrid algorithms. *BMC Bioinformatics*, 2007. *8*: 91.

85. Yeung, K.Y., M. Medvedovic, and R.E. Bumgarner, From co-expression to co-regulation: how many microarray experiments do we need? *Genome Biol*, 2004. *5*(7): R48.

86. Berriz, G.F., et al., Characterizing gene sets with FuncAssociate. *Bioinformatics*, 2003. *19*(18): 2502–2504.

87. De Hoon, M.J., S. Imoto, and S. Miyano, Statistical analysis of a small set of time-ordered gene expression data using linear splines. *Bioinformatics*, 2002. *18*(11): 1477–1485.

88. Spellman, P.T., et al., Comprehensive identification of cell cycle-regulated genes of the yeast *Saccharomyces cerevisiae* by microarray hybridization. *Mol Biol Cell*, 1998. *9*(12): 3273–3297.

89. Whitfield, M.L., et al., Identification of genes periodically expressed in the human cell cycle and their expression in tumors. *Mol Biol Cell*, 2002. *13*(6): 1977–2000.

90. Zhu, H., et al., Analysis of yeast protein kinases using protein chips. *Nat Genet*, 2000. *26*(3): 283–289.

II

EAs FOR GENE EXPRESSION DATA ANALYSIS AND GRN RECONSTRUCTION

4

BICLUSTERING ANALYSIS OF GENE EXPRESSION DATA USING EVOLUTIONARY ALGORITHMS

Alan Wee-Chung Liew
School of Information and Communication Technology, Griffith University, Queensland, Australia

4.1 INTRODUCTION

The major goal of systems biology is to reveal how genes and their products interact to regulate cellular process. To achieve this goal, it is necessary to reconstruct gene regulatory networks, which help us to understand the working mechanisms of the cell. To infer the gene regulatory networks, one often looks at how groups of genes are co-expressed under certain conditions and how they regulate each other. This requires the use of high-throughput technologies such as whole genome expression profiling.

DNA microarray technologies allow us to have an insight into cellular process by simultaneously measuring expression levels of thousands of genes under various conditions. In a typical gene expression matrix, the rows describe genes and the columns describe conditions of the experiments. In DNA microarray experiments, discovering groups of genes that share similar transcriptional characteristics is instrumental in

Evolutionary Computation in Gene Regulatory Network Research, First Edition.
Edited by Hitoshi Iba and Nasimul Noman.

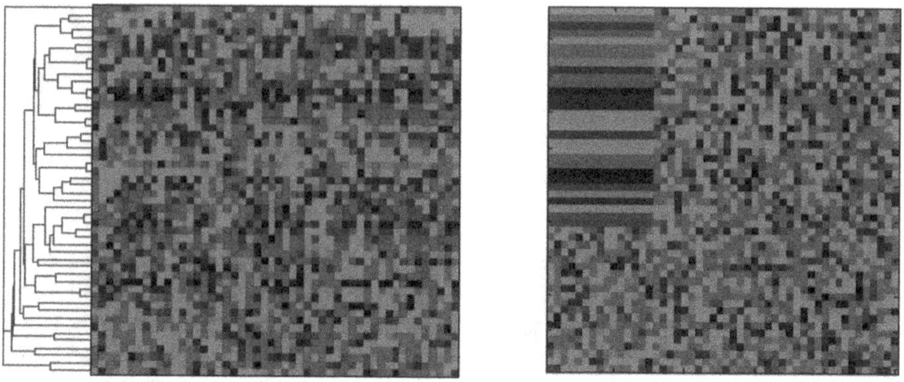

Figure 4.1 An illustrative example where conventional clustering fails but biclustering works. (a) A data matrix, which appears random visually even after hierarchical clustering. (b) A hidden pattern embedded in the data would be uncovered if we permute the rows or columns appropriately [15].

functional annotation, tissue classification, motif identification, and gene regulation [1–3]. Cluster analysis can help elucidate the regulation (or co-regulation) of individual genes, and therefore has been an important tool in gene regulation network study and network reconstruction [3]. However, in many situations, an interesting cellular process is active only under a subset of conditions, or a single gene may participate in multiple pathways that may or may not be co-active under all conditions [4, 5]. In addition, the data to be analyzed often include many heterogeneous conditions from many experiments. In these instances, it is often unrealistic or even undesirable to require that related genes behave similarly across all conditions. Conventional clustering algorithms, such as k-means, hierarchical clustering (HC), and self-organizing maps [6–8], often cannot produce satisfactory solution.

Figure 4.1 illustrates the importance of only grouping the right subset of conditions in clustering. In Figure 4.1a, we see a data matrix clustered using the HC algorithm, where no coherent pattern can be observed by the naked eyes. However, Figure 4.1b indicated that an interesting pattern actually exists within the data if we rearrange the data appropriately. The hidden pattern in Figure 4.1b is called a bicluster, and it shows clearly that only a subset of conditions is relevant in defining this bicluster.

By relaxing the constraint that related genes must behave similarly across the entire set of conditions, "localized" groupings can be uncovered readily. Biclustering allows us to consider only a subset of conditions when looking for similarity between genes. The goal of biclustering is to find submatrices in the dataset, that is, subsets of conditions and subsets of genes, where the subset of conditions exhibits significant homogeneity according to some specific criteria within the subset of genes. Figure 4.2 shows graphically the fundamental difference between clustering and biclustering. Unlike clusters in row-wise or column-wise clustering, biclusters can overlap. In principle, the subsets of conditions for various biclusters can be different. Two biclusters can share some common genes and conditions, and some genes may not belong to

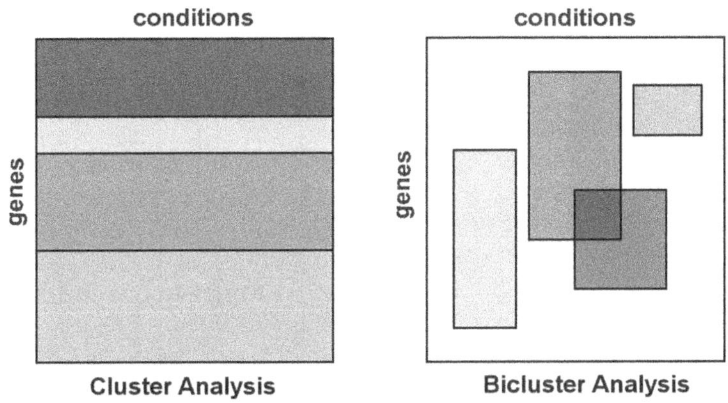

Figure 4.2 Conceptual difference between (a) cluster analysis and (b) bicluster analysis, where biclusters correspond to arbitrary subsets of rows and columns.

any bicluster at all. Since bicluster analysis better reflects the regulatory relationships underlying a cellular process, it has been actively studied for the inference of gene regulatory networks [9].

Biclustering is a very challenging problem computationally. It is an NP hard problem [10]. A bicluster can also have complex coherent patterns. For example, a variety of patterns has been investigated in biclusters such as constant value, linear coherent value, and coherent evolutions. In this chapter, we will discuss the biclustering problem, the different bicluster patterns, existing biclustering techniques, and how evolutionary algorithms (EAs) have been applied to solve the biclustering problem.

4.2 BICLUSTER ANALYSIS OF DATA

Let a dataset of M objects and N attributes be given as a rectangular matrix $D = (a_{ij})_{M\times N}$ where a_{ij} is the value of the ith object in the jth attribute. Denoting the row and column indices of $D_{M\times N}$ as $R = \{1, 2, \ldots, M\}$ and $C = \{1, 2, \ldots, N\}$, we have $D = (R, C) \in \Re^{M\times N}$. Generally, a bicluster is a subset of rows that exhibit similar behaviors across a subset of columns and vice versa. The bicluster $B = (X, Y)$, therefore, appears as a submatrix of D with some similar patterns, where $X = \{M_1, \ldots, M_x\} \subseteq R$ and $Y = \{N_1, \ldots, N_y\} \subseteq C$ are separate subsets of R and C, respectively. Biclustering aims to discover a set of biclusters $B_k = (X_k, Y_k)$ such that each bicluster satisfies some specific characteristics of homogeneity.

Many bicluster patterns have been proposed in the literature [11, 12]. Some of the most common bicluster patterns are as follows: (a) bicluster with constant values, (b) bicluster with constant values in rows or columns, (c) bicluster with coherent values including additive or multiplicative models, and (d) bicluster with coherent evolution. The first three types of biclusters deal with numerical values in the data matrix and they belong to the family of linear bicluster patterns [12, 15]. The bicluster

x	*y*	*z*	*w*
1.3	1.3	1.3	1.3
1.3	1.3	1.3	1.3
1.3	1.3	1.3	1.3
1.3	1.3	1.3	1.3

(a)

x	*y*	*z*	*w*
1.3	1.3	1.3	1.3
2.0	2.0	2.0	2.0
1.5	1.5	1.5	1.5
3.0	3.0	3.0	3.0

(b)

x	*y*	*z*	*w*
1.3	2.0	1.5	3.0
1.3	2.0	1.5	3.0
1.3	2.0	1.5	3.0
1.3	2.0	1.5	3.0

(c)

x	*y*	*z*	*w*
1.2	2.2	0.2	3.2
2.0	3.0	1.0	4.0
1.4	2.4	0.4	3.4
2.4	3.4	1.4	4.4

(d)

x	*y*	*z*	*w*
1.0	2.0	0.5	1.5
2.0	4.0	1.0	3.0
1.4	2.8	0.7	2.1
2.4	4.8	1.2	3.6

(e)

x	*y*	*z*	*w*
1.0	2.1	0.6	1.7
2.0	4.1	1.1	3.2
1.4	2.9	0.8	2.3
2.4	4.9	1.3	3.8

(f)

Figure 4.3 Examples of different bicluster patterns: (a) constant values, (b) constant rows, (c) constant columns, (d) additive coherent values, (e) multiplicative coherent values, and (f) linear coherent values [15].

with coherent evolution aims to find coherent patterns, that is, trends, regardless of the exact numeric values in the data matrix.

Almost all numerical bicluster patterns studied in the literature can be categorized into one of the linear bicluster patterns. This class of bicluster patterns is rich in representation power and their linear structure allows them to be studied as linear geometric primitives embedded in the high-dimensional data space [12–19]. Figure 4.3 enumerates the six possible linear bicluster patterns in a four-dimensional (4D) space: (a) constant values; (b) constant rows; (c) constant columns; (d) additive coherent values, where each row or column is obtained by adding a constant to another row or column; (e) multiplicative coherent values, where each row or column is obtained by multiplying another row or column by a constant value; and (f) linear coherent values, where each column is obtained by multiplying another column by a constant value and then adding a constant. Note that the linear coherent model of (f) subsumes all previous five patterns. In other words, the five patterns (a)–(e) are special cases of the linear coherent model.

Recently, we have shown that the problem of finding linear bicluster patterns has a powerful geometric interpretation, in that detection of linear biclusters in a data matrix can be cast as detecting hyperplanes in a high-dimensional data space [12, 13, 15–19]. Although the six patterns in Figure 4.3 appear to be substantially different from each other visually, if we treat each column as a variable in the 4D space [x, y, z, w] and each row as a point in the 4D space, the six patterns in Figure 4.3 (a) to (f) would correspond to the following six geometric structures: (a) a bicluster at a single point with coordinate $[x, y, z, w] = [1.3, 1.3, 1.3, 1.3]$, (b) a bicluster defined by the line $x = y = z = w$, (c) a bicluster at a single point with coordinate $[x, y, z, w] = [1.3, 2.0, 1.5, 3.0]$, (d) a bicluster defined by the line $x = y - 1 = z + 1 = w - 2$, (e) a bicluster defined by the line $x = 0.5y = 2z = 2w/3$, and (f) a bicluster defined by the line $x = 0.5(y - 0.1) = 2(z - 0.1) = 2(w - 0.2)/3$. Each row in a bicluster is therefore a point lying on one of these points or lines. When a pattern is embedded in a larger

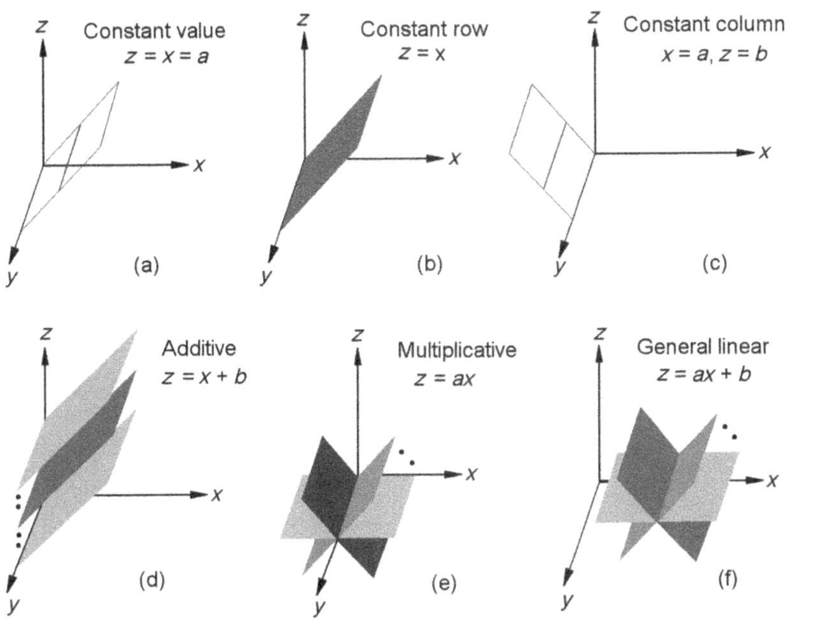

Figure 4.4 Different geometries (lines or planes) in the 3D data space for corresponding bicluster patterns. (a) A bicluster with *constant values* is represented by one of the lines that are parallel to the *y*-axis and lie in the plane $x = z$ (the T-plane). (b) A bicluster with *constant rows* is represented by the T-plane. (c) A bicluster with *constant columns* is represented by one of the lines parallel to the *y*-axis. (d) A bicluster with *additive coherent values* is represented by one of the planes parallel to the T-plane. (e) A bicluster with *multiplicative coherent values* is represented by one of the planes that include the *y*-axis. (f) A bicluster with *linear coherent values* is represented by one of the planes that are parallel to the *y*-axis [15].

data matrix with extra measurements, that is, a bicluster that covers only part of the measurements in the data, the points or lines defined by the bicluster would sweep out a hyperplane in a high-dimensional data space. To visualize this in three-dimensional (3D) space, if we denote the three measurements as x, y, and z, respectively, and assume a bicluster covers x and z only, we can generate 3D geometric views for different patterns as shown in Figure 4.4.

To illustrate the increased complexity of bicluster analysis compared with the traditional cluster analysis, Figure 4.5 shows the different linear bicluster patterns that are hidden in a simple 6×6 data matrix. Note that although we depict the biclustering problem conceptually as cutting the data matrix into submatrices as shown in Figure 4.2 (right), in fact, each bicluster requires a different permutations of the row and column indices. Hence, in general it is not possible to depict all biclusters visually in a single data matrix as in Figure 4.2 (right).

Sometime it is desirable to group data into coherent clusters based on their trends, irrespective of their actual numerical values. For this, we get bicluster patterns of coherent evolution. In this case, the data matrix consists of symbols that reflect trends in the data. The symbols can be purely nominal, of a given order, or encode positive

	A_1	A_2	A_3	A_4	A_5	A_6
O_1	10	10	10	14	7	10
O_2	10	14	10	24	35	10
O_3	15	15	15	12	20	35
O_4	20	20	20	17	25	40
O_5	25	25	25	22	30	45
O_6	10	14	30	27	35	50

(a)

	A_1	A_3	A_6
O_1	10	10	10
O_2	10	10	10

(b)

	A_1	A_2	A_3
O_1	10	10	10
O_3	15	15	15
O_4	20	20	20
O_5	25	25	25

(c)

	A_1	A_2	A_5
O_2	10	14	35
O_6	10	14	35

(d)

	A_3	A_4	A_5	A_6
O_3	15	12	20	35
O_4	20	17	25	40
O_5	25	22	30	45
O_6	30	27	35	50

(e)

	A_5	A_6
O_1	7	10
O_6	35	50

(f)

Figure 4.5 (a) A 6 × 6 data matrix with hidden biclusters, (b) bicluster with constant values, (c) bicluster with constant rows, (d) bicluster with constant columns, (e) bicluster of additive model, where $O_3 = O_4 - 5 = O_5 - 10 = O_6 - 15$ and $A_3 = A_4 + 3 = A_5 - 5 = A_6 - 20$, (f) bicluster of multiplicative model, where $O_1 = 0.2 \times O_6$ and $A_5 = 0.7 \times A_6$.

and negative changes relative to a normal value. Figure 4.6 shows some examples of biclusters with coherent evolution. For this type of bicluster patterns, biclustering strategies different to numerical biclustering are usually used.

Validating the biclustering results is an important part of bicluster analysis. Biclustering results can be validated by using statistics computed from the biclusters or by using domain knowledge [12]. For example, one can compute some statistics of the biclustering solution to assess how well the biclusters found corresponding to the true biclusters in the dataset (i.e., by Jaccard Index or variants of it if ground truth is available, such as in simulated data). In many multi-objective evolutionary-based biclustering algorithms, the quality of the solutions is measured by how well the solutions meet the objectives of the algorithms, such as the homogeneity of the biclusters (i.e., by mean square residual error), size of the biclusters (bigger biclusters are

S1	S1	S1	S1
S1	S1	S1	S1
S1	S1	S1	S1
S1	S1	S1	S1

(a)

S1	S1	S1	S1
S2	S2	S2	S2
S3	S3	S3	S3
S4	S4	S4	S4

(b)

S1	S2	S3	S4
S1	S2	S3	S4
S1	S2	S3	S4
S1	S2	S3	S4

(c)

+1	+1	−1	+1
−1	−1	+1	−1
+1	+1	−1	+1
−1	−1	+1	−1

(d)

Figure 4.6 Types of biclusters with coherent evolution. Considering the entries of a data matrix as symbols, (a) an overall coherent evolution, (b) a coherent evolution on the rows, (c) a coherent evolution on the columns, and (d) a coherent sign change across rows.

preferred), or mean row variance (biclusters with larger mean row variance are assumed to be more interesting).

For gene expression analysis, the preferred validation is to assess whether the biclusters found are biologically meaningful. In general, some domain knowledge about the gene expression dataset is available, and they are used to check for the biological validity of the results. A common way is to check for enrichment in the biclusters by using p-value statistics. The p-value measures the probability of including objects of a given category in a bicluster by chance. Thus, the over-represented bicluster is a bicluster of objects which is very unlikely to be obtained randomly. Currently, Gene Ontology (GO), metabolic pathway maps (MPMs), and protein–protein interaction (PPI) networks can be used to determine the biological functional relevance of genes and conditions in a bicluster. Hence, using the known gene annotation in GO, MPM, and PPI, the p-values of genes associated with the biclusters can be computed for biological validation.

4.3 BICLUSTERING TECHNIQUES

Many biclustering algorithms have been proposed recently and they can be grouped into several categories depending on the bicluster model, the search strategy, and the algorithmic framework used.

4.3.1 Distance-Based Techniques

Distance-based biclustering technique is among the earliest biclustering methods proposed in the literature. This approach measures the quality of the biclusters based on some distance metrics, and searches for the biclusters by minimizing the residual sum of squares cost. In the "direct clustering" algorithm of Hartigan [20], the sum of squares measure "SSQ" is used to evaluate the quality of each bicluster $B_k = (X_k, Y_k)$:

$$\mathrm{SSQ}_k = \sum_{i \in X_k, j \in Y_k} (a_{ij} - a_{X_k Y_k})^2, \tag{4.1}$$

where $a_{X_k Y_k}$ is the average value in the bicluster B_k. Biclusters with lower SSQ are considered to be better than biclusters with higher SSQ. In direct clustering, the number of biclusters is fixed and the solution is reached by minimizing the sum of SSQ_k. Obviously, the direct clustering algorithm only searches for constant biclusters.

Cheng and Church [21] were the first to introduce biclustering into gene expression data analysis. In their well-known δ-bicluster algorithm, they use the H-score, which is a mean squared residue (MSR) score, to measure the degree of coherence of a bicluster. The δ-bicluster algorithm minimizes

$$H(X, Y) = \frac{1}{|X|\,|Y|} \sum_{i \in X, j \in Y} \left(a_{ij} - a_{iY} - a_{Xj} + a_{XY}\right)^2, \tag{4.2}$$

where a_{iY}, a_{Xj}, and a_{XY} are the row mean, column mean, and the mean in the submatrix $B = (X, Y)$, respectively. It can be easily shown that Equation (4.2) is only zero for constant value, constant row or constant column, and additive bicluster patterns, but not multiplicative bicluster pattern. A bicluster is called a δ-bicluster if $H(X, Y) \leq \delta$ for some $\delta > 0$. To find the δ-bicluster, the score H is computed for each possible row/column addition or deletion, and the action that decreases H the most is applied. A bicluster is returned when H cannot be decreased or when $H \leq \delta$. After one δ-bicluster is identified, the elements in the corresponding submatrix are replaced by random numbers before finding the next δ-bicluster. The δ-biclusters are successively extracted from the raw data matrix one at a time until a prespecified number of biclusters have been identified.

Following the work of Cheng and Church, different search strategies were proposed to better detect the δ-bicluster. Bryan et al. [22] proposed a simulated annealing search technique and reported better performance on a variety of datasets. Yang et al. [23] proposed a probabilistic move-based algorithm called FLOC (FLexible Overlapped biClustering) that is able to discover multiple biclusters simultaneously. As a submatrix of a δ-bicluster is not necessarily a δ-bicluster because of outliers, Wang et al. [24] proposed the δ-pCluster model to deal with the outlier problem by further requiring that any 2×2 submatrix in a δ-bicluster has a $pScore \leq \delta$ for some $\delta > 0$, where the *pScore* measures the difference between elements in the 2×2 submatrix.

4.3.2 Factorization-Based Techniques

Factorization-based biclustering algorithm uses spectral decomposition technique to uncover "natural" substructures that are related to the main patterns of the data matrix [25, 26]. The spectral biclustering in Ref. [25] uses singular value decomposition (SVD) and assumes that the data matrix has a checkerboard structure that can be identified in eigenvectors corresponding to characteristic patterns across samples or features. Using SVD, the data matrix $D_{N \times M}$ can be decomposed as $D = U \Lambda V^T$, where Λ is a diagonal matrix with decreasing non-negative entries, and U and V are $N \times \min(N, M)$ and $M \times \min(N, M)$ orthonormal column matrices. If the data matrix has a block diagonal structure (with all elements outside the blocks equal to zero), then each block can be associated with a bicluster. Specifically, if the data matrix is of the form

$$D = \begin{bmatrix} D_1 & 0 & \cdots & 0 \\ 0 & D_2 & \cdots & 0 \\ \vdots & \vdots & \ddots & \vdots \\ 0 & 0 & \cdots & D_r \end{bmatrix}, \tag{4.3}$$

where D_i ($i = 1, \ldots, r$) are arbitrary matrices, then for each D_i there will be a singular vector pair (u_i, v_i) such that a nonzero component of u_i corresponds to rows occupied by D_i and a nonzero component of v_i corresponds to columns occupied by D_i. In a less ideal case, when the elements outside the diagonal blocks are not necessarily zeros but the diagonal blocks still contain dominating values, the SVD is able to reveal the biclusters as dominating components in the singular vector pairs.

Non-negative matrix factorization (NMF) decomposes the data as a product of two matrices that are constrained by having non-negative elements [27]. The NMF is given by $D \approx WH$, where $D \in \Re^{p \times n}$ is a positive data matrix with p variables and n samples, $W \in \Re^{p \times q}$ are the reduced q basis vectors or factors, and $H \in \Re^{q \times n}$ contains the coefficients of the linear combinations of the basis vectors needed to reconstruct the original data (also known as encoding vectors). As both the basis W and encoding vectors H are constrained to be non-negative, only additive combinations are possible. In Ref. [26], non-smooth non-negative matrix factorization algorithm (nsNMF), a variant of the NMF model, has been introduced to identify localized patterns in large datasets. In contrast to NMF, nsNMF produces sparse representation of the factors and encoding vectors by making use of non-smoothness constraints. The sparseness introduced by the algorithm produces more compact and localized feature representation of the data than the standard NMF.

4.3.3 Probabilistic-Based Techniques

The biclustering method in this category typically assumes a probabilistic model of biclusters and applies statistical parameter estimation techniques to search for biclusters [28–31]. In the plaid model of Lazzeoni and Owen [30], the data matrix is viewed as consisting of a series of additive layers, that is, consisting of biclusters or subsets of rows and columns. The model first includes a background layer that accounts for the global effects in the data matrix. Then, any subsequent layer represents additional effects corresponding to biclusters of objects and features that exhibit strong pattern not explained by the background layer. The generalized plaid model is given by

$$a_{ij} = \mu_0 + \sum_{k=1}^{K} \theta_{ijk} \rho_{ik} \kappa_{jk} + \varepsilon_{ij}, \tag{4.4}$$

where μ_0 corresponds to the effect in the global background layer and θ_{ijk} models the effect of layer k. The effect θ_{ijk} can be expressed as a combination of μ_k, α_{ik}, and β_{jk}, where μ_k is the background color in bicluster k, and α and β are row- and column-specific additive constants in bicluster k. The parameter ρ_{ik} (or κ_{jk}) equals 1 when object i (or attribute j) belongs to layer k, and equals 0 otherwise. Any residual, not modeled by the K layers, is accounted for in the noise term ε_{ij}. The biclustering process searches the layers in the dataset one after another, using the expectation maximization algorithm to estimate the model parameters until the variance of expression levels within the current layer is smaller than a threshold.

Gu and Liu [28] proposed a fully generative model called Bayesian biclustering (BBC) algorithm for gene expression data. The data model in BBC is assumed to be

$$a_{ij} = \sum_{k=1}^{K} \left(\left(\mu_k + \alpha_{ik} + \beta_{jk} + \varepsilon_{ijk} \right) \delta_{ik} \kappa_{jk} \right) + e_{ij} \left(1 - \sum_{k=1}^{K} \delta_{ik} \kappa_{jk} \right), \tag{4.5}$$

where K is the total number of clusters (unknown), μ_k is the main effect of cluster k, and α_{ik} and β_{jk} are the effects of sample i and feature j, respectively, in cluster k, ε_{ijk} is the noise term for cluster k, and e_{ij} models the data points that do not belong to any cluster. Here, $\delta_{ik} = 1$ indicates that sample i belongs to cluster k, and $\delta_{ik} = 0$ otherwise. Similarly, $\kappa_{jk} = 1$ indicates that feature j is in cluster k, $\kappa_{jk} = 0$ otherwise. Gibbs sampling method is used for statistical inference in BBC.

Sheng et al. [29] proposed a Bayesian technique for biclustering based on a simple frequency model for the expression pattern of a bicluster and on Gibbs sampling for parameter estimation. The data are discretized and every condition in a bicluster is modeled by a multinomial distribution, where the multinomial distributions for different conditions of a bicluster are assumed to be mutually independent. The Gibbs sampling sets the model in the Bayesian framework, and the Bernoulli posterior distribution is used during Gibbs sampling to find the biclusters.

4.3.4 Geometric-Based Biclustering

Based on a spatial interpretation of biclusters, a geometric-based biclustering framework has recently being introduced [15,16,18,19]. The geometric viewpoint provides a unified mathematical formulation for the simultaneous detection of different types of linear biclusters (i.e., constant, additive, multiplicative, and mixed additive and multiplicative biclusters) and allows biclustering to be done with a generic plane detection algorithm.

The theoretical basis of geometric-based biclustering can be understood as follows: If we consider that the set of columns Y in $B = (X, Y)$ spans a Y-dimensional space, then the data vector in every row of B corresponds to a point in this space. Thus, from a geometric viewpoint, the different biclusters can be considered as different linear geometric patterns in the high-dimensional data space. For example, given a matrix $D_{N\times3}$, a bicluster is represented by a plane in a 3D space as shown in Figure 4.7, where

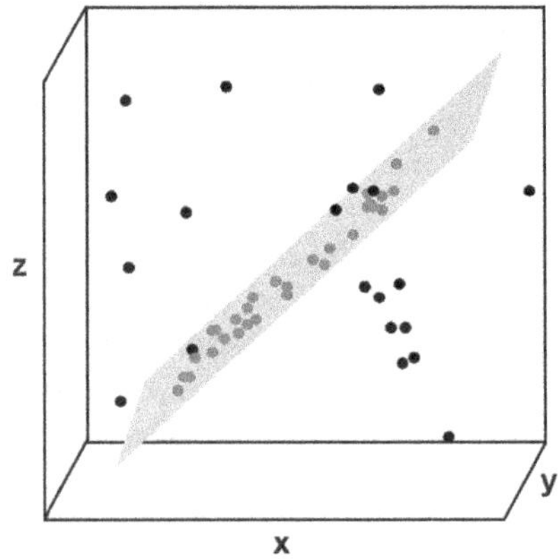

Figure 4.7 A plane formed by points in a bicluster in the three-dimensional data space. The gray dots are data located on the plane.

the N 3D samples are represented by N points. Obviously, a plane can be detected within the 3D data space which provides clues about the hidden bicluster in D. In geometric based biclustering, the problem of identification of coherent submatrices within a data matrix is formulated as the detection of linear geometric patterns (lines, planes, or hyperplanes) in a multidimensional data space [15, 19].

The geometric interpretation of bicluster patterns has important implication in that it unifies the commonly used linear bicluster patterns into a single linear class and allows a unified treatment in detecting these linear biclusters simultaneously. This is in contrast to most existing biclustering algorithms where the cost function implicitly imposes a constraint on the type of bicluster patterns that could be discovered. In principle, any algorithm for detecting linear geometric patterns can be employed in the geometric biclustering framework. In Refs. [15, 16, 18, 19], computer vision techniques that detect linear structures in n-dimensional space based on Hough transform have been employed to find the biclusters. This class of approach has been very successful in detecting linear bicluster patterns from a dataset and is highly robust to noise in the dataset. The major drawback is its high computation cost.

4.3.5 Biclustering for Coherent Evolution

Several biclustering algorithms that find bicluster pattern of coherent evolution have been proposed. Ben-Dor et al. [32] defined a bicluster as an order-preserving submatrix (OPSM). Specifically, a submatrix is order-preserving if there is a permutation of its columns under which the sequence of values in every row is strictly increasing. They define a complete bicluster model as the pair (Y, π) where $\pi = (y_{1,\ldots}, y_s)$ is a linear ordering of the columns in Y. A row supports (Y, π) if the s corresponding values, ordered according to the permutation π, are monotonically increasing. Since an exhaustive algorithm that tries all possible models is not feasible, the algorithm grows partial models iteratively until they become complete models. Similarly, Liu and Wang define a bicluster as an OP-Cluster (order-preserving cluster) [33] which generalizes OPSM to discover biclusters with coherent evolutions on the columns.

Murali and Kasif [34] introduced an algorithm that aims to find the largest xMOTIFs. An xMOTIF is defined as a bicluster with coherent evolutions on its rows. The data are first discretized into a set of symbols by using a list of statistically significant intervals for each row. The motifs are computed starting with a set of randomly chosen columns that act as seeds. For each column, an additional randomly chosen set A of columns is selected, called a discriminating set. The selected bicluster contains all the rows that have states equal to the seed columns and in the columns contained in the discriminating set A. The motif is discarded if less than an α-fraction of the columns matches it. After all the seeds have been used to produce xMOTIFs, the largest xMOTIF, one with the largest number of rows, is returned.

Tanay et al. [35] introduced Statistical-Algorithmic Method for Bicluster Analysis (SAMBA) to detect the biclusters of coherent evolution. The data matrix is modeled as a bipartite graph. Discovering the most significant biclusters under the weighting schemes is equivalent to the selection of the heaviest subgraphs in the bipartite graph.

SAMBA assumes that each a_{ij} can be represented by two symbols S_0 or S_1, where S_1 means change and S_0 means no change. As such, the model graph has an edge between a row and a column when the object is significantly changed with the feature. A large bicluster is the one with a maximum number of rows whose symbol for a_{ij} is expected to be S_1.

Prelic et al. [36] presented a fast divide-and-conquer algorithm called Bimax to detect the inclusion-maximal biclusters in the binary matrix E after a prediscretization procedure. The Bimax algorithm is similar to SAMBA. The idea behind the Bimax algorithm is to partition E into three submatrices, one of which contains only 0 cells and therefore can be disregarded in the results. The algorithm is then recursively applied to the remaining two submatrices U and V. The recursion ends if the current matrix represents a bicluster, that is, contains only 1*s*. If U and V do not share any rows and columns of E, the two matrices can be processed independently from each other. However, if U and V have a set, X, of rows in common, special care is necessary to only generate those biclusters in V that share at least one common column with X. Uitert et al. [37] proposed BicBin (Biclustering Binary data) to find a contiguous block for a large, binary, and sparse genomic data matrix, such as transcription factor binding site, insertional mutagenesis, and gene expression. Assuming that each element in D is the outcome of a Bernoulli trial, a probability-based score function is derived in BicBin to evaluate a submatrix.

4.4 EVOLUTIONARY ALGORITHMS BASED BICLUSTERING

Evolutionary algorithm is a meta-heuristic technique that has been successfully used for various optimization problems because of its excellent exploratory capability in a global search space and its good ability to solve complex problems [38, 39]. Hence, many biclustering algorithms that utilize the EA framework as a search strategy have been proposed. There are two main EA frameworks; the first is based on the genetic algorithm (GA), and the second is based on the artificial immune system (AIS).

Genetic algorithms are a class of EAs which generate solutions to optimization problems using techniques inspired by natural evolution, such as inheritance, mutation, selection, and crossover. In a GA, a population of candidate solutions, called individuals, to an optimization problem is evolved toward better solutions over several generations. Each candidate solution has a set of properties encoded as a chromosome that can be mutated and altered during iteration to form the next generation.

In GA-based biclustering of gene expression data, the row and column indices of the data matrix that correspond to a solution for a bicluster are encoded in a chromosome. The chromosome encoding usually consists of two parts; the first part consists of bit string for genes, and the second part consists of bit string for conditions. For fixed length chromosome, the length of bit string for genes equals the number of rows in the gene expression data matrix, and the length of bit string for conditions equals the number of columns. A bit is set to 1 if the corresponding gene and/or condition is present in the bicluster, and 0 otherwise. At each iteration, the operations of cross-over and mutation are performed to generate the offspring population, and the parent and offspring populations are combined in a pool from which the next generation is selected based on fitness of the individuals.

Artificial immune systems are a class of computationally intelligent systems inspired by the principles and processes of the biological immune system. Depending on the specific immunological theories that AIS is inspired by, the techniques used in AIS can be classified into three types: clonal selection, negative selection, and immune network.

The clonal selection algorithms are inspired by the antigen-driven affinity maturation process of B-cells, with its associated hypermutation mechanism. During the clonal expansion of B-cells, the average antibody affinity increases for the antigen that triggered the clonal expansion. This phenomenon is called affinity maturation, and is responsible for the fact that upon a subsequent exposure to the antigen, the immune response is more effective due to the antibodies having a higher affinity for the antigen. Affinity maturation is caused by a somatic hypermutation and the selection mechanism that occurs during the clonal expansion of B-cells. Somatic hypermutation alters the specificity of antibodies by introducing random changes to the genes that encode for them.

Castro and Zuben [40] proposed a clonal selection algorithm named "CLONALG" for optimization. Two important features of affinity maturation in B-cells are exploited in CLONALG. The first is that the proliferation of B-cells is proportional to the affinity of the antigen that binds it, thus the higher the affinity, the more the clones are produced. Second, the mutations suffered by the antibody of a B-cell are inversely proportional to the affinity of the antigen it binds. CLONALG generates a population of N antibodies, each specifying a random solution for the optimization process. During each iteration, a subset of the best antibodies with the highest affinity to the antigens is selected, cloned, and mutated in order to construct a new candidate population. Finally, a percentage of worst antibodies of previous generation are replaced with new randomly create ones.

Negative selection algorithms are inspired by the main mechanism in the thymus that produces a set of mature T-cells capable of binding only non-self antigens. The process of deleting self-reactive lymphocytes is termed clonal deletion and is carried out via a mechanism called negative selection that operates on lymphocytes during their maturation. For T-cells this mainly occurs in the thymus, which provides an environment rich in antigen presenting cells that present self-antigens. Immature T-cells that strongly bind these self-antigens undergo apoptosis. Thus, the T-cells that survive this process should be unreactive to self-antigens. The property of lymphocytes not to react with the self is called immunological tolerance. The first negative selection algorithm was proposed by Forrest et al. [41] to detect data manipulation caused by a virus in a computer system. The starting point of this algorithm is to produce a set of self-strings S that define the normal state of the system. The task then is to generate a set of detectors D that only bind/recognize the complement of S. These detectors can then be applied to new data in order to classify them as being self or non-self, or whether they have been manipulated.

In the immune network theory [42], in the absence of foreign antigen, the immune system displays behavior or activity resulting from interactions with itself, and from these interactions immunological behavior such as tolerance and memory emerge. The immune network algorithms focus on the network graph structures involved where antibodies (or antibody-producing cells) represent the nodes and the training algorithm involves growing or pruning edges between the nodes based on affinity. Immune network algorithms have been used in clustering, data visualization, control, and optimization domains, and share properties with artificial neural networks [43].

In Ref. [44], a multi-objective GA biclustering algorithm using non-dominated sorting genetic algorithm (NSGA-II) was proposed. In multi-objective optimization [45], the objective function or fitness function consists of two or more conflicting objectives. A solution, which is better with respect to one objective, requires a compromise in other objectives. Hence, there is no single optimum solution. Rather, there exists a set of solutions that are all optimal involving tradeoffs between conflicting objectives. In multi-objective optimization, the set of optimal solution constitute the Pareto front. A solution belongs to the Pareto front (or is Pareto optimal) if there is no other feasible solution capable of reducing the value of one objective without simultaneously increasing at least one of the others. These solutions are called non-dominating. To obtain Pareto optimal solutions that are well spaced out on the Pareto front, that is, with good diversity, the concept of crowding is used. The idea is to retain solutions that are far apart from each other. NSGA-II [46] has three main properties—non-domination, crowding distance, and crowding selection operator—that implement these concepts. In Ref. [44], a local search mechanism, based on the node insertion and deletion heuristics proposed by Cheng and Church [21], was added into NSGA-II. The algorithm employs a binary encoding representation, where each individual in the population is a binary vector with a size equal to the number of rows plus the number of columns of the dataset. A value of 1 in the vector indicates that the corresponding row or column is present in the bicluster. The algorithm starts with a given bicluster. The irrelevant genes or conditions having MSR above (or below) a certain threshold are then selectively eliminated (or added) to the bicluster. Finally, the nodes with low contribution to the residue are added back to the bicluster to increase the volume of the bicluster as long as the residue is smaller than a predefined threshold. These node deletion and insertion procedures are basically greedy procedures similar to Cheng and Church's algorithm. The algorithm uses standard GA operators for crossover and mutation, that is, single-point crossover, single bit mutation, to generate the next generation. Both statistical and biological validations were performed to verify the biclustering results.

In Ref. [47], a multi-objective GA biclustering algorithm which has three objectives and uses NSGA-II optimization with local search based on Cheng and Church's heuristics was proposed. The first objective is the size of the bicluster, the second objective is the MSR score of Cheng and Church, and the third objective is the mean row variance $R\text{var}(I,J)$ given by

$$R\text{var}(I,J) = \frac{1}{|I| \times |J|} \sum_{i \in I, j \in J} \left(a_{ij} - a_{iJ}\right)^2. \tag{4.6}$$

Note that maximizing $R\text{var}(I,J)$ tends to find biclusters with genes that exhibit significant fluctuations in the different conditions, that is, biclusters with significant row variances. Hence, even though both constant value bicluster and additive model bicluster will minimize MSR, only the later maximizes $R\text{var}(I,J)$. The encoding used in this algorithm is based on representing a bicluster as a string composed of two parts: the first part is an ordered row index and the second part is an ordered column index. Single point crossover operator is used in each part of the solution (row part and column part), and each part undergoes crossover separately by exchanging index lists of the two parents after the crossover points obtained as follows. Let the two parents be $P_1 = \{r_1 \ldots r_n\ c_1 \ldots c_m\}$ and $P_2 = \{r'_1 \ldots r'_l c'_1 \ldots c'_k\}$, where $r_n \leq r'_l$. The row part crossover point in P_1 is generated as a random integer in the range $2 \leq \lambda_1 \leq r_n$, then the crossover point in P_2 is obtained by $\lambda_2 = r'_j$ where $r'_j \geq \lambda_1$ and $r'_{j-1} \leq \lambda_1$. Similarly, the column part crossover points are obtained. For the mutation operator, instead of random mutation as in conventional GA, a local search based on Cheng and Church's heuristics is used. Only statistical validation was done to validate the biclustering results.

In Ref. [48], a multi-objective GA biclustering algorithm called "MOGAB" was proposed. MOGAB uses a variable length string that encodes the A gene clusters and B condition clusters to represent the $A{\times}B$ biclusters. It finds biclusters with high mean row variance and an MSR below a threshold δ by optimizing a two-objective function based on the NSGA-II algorithm. MOGAB uses simple single-point crossover, where the gene part and the condition part of the string undergo crossover separately. For the gene part, two crossover points are chosen on the two parent chromosomes, respectively, and the portions of the chromosomes beyond these crossover points are exchanged. Similarly, the crossover for the condition part is performed. Single bit mutation for the gene part and condition part is performed by replacing a randomly chosen position in that part with another randomly chosen index from the corresponding part, respectively. The biclustering results were validated both statistically and biologically.

In Ref. [49], a GA-based multi-objective biclustering algorithm that identifies fuzzy biclusters was introduced. Instead of crisp assignment, fuzzy assignment of genes and conditions was done by using the membership and medoid updating rules adopted from the fuzzy c-mediods algorithm. The biclustering algorithm optimizes three objectives: fuzzy bicluster volume, fuzzy MSR, and fuzzy mean row variance. This algorithm uses a string encoding consisting of a gene part and a condition part, where the indices in each part consist of indices of genes or conditions that act as cluster medoids of genes or conditions. NSGA-II optimization algorithm was used to find the biclusters. Standard GA operators for crossover and mutation, that is, single-point crossover and single bit mutation, were used to generate the next generation. Only statistical validation was used to validate the biclustering results.

In Ref. [50], a multi-objective multi-population artificial immune network (MOM-aiNet) was proposed to detect additive biclusters. The two objectives used in MOM-aiNet are MSR and bicluster volume. These objectives constitute the affinity function that selects the non-dominated individuals in the subpopulation. The algorithm starts with the generation of n subpopulations of one bicluster each, generated by randomly choosing one row and one column of the dataset. Then in the main loop, cloning of each subpopulation occurs, followed by mutation. In the mutation process, each clone undergoes one of the three possible actions, that is, insert a row, insert a column, and remove a row or column, with equal probability. After mutation, all the non-dominated biclusters of this subpopulation (consisting of the original individuals and the mutated clones) are selected to generate the new subpopulation for the next iteration. As each of these subpopulations is cloned and mutated, they converge to distinct promising regions of the search space. MOM-aiNet also performs a suppression operation from time to time, by comparing the degree of overlap of the largest biclusters of each subpopulation. If the overlap is over a set threshold, the two subpopulations are merged into a single subpopulation, and the non-dominated individuals are selective from this new subpopulation. To promote diversity and explore new regions of the search space, random insertion of new subpopulations is also performed from time to time.

One feature of MOM-aiNet is that it keeps not only the best individual of each sub-population, but also several locally (within their subpopulation) non-dominated ones at each generation. The final result returned by MON-aiNet includes not only the final set of non-dominated (i.e., Pareto optimal) individuals, but also all the non-dominated individuals within each subpopulation. Hence, the final set of solutions may provide a higher coverage of the data matrix but may also contain suboptimal biclusters that capture important correlations in the data. The biclustering results on two gene expression datasets and a MovieLens dataset are evaluated based on statistical measures such as MSR, bicluster volume, Pareto front coverage, and overlap. It was shown that MOM-aiNet could provide biclustering solution that maximizes the coverage of the data while minimizing overlap, mainly because of its multi-population aspect and its suppression mechanism, which inhibits two subpopulations from exploring similar areas of the search space.

In Ref. [51], we introduced a technique called condition-based evolutionary biclustering (CBEB) algorithm based on HC and EA search. The general framework of the algorithm is as follows: to search for biclusters in a subspace of conditions, an EA + HC search strategy is used. The chromosome of an individual in the EA method encodes the indices of the conditions in each subspace. All of the rows and the columns corresponding to the selected indices in the chromosome form a submatrix. The conventional HC algorithm is then applied to the submatrix. The clusters in the submatrix obtained from HC can be considered as biclusters in the subspace. Since a bicluster discovered from a single subspace may be a part of a larger bicluster crossing multiple subspaces, after the biclusters are found in each subspace, an expand-and-merge operation is performed to obtain the final biclustering output. The biclustering result obtained from the EA + HC procedure is finally verified using an MSR-based fitness function. As the column dimension of the gene expression data matrix is divided into a number of subsets and EA computation is performed in parallel in each of these subspaces, high computation cost of performing EA search in a space created by a large number of experimental conditions is avoided.

In CBEB, binary encoding is chosen, and each individual (solution) represents one binary chromosome with the bit string length, the same as the column number of the corresponding subspace. For each individual in the population, a bit equal to 1 indicates that the corresponding column is selected to form a submatrix M_s. M_s contains full rows from the original data matrix. Once the submatrix is formed, HC is performed to detect biclusters in this submatrix. The fitness of the solution is evaluated based on a criterion that favors the solution with smaller MSR score and larger bicluster numbers. Solutions with better fitness survive to the next generation and can reproduce offspring. In this work, we employed both the simple crossover method and the binary mutation method for reproducing the offspring. Once EA-based search of the optimal subspaces is performed and a set of biclusters *Bic_Set* is obtained, we first expand each bicluster in *Bic_Set* and then merge small biclusters into larger ones. In expanding the biclusters, a new column is added into a specific bicluster if the newly formed bicluster satisfies the MSR score with a predefined threshold. Similarly, a row is added to the bicluster if the newly expanded bicluster meets the same criterion. In merging biclusters, a smaller bicluster can be merged into a larger one if they mostly overlap.

The CBEB algorithm is evaluated using synthetic data based on the matching score S. Given two sets of biclusters, M_1 and M_2, the match score of M_1 with respect to M_2 is formulated as

$$S\left(M_1, M_2\right) = \frac{1}{|1|} \sum\nolimits_{B(I_1,J_1)\in M_1} \max B\left(I_2, J_2\right) \in M_2 \frac{|I_1 \cap I_2| + |J_1 \cap J_2|}{|I_1 \cup I_2| + |J_1 \cup J_2|}, \tag{4.7}$$

where $B(I, J)$ denotes a bicluster with the set of rows (I) and the set of columns (J). Let M_{op} denote the best set of biclusters and M_t the set of biclusters to be evaluated. The $S(M_{op},M_t)$ can represent the quality of the biclusters in M_t with respect to the best biclusters in M_{op}. For real dataset, the quality of the biclusters is evaluated based on statistical enrichment using GO. Experiments indicated that CBEB significantly outperformed several popular biclustering algorithms on both the synthetic and real datasets. Figure 4.8 shows the result of a comparative study of seven algorithms on the *S. cerevisiae* gene expression dataset from [52].

Note that all the algorithms discussed above detect biclusters with homogeneity defined by the MSR score. It is easy to show that while the MSR for a perfect additive bicluster is 0, it is not true for multiplicative bicluster patterns. As MSR score can only detect additive bicluster patterns, these algorithms are, in principle, only capable of detecting additive bicluster patterns. Additive bicluster patterns are sometime called shifting patterns since the rows or columns in a bicluster are "shift" of other rows or columns in the bicluster.

In order to handle multiplicative biclusters, which are sometime called scaling patterns, Giraldez et al. [53] introduced the maximal standard area (MSA) as another homogeneity criterion besides the MSR in their multi-objective GA biclustering

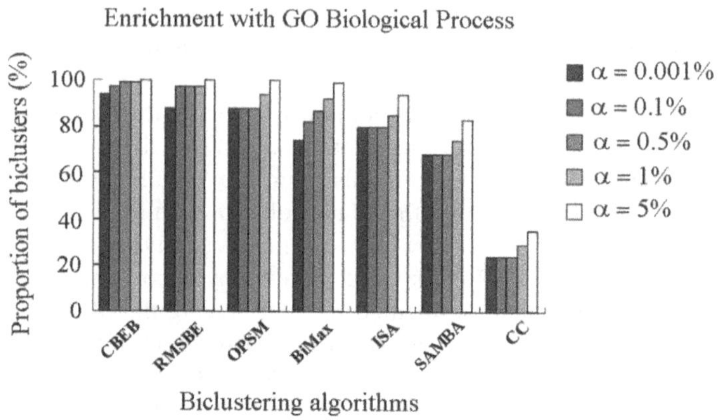

Figure 4.8 Proportion of biclusters significantly enriched by Gene Ontology biological process for different algorithms, where α is the adjusted significant level for the biclusters.

algorithm called sequential multi-objective biclustering (SMOB). The MSA basically measures the area bounded by the maximum and minimum values of standardized expression levels of genes within a bicluster and could handle the non-parallel shifting pattern to some extent as seen in the multiplicative bicluster pattern. SMOB detects biclusters by optimizing four objectives: size of bicluster, MSR, mean row variance, and MSA. Note that although SMOB was able to detect some multiplicative biclusters due to the use of MSA, the MSA does not actually model multiplicative patterns. Noise in the data would create non-parallel shifting patterns, even if the true bicluster pattern is additive. As MSA is not a model for multiplicative bicluster patterns, there is no guarantee that it will find all multiplicative biclusters. A similar attempt to handle both additive and multiplicative bicluster patterns based on data standardization is proposed by Pontes et al. [54] in their algorithm called Evo-Bexpa. In this algorithm, the expression values are standardized by

$$\hat{b}_{ij} = \frac{b_{ij} - \mu_{cj}}{\sigma_{cj}}, \tag{4.8}$$

where μ_{cj} and σ_{cj} are the mean and the standard deviation of column j, respectively. Similar to MSA, the standardization only reduces the column variations (i.e., to 0 mean and 1 standard deviation) but does not actually model multiplicative bicluster patterns. Nevertheless, the standardization scales down the column variations such that multiplicative patterns are more likely to satisfy the MSR score commonly used in many biclustering algorithms.

In Ref. [15, 19], we proposed a class of geometric-based biclustering algorithms that could detect biclusters with general linear patterns. As both additive and multiplicative patterns are the special cases of linear patterns, our algorithms were able to simultaneously detect all additive and multiplicative biclusters. In our approach, a linear bicluster is viewed as a hyperplane in a high-dimensional space. For a submatrix (R, C), a linear bicluster satisfies

$$\mathbf{u}^T\mathbf{x} = \sum_{i=1}^{|C|} u_i x_i = v, \tag{4.9}$$

where $u_1, u_2, \ldots, u_{|C|}, v \in \Re$ and at least one of the u_i is nonzero, and $|C|$ is the cardinality of the column set C. The set of all genes $\mathbf{x} = (x_1, x_2, \ldots, x_{|C|})^T$ that satisfy Equation (4.9) is called a hyperplane of the space $\Re^{|C|}$. In Figure 4.9, all genes $\{\mathbf{x}_1, \mathbf{x}_2, \mathbf{x}_3, \mathbf{x}_4, \mathbf{x}_5\}$ are on the same hyperplane whose equation is given by $c_1 = 0.5c_2 - 1.5c_3 + 2c_4 + 3c_5 - c_6$.

	c_1	c_2	c_3	c_4	c_5	c_6
$\mathbf{x}_1$	0.55	1	2.5	3	0.6	4
$\mathbf{x}_2$	4.95	2.5	3	1.5	2	0.8
$\mathbf{x}_3$	8.75	0.8	0.9	0.6	3	0.5
$\mathbf{x}_4$	5.55	1.2	1.5	1.6	2.1	2.3
$\mathbf{x}_5$	7.95	4.5	2.2	3.1	1.2	0.8

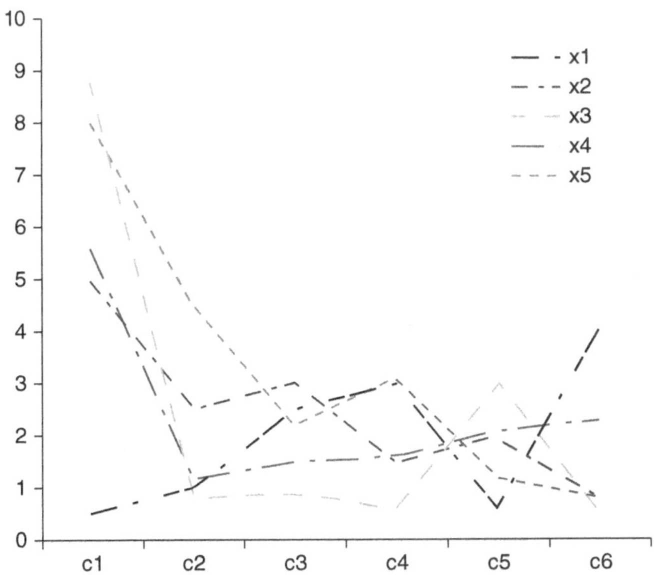

Figure 4.9 Graphical representation of a bicluster containing linear patterns whose relationship is $c_1 = 0.5c_2 - 1.5c_3 + 2c_4 + 3c_5 - c_6$ as shown in the table above.

In Ref. [17], we proposed to detect linear biclusters by searching for hyperplanes using GA. Given a gene $\mathbf{a} = (a_1, a_2, \ldots, a_{|C|})^{\mathrm{T}}$ and a hyperplane $H = \{\mathbf{x} \in \mathrm{R}^{|C|} : \mathbf{u}^T\mathbf{x} = v\}$, the distance from $\mathbf{a}$ to H is given by

$$d(\mathbf{a}, H) = \frac{|\mathbf{u}^{\mathrm{T}}\mathbf{a} - v|}{\|\mathbf{u}\|}. \tag{4.10}$$

In biclustering, we need to find the hyperplane that minimizes the total distances from the hyperplane to all genes in the bicluster. Therefore, the problem for biclustering is formulated as the following GA optimization problem:

Find $\mathbf{z} = (u_1, u_2, \ldots, u_{|C|}, v)^{\mathrm{T}}_{\mathrm{I}}$ and a submatrix $(G, C) \subseteq \mathbf{A}$ which minimizes

$$f(\mathbf{z},G,C) = \sum_{k=1}^{|G|} d(\mathrm{a}_k, H) = \sum_{k=1}^{|G|} \frac{|\mathbf{u}^{\mathrm{T}}\mathrm{a}_k - v|}{\|\mathbf{u}\|}, \tag{4.11}$$

subject to

$$\begin{cases} -1 \le u_i \le 1, \quad i = 1 \ldots |C|, \quad \text{and} \quad \exists u_i \ne 0 \\ -1 \le v \le 1 \end{cases}. \tag{4.12}$$

The above optimization problem is a mixture of discrete and continuous search spaces because the domains of hyperplane parameters and submatrix are continuous and discrete spaces, respectively. We employed GA to find the biclusters. Our algorithm is shown in Figure 4.10.

In our algorithm, each individual which has a dynamic set of biclusters (R, C) is represented by integer string encoding. More specifically, each individual has two subsets containing selected rows and columns and separated by zero values. Figure 4.11 is an illustration of this encoding with three biclusters, and the sets of rows and columns of the first bicluster are $\{3, 2, 5\}$ and $\{1, 8\}$, respectively.

Step 1: Initialize a population of individuals which contain a set of biclusters (G, C).
Step 2: For each bicluster (G, C) of an individual, the steepest descent is used to obtain parameters of hyperplane via minimizing Equation (4.11). The fitness value of each individual is computed using Equation (4.13).
Step 3: Based on the fitness values, a new generation is created via reproduction, crossover and mutation operators.
Step 4: The second and third steps are iterated until maximum number of generation is achieved.

Figure 4.10 Pseudo-code of linear biclusters detection based on the genetic algorithm.

Rows:	3	2	5	**0**	1	6	**0**	5	8	9	15

Columns:	1	8	**0**	1	5	8	9	**0**	20	35	8

Figure 4.11 Individual chromosome encoding.

The parameters of evolutionary computation listed in Table 4.1 are chosen experimentally and the fitness function of each individual is given by

$$\text{fitness} = 1 \Bigg/ \frac{\sum_{i=1}^{h} f_i(\mathbf{z},\mathrm{G},\mathrm{C})}{h}, \tag{4.13}$$

where h is number of biclusters in an individual. In step 2, the iteration of the steepest descent method is stopped if the number of iterations reaches a predefined number or if there are negligible changes in fitness function values between two consecutive iterations.

To validate our algorithm, we run it on simulated datasets where ground truth is known and also on real gene expression datasets. Our algorithm was compared with six other biclustering algorithms, namely FABIA [55], ISA2 [56], xMOTIF [34], Cheng and Church [21], spectral biclustering [25], plaid model [30]. For the simulated datasets, we performed 100 runs by randomly generated 100 datasets whose sizes are 200 rows by 40 columns. Four biclusters of linear model with the number of rows and columns randomly selected within the range from 20 to 40 and 7 to 15, respectively, were embedded into each of these datasets. The Jaccard coefficient was used to measure the similarity between the two biclusters:

$$J(G_1, G_2) = \frac{|G_1 \cap G_2|}{|G_1 \cup G_2|}, \tag{4.14}$$

where G_1 and G_2 are two biclusters. The higher the Jaccard index, the better the performance of the algorithm.

The Jaccard index was computed from the results and the mean and standard deviation of the Jaccard indices over 100 runs are listed in Table 4.2. Note that xMOTIF and spectral biclustering could not find any of the linear biclusters. Our algorithm performed the best with the highest mean Jaccard index and the smallest standard deviation values over the 100 runs among all compared algorithms.

Table 4.1 Parameters of genetic algorithm

Parameter	Value
Population size	500
Number of generations	20
Probability of reproduction	0.1
Probability of crossover	0.7
Probability of mutation	0.2

Table 4.2 Average Jaccard index based on 100 simulated datasets

Methods	Mean	Standard deviation
The proposed algorithm	0.4848	0.0264
xMOTIF	N/A	N/A
Spectral biclustering	N/A	N/A
Plaid model	0.0329	0.1156
Cheng and Church	0.0333	0.0707
ISA2	0.1667	0.3430
FABIA	0.0036	0.0226

We tested our algorithm on the gene expression dataset of diffuse large B-cell lymphoma [57]. The diffuse large-B-cell lymphoma was used to predict the survival after chemotherapy and contains 180 samples of 661 genes. To evaluate the biclustering results, we used GO-TermFinder [58] and ClueGO [59] to assess the biological relevance of the biclusters based on the GO [60] and Kyoto Encyclopedia of Genes and Genomes (KEGG) pathway [61] annotations. All biclusters obtained from biclustering algorithms were enriched to three GO categories: GO biological process (GO BP), GO molecular function (GO MF), GO cellular component (GO CC), and the KEGG pathway. The number of biclusters found by our algorithm and six other biclustering algorithms enriched by three GO categories and KEGG pathway are shown in Tables 4.3, 4.4, 4.5, 4.6, and 4.7. For this dataset, xMOTIF

Table 4.3 Number of biclusters found

Methods	Number of biclusters
The proposed algorithm	13
xMOTIF	0
Spectral biclustering	1
Plaid model	1
Cheng and Church	1
ISA2	10
FABIA	5

Table 4.4 Number of biclusters enriched by GO BP

Methods	p-value = 0.05	p-value = 0.01
The proposed algorithm	13	13
xMOTIF	0	0
Spectral biclustering	0	0
Plaid model	0	0
Cheng and Church	1	1
ISA2	8	7
FABIA	5	5

GO BP, GO biological process.

Table 4.5 Number of biclusters enriched by GO MF

Methods	$p = 0.05$	$p = 0.01$
The proposed algorithm	13	13
xMOTIF	0	0
Spectral biclustering	0	0
Plaid model	0	0
Cheng and Church	1	1
ISA2	2	2
FABIA	3	3

GO MF, GO molecular function.

Table 4.6 Number of biclusters enriched by GO CC

Methods	$p = 0.05$	$p = 0.01$
The proposed algorithm	12	11
xMOTIF	0	0
Spectral biclustering	0	0
Plaid model	0	0
Cheng and Church	1	1
ISA2	5	3
FABIA	3	1

GO CC, GO cellular component.

Table 4.7 Number of biclusters enriched by KEGG pathway

Methods	$p = 0.05$	$p = 0.01$
The proposed algorithm	13	10
xMOTIF	0	0
Spectral biclustering	0	0
Plaid model	0	0
Cheng and Church	1	1
ISA2	5	5
FABIA	4	3

KEGG, Kyoto Encyclopedia of Genes and Genomes.

did not give any bicluster, and Cheng and Church considered the whole dataset as a bicluster, while spectral clustering and plaid model gave only one bicluster that was not significantly enriched by the three GO categories and KEGG pathway.

4.5 CONCLUSION

Bicluster analysis has recently emerged as a powerful tool for unsupervised pattern discovery, especially for the analysis of gene expression data. However, biclustering is an NP-hard problem, and there is a need for highly effective and efficient search algorithms to find the biclusters in the dataset. Evolutionary algorithms have generated a lot of interest in the optimization community due to its ability to find near optimal solution in many hard optimization problems. In this chapter, we provided a comprehensive review of the biclustering of gene expression data and the application of EAs to solve the biclustering problem. We first described the different types of bicluster patterns and highlighted an interesting and important viewpoint of the biclustering problem: the detection of linear hyperplanes in a high-dimensional data space. As different types of algorithms have been proposed for gene expression data biclustering depending on the objective functions, search strategies, and algorithmic frameworks used, we also briefly described some representative algorithms in each category. We reviewed recent application of GAs and artificial immunity system algorithms to the biclustering problem and presented some of our own results. To handle multiple (possibly conflicting) objectives in the optimization, many of these EAs employ the concept of Pareto optimality and search for a set of solutions, called non-dominating solutions, that constitute the Pareto front. We stressed that the majority of these evolutionary biclustering algorithms detect only the additive bicluster pattern and there is a need for algorithms that can handle the more general linear bicluster pattern.

REFERENCES

1. D. A. Rew, "DNA microarray technology in cancer research", *European Journal of Surgical Oncology*, *27*(5): 504–508, 2001.
2. T. R. Golub, et al., "Molecular classification of cancer: class discovery and class prediction by gene expression monitoring", *Science*, *286*(5439): 531–537, 1999.
3. P. D'haeseleer, S. Liang, R. Somogyi, "Genetic network inference: from co-expression clustering to reverse engineering", *Bioinformatics*, *16*(8): 707–726, 2000.
4. M. T. Laub, H. H. McAdams, T. Feldblyum, C. M. Fraser, L. Shapiro, "Global analysis of the genetic network controlling a bacterial cell cycle", *Science*, *290*(5499): 2144–2148, 2000.
5. J. L. DeRisi, V. R. Iyer, P. O. Brown, "Exploring the metabolic and genetic control of gene expression on a genomic scale", *Science*, *278*(5338): 680–686, 1997.
6. S. Tavazoie, J. D. Hughes, M. J. Campbell, R. J. Cho, G. M. Church, "Systematic determination of genetic network architecture", *Nature Genetics*, *22*(3): 281–285, 1999.
7. M. B. Eisen, P. T. Spellman, P. O. Brown, D. Botstein, "Cluster analysis and display of genome-wide expression patterns", *Proceedings of the National Academy of Sciences of the United States of America*, *95*(25): 14863–14868, 1998.
8. P. Tamayo, D. Slonim, J. Mesirov, Q. Zhu, S. Kitareewan, E. Dmitrovsky, E. S. Lander, T. R. Golub, "Interpreting patterns of gene expression with self-organizing maps: methods and application to hematopoietic differentiation", *Proceedings of the National Academy of Sciences of the United States of America*, *96*(6): 2907–2912, 1999.
9. D. Reiss, N. Baliga, R. Bonneau, "Integrated biclustering of heterogeneous genome-wide datasets for the inference of global regulatory networks", *BMC Bioinformatics*, *7*(1): 280, 2006.

10. S. Busygin, O. Prokopyev, P. M. Pardalos, "Biclustering in Data Mining", *Computers and Operation Research, 35*: 2964–2987, 2008.
11. S. C. Madeira, A. L. Oliveira. Biclustering algorithms for biological data analysis: a survey. *IEEE/ACM Transactions on Computational Biology and Bioinformatics, 1*: 24–45, 2004.
12. H. Zhao, A. W. C. Liew, D. Z. Wang, H. Yan, "Biclustering analysis for pattern discovery: current techniques, comparative studies and applications", *Current Bioinformatics, 7*(1): 43–55, 2012.
13. A. W. C. Liew, N. F. Law, H. Yan, " Recent Patents on Biclustering Algorithms for Gene Expression Data Analysis", *Recent Patents on DNA and Gene Sequences, 5*(2): 117–125, 2011.
14. K. O. Cheng, N. F. Law, W. C. Siu, A. W. C. Liew, "Identification of coherent patterns in gene expression data using an efficient biclustering algorithm and parallel coordinate visualization", *BMC Bioinformatics, 9*: 210, 2008.
15. X. Gan, A. W. C. Liew, H. Yan, "Discovering biclusters in gene expression data based on high-dimensional linear geometries", *BMC Bioinformatics, 9*: 209, 2008.
16. H. Zhao, A. W. C. Liew, X. Xie, H. Yan, "A new geometric biclustering algorithm based on the Hough transform for analysis of large-scale microarray data", *Journal of Theoretical Biology, 251*(2): 264–274, 2008.
17. C. To, A. W. C. Liew, "Genetic algorithm based detection of general linear biclusters", 2014 International Conference on Machine Learning and Cybernetics, ICMLC2014, Lanzhou, China, July 13–16, 2014.
18. H. Zhao, A. W. C. Liew, H. Yan, "A new strategy of geometrical biclustering for microarray data analysis" Proceedings of the Fifth Asia Pacific Bioinformatics Conference, APBC2007, Hong Kong, January 15–17, pp. 47–56, 2007.
19. X. Gan, A. W. C. Liew, H. Yan, "Biclustering gene expression data based on a high dimensional geometric method", Proceedings of the Fourth International Conference on Machine Learning and Cybernetics (ICMLC 2005), Guangzhou, China, August 18–21, 2005.
20. J. A. Hartigan, "Direct clustering of a data matrix", *Journal of the American Statistical Association, 67*(337): 123–129, 1972.
21. Y. Cheng, G. M. Church, "Biclustering of expression data", *Proceedings of 8th International Conference on Intelligent Systems for Molecular Biology*, Menlo Park, CA, pp. 93–103, 2000.
22. K. Bryan, P. Cunningham, N. Bolshakova, "Biclustering of expression data using simulated annealing", *Proceedings of the 18th IEEE Symposium on Computer-Based Medical Systems*, Washington, DC, pp. 383–388, 2005.
23. J. Yang, W. Wang, H. Wang, P. Yu, "Enhanced biclustering on expression data", *Proceedings of the Third IEEE Conference Bioinform Bioeng*, Bethesda, Maryland, pp. 321–327, 2003.
24. H. Wang, W. Wang, J. Yang, P. Yu, "Clustering by pattern similarity in large data sets", *Proceedings of the 2002 ACM SIGMOD Int'l Conference Management of Data*, Madison, Wisconsin, pp. 394–405, 2002.
25. Y. Klugar, R. Basr, J. T. Chang, M. Gerstein, "Spectral biclustering of microarray data: Coclustering genes and conditions", *Genome Research, 13*: 703–716, 2003.

26. P. Carmonan-Saez, R. D. Pascual-Marqui, F. Tirado, J. M. Carazo, A. Pascual-Montano, "Biclusteing of gene expression data by non-smooth non-negative matrix factorization", *BMC Bioinformatics*, *7*: 78, 2006.
27. D. D. Lee, H. S. Seung, "Learning the parts of objects by nonnegative matrix factorization", *Nature*, *401*: 788–791, 1999.
28. J. Gu, J. S. Liu, "Bayesian biclustering of gene expression data", *BMC Genomics*, *9*: 4, 2008.
29. Q. Sheng, Y. Moreau, B. De Moor, "Biclustering microarray data by Gibbs sampling", *Bioinformatics*, *19*: ii196–ii205, 2003.
30. L. Lazzeroni, A. Owen, "Plaid models for gene expression data", Technical report, Stanford University, Stanford, CA, 2000.
31. G. Govaert, M. Nadif, "Clustering with block mixture models", *Pattern Recognition*, *36*(2): 463–473, 2003.
32. A. Ben-Dor, B. Chor, R. Karp, Z. Yakhini, "Discovering local structure in gene expression data: The order-preserving submatrix problem", *Journal of Computational Biology*, *10*(3–4): 373–384, 2003.
33. J. Liu, W. Wang, "OP-Cluster: Clustering by tendency in high dimensional space", *Proceedings of the 3rd IEEE Int'l Conference Data Mining*, Florida, USA, pp. 187–194, 2003.
34. T. M. Murali, S. Kasif, "Extracting conserved gene expression motifs from gene expression data", *Proceedings of Pacific Symposium on Biocomputing*, Lihue, Hawaii, *8*: 77–88, 2003.
35. A. Tanay, R. Sharan, R. Shamir, "Discovering statistically significant biclusters in gene expression data", *Bioinformatics*, *18*: S136–S144, 2002.
36. A. Prelic, et al., "A systematic comparison and evaluation of biclustering methods for gene expression data", *Bioinformatics*, *22*: 1122–1129, 2006.
37. M. Uitert, W. Meuleman, L. Wessels, "Biclustering sparse binary genomic data", *Journal of Comparative Biology*, *15*(10): 1329–1345, 2008.
38. M. Mitchell. *An Introduction to Genetic Algorithm*, MIT Press, London, 2001.
39. K. A. De Jong, *Evolutionary Computation: a Unified Approach*, MIT Press, Cambridge, MA, 2006.
40. L. N. de Castro, F. J. Von Zuben, "Learning and optimization using the clonal selection principle", *IEEE Transactions on Evolutionary Computation, Special Issue on Artificial Immune Systems*, *6*(3): 239–251, 2002.
41. S. Forrest, A. S. Perelson, L. Allen, R. Cherukuri, "Self-Nonself discrimination in a computer", *Proceedings of IEEE Symposium on Research in Security and Privacy*, Oakland, CA, pp. 202–212, 1994.
42. N. K. Jerne, "Towards a network theory of the immune system", *Annals of Immunology*, *125C*: 373–389, 1974.
43. J. Timmis, M. Neal, J. Hunt, "An artificial immune system for data analysis", *Biosystems*, *55*(1): 143–150, 2000.
44. S. Mitra, H. Banka, "Multi-objective evolutionary biclustering of gene expression data", *Pattern Recognition*, *39*: 2464–2477, 2006.
45. K. Deb, *Multi-Objective Optimization Using Evolutionary Algorithms*. Wiley, London, 2001.

46. K. Deb, A. Pratap, S. Agarwal, T. Meyarivan, "A fast and elitist multiobjective genetic algorithm: NSGA-II", *IEEE Transactions on Evolutionary Computation*, *6*(2): 182–197, 2002.
47. K. Seridi, L. Jourdan, E. Talbi, "Multi-objective evolutionary algorithm for biclustering in microarrays data", *Proceeding of IEEE Congress on Evolutionary Computation*, New Orleans, USA, pp. 2593–2599, 2011.
48. U. Maulik, S. Bandyopadhyay, "Finding multiple coherent biclusters in microarray data using variable string length multi-objective genetic algorithm", *IEEE Transactions on Information Technology in Biomedicine*, *13*: 969–975, 2009.
49. U. Maulik, A. Mukhopadhyay, S. Bandyopadhyay, M. Q. Zhang, X. Zhang, "Multiobjective fuzzy biclustering in microarray data: method and a new performance measure", *Proceedings of the IEEE Congress on Evolutionary Computation*, Hong Kong, 2008, pp. 1536–1543,.
50. G. P. Coelho, F. O. de Franca, F. J. Von Zuben, "Multi-objective biclustering: when non-dominated solutions are not enough", *Journal of Mathematical Modelling and Algorithms*, *8*: 175–202, 2009.
51. Q. Huang, D. Tao, X. Li, A. W. C. Liew, "Parallelized evolutionary learning for detection of biclusters in gene expression data", *IEEE/ACM Transactions on Computational Biology and Bioinformatics*, *9*:560–570, 2012.
52. A. P. Gasch, P. T. Spellman, C. M. Kao, O. Carmel-Harel, M. B. Eisen, G. Storz, D. Botstein, P. O. Brown, "Genomic expression programs in the response of yeast cells to environmental changes", *Molecular Biology of the Cell*, *11*: 4241–4257, 2000.
53. R. Giraldez, F. Divina, B. Pontes, J. S. Aguilar-Ruiz, "Evolutionary search of biclusters by minimal intrafluctuation", *Proceedings of the IEEE International Fuzzy Systems Conference*, London, UK, pp. 1–6, 2007.
54. B. Pontes, R. Giraldez, J. S. Aguilar-Ruiz, "Configurable pattern-based evolutionary biclustering of gene expression data", *Algorithms for Molecular Biology*, *8*: 4, 2013.
55. S. Hochreiter, U. Bodenhofer, M. Heusel, A. Mayr, A. Mitterecker, A. Kasim, T. Khamiakova, S. Van Sanden, D. Lin, W. Talloen, L. Bijnens, H. W. Göhlmann, Z. Shkedy, D. A. Clevert, "FABIA: factor analysis for bicluster acquisition", *Bioinformatics*, *26*: 1520–1527, 2010.
56. J. Ihmels, S. Bergmann, N. Barkai, "Defining transcription modules using large-scale gene expression data", *Bioinformatics*, *20*: 1993–2003, 2004.
57. A. Rosenwald, et al., "The use of molecular profiling to predict survival after chemotherapy for diffuse large-B-cell lymphoma", *New England Journal of Medicine*, *346*: 1937–1947, 2002.
58. E. I. Boyle, S. Weng, J. Gollub, H. Jin, D. Botstein, J. M. Cherry, G. Sherlock, "GO: TermFinder—open source software for accessing Gene Ontology information and finding significantly enriched Gene Ontology terms associated with a list of genes", *Bioinformatics*, *20*: 3710–3715, 2004.
59. G. Bindea, et al., ClueGO: a Cytoscape plug-in to decipher functionally grouped gene ontology and pathway annotation networks", *Bioinformatics*, *25*: 1091–1093, 2009.
60. M. Ashburner, et al., "Gene Ontology: tool for the unification of biology", *Nature Genetics*, *25*: 25–29, 2000.
61. M. Kanehisa, S. Goto, "KEGG: kyoto encyclopedia of genes and genomes", *Nucleic Acids Research*, *28*: 27–30, 2000.

5

INFERENCE OF VOHRADSKÝ'S MODELS OF GENETIC NETWORKS USING A REAL-CODED GENETIC ALGORITHM

Shuhei Kimura

Graduate School of Engineering, Tottori University, Tottori, Japan

5.1 INTRODUCTION

Proteins play major roles in cells, and are synthesized using information from genes. Gene expression is a process by which information from a gene is used in the synthesis of a protein. The analysis of gene expression is regarded as an important means to understand biological systems.

Advancements in biological technologies, such as deoxyribonucleic acid (DNA) microarrays and ribonucleic acid sequencing (RNA-seq), allow us to measure the expression levels of thousands of genes. However, in order to utilize these technologies, we must find a way to extract hidden information from the measured gene expression data. The inference of genetic networks is considered a promising approach for extracting useful information from these data. A genetic network, which is also referred to as a gene regulatory network, is a functioning circuit in living cells. The inference of genetic networks is a problem in which mutual interactions among

Evolutionary Computation in Gene Regulatory Network Research, First Edition.
Edited by Hitoshi Iba and Nasimul Noman.

genes are deduced from time series of gene expression data. The inferred model of the genetic network is conceived of as a tool to help biologists generate hypotheses and facilitate the design of their experiments. A number of researchers have taken an interest in the inference of genetic networks, and, therefore, a number of genetic network inference methods have been proposed [2,4,8,10–12,21,23,25,36, 37,44,45].

The inference of genetic networks is often defined as nonlinear function optimization problems. In order to infer reasonable genetic networks, therefore, some inference methods use evolutionary algorithms as a function optimizer (e.g., Refs. [6,16,18,20,24,27,28,34,35,42,43]). The dimensions of the objective functions defined in these methods generally depend on the number of genes contained in a target genetic network. Therefore, when trying to analyze a genetic network consisting of a large number of genes, high-dimensional nonlinear function optimization problems must be solved. Among these inference methods, this chapter describes a method proposed by Kimura et al. that resolves this high dimensionality by defining the inference of a genetic network as several two-dimensional function optimization problems [22].

5.2 MODEL

Numerous models have been used to describe genetic networks, each of which has its own merits and demerits [13,15,17,39,44]. Our group has focused on models based on sets of differential equations, since they have an ability to capture the dynamic behavior of gene expression. When we use a set of differential equations, a genetic network is described as

$$\frac{dX_n}{dt} = F_n(X_1, X_2, \ldots, X_N), \qquad (n = 1, 2, \ldots, N), \tag{5.1}$$

where X_n is the expression level of the nth gene, N is the number of genes contained in the target network, and F_n is a function of arbitrary form. The purpose of a genetic network inference problem based on a set of differential equations is to identify the function F_n from the observed gene expression levels.

In order to identify the function F_n, Sakamoto and Iba [31] have proposed to use a genetic programming, a kind of evolutionary algorithm. The scope of their method has been limited, however, as the difficulty of identifying arbitrary functions restricts application to small-scale genetic network inference problems. It is generally better to approximate the function F_n than to try to identify it. In many earlier studies, the function F_n was approximated using a set of differential equations of a fixed form. When we use a set of differential equations of a fixed form to describe genetic networks, the purpose of the genetic network inference problem is to estimate all of the parameters contained in the model. A linear model is a well-studied model based on a set of differential equations of a fixed form, and several inference methods based on the linear model have been proposed [14, 44]. The computation times of these methods are reportedly very short. As the linear model requires that the system is operating near a steady state, however, it is unsuitable for analyzing the time series of gene expression levels [44]. An S-system model is another well-studied model based on a set of differential equations of a fixed form [39]. This model has a rich structure capable of capturing various dynamics and can be analyzed by several available methods. A number of inference methods based on the S-system model have thus been proposed [5, 6, 18–20, 24, 27, 35, 40]. However, there are a large number of parameters in the S-system model. For example, the number of parameters in the linear model is $N(N+1)$, where N is the number of genes contained in the target network. On the other hand, the number of parameters in the S-system model is $2N(N+1)$. Therefore, we must give the inference methods based on the S-system model more gene expression data to obtain reasonable results.

When trying to infer genetic networks, we should use a model that has an ability to approximate actual biochemical reactions. Moreover, as it is generally difficult to measure a sufficient amount of gene expression data, the model should contain fewer parameters. A model proposed by Vohradský seems to satisfy the requirements above, since the number of parameters is comparable to that of the linear model and it is reportedly capable of capturing the process of gene expression [38]. Because of its preferable features, several inference methods based on this model have been proposed [28, 42, 43], and the inference method described in this chapter [22] is also based on this model. The Vohradský's model is a set of nonlinear differential equations of the form

$$\frac{dX_n}{dt} = \alpha_n f\left(\sum_{m=1}^{N} w_{n,m}X_m + b_n\right) - \beta_n X_n, \qquad (n = 1, 2, \dots, N), \tag{5.2}$$

where

$$f(x) = \frac{1}{1+e^{-x}},$$

and α_n (> 0), β_n (> 0), b_n and $\boldsymbol{w}_n = (w_{n,1}, w_{n,2}, \dots, w_{n,N})$ $(n = 1, 2, \dots, N)$ are model parameters. The first and the second terms on the right-hand side of Equation (5.2) represent processes that contribute to the increase and decrease, respectively, in X_n. When the mth gene positively or negatively regulates the nth gene, the value for the parameter $w_{n,m}$ is positive or negative, respectively. On the other hand, when the nth gene is not regulated by the mth gene, the value for $w_{n,m}$ is 0.

In the inference of a Vohradský's model of a genetic network, the purpose is to estimate all of the model parameters that produce time-series data consistent with the observed gene expression levels. The numbers of the parameters α_n, β_n, b_n, and $w_{n,m}$ $(n, m = 1, 2, \dots, N)$ are N, N, N, and N^2, respectively. The total number of model parameters to be estimated is therefore $N(N+3)$.

5.3 INFERENCE BASED ON BACK-PROPAGATION THROUGH TIME

The Vohradský's model can be viewed as a recurrent neural network. Therefore, in order to estimate the parameters of this model, we can use learning algorithms for recurrent neural networks, such as a back-propagation through time (BPTT) [41]. When we use the BPTT, the inference of a Vohradský's model of a genetic network consisting of N genes is defined as a minimization problem of the following function:

$$F(\boldsymbol{\alpha}, \boldsymbol{\beta}, \boldsymbol{b}, \boldsymbol{w}) = \sum_{n=1}^{N} \sum_{k=2}^{K} \left(X_n|_{t_k} - X_n|_{t_k}^{\text{cal}} \right)^2, \tag{5.3}$$

where $\boldsymbol{\alpha} = (\alpha_1, \alpha_2, \dots, \alpha_N)$, $\boldsymbol{\beta} = (\beta_1, \beta_2, \dots, \beta_N)$, $\boldsymbol{b} = (b_1, b_2, \dots, b_N)$, and $\boldsymbol{w} = (w_{1,1}, w_{1,2}, \dots, w_{N,N})$ are the model parameters, and K is the number of measurements. $X_n|_{t_k}$ and $X_n|_{t_k}^{\text{cal}}$ are the measured and the computed expression levels of the nth gene at time t_k, respectively. However, $X_n|_{t_k}^{\text{cal}}$ is not computed from a set of differential equations (5.2), but from its discrete form, that is,

$$X_n|_{t_k} = \alpha_n f \left(\sum_{m=1}^{N} w_{n,m} X_m|_{t_{k-1}} + b_n \right) \Delta t + (1 - \beta_n \Delta t) X_n|_{t_{k-1}}, \tag{5.4}$$

where $\Delta t = t_k - t_{k-1}$.

The existing inference methods based on the Vohradský's model have been designed according to this problem definition [28, 42, 43]. As the training of recurrent neural networks is not always easy, evolutionary algorithms are used as function optimizers. The method proposed by Xu et al. uses a hybrid of a differential evolution and a particle swarm optimization to optimize the objective function above [42]. Xu et al. [43] also proposed another inference method that uses two different particle swarm optimizations. This method uses one particle swarm optimization to infer the structure of the target genetic network and the other to estimate the parameters of a Vohradský's model. It is known that a gene is regulated by a small number of genes [33], and the inference method [43] was developed to introduce this knowledge into the genetic network inference. On the other hand, in order to optimize the objective function (5.3), Palafox and Iba [28] have used population-based incremental learning [3], a kind of estimation of distribution algorithm. Although their method has an ability to infer genetic networks with very short computation times, it is unable to estimate model parameters precisely.

In contrast to these inference methods, the method proposed by Kimura et al. estimates the model parameters by solving simultaneous equations, as described in Section 5.4 [22].

5.4 INFERENCE BY SOLVING SIMULTANEOUS EQUATIONS

As mentioned in Section 5.3, several genetic network inference methods based on the Vohradský's model have been proposed [28, 42, 43]. As these methods are designed on the basis of a learning algorithm for recurrent neural networks, they estimate all of the model parameters simultaneously. The number of parameters in the Vohradský's model is $N(N+3)$, where N is the number of genes contained in the network. When trying to analyze genetic networks with a large number of genes, high-dimensional nonlinear function optimization problems must be solved.

In order to overcome this high dimensionality, the inference method proposed by Kimura et al. [22] first divides the inference problem of a Vohradský's model of a genetic network consisting of N genes into N subproblems, each of which corresponds to each gene. Our method then defines each subproblem as a two-dimensional function optimization problem. This section describes the nth subproblem corresponding to the nth gene.

5.4.1 Problem Definition

The nth subproblem is defined as the problem of solving the following simultaneous equations:

$$\begin{aligned}
\left.\frac{dX_n}{dt}\right|_{t_1} &= \alpha_n f\left(\sum_{m=1}^{N} w_{n,m}\, X_m\big|_{t_1} + b_n\right) - \beta_n\, X_n\big|_{t_1}, \\
\left.\frac{dX_n}{dt}\right|_{t_2} &= \alpha_n f\left(\sum_{m=1}^{N} w_{n,m}\, X_m\big|_{t_2} + b_n\right) - \beta_n\, X_n\big|_{t_2}, \\
&\vdots \\
\left.\frac{dX_n}{dt}\right|_{t_K} &= \alpha_n f\left(\sum_{m=1}^{N} w_{n,m}\, X_m\big|_{t_K} + b_n\right) - \beta_n\, X_n\big|_{t_K},
\end{aligned} \tag{5.5}$$

where $X_m\big|_{t_k}$ is the expression level of the mth gene at time t_k, $\left.\frac{dX_n}{dt}\right|_{t_k}$ is the time derivative of the expression level of the nth gene at time t_k, and K is the number of measurements. Note that $X_m\big|_{t_k}$ is measured using some biological technology. $\left.\frac{dX_n}{dt}\right|_{t_k}$ is, on the other hand, estimated directly from the observed time series of the gene expression levels using a smoothing technique, such as spline interpolation [29], local linear regression [9], and so on. By solving the simultaneous equations (5.5) in the nth subproblem, the method estimates the parameters corresponding to the nth gene, that is, α_n, β_n, b_n, and $\boldsymbol{w}_n = (w_{n,1}, w_{n,2}, \ldots, w_{n,N})$. Based on a similar idea, several genetic network inference methods have been proposed [7, 19, 23, 40, 44].

When we simply use a least-squares method to solve these simultaneous equations, for example, the estimation of the model parameters is defined as a minimization problem of the function

$$G_n(\alpha_n, \beta_n, b_n, \boldsymbol{w}_n) = \sum_{k=1}^{K} \left[\left.\frac{dX_n}{dt}\right|_{t_k} - R_k(\alpha_n, \beta_n, b_n, \boldsymbol{w}_n) \right]^2, \tag{5.6}$$

where

$$R_k(\alpha_n, \beta_n, b_n, \mathbf{w}_n) = \alpha_n f\left(\sum_{m=1}^{N} w_{n,m} X_m|_{t_k} + b_n\right) - \beta_n X_n|_{t_k}.$$

It is not always easy to solve this problem, however, since the objective function (5.6) is nonlinear. Moreover, as the dimension of this function is $N + 3$, the difficulty in optimizing it increases with an increase in the number of genes contained in the network, that is, N. In order to cope with this difficulty, our group proposed an efficient technique for solving the simultaneous equations (5.5), as mentioned in Section 5.4.2.

5.4.2 Efficient Technique for Solving Simultaneous Equations

5.4.2.1 *Concept* To solve the simultaneous equations (5.5), our method uses a feature arising from the transformation of the equations, described below.

By rearranging the kth member of Equations (5.5), we have

$$Y_k = f\left(\sum_{m=1}^{N} w_{n,m} X_m|_{t_k} + b_n\right), \tag{5.7}$$

where

$$Y_k = \frac{\frac{dX_n}{dt}\Big|_{t_k} + \beta_n X_n|_{t_k}}{\alpha_n}.$$

By applying $f(x) = \frac{1}{1+e^{-x}}$ to Equation (5.7), we obtain

$$\begin{aligned} Y_k &= \frac{1}{1 + \exp\left[-\left(\sum_{m=1}^{N} w_{n,m} X_m|_{t_k} + b_n\right)\right]}, \\ \frac{1 - Y_k}{Y_k} &= \exp\left[-\left(\sum_{m=1}^{N} w_{n,m} X_m|_{t_k} + b_n\right)\right], \\ \frac{Y_k}{1 - Y_k} &= \exp\left(\sum_{m=1}^{N} w_{n,m} X_m|_{t_k} + b_n\right). \end{aligned} \tag{5.8}$$

By taking logarithms of both sides of the above equation, we then have

$$\log\left(\frac{Y_k}{1 - Y_k}\right) = \sum_{m=1}^{N} w_{n,m} X_m|_{t_k} + b_n. \tag{5.9}$$

All of the members of the simultaneous equations (5.5) can be converted in the same way. We therefore have new simultaneous equations:

$$
\begin{aligned}
\log\left(\frac{Y_1}{1-Y_1}\right) &= \textstyle\sum_{m=1}^{N} w_{n,m}\, X_m\big|_{t_1} + b_n,\\
\log\left(\frac{Y_2}{1-Y_2}\right) &= \textstyle\sum_{m=1}^{N} w_{n,m}\, X_m\big|_{t_2} + b_n,\\
&\;\;\vdots\\
\log\left(\frac{Y_K}{1-Y_K}\right) &= \textstyle\sum_{m=1}^{N} w_{n,m}\, X_m\big|_{t_K} + b_n.
\end{aligned}
\tag{5.10}
$$

Note that, although these transformed simultaneous equations are nonlinear with respect to the parameters α_n and β_n, they are linear with respect to the parameters b_n and $\boldsymbol{w}_n = (w_{n,1}, w_{n,2}, \dots, w_{n,N})$. This fact suggests that, when the parameters α_n and β_n are given, the other parameters b_n and $\boldsymbol{w}_n$ are easily estimated. Our method uses this feature for solving the simultaneous equations (5.5).

5.4.2.2 *Objective Function* Our inference method uses a least-squares method to solve the simultaneous equations (5.5). As mentioned in Section 5.4.2.1, however, we can easily estimate the parameters b_n and $\boldsymbol{w}_n$, when the parameters α_n and β_n are given. Therefore, in order to solve the simultaneous equations, our method searches for optimum values for the parameters α_n and β_n only. The objective function of this problem is thus defined as

$$
H_n(\alpha_n, \beta_n) = \sum_{k=1}^{K} \left[\left.\frac{dX_n}{dt}\right|_{t_k} - R_k(\alpha_n, \beta_n, b_n^*, \boldsymbol{w}_n^*) \right]^2, \tag{5.11}
$$

where

$$
R_k(\alpha_n, \beta_n, b_n^*, \boldsymbol{w}_n^*) = \alpha_n f\left(\sum_{m=1}^{N} w_{n,m}^*\, X_m\big|_{t_k} + b_n^* \right) - \beta_n\, X_n\big|_{t_k},
$$

and b_n^* and $\boldsymbol{w}_n^* = (w_{n,1}^*, w_{n,2}^*, \dots, w_{n,N}^*)$ are the optimal values for b_n and $\boldsymbol{w}_n = (w_{n,1}, w_{n,2}, \dots, w_{n,N})$, respectively, under given α_n and β_n. In Section 5.4.2.3, we will describe a method for obtaining b_n^* and $\boldsymbol{w}_n^*$.

Note here that the shapes of the objective functions (5.6) and (5.11) are the same. The objective function (5.11) is therefore still nonlinear. Its low dimensionality however makes the problem much easier to solve.

5.4.2.3 *Estimation of b_n^* and w_n^** When trying to compute a value for the objective function (5.11), we must always provide values for b_n^* and $\boldsymbol{w}_n^*$. In our approach, they serve as the solution of a set of the transformed equations (5.10) under given α_n and β_n. Note that, when the parameters α_n and β_n are given, these equations are linear with respect to the unknown parameters, that is, b_n and $\mathbf{w}_n$. We can thus easily estimate b_n^* and $\mathbf{w}_n^*$. In our approach, we define the estimation of these parameters as the following constrained function minimization problem:

$$\underset{b_n, \mathbf{w}_n, \xi_k^+, \xi_k^-}{\text{minimize}} \frac{C}{\Gamma} \sum_{k=1}^{K} \left(\gamma_k^+ \xi_k^+ + \gamma_k^- \xi_k^- \right) + \sum_{m=1}^{N} |w_{n,m}|, \tag{5.12}$$

subject to

$$\begin{cases} \log\left(\frac{Z_k}{1-Z_k}\right) - \sum_{m=1}^{N} w_{n,m} X_m|_{t_k} - b_n \geq -\xi_k^+, & (k = 1, 2, \ldots, K), \\ \xi_k^+ \geq 0, & (k = 1, 2, \ldots, K), \\ \log\left(\frac{Z_k}{1-Z_k}\right) - \sum_{m=1}^{N} w_{n,m} X_m|_{t_k} - b_n \leq \xi_k^-, & (k = 1, 2, \ldots, K), \\ \xi_k^- \geq 0, & (k = 1, 2, \ldots, K), \end{cases}$$

where

$$Z_k = \begin{cases} Y_k, & (\text{if } \delta \leq Y_k \leq 1 - \delta), \\ \delta, & (\text{if } Y_k < \delta), \\ 1 - \delta, & (\text{otherwise}), \end{cases}$$

$$Y_k = \frac{\left.\frac{dX_n}{dt}\right|_{t_k} + \beta_n X_n|_{t_k}}{\alpha_n},$$

$$\Gamma = \frac{1}{2} \sum_{k=1}^{K} (\gamma_k^+ + \gamma_k^-),$$

ξ_k^+ and ξ_k^- are slack variables, and γ_k^+, γ_k^-, δ and C are constant parameters. In the problem (5.12), we treat the parameters α_n and β_n as constants. Note that, whenever trying to compute a value for the objective function (5.11), we must always solve the function optimization problem (5.12) (see Figure 5.1).

ξ_k^+ and ξ_k^- represent the differences between the left- and the right-hand sides of the kth member of the simultaneous equations (5.10). The first term of the objective function of the problem (5.12), that is, $\frac{C}{\Gamma} \sum_{k=1}^{K} \left(\gamma_k^+ \xi_k^+ + \gamma_k^- \xi_k^- \right)$, is thus a weighted sum of the differences between the left- and right-hand sides of Equations (5.10). As described in Section 5.2, on the other hand, when the nth gene is not regulated by the mth gene, the parameter corresponding to this regulation, that is, $w_{n,m}$, is 0 in the Vohradský's model. Therefore, most $w_{n,m}$ values should be 0 because genetic

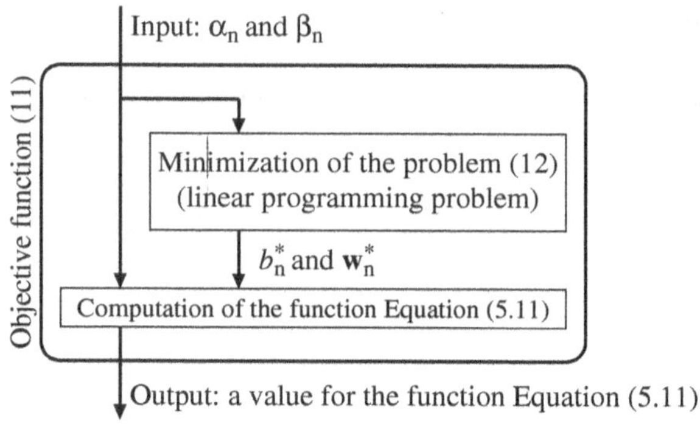

Figure 5.1 The computation of the objective function (5.11).

networks are known to be sparsely connected [33]. The second term of the objective function, that is, $\sum_{m=1}^{N} |w_{n,m}|$, introduces this knowledge into the parameter estimation.

The transformation of the simultaneous equations (5.5), described in Section 5.4.2.1, can be done, only when the condition $0 < Y_k < 1$ is satisfied. Even when the optimum values are set for α_n and β_n, however, the noise contained in the observed gene expression data may mean that the condition above is not satisfied. Our approach thus introduces a threshold parameter δ, and sets its value to 10^{-10}. On the other hand, we should note that, when Y_k approaches 0 or 1, the term $\log\left(\frac{Y_k}{1-Y_k}\right)$ contained in the kth member of the transformed simultaneous equations (5.10) approaches $-\infty$ or $+\infty$, respectively. Therefore, when Y_k is approximately equal to 0 or 1, the transformation of the kth member of Equations (5.5) would amplify the noise contained in the measurement data. We should not rely too much on the equations transformed under these conditions. In order to introduce this notion into the parameter estimation, we set the constant parameters γ_k^+ and γ_k^- to

$$\gamma_k^+ = \begin{cases} \left[1 - 4\left(Z_k - \frac{1}{2}\right)^2\right]^{\frac{1}{2}}, & (\text{if } 0 \le Z_k \le \frac{1}{2}), \\ 1 - 4\left(Z_k - \frac{1}{2}\right)^2, & (\text{otherwise}), \end{cases}$$

$$\gamma_k^- = \begin{cases} 1 - 4\left(Z_k - \frac{1}{2}\right)^2, & (\text{if } 0 \le Z_k \le \frac{1}{2}), \\ \left[1 - 4\left(Z_k - \frac{1}{2}\right)^2\right]^{\frac{1}{2}}, & (\text{otherwise}). \end{cases}$$

The problem (5.12) can be converted into a linear programming problem. Therefore, we can always find an optimum solution to the problem. In order to solve this problem, our inference method uses an interior point method [26].

5.5 REX^{STAR}/JGG

Any function optimization algorithm can be used to minimize the objective function (5.11). The defined functions are expected to be easily optimized, since they are only two dimensional. When we applied a local search to the optimization problem of this function, however, we found several local optima. Therefore, in order to increase the probability of obtaining a reasonable solution, our inference method uses REX^{star}/JGG [1] as a function optimizer. REX^{star}/JGG, which is a real-coded genetic algorithm, a sort of evolutionary algorithm, reportedly shows excellent performance for many benchmark problems. REX^{star}/JGG uses just generation gap (JGG) as a generation alternation model and real-coded ensemble crossover^{star} (REX^{star}) as a recombination operator. This section describes each of the operators in detail.

5.5.1 JGG

JGG is a generation alternation model developed chiefly for multiparental recombination operators. The generation alternation model is a procedure for selecting individuals to breed and for selecting individuals to form a new population in the next generation. The following is an algorithm of JGG.

[Algorithm: JGG]

1. **Initialization**
   As an initial population, create n_p individuals. As REX^{star}/JGG is a real-coded genetic algorithm, these individuals are represented as s-dimensional real number vectors, where s is the dimension of the search space. Set *Generation* $= 0$.
2. **Selection for Reproduction**
   Randomly select m individuals without replacement from the population. The selected individuals, expressed here as $\boldsymbol{p}_1, \boldsymbol{p}_2, \dots, \boldsymbol{p}_m$, are used as the parents for the recombination operator in the next step. As mentioned in the following section, REX^{star} uses $s + 1$ parents, that is, $m = s + 1$.
3. **Generation of Offsprings**
   Generate n_c children by applying the recombination operator to the parents selected in the previous step. Our inference method uses REX^{star} as the recombination operator.
4. **Selection for Survival**
   Select the best m individuals from the family containing the m parents, that is, $\boldsymbol{p}_1, \boldsymbol{p}_2, \dots, \boldsymbol{p}_m$, and their children. Then, replace the m parents with the selected individuals. In the original JGG, the best m individuals are selected only from

the children. However, we slightly modified its algorithm as this optimization process seemed to be unstable.

5. **Termination**
   Stop if the halting criteria are satisfied. Otherwise, *Generation* ← *Generation* +1, and then return to step 2.

5.5.2 REX^{star}

REX^{star} is a multiparental recombination operator. It uses $s+1$ parents, where s is the dimension of the search space, and generates n_c ($> s+1$) children according to the following algorithm:

[Algorithm: REX^{star}]

1. Generate reflection points $\underline{\boldsymbol{p}_1}, \underline{\boldsymbol{p}_2}, \dots, \underline{\boldsymbol{p}_{s+1}}$ of the parents $\boldsymbol{p}_1, \boldsymbol{p}_2, \dots, \boldsymbol{p}_{s+1}$, that is,

$$\underline{\boldsymbol{p}_i} = 2\boldsymbol{G} - \boldsymbol{p}_i, \tag{5.13}$$

   where

$$\boldsymbol{G} = \frac{1}{s+1} \sum_{i=1}^{s+1} \boldsymbol{p}_i.$$

2. Compute the objective values of the $s+1$ reflection points generated in the previous step. In REX^{star}, these reflection points are treated as the children.
3. From the parents and their reflection points, select the best $s+1$ individuals with respect to the objective value, and then compute the center of the gravity of the selected $s+1$ individuals. This study represents this as $\boldsymbol{G}^b$.
4. Generate $n_c - s - 1$ children by applying the following equation $n_c - s - 1$ times. Note that the $s+1$ reflection points generated in step 1 are treated as the children. The total number of the children generated by REX^{star} is therefore n_c:

$$\boldsymbol{c} = \boldsymbol{G} + \text{diag}(\xi_1^t, \xi_2^t, \dots, \xi_s^t)(\boldsymbol{G}^b - \boldsymbol{G}) + \sum_{i=1}^{s+1} \xi^i (\boldsymbol{p}_i - \boldsymbol{G}), \tag{5.14}$$

   where $\boldsymbol{c}$ represents the child, and

$$\text{diag}(d_1, d_2, \dots, d_s) = \begin{pmatrix} d_1 & 0 & \cdots & 0 \\ 0 & d_2 & \cdots & 0 \\ \vdots & \vdots & \ddots & \vdots \\ 0 & 0 & \cdots & d_s \end{pmatrix}.$$

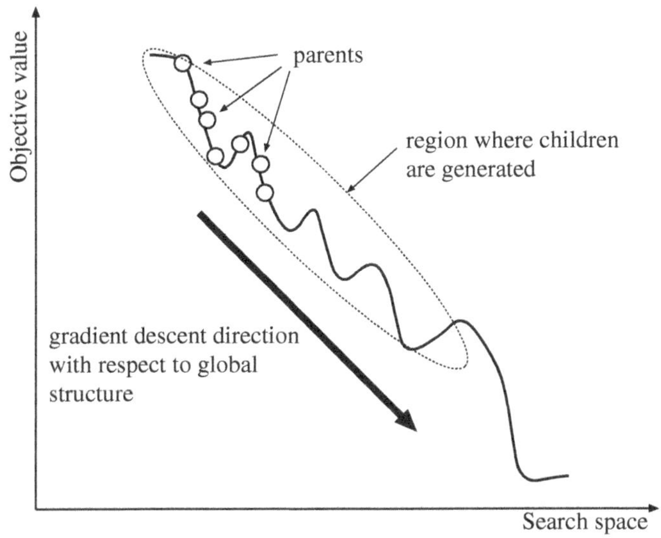

Figure 5.2 Generation of offspring by REX^{star}.

ξ_i^ts and ξ^is are random numbers drawn from uniform distributions $[0, t]$ and $\left[-\sqrt{\frac{3}{s+1}}, \sqrt{\frac{3}{s+1}}\right]$, respectively, where t is a hyper-parameter named a step-size parameter.

REX^{star} estimates a global structure of a landscape of an objective function, disregarding its local roughness. The estimated global structure is used to generate offspring, that is, the region where children are generated is enlarged toward the gradient descent direction with respect to the estimated global structure (Figure 5.2). Because of this feature, REX^{star}/JGG has an ability to find optimum solutions with a smaller number of function evaluations.

In the original paper of REX^{star}/JGG, the following settings are recommended for its hyper-parameters: the population size n_p is set between $2s$ and $20s$; the number of children generated per selection n_c is set between $2s$ and $3s$; and the step-size parameter t is set between 2.5 and 15.

5.6 INFERENCE OF AN ARTIFICIAL NETWORK

In order to confirm that our method is capable of estimating reasonable values for the parameters of the Vohradský's model, we applied it to an artificial genetic network inference problem.

5.6.1 Experimental Setup

We used a Vohradský's model with four genes ($N = 4$) as a target network [43]. The model parameters and the network structure of this system are given in Table 5.1 and

Table 5.1 The model parameters of the artificial network

n	$w_{n,1}$	$w_{n,2}$	$w_{n,3}$	$w_{n,4}$	b_n	α_n	β_n
1	20.0	−20.0	0.0	0.0	0.0	0.1	0.1
2	15.0	−10.0	0.0	0.0	−5.0	0.2	0.2
3	0.0	−8.0	12.0	0.0	0.0	0.2	0.2
4	0.0	0.0	8.0	−12.0	0.0	0.2	0.2

Figure 5.3, respectively. Note that, as this network consists of four genes, our inference method solves four individual two-dimensional function optimization problems to estimate all of the model parameters.

As the observed gene expression patterns, three sets of time-series data, each covering four genes, were computed from the differential equations (5.2) on the target model. The sets began from initial values randomly generated in [0.0, 1.0], and 50 sampling points for the time-series data were assigned to each gene in each set. The number of observations K is therefore $3 \times 50 = 150$. A sample of the time-series data we generated is shown in Figure 5.4. In a practical application, these sets would be obtained by actual biological experiments under different experimental conditions. This experiment simulated no measurement noise in the computed data. The time derivatives of the gene expression levels were thus directly computed from the model (5.2) of the target network. In this experiment, we estimated the parameters of the target model only from the gene expression levels and their derivatives.

We performed 10 trials, each with different sets of gene expression data. We considered the model parameters to be successfully estimated only when the value of the objective function (5.11) dropped to less than 1.0×10^{-6}. As the parameters α_n and β_n contained in the objective function (5.11) are both positive, this study searched for them in a logarithmic space. Their search area was set to $[-3.0, 3.0]^2$. Based on the preliminary experiments, we set the constant parameter C contained in the constrained function minimization problem (5.12) to 2000. In accordance with the

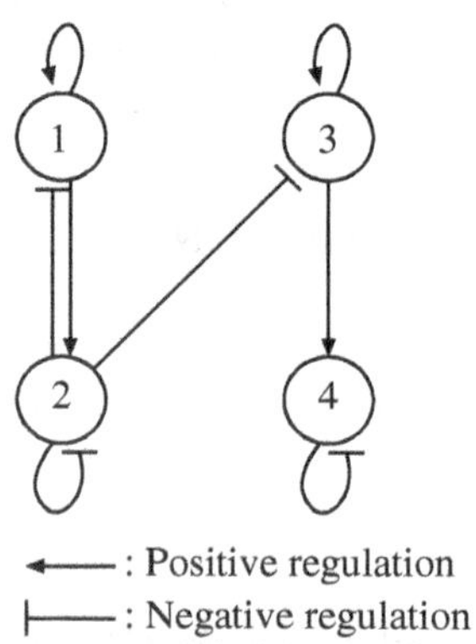

Figure 5.3 The structure of the artificial network.

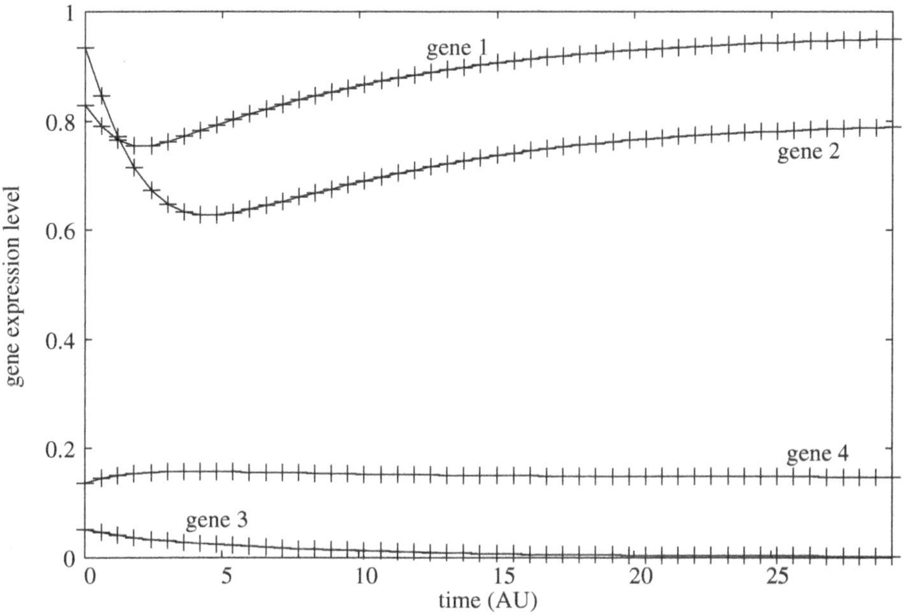

Figure 5.4 A sample of time-series data used for inferring the artificial genetic network.

recommended values, this study used the following values for the hyper-parameters of REX^{star}/JGG [1]: the population size n_p is 40; the number of children generated per selection n_c is 6; and the step-size parameter t is 2.5. Each run of REX^{star}/JGG was continued until the maximum number of generation alternations reached 250.

5.6.2 Results

The inference method described here succeeded in estimating the parameter values with precision in seven trials. In these trials, the parameters were estimated with high precision (Table 5.2). Even in the rest of the trials, most of the parameters were correctly estimated. Table 5.3 shows a sample of the model parameters estimated in one of the failed trials. As described previously, the inference method divided the parameter estimation problem of the target network here into four subproblems,

Table 5.2 A typical sample of the parameters correctly estimated

n	$w_{n,1}$	$w_{n,2}$	$w_{n,3}$	$w_{n,4}$	b_n	α_n	β_n
1	20.003	−20.002	0.000	−0.001	0.000	0.100	0.100
2	15.000	−10.000	0.000	0.000	−5.000	0.200	0.200
3	−0.016	−7.991	11.986	−0.007	0.008	0.200	0.200
4	0.000	0.000	8.000	−12.000	0.000	0.200	0.200

Table 5.3 A sample of the parameters erroneously estimated

n	$w_{n,1}$	$w_{n,2}$	$w_{n,3}$	$w_{n,4}$	b_n	α_n	β_n
1	1.405	−0.043	0.552	0.068	−2.424	4.004	1.561
2	15.000	−10.001	0.000	0.000	−5.000	0.200	0.200
3	0.000	−7.854	11.998	−0.275	0.031	0.200	0.200
4	0.000	0.000	8.000	−12.000	0.000	0.200	0.200

each of which is defined as a two-dimensional function optimization problem. In this experiment, our method therefore solved $4 \times 10 = 40$ subproblems and failed to find the optimum solutions for only 3 of these 40 subproblems. While the averaged objective value (5.11) of the three failed subproblems was $1.590 \times 10^{-3} \pm 2.120 \times 10^{-3}$, that of the other subproblems was $4.607 \times 10^{-9} \pm 1.622 \times 10^{-8}$. When we gave the correct parameter values to the objective functions of the three failed subproblems, we obtained better objective values. The solutions found in the three failed subproblems are therefore local optima. To estimate all of the model parameters for this network, our method took about 10.4 ± 0.1 min on a single-CPU personal computer (Pentium IV 2.8 GHz).

As described previously, the inference method proposed by Kimura et al. uses an evolutionary algorithm, REX^{star}/JGG, to optimize the objective function (5.11). When we use a local search, instead of the evolutionary algorithm, we can shorten the computation time required for optimizing it. As the objective function seems to be multimodal, however, the use of the local search makes it difficult to infer reasonable genetic networks. When we used the modified Powell's method [29], a local search algorithm, the computation time to estimate all of the model parameters was 3.4 ± 0.6 min on the personal computer (Pentium IV 2.8 GHz). The inference method with the modified Powell's method, on the other hand, failed in finding the optimum solutions for 19 of the 40 subproblems. The averaged objective values (5.11) of the 19 failed subproblems and the other subproblems were $7.052 \times 10^{-2} \pm 2.205 \times 10^{-1}$ and $1.653 \times 10^{-8} \pm 7.392 \times 10^{-8}$, respectively. When we applied the inference method with the modified Powell's method to artificial genetic network inference problems consisting of 30 genes, the method failed in finding reasonable solutions for 145 of $30 \times 10 = 300$ subproblems. This fact indicates that the difficulty in solving each subproblem defined in our approach is independent of the number of genes contained in the network. This is a preferable feature when we try to analyze larger-scale genetic networks.

As mentioned in Section 5.3, several genetic network inference methods based on the Vohradský's model have already been proposed [28, 42, 43]. However, for computational simplicity, they substantially limit the search space to $\alpha_n = \beta_n$. In order to make a fair comparison, we therefore constructed an inference method based on the BPTT, referred to here as BPTTGA, that does not limit the search space. BPTTGA uses REX^{star}/JGG to optimize the objective function (5.3). Although the computation time of BPTTGA was comparable to that of the proposed inference method, it was unable to estimate the model parameters precisely. When we used

Table 5.4 A sample of the parameters estimated by BPTTGA

n	$w_{n,1}$	$w_{n,2}$	$w_{n,3}$	$w_{n,4}$	b_n	α_n	β_n
1	20.847	−21.373	−0.719	0.626	0.750	0.097	0.098
2	13.222	−8.787	0.092	−0.110	−4.421	0.194	0.189
3	−0.821	−4.595	11.752	−4.166	1.093	0.189	0.189
4	−0.023	0.024	6.553	−9.815	0.084	0.189	0.189

BPTTGA to infer the target network here, the averaged computation time on the personal computer (Pentium IV 2.8 GHz) and the averaged objective value were 9.3 ± 7.5 min and $5.987 \times 10^{-3} \pm 1.062 \times 10^{-2}$, respectively. A typical sample of the model parameters estimated by BPTTGA is shown in Table 5.4. Note that, although the existing inference methods [28, 42, 43] substantially limit the search space, they were reportedly still unable to estimate the model parameters with precision.

5.7 INFERENCE OF AN ACTUAL GENETIC NETWORK

We then used our inference method to analyze actual gene expression data.

5.7.1 Experimental Setup

We applied our inference method to the actual inference problem of the SOS DNA repair regulatory network in *Escherichia coli* (Figure 5.5) [32]. More than 30 genes, including *lexA* and *recA*, are known to be involved in this system. These genes are regulated by *lexA* and *recA*. In a basal state, LexA, a master repressor, is bound to the interaction site in the promoter regions of these genes. When DNA is damaged, RecA, another SOS protein, senses the damage and mediates LexA autocleavage. The decrease in LexA protein level halts the repression of the SOS genes, and they

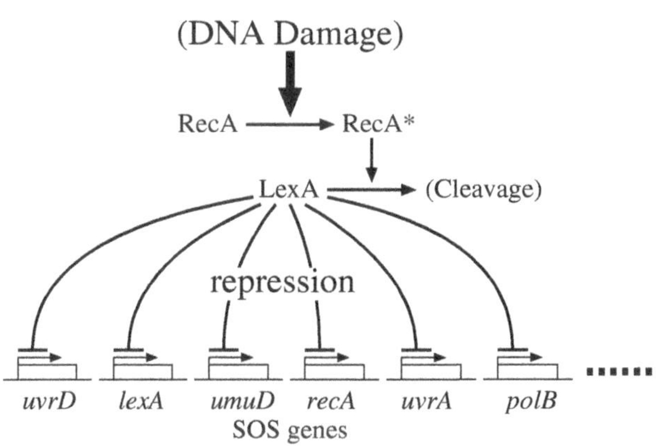

Figure 5.5 The SOS DNA repair system in *Escherichia coli*.

then start DNA repair. Once the damage has been repaired, RecA stops mediating LexA autocleavage, LexA accumulates and represses the SOS genes, and the cells return to their basal state.

This experiment analyzed the expression data of six genes, that is, *uvrD*, *lexA*, *umuD*, *recA*, *uvrA*, and *polB*, which had been measured by Ronen et al. [30] ($N = 6$). In order to infer the genetic network, therefore, we must solve six individual two-dimensional function optimization problems. These expression data have often been used to confirm the performances of inference methods [5,6,16,19,21,23,42,43]. The original expression data contain four sets of time-series data. This experiment however used only two sets (the third and fourth sets), since these two had been measured under the same experimental conditions. Each set of time-series data consisted of 50 measurement values including the initial concentrations of 0. In the experiment, we removed the initial concentrations from both sets as models based on a set of differential equations cannot produce different time courses from the same initial conditions. The number of measurements K is thus $2 \times 49 = 98$. We normalized the data corresponding to each gene against its maximum expression level. The actual gene expression data are generally polluted by noise. This experiment therefore smoothed the normalized gene expression data using the local linear regression [9], a smoothing technique. We assigned a value of 10^{-6} to expression levels with values of less than 10^{-6}, as the gene expression levels must not be negative. The time derivatives of the gene expression levels were estimated from the smoothed data. We performed 10 trials by changing a seed for pseudo-random numbers. All of the other experimental conditions were the same as those described previously.

5.7.2 Results

Although we performed 10 trials in this experiment, the inferred models had parameters similar to each other. A sample of the parameters of the inferred model is shown in Table 5.5. The computation time required for analyzing this system was approximately 13.9 ± 0.2 min on the personal computer (Pentium IV 2.8 GHz). We extracted the structures of the network from the estimated model parameters. The structures of the networks were extracted according to the following rules: when $w_{n,m} \geq T_n$ or $w_{n,m} \leq -T_n$, we conclude that the mth gene positively or negatively, respectively,

Table 5.5 A sample of the estimated parameters in the analysis of the SOS DNA repair network in *Escherichia coli*

n	$w_{n,1}$	$w_{n,2}$	$w_{n,3}$	$w_{n,4}$	$w_{n,5}$	$w_{n,6}$	b_n	α_n	β_n
1	4.412	−6.461	−0.355	0.000	5.901	−0.028	−4.185	0.337	0.055
2	6.054	−14.281	0.000	−7.562	19.762	1.667	−4.771	0.164	0.050
3	3.292	−11.752	7.891	−5.102	9.330	0.476	−4.997	0.604	0.095
4	4.177	−14.748	6.997	−3.644	11.519	0.000	−5.194	0.463	0.069
5	5.220	−20.101	9.979	−7.332	19.523	0.159	−4.301	0.265	0.252
6	−0.372	−4.863	−17.110	10.927	15.720	7.347	−2.535	0.094	0.110

Genes 1, 2, 3, 4, 5, and 6 represent *uvrD*, *lexA*, *umuD*, *recA*, *uvrA* and *polB*, respectively.

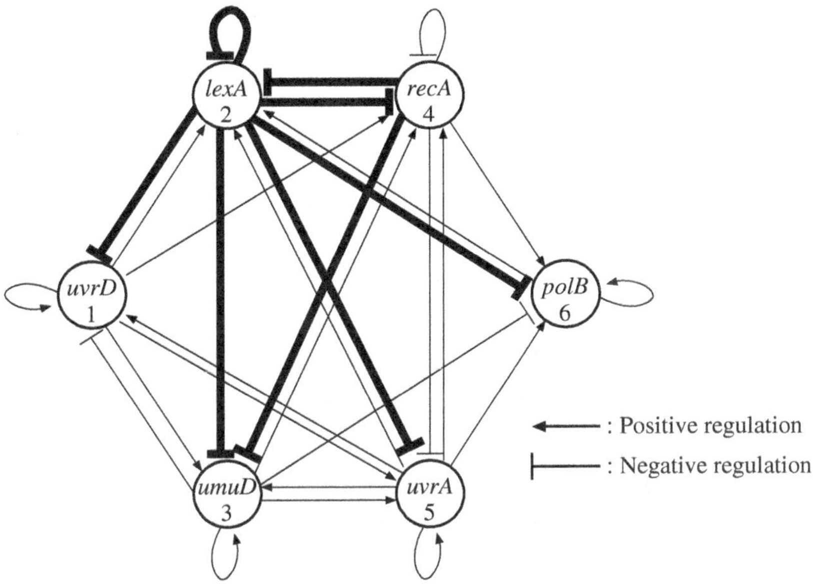

Figure 5.6 The network structure extracted from the estimated parameters in the analysis of the SOS DNA repair network in *Escherichia coli*. Bold lines represent biologically plausible regulations.

regulates the nth gene, where T_n is a threshold; otherwise, we infer no regulation from the mth gene to the nth gene. This experiment set the threshold T_n to

$$T_n = 0.05 \times \max \left\{ |w_{n,1}|, |w_{n,2}|, \dots, |w_{n,N}|, 10^{-3} \right\}.$$

All 10 of the inferred networks had the same structure. Figure 5.6 shows the structure of the network inferred by our inference method.

As mentioned before, LexA is known to repress the SOS genes. Therefore, the negative regulations of all of the genes by *lexA*, inferred by our method, are reasonable. Similarly, the negative regulation of *lexA* by *recA* would be reasonable, as RecA senses the damage of DNA and mediates LexA autocleavage. The regulation of *umuD* by *recA*, contained in the inferred network, also appears to be reasonable, since it is contained in a network now known [14]. As shown in the figure, the network inferred by our method still contains a number of regulations that are not mentioned above. Some of these regulations might be new findings. However, the rest should be false positives. In order to analyze larger-scale genetic networks using our inference method, therefore, we must find a way to reduce these erroneous results.

5.8 CONCLUSION

This chapter described the genetic network inference method proposed by Kimura et al. [22]. This method uses the Vohradský's model to describe genetic networks.

In order to infer a Vohradský's model of a genetic network consisting of N genes, $N(N+3)$ model parameters must be estimated. The estimation of these parameters is generally defined as function optimization problems whose dimensions depend on the number of genes. Therefore, it is not always easy to infer large-scale genetic networks consisting of many genes. The inference method described in this chapter resolves the difficulty in the estimation of these parameters by dividing the inference problem into N subproblems first. The method then defines each subproblem as a two-dimensional function optimization problem. As the defined subproblems seem to be multimodal, it uses the evolutionary algorithm, REX^{star}/JGG, as a function optimizer. The experimental results showed that the inference method has an ability to infer reasonable genetic networks. The network inferred by the method, however, seemed to still contain a number of false-positive regulations. In addition, its computation times were not always sufficiently short. In future work, the author must deal with these drawbacks.

A variety of inference methods based on a variety of mathematical models have been proposed. However, we still do not know which method is most suitable for the inference of genetic networks. The author thinks that, in order to obtain a reliable network, it is important to analyze the measurement data using multiple inference methods based on different models. Kimura et al. have therefore proposed several inference methods based on several mathematical models [20, 21, 23]. The development of a technique to integrate the results obtained from multiple inference methods is also a future goal.

ACKNOWLEDGEMENTS

The author thanks Dr Mariko Okada-Hatakeyama of RIKEN Center for Integrative Medical Sciences and Dr Masanao Sato of National Institute for Natural Sciences for their useful suggestions and comments, and the editors for reviewing the draft of this chapter. Our inference method used a software, BPMPD, developed by Dr Csaba Mészáros at MTA SZTAKI to solve the linear programming problems. This work was partially supported by JSPS KAKENHI Grant Number 26330275.

REFERENCES

1. Y. Akimoto, J. Sakuma, I. Ono, and S. Kobayashi. Adaptation of expansion rate for real-coded crossovers. In: *Proceedings of 2009 Genetic and Evolutionary Computation Conference*, ACM Press, New York, NY, 739–746 (2009).
2. T. Akutsu, S. Miyano, and S. Kuhara. Inferring qualitative relations in genetic networks and metabolic pathways, *Bioinformatics*, 16: 727–734 (2000).
3. S. Baluja. Population-based incremental learning, Technical Report CMU-CS-94-163, Carnegie Mellon University, Pittsburgh, PA (1994).
4. M. Bansal and D. di Bernardo. Inference of gene networks from temporal gene expression profiles. *IET Syst. Biol.*, 1: 306–312 (2007).

5. N. Chemmangattuvalappil, K. Task, and I. Banerjee. An integer optimization algorithm for robust identification of non-linear gene regulatory networks. *BMC Systems Biology*, 6: 119 (2012).
6. D.Y. Cho, K.H. Cho, and B.T. Zhang. Identification of biochemical networks by s-tree based genetic programming. *Bioinformatics*, 22: 1631–1640 (2006).
7. I.C. Chou, H. Martens, and E.O. Voit. Parameter estimation in biochemical systems models with alternating regression. *Theoretical Biology and Medical Modelling*, 3: 25 (2006).
8. I.C. Chou and E.O. Voit. Recent development in parameter estimation and structure identification of biochemical and genomic systems. *Mathematical Biosciences*, 219: 57–83 (2009).
9. W.S. Cleveland. Robust locally weight regression and smoothing scatterplots. *Journal of American Statistical Association*, 79: 829–836 (1979).
10. P. D'haeseleer, S. Liang, and R. Somogyi. Genetic network inference: from co-expression clustering to reverse engineering. *Bioinformatics*, 16: 707–726 (2000).
11. A. Ergün, C.A. Lawrence, M.A. Kohanski, T.A. Brennan, and J.J. Collins. A network biology approach to prostate cancer. *Molecular Systems Biology*, 3: 82 (2007).
12. J.J. Faith, B. Hayete, J.T. Thaden, I. Mogno, J. Wierzbowski, G. Cottarel, S. Kasif, J.J. Collins, and T.S. Gardner. Large-scale mapping and validation of *Escherichia coli* transcriptional regulation from a compendium of expression profiles. *PLoS Biology*, 5: e8 (2007).
13. N. Friedman, M. Linial, I. Nachman, and D. Pe'er. Using Bayesian networks to analyze expression data. *Journal of Computational Biology*, 7: 601–620 (2000).
14. T.S. Gardner, D. di Bernardo, D. Lorenz, and J.J. Collins. Inferring genetic networks and identifying compound mode of action via expression profiling. *Science*, 301: 102–105 (2003).
15. P.J.E. Goss and J. Peccoud. Quantitative modeling of stochastic systems in molecular biology by using stochastic petri nets. *Proceedings of National Academy of Sciences of the United States of America*, 95: 6750–6755 (1998).
16. S. Kabir, N. Noman, and H. Iba. Reverse engineering gene regulatory network from microarray data using linear time-variant model. *BMC Bioinformatics*, 11: S56 (2010).
17. S.A. Kauffman. Metabolic stability and epigenesis in randomly constructed genetic nets. *Journal of Theoretical Biology*, 22: 437–467 (1969).
18. S. Kikuchi, D. Tominaga, M. Arita, K. Takahashi, and M. Tomita. Dynamic modeling of genetic networks using genetic algorithm and S-system. *Bioinformatics*, 19: 643–650 (2003).
19. S. Kimura, D. Araki, K. Matsumura, and M. Okada-Hatakeyama. Inference of S-system models of genetic networks by solving one-dimensional function optimization problems *Mathematical Biosciences*, 235: 161–170 (2012).
20. S. Kimura, K. Ide, A. Kashihara, M. Kano, M. Hatakeyama, R. Masui, N. Nakagawa, S. Yokoyama, S. Kuramitsu, and A. Konagaya. Inference of S-system models of genetic networks using a cooperative coevolutionary algorithm. *Bioinformatics*, 21: 1154–1163 (2005).
21. S. Kimura, S. Nakayama, and M. Hatakeyama. Genetic network inference as a series of discrimination tasks. *Bioinformatics*, 25: 918–925 (2009).

22. S. Kimura, M. Sato, and M. Okada-Hatakeyama. Inference of Vohradský's models of genetic networks by solving two-dimensional function optimization problems. *PLoS One*, 8: e83308 (2013).
23. S. Kimura, K. Sonoda, S. Yamane, H. Maeda, K. Matsumura, and M. Hatakeyama. Function approximation approach to the inference of reduced NGnet models of genetic networks. *BMC Bioinformatics*, 9: 23 (2008).
24. P.K. Liu and F.S. Wang. Inference of biochemical network models in S-system using multiobjective optimization approach. *Bioinformatics*, 24: 1085–1092 (2008).
25. A.A. Margolin, I. Nemenman, K. Basso, C. Wiggins, G. Stolovitzky, R. Dalla Favera, and A. Califano. ARACNE: an algorithm for the reconstruction of gene regulatory networks in a mammalian cellular context. *BMC Bioinformatics*, 7: S7 (2006).
26. S. Mehrotra. On the implementation of a primal-dual interior point method. *SIAM Journal on Optimization*, 2: 575–601 (1992).
27. M. Nakatsui, T. Ueda, Y. Maki, I. Ono, and M. Okamoto. Method for inferring and extracting reliable genetic interactions from time-series profile of gene expression. *Mathematical Biosciences*, 215: 105–114 (2008).
28. L. Palafox and H. Iba. On the use of population based incremental learning to do reverse engineering on gene regulatory networks. In: *Proceedings of 2012 Congress on Evolutionary Computation*, IEEE Publishing, Piscataway, NJ, 1865–1872 (2012).
29. W. Press, S. Teukolsky, W. Vetterling, and B. Flannery. *Numerical Recipes in C*, 2nd edn, Cambridge University Press, Cambridge (1995).
30. M. Ronen, R. Rosenberg, B.I. Shraiman, and U. Alon. Assigning numbers to the arrows: parameterizing a gene regulation network by using accurate expression kinetics. *Proceedings of National Academy of Sciences of the United States of America*, 99: 10555–10560 (2002).
31. E. Sakamoto and H. Iba. Inferring a system of differential equations for a gene regulatory network by using genetic programming. In: *Proceedings of 2001 Congress on Evolutionary Computation*, IEEE Publishing, Piscataway, NJ, 720–726 (2001).
32. M.D. Sutton, B.T. Smith, V.G. Godoy, and G.C. Walker. The SOS response: recent insights into umuDC-Dependent mutagenesis and DNA damage tolerance. *Annual Review of Genetics*, 34: 479–497 (2000).
33. D. Thieffry, A.M. Huerta, E. Pérez-Rueda, and J. Collado-Vides. From specific gene regulation to genomic networks: a global analysis of transcriptional regulation in *Escherichia coli*. *BioEssays*, 20: 433–440 (1998).
34. D. Tominaga and P. Horton. Inference of scale-free networks from gene expression time series. *Journal of Bioinformatics and Computational Biology*, 4: 503–514 (2006).
35. K.Y. Tsai and F.S. Wang. Evolutionary optimization with data collocation for reverse engineering of biological networks. *Bioinformatics*, 21: 1180–1188 (2005).
36. W. Tucker, Z. Kutalik, and V. Moulton. Estimating parameters for generalized mass action models using constraint propagation. *Mathematical Biosciences*, 208: 607–620 (2007).
37. S.R. Veflingstad, J. Almeida, and E.O. Voit. Priming nonlinear searches for pathway identification. *Theoretical Biology and Medical Modelling*, 1: 8 (2004).
38. J. Vohradský. Neural network model of gene expression. *FASEB Journal*, 15: 846–854 (2001).

39. E.O. Voit. *Computational Analysis of Biochemical Systems*. Cambridge University Press, Cambridge (2000).
40. E.O. Voit and J. Almeida. Decoupling dynamical systems for pathway identification from metabolic profiles. *Bioinformatics*, 20: 1670–1681 (2004).
41. R.J. Williams and J. Peng. An efficient gradient-based algorithm for on-line training of recurrent network trajectories. *Neural Computation*, 2: 490–501 (1990).
42. R. Xu, G.K. Venayagamoorthy, and D.C. Wunsch II. Inference of gene regulatory networks with hybrid differential evolution and particle swarm optimization. *Neural Networks*, 20: 917–927 (2007).
43. R. Xu, D.C. Wunsch II, and R.L. Frank. Inference of genetic regulatory networks with recurrent neural network models using particle swarm optimization. *IEEE/ACM Transactions on Computational Biology and Bioinformatics*, 4: 681–692 (2007).
44. M.K.S. Yeung, J. Tegnér, and J.J. Collins. Reverse engineering gene networks using singular value decomposition and robust regression. *Proceedings of National Academy of Sciences of the United States of America*, 99: 6163–6168 (2002).
45. J. Yu, V.A. Smith, P.P. Wang, J. Hartemink, and E.D. Jarvis. Advances to Bayesian network inference for generating causal networks from observational biological data. *Bioinformatics*, 20: 3594–3603 (2004).

6

GPU-POWERED EVOLUTIONARY DESIGN OF MASS-ACTION-BASED MODELS OF GENE REGULATION

Marco S. Nobile

Dipartimento di Informatica, Sistemistica e Comunicazione,
Università degli Studi di Milano-Bicocca, Milano, Italy
SYSBIO Centre for Systems Biology, Milano, Italy

Davide Cipolla

Dipartimento di Informatica, Sistemistica e Comunicazione,
Università degli Studi di Milano-Bicocca, Milano, Italy

Paolo Cazzaniga

Dipartimento di Scienze Umane e Sociali, Università degli Studi di Bergamo,
Bergamo, Italy
SYSBIO Centre for Systems Biology, Milano, Italy

Daniela Besozzi

Dipartimento di Informatica, Università degli Studi di Milano, Milano, Italy
SYSBIO Centre for Systems Biology, Milano, Italy

Evolutionary Computation in Gene Regulatory Network Research, First Edition.
Edited by Hitoshi Iba and Nasimul Noman.

6.1 INTRODUCTION

The goal of synthetic biology (SB) is to design and construct novel biological circuits—in particular, gene regulatory networks (GRNs)—that are able to reproduce a desired behavior. This task is similar to the reverse engineering (RE) problem [10], whose purpose is to identify the network of interactions among the components of a real biological system, that fits an experimentally observed dynamics. The main difference between RE and the design of a synthetic circuit is that, in RE, the target dynamics is not a specifically chosen behavior, but it usually consists in laboratory measurements of some chemical species. In this context, mathematical models and computational analysis of GRNs are needed to facilitate the experimental research and to provide useful insights for the control of gene interactions.

GRNs are traditionally modeled by means of high-level formalisms, in which the interaction of genes is expressed in terms of promotion and inhibition mechanisms. An example is represented by S-systems [65], namely, systems of nonlinear ordinary differential equations (ODEs) in which gene expression is modeled by power-law functions. S-systems models of GRNs are able to capture the intrinsic nonlinearity of gene expression, providing a good description of the behavior of the corresponding artificial gene regulation system [53]. In the context of SB, the inference of S-systems was tackled by means of evolutionary computation (EC) [3,9,26,53], where a population of candidate solutions iteratively evolves under the pressure of a fitness function. The EC technique traditionally applied to this problem is genetic programming (GP) [31], in which candidate GRNs are encoded by complex recursive data structures like derivation trees or LISP s-expressions. Another EC methodology that was exploited for the inference of S-systems is differential evolution [66]; however, this method suffers from the additional problem related to overfitting solutions, which must be tackled during the evolution of the typical sparse structure of GRNs.

S-systems provide a valuable formalism for the modeling of GRNs, but they are not powerful enough to describe the actual mechanisms allowing a GRN to express a certain dynamics, so that these models have a low predictive capability. In order to provide biologists with additional information regarding gene regulation mechanisms, GRNs can be modeled as mechanistic reaction-based models (RBM), which describe in detail the molecular interactions among the chemical species. An advantage of using RBMs is that they can be easily exploited to analyze and predict emerging dynamics of GRNs in different conditions, thanks to several existing simulation tools [25,50,52].

In this chapter, we call *evolutionary design* (ED) the problem of automatically deriving an RBM that is able to reproduce a desired behavior, by exploiting EC methods only. The ED of RBMs can be more complicated than the ED of S-systems, because RBMs also require a proper description of the stoichiometry of reagents and products in each biochemical reaction. Furthermore, as in the case of S-systems, a correct kinetic parameterization of all reactions is needed, in order to produce a reliable simulation of the system dynamics. The identification of a plausible set of kinetic parameters, which is known as the parameter estimation (PE) problem

[10], can be tackled by means of EC as well [44, 49]. Many EC algorithms were successfully applied to PE: in Ref. [13], it was empirically shown that particle swarm optimization (PSO) [29] is the most efficient algorithm, as also suggested in Ref. [4], where PSO and genetic algorithms (GAs) were compared.

In this chapter, we introduce a computational strategy to infer an RBM that specifically represents a gene regulation model (GRM) characterized by some predefined behavior. In particular, we present a two-level ED methodology, named cuGENED, which integrates two EC algorithms: Cartesian genetic programming (CGP) [43] and PSO [29]. CGP exploits individuals encoded as fixed-length vectors of integer numbers which, in contrast to standard GP, are mapped onto directed graphs rather than derivation trees [43]. The choice of CGP for the ED process is motivated by the fact that graphs are suitable for the representation of a network, and the mapping between a CGP individual and the corresponding GRM allows a direct translation of the individual into a human-comprehensible set of chemical reactions. Moreover, since CGP exploits fixed-length individuals, it does not need any explicit strategy to avoid bloating, that is, the uncontrolled growth of the size of solutions that usually occurs in GP.

cuGENED exploits CGP to derive the network of biochemical reactions which describe the genetic regulation mechanisms in a GRM. Then, PSO is exploited to estimate the kinetic parameters of the GRMs. The target of the overall evolutionary process is represented by some representative dynamics of a predefined number of genes that participate in the GRN. Eventually, the effective behavior of the inferred model has to be validated after the synthetic engineering of the genetic circuit. Such validation might be done by measuring, for example, the transcription levels of the genetic components of the circuit.

It is worth noting that, for the ED of synthetic GRNs, it is not possible to consider any network structure as target of the optimization process, so that the quality of a candidate GRM can be only evaluated by comparing its simulated dynamics against the desired target behavior. Indeed, an absolute novelty of cuGENED with respect to the state-of-the-art is that it allows 0-knowledge inference since the number of intermediate chemical species occurring in the GRM is usually unknown a priori.

In cuGENED, every generation of CGP requires the execution of PSO, which realizes the PE for each individual. Since PSO is a population-based algorithm as well, the whole methodology is computationally expensive. Nevertheless, all fitness evaluations in each iteration of the PSO are independent and can be accelerated by means of a parallel architecture. Among the existing technologies, one of the most efficient is the *general-purpose GPU computing*, in which the huge computational power of modern video cards is exploited for general-purpose computation. In the recent years, the adoption of graphics engines experienced a great boost in bioinformatics, systems biology and computational biology [12,24,57], where central processing units (CPUs) traditionally represented the standard workhorses. As a matter of fact, when several batches of simulations need to be executed—as in the case of the ED process—the necessary computing power can rapidly overtake the capabilities of standard desktop computers, therefore requiring high-performance computing solutions. To solve this problem, in this chapter, we consider cupSODA [50], a GPU-powered deterministic simulator of biochemical systems modeled by means of RBMs. cupSODA, exploited during the PE phase, is fully integrated in cuGENED and allows a strong reduction of the overall running time.

The chapter is structured as follows. In Section 6.2, we briefly present some EC methods previously used for the inference of biochemical models, using different assumptions for the description of kinetic rates. In Section 6.3, we describe the formalization of GRMs by means of mass-action-based models, and give a brief introduction of the EC techniques (CGP and PSO) used in the ED of GRNs. Then, we briefly explain the GPU computing framework exploited to speed up the optimization process. In Section 6.4, we provide a detailed description of our ED methodology to automatically derive GRMs. In Section 6.5, we present the results of the application of cuGENED for the automatic design of GRNs consisting in two and three genes. We discuss the whole methodology in Section 6.6, and in Section 6.7, we conclude with some final remarks and future developments.

6.2 EVOLUTIONARY COMPUTATION FOR THE INFERENCE OF BIOCHEMICAL MODELS

To define a mathematical model of a biochemical system, the inference of the network of unknown reactions can be performed using EC. Historically, the first example was proposed in 2001 by Koza *et al.*, who used GP to evolve a huge population (100,000 individuals) of putative biochemical systems [32]. Specifically, each individual is composed of multiple trees representing chemical reactions. In order to assess the fitness values, individuals are converted into analog electrical circuits (e.g., chemical concentrations are represented by voltages, reactions are represented by analog components) and simulated by means of an extended version of the SPICE simulation tool kit for electrical circuits [61]. Authors showed the feasibility of this methodology by reverse engineering metabolic pathways. The huge computational complexity of the method was then discussed in Ref. [33], concerning the parallelization of the computations on a 1000-nodes cluster system.

Similarly, Sugimoto *et al.* proposed a GP-based methodology for the RE of biochemical systems [67], which derives ODE-based models not restricted to mass-action kinetics. Indeed, their approach allows the inference of any arbitrary kinetics (e.g., Hill functions or Michaelis-Menten [46]). Because of this freedom, the method is prone to overfitting. In order to mitigate this problem, the fitness function was extended with a "parsimony" term: it is proportional to the complexity of the derivation tree and introduces a selective pressure on the population. Their results showed the successful inference of a single reaction in a biochemical system.

A similar GP-based approach was proposed by Ando *et al.* [3]. In this work, each individual is defined as a set of trees, which represent the right-hand side of the ODEs associated with each chemical species in the system. Differently from Ref. [67], they showed a successful RE of a metabolic pathway characterized by two reactions and three chemical species. Similar to Refs. [32, 67], this method does not allow the inference of models that are strictly based on mass-action kinetics.

A different approach for RE was proposed by Nummela and Briant [54], in which a population of individuals modeled as Petri nets [60] is evolved by means of GAs, extending the previous work by Kitagawa and Iba [30]. Nummela and Briant exploited a pseudo-first-order approximation, so that the Michaelis–Menten kinetics was simplified to a first-order rate equation. This simplification can be done thanks to the steady-state approximation on the concentrations of the intermediate complexes, and under the assumption that the equilibrium constant for the reversible formation of the complex is larger than the substrate concentration. Similar to Ref. [67], the bloating of the solutions is mitigated by introducing an additional factor in the fitness function, proportional to the number of transitions in the putative Petri net. Since the method is based on a GA, it is not able to simultaneously evolve the network of reactions and their kinetic parameters: the latter were optimized by means of stochastic hill-climbing during the fitness evaluation phase. Even though the method failed in identifying the correct target networks, it was shown to propose putative pathways whose dynamics is similar to the target dynamics. Nobile *et al.* proposed an alternative EC methodology for the robust evolution of populations of Petri nets: the evolutionary Petri nets (EPNs) [47]. Conceptually similar to GP, EPNs are characterized by novel genetic operators for the crossover and mutation of individuals. For the specific application to the RE of biochemical systems, though, EPNs require a separate PE methodology for the inference of kinetic parameters.

The strict decoupling of the network inference and PE is also exploited by Lenser *et al.* [35] and Nobile *et al.* [51]. In the first work, the authors exploited a custom evolutionary algorithm that separates the evolution of the network (performed with a GP-like approach) and the parameters inference (executed with an evolution strategy algorithm), showing that this methodology prevents a premature convergence. In the second work, RE was performed by means of a specific class of GP, the CGP [43], and parameters were estimated using PSO [29]. CGP allows to evolve individuals that can be represented as sets of expressions: in the case of RE, they can be interpreted as chemical reactions. Thus, this approach allows to directly evolve systems of reactions describing the biochemical network, under the assumption that all reactions have mass-action kinetics. In so doing, no arbitrary kinetics function can be introduced, avoiding the overfitting of solutions. In Ref. [51], PSO was used for PE, as it was previously shown to represent the best option to tackle this kind of problem [4, 13].

6.3 METHODS

6.3.1 Mass-Action-Based Modeling of Gene Regulation

Given a biochemical system η consisting of some molecular species (e.g., genes, proteins, metabolites) and their mutual interactions, a mechanistic reaction-based model (RBM) of η can be formally defined by specifying the set $S = \{S_1, \dots, S_N\}$ of different molecular species occurring in η and the set $\mathcal{R} = \{R_1, \dots, R_M\}$ of biochemical reactions. A reaction $R_\mu \in \mathcal{R}$, $\mu = 1, \dots, M$, is usually given in the form $R_\mu : a_1 S_1 + \cdots + a_N S_N \rightarrow b_1 S_1 + \cdots + b_N S_N$, where $a_i, b_i \in \mathbb{N}$ are called the stoichiometric coefficients of R_μ. The species occurring on the left-hand (right-hand) side of R_μ are called reagents (products, respectively). A numerical value $k_\mu \in \mathbb{R}^+$ is associated with R_μ, representing the rate constant of that reaction.

By considering the values of the stoichiometric coefficients of the reactions in $\mathcal{R}$, we can identify some particular reactions:

- if the stoichiometric coefficients a_i of the reagents are null for all $i = 1, \dots, N$, then R_μ is called a *source* reaction and it will be denoted as $\emptyset \rightarrow$ *products*. Similarly, if the stoichiometric coefficients b_i of the products are null for all $i = 1, \dots, N$, then R_μ is called a *sink* (or degradation) reaction and it will be denoted as *reagents* $\rightarrow \emptyset$. Reactions of the form $\emptyset \rightarrow \emptyset$, where $a_i = 0$ and $b_i = 0$ for all $i = 1, \dots, N$, are not considered in RBMs;
- meaningless reactions of the form $a_i S_i \rightarrow b_i S_i$ are excluded from any RBM since they correspond to unfeasible biochemical processes where a_i molecules of species S_i are converted into b_i molecules of the same species;
- at most second-order reactions (i.e., chemical reactions where at most two molecules occur as reactants) are considered in RBMs since third-order (or any higher order) reactions have a probability to occur almost equal to zero, as they would require the simultaneous collision of three (or more) reactant molecules.

In this chapter, we are interested, in particular, in reaction-based gene regulation models (GRMs), which formally describe the biochemical reactions involved in gene transcription and translation. To properly define these reactions, we assume that the set $\mathcal{S}$ of molecular species is given by the union of two disjoint sets of species, which we denote as $\Gamma = \{\gamma_0, \dots, \gamma_{N_\Gamma}\}$ and $\Sigma = \{\sigma_0, \dots, \sigma_{N_\Sigma}\}$, for some $N_\Gamma, N_\Sigma \in \mathbb{N}$. We assume that $|\Gamma| \leq |\Sigma|$, where $|\cdot|$ represents the cardinality of the set.

The set Γ represents the genes that are available for the synthetic engineering of gene regulation circuits. We assume that each element γ_i in Γ is strictly associated with a messenger RNA (mRNA), the product of gene transcription, whose expression can be experimentally evaluated through cutting-edge technologies (e.g., microarray [20], qRT-PCR [34]). Thus, the cardinality of Γ is equivalent to the number of target time series used in cuGENED. By abuse of notation, we will identify γ_i either with the "gene" or with its "transcription product", according to the context.

The set Σ represents a set of generic species, as well as their mutual chemical complexes, that are related to the processes of gene expression. These species are the actual effectors of gene regulation, and need to be included in any GRM to evaluate their influence on the emerging behavior of the circuit. An element $\sigma_i \in \Sigma$ might represent any kind of gene product (e.g., protein) related to γ_i, or the molecular complex formed by the interaction of the gene product (acting as promoting/inhibiting transcription factor) with another gene $\gamma_j \in \Gamma$.

To better clarify the meaning of the sets Γ and Σ, we illustrate some types of reactions that formally describe the processes of gene regulation and that might be present in a GRM, along with other generic reactions mentioned above.

Reactions that represent gene expression (transcription/translation) can be written in the form $\gamma_i \rightarrow \gamma_i + \sigma_i$. Given a gene $\gamma_j \in \Gamma$, its regulation by means of gene γ_i can be described either as $\gamma_i + \gamma_j \rightarrow \sigma_k$ or as $\sigma_i + \gamma_j \rightarrow \sigma_k$, with i not necessarily distinct from j. Here, the species σ_k might represent two different elements: (1) the product of the regulation of γ_j (e.g., the protein encoded by γ_j); in this case we can simply say that σ_k is equal to σ_j; (2) the chemical complex between gene γ_j and its own regulator; in this case σ_k is a compact formalization for the species $\gamma_i\gamma_j$ or $\sigma_i\gamma_j$. The inverse reactions of the type $\sigma_k \rightarrow \sigma_i + \gamma_j$ can then be used to describe the dissociation of the regulator σ_i from gene γ_j. The effective expression of a regulated gene γ_j—that takes place after the occurrence of some reactions of type $\sigma_i + \gamma_j \rightarrow \sigma_k$—can be synthetically written in the form $\sigma_k \rightarrow \sigma_j$. This reaction states that species σ_j, related to gene γ_j, is derived from species σ_k, which represents an intermediate molecular complex having γ_j as element (e.g., $\sigma_k = \gamma_i\gamma_j$ or $\sigma_k = \sigma_i\gamma_j$, as mentioned above). Finally, reactions of the form $\sigma_h \rightarrow \sigma_g$ can also be used to represent the degradation, or any other generic transformation reaction (e.g., post-translational modification of a protein) of the species occurring in Σ.

The RE process of a given biological system consists in the automatic identification of the network of interactions among the molecular components of that system. In this task, a set of experimental time series measurements of some species occurring in the system can be exploited as target to drive the evolutionary inference of the network [51]. On the contrary, in this chapter, we assume that only an expected dynamical behavior of the network is given as target. Quite obviously, we assume that no specific laboratory measurements of the temporal evolution of chemicals that will constitute the circuit can be given *before* the circuit itself has been engineered. For this reason, in what follows, we do not formally specify the type and the unit measurements of the amount of chemicals, but we use the general term "level". The level can indicate either the number of molecules or the concentration values that might then be actually measured by *ad hoc* laboratory methodologies *after* the construction of the circuit. For instance, if qRT-PCR is exploited to validate the functioning of the synthetically engineered circuit, then the level of the species in Γ represents the concentrations of the corresponding mRNA, measured by relative quantification with respect to the amount of some control housekeeping gene [37].

According to the law of mass-action[1], given a set of biochemical reactions $\mathcal{R}$—each one characterized by its own kinetic constant k_μ[2] —for each molecular species appearing either as reagent or product in some reaction in $\mathcal{R}$, it is possible to derive a rate equation that describes the variation in its concentration with respect to time. In other words, any given RBM or GRM can be formalized as an equivalent system of coupled (nonlinear) first-order ODEs [71]. This formalization can be exploited to determine the temporal evolution of the network (that is, its dynamics) by means of

[1]The mass-action law states that, in a dilute solution, the rate of an elementary reaction is proportional to the product of the concentration of its reactants raised to the power of the corresponding stoichiometric coefficient [46].

[2]The unit of measurement of k_μ, that we omit for simplicity, depends on the order of each reaction: it is equal to $(\text{mol/L})^{1-n}(\text{time})^{-1}$ for reactions of order n, $n \geq 1$, and equal to $(\text{time})^{-1}$ if $n = 0$.

different numerical integration methods [7]. These algorithms require as input the set of ODEs, along with the set of kinetic constants and the initial concentrations of the chemical species. To this aim, in this chapter, we exploit cupSODA [48,50], a GPU-powered tool that automatically derives a set of ODEs from RBMs defined according to the mass-action kinetics, and then exploits the numerical integration algorithm LSODA [59] to perform the deterministic simulation of the system dynamics. In Section 6.4, we show how to exploit CGP to carry out the ED of a GRM, which is then automatically converted to an ODEs system and simulated by means of cupSODA, considering the kinetic constants that are inferred by PSO.

6.3.2 Cartesian Genetic Programming

CGP is an EC method whose individuals are described by means of indexed graphs, rather than trees, having (sequentially numbered) nodes arranged in a Cartesian coordinate grid [43]. The genotype of each individual is represented by a sequence of node connections and functions in the grid, called *Cartesian program* (CP).

Formally, a CP is a 9-tuple $\{G, n_i, n_o, F, FN, n_r, n_c, n_n, l\}$ where

- G is the genotype, encoded as fixed-length vector of integer numbers representing the connections from the input nodes to the outputs nodes of the grid;
- $n_i, n_o \in \mathbb{N}$ are the number of input and output nodes, respectively;
- F is a finite set of functions (for instance, elementary arithmetic operations as $\{+, -, \times, /\}$);
- FN is a grid of functional nodes, sequentially indexed by rows and columns, each one containing a function from F;
- $n_r, n_c \in \mathbb{N}$ are the number of functional nodes appearing in each row and in each column of the grid, respectively, so that $|FN| = n_r n_c$;
- $n_n \in \mathbb{N}$ is the number of input connections in each functional node;
- $l \in \mathbb{N}$ is the so-called *"levels back"* parameter, a measure of the CP inter-connectivity which determines how many preceding columns (in the grid of functional nodes) can have their output connected to the functional nodes appearing in any given column of the grid.

The connections of a CP start from the input nodes and pass through the functional nodes, each one having a fixed number n_n of input connections. The length of the genotype G is equal to $n_r n_c (n_n + 1) + n_o$, that is, an integer number is assigned to each input connection and to the output connection of every functional node, as well as to each output node of the grid. The inter-connectivity of nodes generally exploited in CGP, and specifically considered in this chapter, is strictly feed-forward, meaning that nodes belonging to the same column of the grid cannot be connected to each other. In addition, any node can be either connected or disconnected; disconnected nodes represent non-coding genes in the genotype and are ignored in the phenotype. The phenotype of a given CP is given by the actual graph that the CP represents. In this chapter, the semantic of the phenotype corresponds to the set of reactions derived from the graph, that will constitute the GRM. An example of CP genotype and phenotype, in the context of GRMs, is given in Section 6.4.

In a CGP population, after the fitness function of each CP has been evaluated, the best candidate solutions are selected to generate the offspring by means of a mutation operator. During the iterative process, CGP evolves a population of individuals characterized by a set of expressions, each one formed by the composition of input nodes and functional nodes. The total number of possible expressions represented by a given CP is upper bounded by the number n_o of output nodes. If an output node is linked to a disconnected functional node, some internal nodes will not be part of any path connecting an input node to that output node, a circumstance that leads to the possible existence of different genotypes mapping to the same phenotype. These non-coding regions are important in CGP for three reasons: (1) they reduce the size of the phenotype, as stated above; (2) they reduce the effect of mutations since a mutation acting on a disconnected functional node will not contribute to the variation of the phenotype; (3) even if they are unused in the current CP, a mutation could suddenly connect some disconnected node to the rest of the graph, resulting in a possibly relevant change in the phenotype.

In this chapter, the candidate solutions evolve by using a $(1+\lambda)$ ES [5], as described in Ref. [43]: all individuals are evaluated and the best one is selected as a parent for the next generation. Then, λ offspring are produced by means of random mutations, that is, random modifications of the integers which constitute the genotype of the parent individual. The proportion of genes that are mutated is determined by the mutation rate parameter $\rho \in (0, 1)$. The ES methodology does not exploit any crossover mechanism.

In Section 6.4, we show how to exploit CGP to perform the ED of GRMs. In order to evaluate the GRM that each CP represents, we need to define a proper fitness function. As target of the optimization process, we exploit a set of artificial temporal data that represent the desired behavior of the synthetic gene regulation circuit. These artificial data informally correspond to the concentrations of the molecular species that might be experimentally measured after the construction of the circuit itself. Therefore, the fitness function is based on the comparison of the target series against a simulated dynamics of the model, which is generated by exploiting the cupSODA tool [48, 50] explained in Section 6.3.4. Since the GRM derived with CGP is not complete until a proper kinetic parameterization is given, the fitness evaluation of each candidate solution embeds a PE phase, which is performed by means of PSO.

6.3.3 Particle Swarm Optimization

PSO is a population-based meta-heuristic inspired by the collective movement of flock of birds and school of fish [29]. In PSO, a *swarm* of P candidate solutions (named *particles*) moves inside an M-dimensional Euclidean search space, cooperating in the identification of the optimal solution, whose quality is measured using a specified fitness function. The movement of the swarm is bounded inside a subspace of $\mathbb{R}^M$: if a particle is driven outside the boundaries of this subspace as a consequence of its movement, then it is relocated inside the feasible region by means of the *damping* mechanism described in Ref. [72].

The i-th particle of the swarm is defined by two vectors: the position of the particle $\varphi_i \in \mathbb{R}^M$ and its velocity $\mathbf{v}_i \in \mathbb{R}^M$. During the iterative optimization phase, the velocity is updated as a result of two attractors: the best position $\mathbf{b}_i \in \mathbb{R}^M$ ever found by the particle itself, and the best position $\mathbf{g} \in \mathbb{R}^M$ identified by the whole swarm. These attractors are balanced by means of two settings specified by the user: the cognitive factor $c_{cog} \in \mathbb{R}^+$ and the social factor $c_{soc} \in \mathbb{R}^+$. The components of both attractors are further modulated by two vectors $\mathbf{R}_1$ and $\mathbf{R}_2$ of random numbers uniformly sampled in $[0, 1)$, which allows the particles to move stochastically and to avoid the entrapment into local minima. Moreover, to reduce the possibility of chaotic movements of particles, the change of velocity is weighted by an inertia factor $w \in \mathbb{R}^+$. The final definition of the velocity update formula for the i-th particle is:

$$\mathbf{v}_i = w \cdot \mathbf{v}_i + c_{soc} \cdot \mathbf{R}_1 \circ \left(\varphi_i - \mathbf{g}\right) + c_{cog} \cdot \mathbf{R}_2 \circ \left(\varphi_i - \mathbf{b}_i\right), \qquad (6.1)$$

where $\circ$ represents the component-wise multiplication operator between vectors. Once the velocity is updated, the new position of each particle is determined by calculating $\varphi_i = \varphi_i + \mathbf{v}_i$.

In this chapter, a particle corresponds to a candidate kinetic parameterization of each GRM determined by CGP. Therefore, M is equal to the number of connected outputs in the CP, that is, the total number of reactions in the GRM.

6.3.4 General-Purpose GPU Computing

The Compute Unified Device Architecture (CUDA) is a parallel computing platform and programming model introduced by Nvidia in 2006, consisting in a framework suitable to exploit GPUs in general-purpose computational tasks (GPGPU computing). GPGPU computing represents a valuable alternative to the traditional high-performance computing infrastructures (e.g., clusters of machines) since GPUs are characterized by low costs and a reduced energy consumption, allowing the access to tera-scale computing on common workstations of mid-range price. Nevertheless, a direct porting of sequential code on the GPU is most of the times unfeasible, due to the innovative architecture and the intrinsic limitations of this technology. Therefore, a fully exploitation of the computational power and massive parallelism of the GPU is usually challenging [17].

CUDA combines the single instruction multiple data (SIMD) architecture with multi-threading, and automatically handles the conditional divergence between threads. However, this flexibility has a drawback since any divergence of the execution flow among threads results in a serialization of the execution, affecting the overall performances. Following the naming conventions used in CUDA, a C/C++ function, called *kernel*, is loaded from the host (the CPU) to the devices (one or more GPUs) and replicated in many copies named *threads*. Threads can be organized in three-dimensional structures named *blocks* which, in turn, are contained in three-dimensional *grids* (as schematized in Figure 6.1, left side). Whenever the host computer runs a kernel function, the GPU creates the corresponding grid and automatically

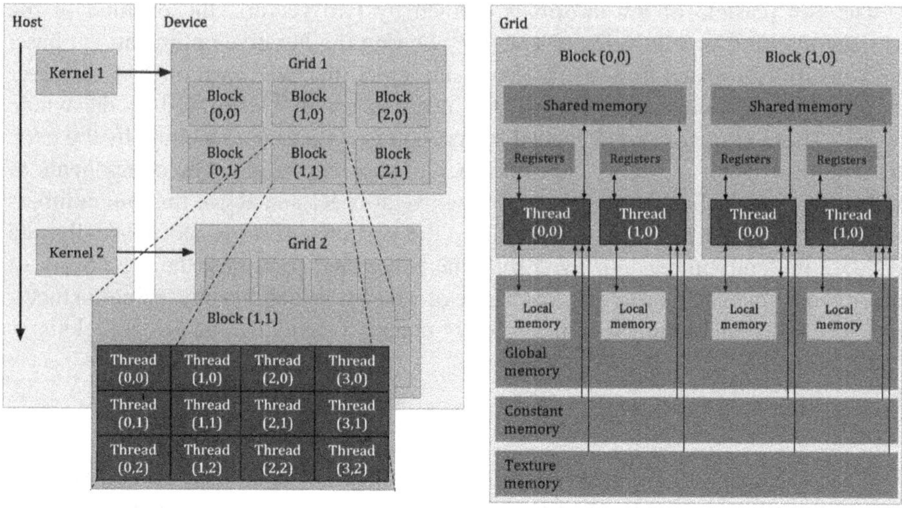

Figure 6.1 Architecture of CUDA's threads and memory hierarchy. *Left side*. Threads organization: a kernel is invoked from the CPU (the host) and is executed in multiple threads on the GPU (the device). Threads are organized in three-dimensional structures named blocks which are, in turn, organized in three-dimensional grids. The programmer must decide the dimensions of blocks and grids before the kernel launch. *Right side*. Memory hierarchy: threads can access data from multiple kinds of memories, all with different scopes and characteristics. Registers and local memories are private for each thread; shared memory lets threads belonging to the same block communicate, and has low access latency; all threads can access the global memory, which suffers from high latencies, but it is cached since the introduction of the Fermi architecture; texture and constant memory can be read from any thread and are equipped with a cache as well. cupSODA takes advantage of the memory hierarchy of the GPU, by allocating the most frequently updated data (e.g., the state of the system) in the shared memory, and the unvarying data (e.g., the set of reactions) in the constant memory. Adapted from Nvidia's CUDA programming guide [55].

schedules each block of threads on an available streaming multi-processor of the GPU, thus allowing a transparent scaling of performances on different devices.

GPUs are equipped with different types of memory. As described in Figure 6.1, right side, the GPU memory hierarchy consists in the *global memory* (accessible from all threads), the *shared memory* (accessible from threads of the same block), the *local memory* (registers and arrays, accessible from owner thread), and the *constant memory* (cached and not modifiable). The best performances in the execution of CUDA code can be achieved by exploiting the shared memory as much as possible. Unfortunately, the shared memory is a very limited resource (49,152 bytes for each multi-processor, since the introduction of the Fermi architecture) that brings about restrictions on the blocks' size. On the contrary, the global memory is very large (thousands of megabytes) but suffers from high latencies. In order to mitigate this issue, starting from the Fermi architecture, the global memory has been equipped with a L2 cache (see Figure 6.2). Moreover, with the Fermi architecture, the programmer can balance 64 KB of fast on-chip memory between the shared memory and L1 cache,

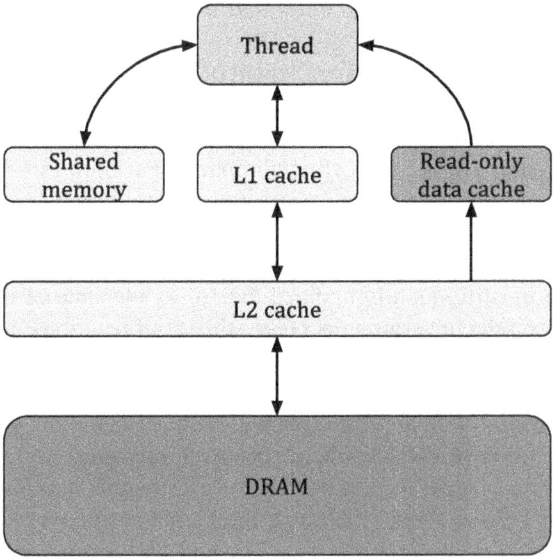

Figure 6.2 Schematic description of the memory hierarchy in Fermi and Kepler architectures. GPUs based on these architectures are equipped with a two-level data cache and a read-only data cache. The shared memory and the L1 cache share the same on-chip 64 KB memory banks. The amount of memory can be reconfigured by the user, according to the specific needs of the application. Adapted from Nvidia's Kepler GK110 whitepaper [56].

specifying two different configurations: 48 KB for the shared memory and 16 KB for L1 cache, or 16 KB for the shared memory and 48 for L1 cache. In addition, using the Kepler architecture (exploited in our work), a third and perfectly balanced configuration can be specified by assigning the same amount of memory (32 KB) both to the shared memory and L1 cache.

Despite the remarkable advantages concerning the computational speedup, computing with GPUs usually requires the development and the implementation of *ad hoc* algorithms since GPU-based programming substantially differs from CPU-based computing. As a consequence, scientific applications of GPUs might undergo the risk of remaining a niche for few specialists. To avoid such limitations, several packages and software tools were recently released (see, e.g., Refs. [24, 50, 69, 73]), so that users with no knowledge of GPUs hardware and programming can also access the high-performance computing power of graphics engines.

In this chapter, the simulations needed to execute the ED of GRMs were carried out by means of cupSODA, a GPU-powered simulator for biological systems that allows the efficient execution of high numbers of parallel deterministic simulations [48,50]. cupSODA relies on a C version of the numerical integration algorithm LSODA [59], ported and adapted to the CUDA architecture. This tool requires an Nvidia video card and can be executed on Microsoft Windows, Linux, and Apple OS X-based operating systems. cupSODA was designed to the purpose of being a *black-box* simulator, that can be easily used without any programming skills. In particular, cupSODA automatically converts any mass-action-based RBM of a biological system into the corresponding system of ODEs.

cupSODA exploits the massive parallelism of CUDA architecture for the parallel execution of independent simulations—each one characterized by a different initial parameterization of the model—in each thread. cupSODA was designed to speed up the time-consuming tasks typical of computational biology [2, 10], which rely on the repetition of large numbers of simulations in perturbed conditions, generally realized by varying the initial concentrations of chemical species or the value of the kinetic constants. cupSODA takes advantage of the memory hierarchy of the GPU, by allocating the most frequently updated data (e.g., the state of the system) in the shared memory, and the unvarying data (e.g., the set of reactions in the RBM) in the constant memory.

cupSODA integrates a specific functionality for the evaluation of the fitness function since it allows to compare the outcome of the simulations with any available target data (e.g., the levels of chemical species at chosen time instants $t_0, \ldots, t_F$). This is useful, for instance, to perform a PE task. To this aim, cupSODA invokes the LSODA kernel $F - 1$ times: each time the kernel is run over a (simulated) time interval of length $\Delta t = t_i - t_{i-1}$, $i = 1, \ldots, F$, and the concentration values of the output species are stored at the end of each Δt. Once the concentrations are stored, cupSODA provides a set of metrics (e.g., Equation 6.2) that can be exploited to evaluate in parallel the fitness function of each particle.

6.4 DESIGN METHODOLOGY OF GENE REGULATION MODELS BY MEANS OF CGP AND PSO

In this section, we describe our ED methodology, based on the integration of (1) CGP for the design of the synthetic circuit, (2) PSO for the estimation of the kinetic parameters of the corresponding GRM, and (3) cupSODA for the GPU-accelerated fitness evaluation.

To the best of our knowledge, this is the first time that the ED is performed by means of a hybrid approach based on CGP and PSO. The choice of PSO to perform the PE task is motivated by empirical studies in the field, highlighting its higher performances with respect to other popular global optimization techniques like GAs, differential evolution and evolution strategy [4, 13].

CGP phase. In cuGENED, CGP is implemented using $F = \{+, -\}$ as the set of functions. The composition of input nodes and functional nodes allows CGP to encode complex expressions for each genotype G. The expressions can be then converted, by means of symbolic manipulation, into arithmetical equations as described hereafter. In the context of ED of GRMs, the set of input nodes is determined by the species occurring in Γ and Σ, so that $n_i = |\Gamma| + |\Sigma|$. Since the exact regulatory interactions necessary to reproduce the desired behavior of the synthetic circuit are to be determined, it follows that the number of chemical species belonging to the set

Σ is unknown as well. Therefore, also the cardinality of Σ must be inferred by means of some heuristics. As a matter of fact, the value $|\Sigma|$ has a relevant impact on the ED process since it determines the space of the possible GRMs that CGP can explore. If $|\Sigma|$ is too small, then a circuit that is able to achieve a perfect fit of the desired dynamics might be impossible to be designed. On the contrary, if $|\Sigma|$ is too large, then the evolutionary algorithm can take a longer time to converge to an optimal solution, or might suffer from overfitting. Nevertheless, in the second case, CGP is able to automatically exclude the unnecessary chemical species from the set of candidate solutions if they have no impact on the fitness value. In this work, we always use the value $|\Sigma| = 4 \cdot |\Gamma|$ as heuristic, because we assume that, for each target species in Γ, four intermediates are enough to model gene regulation mechanisms.

The number of input connections for each functional node is $n_n = 2$, because we only exploit addition and subtraction as functions for the construction of the expressions. The set of output nodes corresponds to the biochemical reactions that will be part of the inferred GRM. In these reactions, the species corresponding to the input nodes appear either as reactants, if a functional node containing function + is crossed, or as products, if a functional node containing function is crossed. In so doing, each expression encodes a single reaction of the GRM.

Generally, in CGP, the number of output nodes n_o is fixed. In the context of GRMs, this implicitly means that the actual number of reactions should be known before the optimization takes place, which is clearly unreasonable. Since the GRM itself is the goal of the optimization, in this chapter, the value n_o represents an upper bound to the number of reactions that will appear in the model. Hence, a proper choice for n_o is fundamental: it should be large enough to include all the necessary regulatory mechanisms, and small enough to avoid bloating and overfitting. As heuristic for the selection of n_o, we suggest to fix its value equal to the number of all possible reactions that can involve the set of genes and the set of generic chemical species, that is, $n_o = |\Gamma| \cdot |\Sigma| + |\Gamma| + |\Sigma|$. Furthermore, in this implementation, we set $l = n_c$, meaning that output nodes can be connected to any functional or input node, and each functional node belonging to column j can be connected only to nodes between column $j - 1$ and the input nodes.

An example of CP related to the ED problem of GRMs is explained in Example 6.1 and graphically represented in Figure 6.3.

■ EXAMPLE 6.1

There exists a strict correspondence between the genotype of the CP and the connectivity of each node in the Cartesian coordinate grid. Functions + and − in the grid are represented by 0 and 1 in the genotype, respectively (the first number in each of the four triplets). Since + and − are binary operators, we use $n_n = 2$ input connections to each functional node. The grid of functional nodes consists in $|FN| = n_r n_c = 4$ nodes since the grid is composed of two rows and two columns. The genotype is generated by randomly drawing values related to the chemical species (0 for γ_0, 1 for σ_0, 2 for σ_1) and to the output of each functional

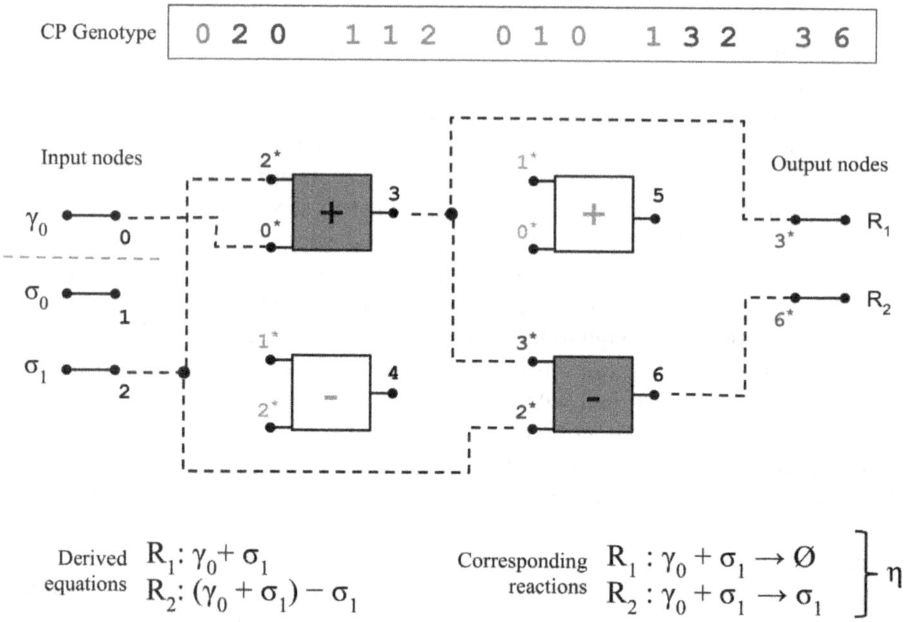

Figure 6.3 Example of the conversion of a CP genotype into the corresponding GRM.

node (3, … , 6). Then, these values are assigned to the input of the functional nodes and to the output nodes (numerical values with asterisk). Note that, in the construction of the genotype, given a functional node, only the values of previous input and functional nodes can be randomly selected and assigned to its input connections since $l = n_c$. The grid is parsed backwards, from the $n_o = 2$ output nodes R_1, R_2 to the $n_i = 3$ input nodes $\gamma_0, \sigma_0, \sigma_1$. The resulting equations are then automatically converted into an equivalent set of biochemical reactions. The grey functional nodes are involved in the equations since a path from the input nodes to the output nodes exists here. The white functional nodes, instead, are not connected in the grid and do not participate in any equation in this CP. Thus, their corresponding genes (gray numbers) are non-coding sequences.

The pseudo-code of our ED methodology based on CGP and PSO is reported in Algorithm 6.1. The evolutionary process begins with the creation of a population of $I = 1 + \lambda$ random CPs (lines 2–7). The population evolves by means of an ES process which can be decomposed in three main steps:

1. in **Step 1**, the representation of each CP is converted into a GRM. Then, PSO is used to estimate the values of the kinetic constants and to assess the fitness value of the parameterized candidate network (lines 10–14);
2. in **Step 2**, the CPs are ranked according to the fitness values, to identify the best CP in the population (line 15);

3. in **Step 3**, a brand new population is formed by considering the best CP, together with the $I - 1$ offspring created by applying the mutation operator on the best CP (lines 16–30).

Algorithm 6.1

```
1  CGP+PSO algorithm {
2          CGP_population ← create_empty_population()
3          for 1 to I
4          {
5              CP = create_CP(S, n_0, n_r, n_c)
6              CGP_population.add(CP)
7          }
8          for 1 to CGP_generations
9          {
10             for each CP in CGP_population
11             {
12                 network ← CP.convert_to_network()
13                 CP.fitness, CP.parameters ← PSO(network)
14             }
15             best_CP ← find_best_individual(CGP_population)
16             CGP_population ← create_empty_population()
17             CGP_population.add(best_CP)
18             for 1 to I − 1
19             {
20                 repeat
21                 {
22                     candidate_CP ← best_CP.mutation(ρ)
23                 } until candidate_CP not in CGP_population
24                 for each reaction in candidate_CP
25                 {
26                     if not consistent(reaction)
27                         candidate_CP.remove(reaction)
28                 }
29                 CGP_population.add(candidate_CP)
30             }
31             if fitness(elite_CP) < fitness(best_CP)
32                 elite_CP ← best_CP
33             CGP_population.add(elite_CP)
34         }
35         return best_CP
36 }
```

Figure 6.3 schematizes the conversion of a CP into the corresponding GRM (**Step 1**). The connections of each output node are followed backwards by recursively passing through the functional nodes. This process yields a derivation tree that is translated into an arithmetical equation, where positive terms (respectively, negative) are considered as reactants (products). Each equation produces a single candidate chemical reaction for the network. A candidate GRM η for the synthetic circuit is then obtained by repeating the algorithm for all output nodes.

The fitness values of all candidate solutions are calculated at the end of the PE process by means of PSO, so that a ranking of the CPs can be assessed and the best solution ($best_{CP}$) in the CGP population is identified (**Step 2**).

During each generation of the CGP, the new offspring solutions are obtained by applying a mutation operator to $best_{CP}$. The GRM η is compared to the rest of the population: if it is identical to some other CP, it is mutated again in order to achieve a heterogeneous population, and the process is repeated. In this chapter, we do not verify that η consists of a single connected component since not all chemical species in Σ necessarily need to be present in the GRM. The rationale behind this choice is that, in the ED methodology presented here, we do not specify a priori the exact number of chemical species that should occur in the system (we only provide an upper bound). Therefore, in this methodology also the number of chemical species in Σ undergoes the optimization process. This approach is different from typical methods used for the RE problem. In particular, it differs from the RE method based on CGP and PSO that we previously proposed in Ref. [51]: in that case, the number of chemical species participating in the system was known a priori; therefore, it was mandatory for the candidate solutions to consist in a single connected component.

When a new population of CGP is generated, for each candidate solution we perform a consistency check (lines 24–28 in Algorithm 6.1). Specifically, we verify that each reaction in the GRM η obeys the conditions described in Section 6.3.1, together with the additional condition that no identical reactions appear in η. In particular, since we consider at most second-order reactions, we can limit the length of CP expressions and reduce the bloating by setting $n_c \leq 3$. In general, if a reaction R is not consistent, we remove it from the network, so that $\eta = \eta \setminus \{R\}$. Eventually, when a new consistent CP is produced, the individual is inserted into the new population.

To improve the convergence speed, at the end of **Step 3** we apply the elitist selection by adding a non-mutated copy of the best candidate network found so far. This solution becomes the $(I + 1)$th individual of the population (lines 31–33 in Algorithm 6.1). When the new population is formed, we calculate the fitness value of each individual by executing a PE by means of PSO, as described hereby. This iterative process is repeated for $GEN_{\max} = 100$ generations. Finally, the GRM with the best fitness, along with its parameterization, is chosen as the result of the ED problem.

PSO phase. For the execution of the PE task, each particle in the PSO corresponds to a candidate kinetic parameterization of a GRM inferred by CGP. Namely, for each candidate GRM, we exploit a whole swarm of particles to determine the reaction constants that best fit the desired target behavior for that GRM. To this aim, the PSO adopts the following settings: $P = 64$ particles, randomly generated using a logarithmic distribution to better distribute the values of kinetic constants over different orders of magnitude; $c_{soc} = c_{cog} = 2.05$, as suggested in Ref. [13]; inertia linearly decremented from $w = 0.9$ to $w = 0.4$; velocity vector automatically clamped to a maximum intensity, equal to 10% of the maximum point-to-point distance in the search space; search space bounded between 10^{-5} and 10^{2} (for each kinetic parameter) with damping boundary conditions [72]. The choice of the upper bound of the search space for the kinetic constants is fundamental for the entire optimization process since a limited range of variation might exclude the global optimum. However, to perform PE of biochemical systems consisting in a small set of reactions, and characterized by molecular species with low concentrations, limiting the search space of parameters is a good practice for the following reason. Very high values of kinetic constants could lead to undesired dynamics in which chemical species are entirely consumed in the first time instants of the simulation.

The movement of the swarm during the PE phase is determined according to the fitness function. In this chapter, for each particle φ_i, $i = 1, \dots, P$, which encodes the kinetic parameterization of a GRM η, the fitness is defined as the normalized distance between the values of the simulated dynamics of gene expression levels in η and the desired target dynamics. Formally, given the set Γ of genes that are available for the synthetic engineering of the gene regulation circuit and the set $\mathcal{R}$ of reactions of the GRM inferred by CGP, we denote by $X_h(t_c)$ the expected measurement at time t_c of the hth chemical species in Γ and by $Y_h^{\varphi_i}(t_c)$ its simulated value sampled at time t_c. The value $Y_h^{\varphi_i}(t_c)$ is obtained by means of cupSODA, using the kinetic constants contained in particle φ_i. The fitness function $\mathcal{F}$ of particle φ_i is therefore given by:

$$\mathcal{F}(\varphi_i) = \sum_{c=1}^{C} \sum_{h=0}^{N_\Gamma} \frac{|Y_h^{\varphi_i}(t_c) - X_h(t_c)|}{X_h(t_c)}, \qquad (6.2)$$

where $C \in \mathbb{N}$ corresponds to the number of time instants that are arbitrarily sampled in the target dynamics of the GRM.

In this chapter, the PSO algorithm is halted after $IT_{\max} = 300$ iterations. Then, for each GRM η, the value $\mathcal{F}(\mathbf{g}_\eta)$—that is, the fitness of the best solution $\mathbf{g} = (k_1, \dots, k_M)$ determined within the swarm of all candidate parameterizations of η—is taken as the fitness of the inferred GRM η. The GRM characterized by the minimum fitness value among all GRMs inferred by CGP is eventually chosen as the best solution of the whole ED problem.

GPU implementation. Fitness evaluations are computationally expensive since multiple simulations (one for each particle) must be performed during all iterations of the PE on each candidate network inferred by CGP. More precisely, it takes $O(GEN_{\max} \cdot IT_{\max} \cdot I \cdot P)$ fitness evaluations to perform a whole ED process. Anyway, all simulations during each PE phase are mutually independent and can be straightforwardly accelerated by means of a parallel architecture. In order to take advantage of the parallelism of modern GPUs, we exploit the GPU-accelerated simulator cupSODA [48,50] and launch P threads, which perform the simulations and calculate the fitness functions defined in Equation 6.2. This way, the impact of fitness evaluations on the overall running time decreases to $O(GEN_{\max} \cdot IT_{\max} \cdot I)$.

The rest of the methodology (CGP and PSO) is implemented using the Python language (version 2.7) and is executed in a strictly sequential fashion. PSO invokes cupSODA to assess the fitness values by means of synchronous subprocess calls.

6.5 RESULTS

To test the feasibility and the effectiveness of cuGENED, we performed the ED of synthetic circuits composed by two and three genes, using as target of the optimization process a desired temporal dynamics of a small set of (*in silico* generated) mRNA levels. The target dynamics were created from scratch to be complex enough to require the ED of GRMs possibly characterized by multiple regulation mechanisms and intermediate species.

As previously mentioned, the results of the ED process can be only evaluated by considering the desired dynamics. No target GRM can be exploited to assess the quality of the inferred network since the goal of the ED process is to determine a synthetic circuit that still has to be constructed in laboratory. Therefore, to prove the effectiveness of cuGENED, only comparisons between the expected and the simulated dynamics of the inferred GRM are presented hereby.

All the following tests were performed on a workstation with a CPU Intel Core i7-2600, with a clock frequency of 3.4 GHz, and with a GPU Nvidia GeForce GTX 590, running OS Windows 7 64 bit. The settings of the LSODA integrator used for the simulation of the dynamics of the candidate solutions are: relative error equal to $1 \cdot 10^{-10}$, absolute error equal to $1 \cdot 10^{-10}$, maximum number of integration steps equal to 10,000.

6.5.1 ED of Synthetic Circuits with Two Genes

For the ED of synthetic circuits consisting of two genes, we set $|\Gamma| = 2$ and $|\Sigma| = 8$, so that the inferred GRM can contain at most $n_i = 10$ chemical species and $n_o = 26$ reactions (according to the heuristics described in Section 6.4). The desired target dynamics of the species in Γ are described by two sigmoidal curves, characterized by different slopes. This behavior corresponds to a down-regulation of both genes, whose constitutive expression decreases in time by the action of some mutual regulatory mechanism, that has to be inferred by cuGENED.

Some preliminary tests were executed to analyze the influence of cuGENED settings on the ED process. In the first test, we investigated the impact of the parameter n_r—that is, the number of rows in the grid of functional nodes—on the convergence speed (Figure 6.4). According to our results for the two-genes system under investigation, too small ($n_r = 8$) or too large ($n_r = 14$) values of n_r yield a worse convergence of cuGENED, while the best results are achieved with $n_r = 10$. Moreover, by analyzing the early phases of the CGP process we can observe that the higher the n_r value the faster the convergence. Nevertheless, it is the intermediate value $n_r = 10$ that achieves the best convergence speed, confirming a similar result

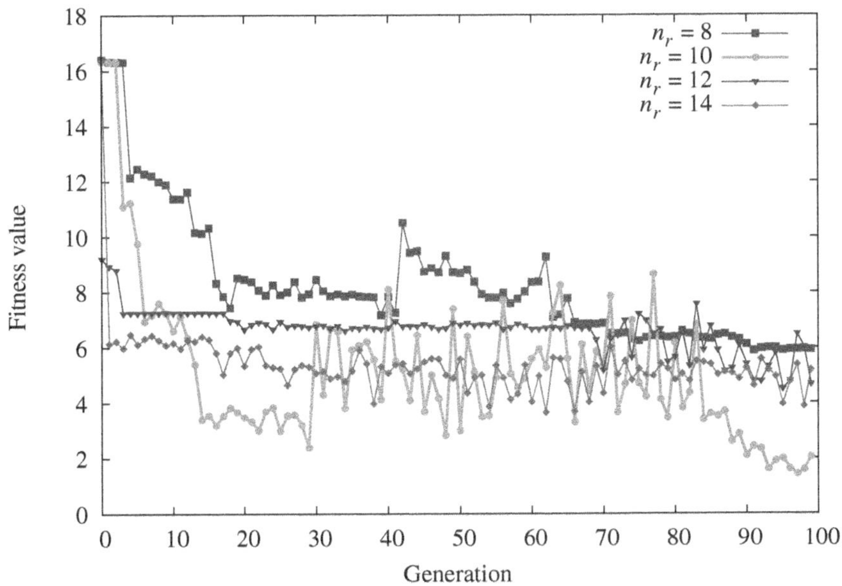

Figure 6.4 Fitness value of the best solution of the ED process obtained by varying the settings for n_r in CGP. The best setting for the two-genes system under investigation is $n_r = 10$, while smaller and larger values yield worse results. The plot highlights the presence of fluctuations due to the stochasticity of PSO, emphasizing how a proper choice for the grid size is fundamental for the convergence of cuGENED.

presented in Ref. [51]. In addition, Figure 6.4 highlights a feature that is typical of methodologies embedding PSO into CGP, as previously discussed in Ref. [51]: due to the stochasticity of PSO, two different PE executions on the same network usually lead to two different kinetic parameterizations which, in turn, are likely to have two different fitness values, explaining the large fluctuations in the convergence speed plots.

As a second test, we analyzed the impact of the parameter ρ, which determines how many elements of a CP genotype are mutated during the offspring generation. Figure 6.5 shows the convergence of cuGENED with different values of ρ, highlighting that the best choice is $\rho = 0.2$. This setting represents the best trade-off between exploration and exploitation of the search space. Indeed, if the value of ρ is too small, CGP is not able to properly explore the search space and cannot converge to individuals characterized by a good fitness value. On the contrary, if the value of ρ is too large, the mutation easily disrupts the structure of the best individual, yielding random individuals with a large fitness value and reducing the advantage of an evolution-guided exploration of the search space.

The best setting identified for cuGENED (i.e., $n_r = 10$ and $\rho = 0.2$) was then exploited to derive the GRM $\eta_{2\gamma}$ that achieves the best fit with the expected behavior of the two-genes system. Figure 6.6 shows the comparison between the desired

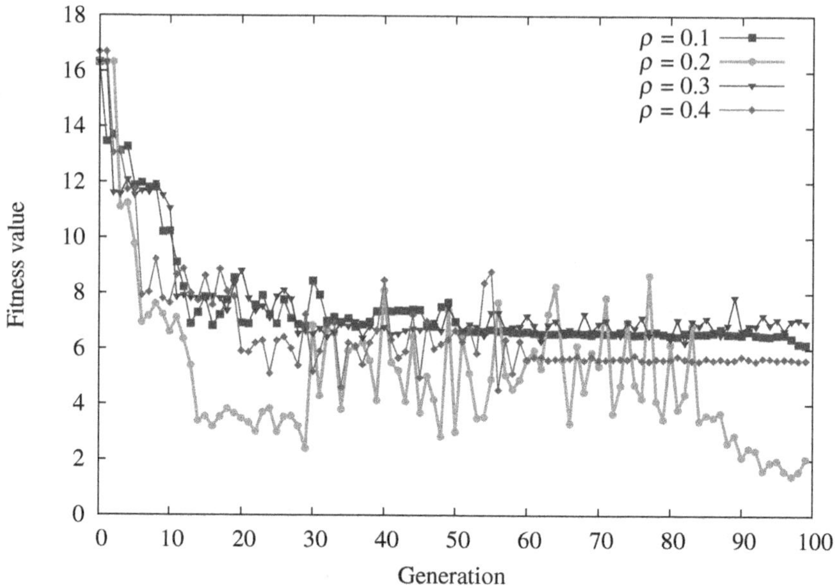

Figure 6.5 Fitness value of the best solution of the ED process obtained by varying the settings for ρ in CGP. The best setting for the two-genes system under investigation is $\rho = 0.2$, that is, about 20% of the genome must be modified when producing new offspring during each generation of the CGP.

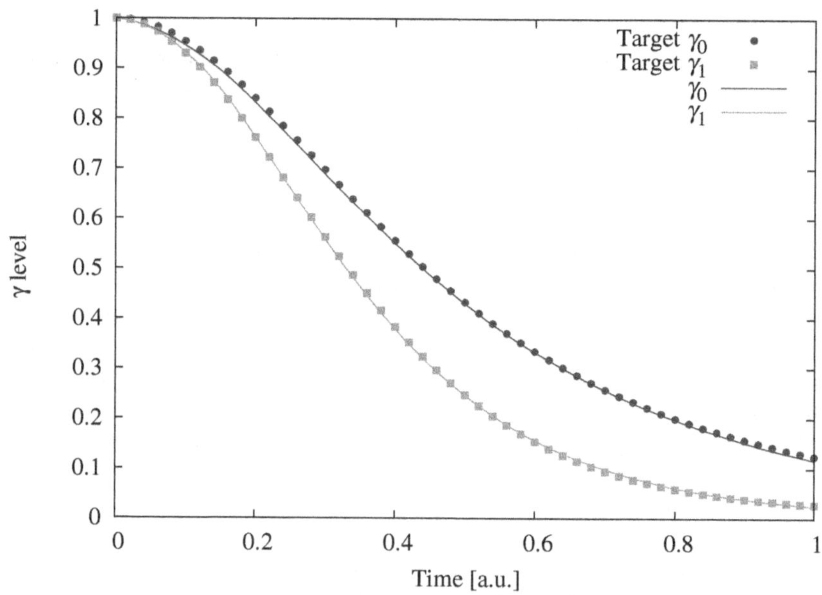

Figure 6.6 Comparison of the target dynamics (dots) with the simulated dynamics of the GRM inferred by cuGENED (lines). The GRM $\eta_{2\gamma}$ achieves a perfect fitting of the desired behavior.

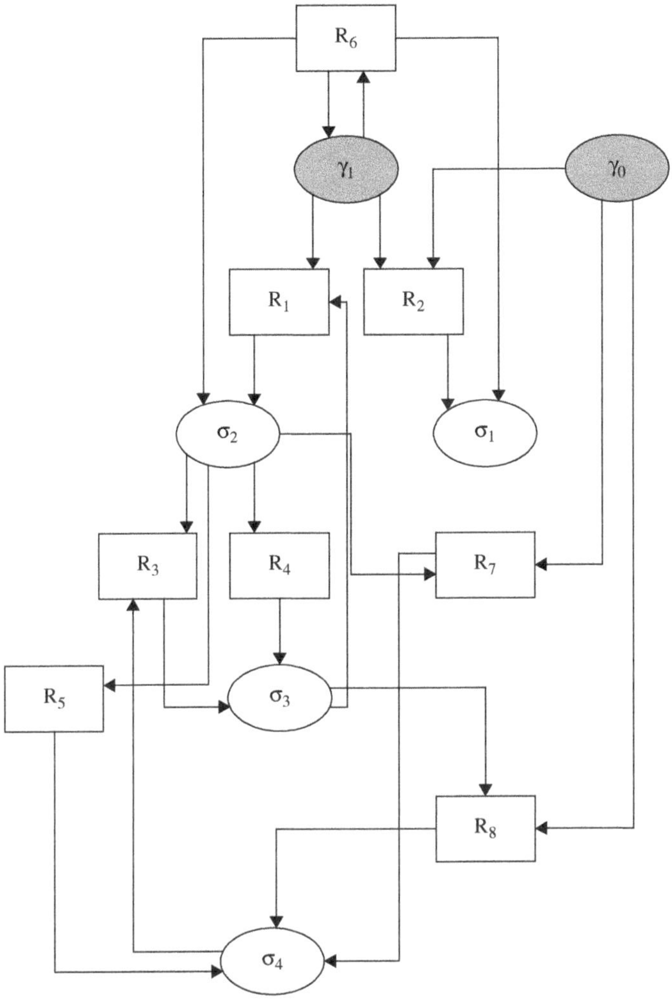

Figure 6.7 The interaction diagram shows the best GRM $\eta_{2\gamma}$ inferred by cuGENED to reproduce the desired behavior shown in Figure 6.6. Circular nodes represent the chemical species involved in the network, while rectangular nodes represent the reactions. The gray nodes denote the chemical species (i.e., mRNA) whose dynamics is considered as target.

target dynamics (dots) and the simulated dynamics (lines) of $\eta_{2\gamma}$. This network is represented in Figure 6.7 and consists of the following reactions:

$$R_1 : \gamma_1 + \sigma_3 \xrightarrow{27.535} \sigma_2,$$

$$R_2 : \gamma_1 + \gamma_0 \xrightarrow{2.390} \sigma_1,$$

$$R_3 : \sigma_2 + \sigma_4 \xrightarrow{13.753} \sigma_3,$$

$$R_4 : \sigma_2 \xrightarrow{19.131} \sigma_3,$$

$$R_5 : \sigma_2 \xrightarrow{7.697} \sigma_4,$$

$$R_6 : \gamma_1 \xrightarrow{10.014} \sigma_2 + \sigma_1 + \gamma_1,$$

$$R_7 : \sigma_2 + \gamma_0 \xrightarrow{2.303} \sigma_4,$$

$$R_8 : \sigma_3 + \gamma_0 \xrightarrow{15.420} \sigma_4,$$

where the numerical values above the arrows correspond to the best kinetic parameterization determined by PSO.

All the reactions in $\eta_{2\gamma}$ are biologically consistent and contribute well to the definition of a network of plausible genetic interactions, which could be implemented by means of synthetic biology techniques. For instance, in $\eta_{2\gamma}$ we can observe the presence of a multi-product reaction (R_6) [14] and alternative splicing reactions (R_4 and R_5) [40]. Furthermore, $\eta_{2\gamma}$ is characterized by a low fitness value, which indicates a consistent similarity with the expected target dynamics obtained by using the kinetic constants ($k_1, \dots, k_8$) inferred by PSO (Figure 6.6). Finally, it is worth noting that this optimal solution does not exploit all the chemical species in the set Σ: the CGP avoided the bloating and evolved a well-fitting GRM that only contains 4 out of the 8 possible species in Σ.

6.5.2 ED of Synthetic Circuits with Three Genes

As a further test, we performed the ED on a system consisting of three genes, using the best settings of cuGENED identified during the preliminary tests (Section 6.5.1). For the ED of this synthetic circuit, we set $|\Gamma| = 3$ and $|\Sigma| = 12$, so that the inferred GRM can contain at most $n_i = 15$ chemical species and $n_o = 51$ reactions (according to the heuristics described in Section 6.4).

The desired target dynamics of the species in Γ are described by two sigmoidal curves, characterized by different slopes, and a monotonically increasing curve (Figure 6.8, dots). This behavior corresponds to a down-regulation of two genes and an up-regulation of a third gene, whose underlying regulatory interplay is to be inferred by means of cuGENED.

The best GRM $\eta_{3\gamma}$ inferred by cuGENED is represented in Figure 6.9 and consists in the following reactions:

$$R_1 : \gamma_0 + \sigma_4 \xrightarrow{4.679} \sigma_3,$$

$$R_2 : \gamma_2 + \sigma_{10} \xrightarrow{20.491} \gamma_2,$$

$$R_3 : \gamma_1 + \sigma_4 \xrightarrow{8.416} \sigma_3 + \sigma_{10},$$

$$R_4 : \sigma_{10} \xrightarrow{4.654} \sigma_0,$$

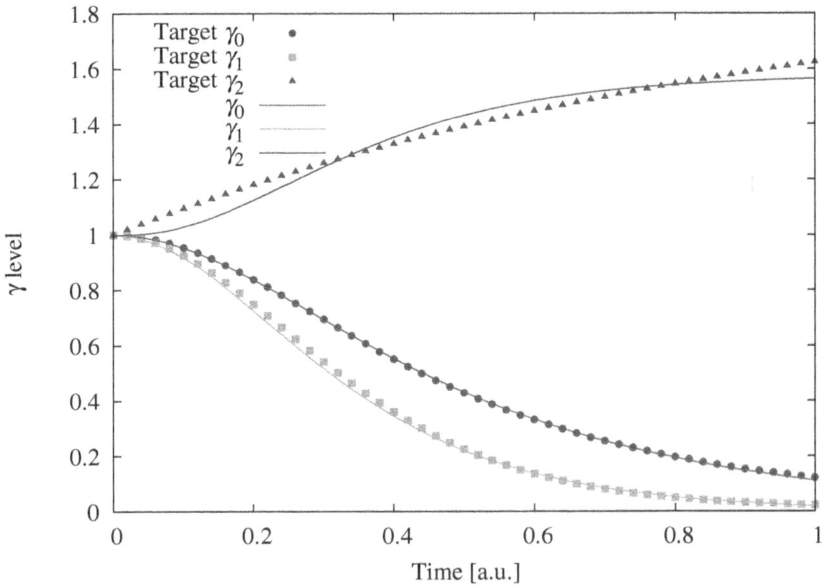

Figure 6.8 Comparison of the target dynamics (dots) with the simulated dynamics of the GRM inferred by cuGENED (lines). The GRM η_{3_γ} achieves an almost perfect fitting of the desired behavior.

$$R_5 : \gamma_1 + \sigma_2 \xrightarrow{21.831} \gamma_1 + \sigma_4,$$

$$R_6 : \sigma_{10} \xrightarrow{10.321} \sigma_{11},$$

$$R_7 : \sigma_6 + \sigma_{10} \xrightarrow{14.681} \sigma_{11},$$

where the numerical values above the arrows correspond to the best kinetic parameterization determined by PSO.

The result in Figure 6.8 shows that η_{3_γ} almost perfectly fits the desired dynamics. In particular, the target and simulated dynamics for γ_0 and γ_1 are perfectly overlapped, while the curve of γ_2 slightly diverges from the expected behavior. Nevertheless, the simulated behavior of γ_2 is qualitatively and quantitatively similar to the target behavior, which means that, from a biological standpoint, the ED goal is fully achieved.

The interaction network in Figure 6.9 has two important characteristics, highlighted by the dashed lines, which represent side effects of cuGENED:

- the chemical species σ_2, which is a product of reaction R_5, is not involved in the rest of the network. Hence, this species can be removed from η_{3_γ} without any impact on the system dynamics. We highlight that the role of the chemical species σ_3 is different from σ_2, because it is fundamental for the degradation of γ_0 as a consequence of reaction R_1;

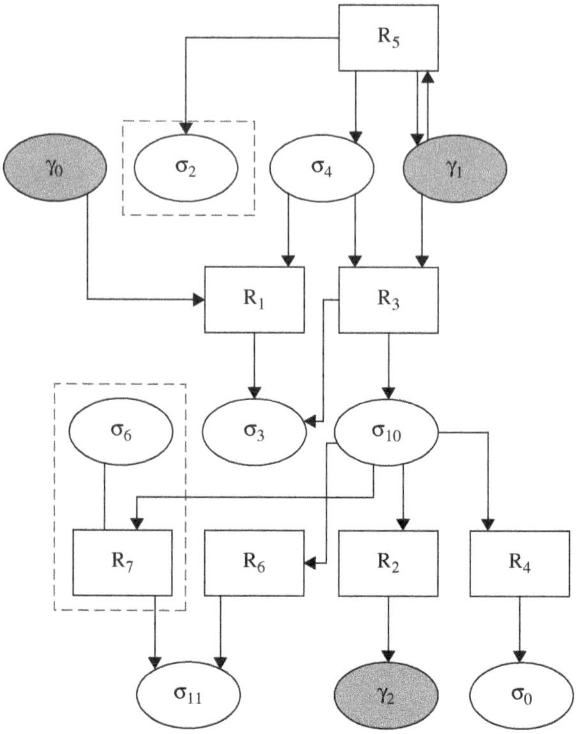

Figure 6.9 The interaction diagram shows the best GRM $\eta_{3\gamma}$ inferred by cuGENED to reproduce the desired behavior shown in Figure 6.8. Circular nodes represent the chemical species involved in the network, while rectangular nodes represent the reactions. The gray nodes denote the chemical species (i.e., mRNA) whose dynamics is considered as target for the ED. Dashed lines highlight reactions and chemical species that have no effective role in the system dynamics.

- the reactants of reaction R_7 are species σ_6 and σ_{10}: while the latter is produced by reaction R_3, the former is not produced by any other reaction, so that the level of σ_6 remains fixed to 0 during the simulations. The consequence is that reaction R_7 will never take place and hence it can be removed from $\eta_{3\gamma}$, along with σ_6. On the contrary, species σ_{11}, the product of reaction R_7, cannot be removed from the network, being involved also in reaction R_6.

6.5.3 Computational Results

Using a GPU Nvidia GeForce GTX 590 equipped with 1024 cores, the ED of $\eta_{2\gamma}$ required 70136 seconds (about 19 hours), while $\eta_{3\gamma}$ required 72645 seconds (about 20 hours). Thus, the running time of cuGENED presumably scales less than linearly with the number of genes, making it feasible for the ED of systems characterized by a large number of genes. The increased running time is probably due to the different size of the candidate networks: the GRMs for the three-genes system are indeed characterized by a larger number of species and reactions, a circumstance that inevitably slows down the simulations. On the other hand, thanks to the use of the cupSODA deterministic simulator [48], the fitness evaluations are computed in parallel, so that the running time is not affected by the number of particles used in the PSO phase.

In order to compare the performances of cuGENED using CPU and GPU implementations of LSODA, we run 640 simulations—corresponding to 64 random parameterizations of 10 random GRMs, each one determined by a CP—which are equivalent to a single CGP iteration. As a reference CPU implementation we used Scipy's `odeint` integrator [27], based on LSODA from the FORTRAN library odepack. The running time to perform the simulations was 1.5 s on the CPU, and 0.061 s on the GPU by using cupSODA, corresponding to a 24.5× speedup achieved by cuGENED.

6.6 DISCUSSION

cuGENED consists in the integration of two evolutionary algorithms which work at two different levels: CGP to infer the structure of a GRN, and PSO to estimate the kinetic parameters of the reactions involved in the network. To the best of our knowledge, cuGENED represents the first attempt to tackle the ED of GRMs by means of CGP combined with PSO. This methodology is also the first attempt to automatically model a GRN assuming 0-knowledge on the set of molecular species. In this context, there are some issues that are worth considering.

In cuGENED, the ED problem is formalized without any a priori information on the structure of the system, defined in terms of molecular interactions, except for a desired target dynamics of a small subset of species. On the one hand, this information is sufficient to properly reconstruct a network whose dynamics fits the target data (see, for instance, Figure 6.6). On the other hand, this information alone is usually not enough to discriminate between possibly different network topologies [11], which could all be able to reproduce the same behavior used as target of the optimization. This well known issue in ED and RE is named *indistinguishability* of equivalent networks [68]. To lessen this computational difficulty, a good practice should be the multiple execution of the ED process, followed by an analysis of the obtained solutions to assess their laboratory feasibility. This analysis could as well be performed by means of automatic algorithms, which exploit a knowledge base of feasible reactions specifically defined by synthetic biologists.

With respect to other existing methodologies for the inference of GRNs, an advantage of cuGENED is that, thanks to the use of CGP, it allows to exploit a simple representation of the candidate GRMs. The genotype of CGP, indeed, is a fixed-length vector of integer values, whose corresponding phenotype is a set of human-readable chemical reactions. Moreover, cuGENED requires the estimation of only M kinetic parameters associated with the reactions, where, in general, $M \ll (N^2 + 1)/2$. On the contrary, other existing GP methodologies generally necessitate complex parameterizations. For instance, S-tree-based methods require the estimation of $2N(N + 1)$ kinetic parameters [9], thus making this method hardly exploitable for systems involving many genes. Moreover, thanks to the explicit limitation to second-order reactions in GRMs, cuGENED does not generate families of candidate solutions having an identical structure but a different stoichiometry, which still represents an open issue in GP-based methodologies [39].

Another relevant strength of cuGENED is the ED of synthetic circuits formally defined as GRMs, which, in contrast to most existing approaches [8], are not based on arbitrary kinetics functions. The GRNs evolved by CGP and PSO consist in mass-action reaction-based models, which describe the biological processes in terms of simple molecular interactions, and not as reaction rate approximations based on some kind of chemical assumption. As a matter of fact, Hill functions were shown not to represent physically realistic reaction schemes [70], especially for gene expression processes [63], while the equilibrium or quasi steady-state assumptions at the basis of Michaelis–Menten constants are valid only in specific conditions [46].

A current drawback of cuGENED is that the choice of its initial setting is relevant to converge to an optimal solution since both CGP and PSO are not settings-free algorithms. We relied on common literature settings for the PSO, while we analyzed the impact of different parameter values on the performances of CGP (i.e., the mutation rate ρ and the number of rows n_r). As a future improvement of cuGENED, we will investigate the relationship between these two values, the number of genes and the complexity of the desired target dynamics, in order to derive a heuristic for the *automatic* selection of the most appropriate settings.

cuGENED relies on the nested execution of two evolutionary algorithms. As a consequence, its overall computational cost can be relevant, being proportional to the number of generations of CGP, to the number of iterations of PSO, to their population size, to the number of reactions involved in the candidate networks, and to the time length of the simulations. The largest part of the running time is due to the simulation of the dynamics of the candidate solutions, which is a fundamental step for the fitness computation. However, the computational cost can be strongly reduced by exploiting the GPU-powered cupSODA simulator [50]. In so doing, we realized a parallel execution of the simulations and the fitness calculations of all candidate solutions, achieving a 24.5× speedup with respect to a strictly sequential execution of the same tasks. It is worth noting that the computational time could be further reduced by executing parallel ED instances. To the best of our knowledge, cuGENED also represents the first attempt ever in exploiting GPUs to perform an accelerated ED of GRMs.

6.7 CONCLUSIONS AND FUTURE PERSPECTIVES

The idea of a circuit-like connectivity between biological parts was postulated for the first time in the 1960s [45]. This intuition lead to several attempts to properly formalize biological regulation systems through mathematical models [22,23,28,64] and analyze the cellular pathways under investigation by exploiting electrical circuit analogies [41,42]. This research field was favored by the progression of molecular biology and genomics, which provided several methods and the necessary knowledge to physically assemble biomolecular components. As a matter of fact, it is nowadays possible to properly evaluate the interaction among different genes through expression data coming from high-throughput techniques like microarrays [20] and real-time PCR [34]. It is also possible to engineer *in vitro* or *in vivo* customized signaling circuits [36]. The proof of concept that a computing-like behavior could be applied to biological systems was the design of the first synthetic gene networks, realized by using engineering-based methodologies [16,19]. Lately, several network models were proposed, all of them integrating biochemical pathway information and expression data [1,6,38,58].

In this context, we developed cuGENED, a computational methodology for the automatic design of GRNs characterized by a predefined dynamical behavior. In particular, cuGENED allows to derive a mechanistic model of a gene regulation system, modeled as a parameterized set of biochemical reactions. The reactions describe the processes related to gene expression, and they involve a set of molecular species (e.g., genes, mRNA) whose dynamics represent the target behavior of the optimization.

It is often the case that a gene can regulate either itself (autoregulation) or a single other gene. This is due to the fact that a generic regulatory species might be able to interact with a single component of the gene network through a *unique* DNA sequence. This means that all species could be extremely selective for their molecular target (i.e., each species can bind to a unique DNA sequence), so that no unknown interactions can occur between the regulators and their corresponding molecular targets [18]. Therefore, the molecular target of a regulator could be either the gene itself or another gene, for example, in a very selective feedback system. Thus, as future developments of this work, we plan to improve the ED methodology as follows:

- we will include the verification of the requirements about selective regulation to narrow the search space of candidate solutions, possibly obtaining GRMs that could be easier to implement with laboratory techniques;
- we will exploit the massive parallelism provided by GPUs to perform many parallel ED processes, in order to collect a set of optimal GRMs. This set can then be analyzed to identify the GRMs that show a better selective regulation (i.e., the GRMs that better fulfill the requirements stated above).

We also plan to introduce an automatic mechanism to detect the unessential species (e.g., products that have no impact on the system dynamics), as well as the reactions that cannot take place (i.e., reactions whose reactants have zero concentration, like those highlighted in Figure 6.9). In addition, we will consider the fact that the inference process might converge to a GRM whose simulation perfectly fits the desired behavior, though some reactions appearing in the GRM might not be biologically plausible. As mentioned in Section 6.6, a solution to this problem might come from the inclusion of some knowledge domain constraints in the ED, in order to automatically remove such reactions from the candidate GRMs.

Finally, in cuGENED we have not explicitly taken into account the biological noise that is typical of gene expression [15]. Actually, some species involved in GRNs can have very low intracellular amounts (in the order of tens or a few hundreds of molecules), so that a stochastic simulation methodology could be preferable with respect to deterministic simulations to correctly reproduce any noise-induced emergent phenomena. cuGENED can be straightforwardly extended to support also stochastic simulation algorithms for the simulation of the temporal dynamics of candidate solutions [21,52,62]. Indeed, as a further development, we will investigate the possibility of the automatic reconstruction of networks in presence of intrinsic noise, oscillations, or multistability phenomena.

REFERENCES

1. T. Akutsu, S. Miyano, and S. Kuhara. Identification of genetic networks from a small number of gene expression patterns under the Boolean network model. In *Pacific Symposium on Biocomputing*, volume 4, pages 17–28, 1999.
2. B. B. Aldridge, J. M. Burke, D. A. Lauffenburger, and P. K. Sorger. Physicochemical modelling of cell signalling pathways. *Nature Cell Biology*, 8:1195–1203, 2006.
3. S. Ando, E. Sakamoto, and H. Iba. Evolutionary modeling and inference of gene network. *Information Sciences*, 145:237–259, 2002.
4. D. Besozzi, P. Cazzaniga, G. Mauri, D. Pescini, and L. Vanneschi. A comparison of genetic algorithms and particle swarm optimization for parameter estimation in stochastic biochemical systems. In C. Pizzuti, M.D. Ritchie, and M. Giacobini, editors, *Evolutionary Computation, Machine Learning and Data Mining in Bioinformatics (Proceedings EvoBIO 2009)*, volume 5483 of *LNCS*, pages 116–127. Springer, 2009.
5. H. Beyer and H. Schwefel. Evolution strategies—a comprehensive introduction. *Natural Computing*, 1:3–52, 2002.
6. J. M. Bower and H. Bolouri. *Computational Modeling of Genetic and Biochemical Networks*. MIT Press, 2001.
7. J. C. Butcher. *Numerical Methods for Ordinary Differential Equations*. John Wiley & Sons, 2003.
8. H. Cao, L. Kang, Y. Chen, and J. Yu. Evolutionary modeling of systems of ordinary differential equations with genetic programming. *Genetic Programming and Evolvable Machines*, 1:309–337, 2000.
9. D. Cho, K. Cho, and B. Zhang. Identification of biochemical networks by S-tree based genetic programming. *Bioinformatics*, 22(13):1631–1640, 2006.
10. I. C. Chou and E. O. Voit. Recent developments in parameter estimation and structure identification of biochemical and genomic systems. *Mathematical Biosciences*, 219:57–83, 2009.
11. G. Craciun and C. Pantea. Identifiability of chemical reaction networks. *Journal of Mathematical Chemistry*, 44(1):244–259, 2008.
12. L. Demattè and D. Prandi. GPU computing for systems biology. *Briefings in Bioinformatics*, 11(3):323–333, 2010.

13. A. Dräger, M. Kronfeld, M. J. Ziller, J. Supper, H. Planatscher, and J. B. Magnus. Modeling metabolic networks in *C. glutamicum*: a comparison of rate laws in combination with various parameter optimization strategies. *BMC Systems Biology*, 3(5), 2009.
14. X. Du, J. Wang, H. Zhu, L. Rinaldo, K. Lamar, A. C. Palmenberg, C. Hansel, and C. M. Gomez. Second cistron in *CACNA1A* gene encodes a transcription factor mediating cerebellar development and SCA6. *Cell*, 154(1):118–133, 2013.
15. A. Eldar and M. B. Elowitz. Functional roles for noise in genetic circuits. *Nature*, 467(7312):167–173, 2010.
16. M. B. Elowitz and S. Leibler. A synthetic oscillatory network of transcriptional regulators. *Nature*, 403(6767):335–338, 2000.
17. R. M. Farber. Topical perspective on massive threading and parallelism. *Journal of Molecular Graphics and Modelling*, 30:82–89, 2011.
18. N. C. Garbett and J. B. Chaires. Binding: a polemic and rough guide. *Methods in Cell Biology*, 84:1–23, 2008.
19. T. S. Gardner, C. R. Cantor, and J. J. Collins. Construction of a genetic toggle switch in *Escherichia coli*. *Nature*, 403(6767):339–342, 2000.
20. G. Gibson. Microarray analysis. *PLoS Biology*, 1(1):e15, 2003.
21. D. T. Gillespie. Stochastic simulation of chemical kinetics. *Annual Review of Physical Chemistry*, 58:35–55, 2007.
22. L. Glass. Classification of biological networks by their qualitative dynamics. *Journal of Theoretical Biology*, 54(1):85–107, 1975.
23. L. Glass and S. A. Kauffman. The logical analysis of continuous, non-linear biochemical control networks. *Journal of Theoretical Biology*, 39(1):103–129, 1973.
24. M. J. Harvey and G. De Fabritiis. A survey of computational molecular science using graphics processing units. *Wiley Interdisciplinary Reviews: Computational Molecular Science*, 2(5):734–742, 2012.
25. S. Hoops, S. Sahle, R. Gauges, C. Lee, J. Pahle, N. Simus, M. Singhal, L. Xu, P. Mendes, and U. Kummer. COPASI—a COmplex PAthway SImulator. *Bioinformatics*, 22:3067–3074, 2006.
26. H. Iba. Inference of differential equation models by genetic programming. *Information Sciences*, 178:4453–4468, 2008.
27. E. Jones, T. Oliphant, and P. Peterson. SciPy: Open source scientific tools for Python, 2001.
28. S. A. Kauffman. The large scale structure and dynamics of gene control circuits: An ensemble approach. *Journal of Theoretical Biology*, 44(1):167–190, 1974.
29. J. Kennedy and R. C. Eberhart. Particle swarm optimization. In *Proceedings of the IEEE International Conference on Neural Networks*, volume 4, pages 1942–1948, Piscataway, NJ, 1995.
30. J. Kitagawa and H. Iba. Identifying metabolic pathways and gene regulation networks with evolutionary algorithms. In Gary B. Fogel and David W. Corne, editors, *Evolutionary Computation in Bioinformatics*, pages 255–278. San Francisco: Morgan Kaufman, 2003.
31. J. R. Koza. *Genetic Programming: On the Programming of Computers by Means of Natural Selection*. The MIT Press, 1992.
32. J. R. Koza, W. Mydlowec, G. Lanza, J. Yu, and M. A. Keane. Reverse engineering of metabolic pathways from observed data using genetic programming. In *Pacific Symposium on Biocomputing*, volume 6, pages 434–445, 2001.

33. J. R. Koza, W. Mydlowec, G. Lanza, J. Yu, and M. A. Keane. Reverse engineering of metabolic pathways from observed data using genetic programming. Technical Report SMI-2000-0851, Stanford University, Stanford, California (USA), 2001.

34. M. Kubista, J. M. Andrade, M. Bengtsson, A. Forootan, J. Jonák, K. Lind, R. Sindelka, R. Sjöback, B. Sjögreen, L. Strömbom, A. Ståhlberg, and N. Zoric. The real-time polymerase chain reaction. *Molecular Aspects of Medicine*, 27(2–3):95–125, 2006.

35. T. Lenser, T. Hinze, B. Ibrahim, and P. Dittrich. Towards evolutionary network reconstruction tools for Systems Biology. In E. Marchiori, J. H. Moore, and J. C. Rajapakse, editors, *Evolutionary Computation, Machine Learning and Data Mining in Bioinformatics (Proceedings EvoBIO 2007)*, volume 4447 of *LNCS*, pages 132–142. Springer, 2007.

36. W.A. Lim. Designing customized cell signalling circuits. *Nature Reviews Molecular Cell Biology*, 11(6):393–403, 2010.

37. J. Logan, K. Edwards, and N. Saunders, editors. *Real-Time PCR: Current Technology and Applications*. Caister Academic Press, 2009.

38. Y. Maki, D. Tominaga, M. Okamoto, S. Watanabe, and Y. Eguchi. Development of a system for the inference of large scale genetic networks. In *Pacific Symposium on Biocomputing*, volume 6, pages 446–458, 2001.

39. D. Marco, C. Shankland, and D. Cairns. Evolving Bio-PEPA algebra models using Genetic Programming. In *Proceedings of the 2012 Annual Conference on Genetic and Evolutionary Computation*, pages 177–183. ACM, 2012.

40. A. J. Matlin, F. Clark, and C. W. J. Smith. Understanding alternative splicing: towards a cellular code. *Nature Reviews Molecular Cell Biology*, 6(5):386–398, 2005.

41. H. H. McAdams and A. Arkin. Gene regulation: Towards a circuit engineering discipline. *Current Biology*, 10(8):318–320, 2000.

42. H. H. McAdams and L. Shapiro. Circuit simulation of genetic networks. *Science*, 269(5224):650–656, 1995.

43. J. Miller and P. Thomson. Cartesian genetic programming. In R. Poli , W. Banzhaf, W. B. Langdon, J. Miller, P. Nordin, T. C. Fogarty, editors, *Proceedings of the Third European Conference on Genetic Programming (EuroGP2000)*, volume 1802 of *LNCS*, pages 121–132. Springer, 2000.

44. C. G. Moles, P. Mendes, and J. R. Banga. Parameter estimation in biochemical pathways: a comparison of global optimization methods. *Genome Research*, 13:2467–2474, 2003.

45. J. Monod and F. Jacob. General conclusions: teleonomic mechanisms in cellular metabolism, growth, and differentiation. In *Cold Spring Harbor Symposia on Quantitative Biology*, volume 26, pages 389–401. Cold Spring Harbor Laboratory Press, 1961.

46. D. L. Nelson and M. M. Cox. *Lehninger Principles of Biochemistry*. W. H. Freeman Company, 2004.

47. M. S. Nobile, D. Besozzi, P. Cazzaniga, and G. Mauri. The foundation of evolutionary Petri nets. In G. Balbo and M. Heiner, editors, *Proceedings of the 4th International Workshop on Biological Processes & Petri Nets (BioPPN 2013)*, volume 988, pages 60–74. CEUR Workshop Proceedings, 2013.

48. M. S. Nobile, D. Besozzi, P. Cazzaniga, and G. Mauri. GPU-accelerated simulations of mass-action kinetics models with cupSODA. *The Journal of Supercomputing*, 69(1):17–24, 2014.

49. M. S. Nobile, D. Besozzi, P. Cazzaniga, G. Mauri, and D. Pescini. A GPU-based multi-swarm PSO method for parameter estimation in stochastic biological systems exploiting

discrete-time target series. In M. Giacobini, L. Vanneschi, and W. S. Bush, editors, *Evolutionary Computation, Machine Learning and Data Mining in Bioinformatics (Proceedings EvoBIO 2012)*, volume 7246 of *LNCS*, pages 74–85. Springer, 2012.

50. M. S. Nobile, D. Besozzi, P. Cazzaniga, G. Mauri, and D. Pescini. cupSODA: a CUDA-powered simulator of mass-action kinetics. In V. Malyshkin, editor, *Proceedings of 12th International Conference on Parallel Computing Technologies (PaCT 2013)*, volume 7979 of *LNCS*, pages 344–357. Springer, 2013.
51. M. S. Nobile, D. Besozzi, P. Cazzaniga, D. Pescini, and G. Mauri. Reverse engineering of kinetic reaction networks by means of cartesian genetic programming and particle swarm optimization. In *Evolutionary Computation (CEC), 2013 IEEE Congress on*, pages 1594–1601. IEEE, 2013.
52. M. S. Nobile, P. Cazzaniga, D. Besozzi, D. Pescini, and G. Mauri. cuTauLeaping: A GPU-powered tau-leaping stochastic simulator for massive parallel analyses of biological systems. *PLoS ONE*, 9(3):e91963, 2014.
53. N. Noman and H. Iba. Inference of gene regulatory networks using S-system and differential evolution. In *Proceedings of the 2005 Conference on Genetic and Evolutionary Computation*, pages 439–446. ACM, 2005.
54. J. Nummela and J. A. Bryant. Evolving Petri nets to represent metabolic pathways. In *Proceedings of the 2005 Conference on Genetic and Evolutionary Computation*, pages 2133–2139. ACM, 2005.
55. Nvidia. Nvidia CUDA C Programming Guide v5.0, 2012.
56. Nvidia. NVIDIA's Next Generation CUDA Compute Architecture: Kepler GK110, 2012.
57. J. L. Payne, N. A. Sinnott-Armstrong, and J. H. Moore. Exploiting graphics processing units for computational biology and bioinformatics. *Interdisciplinary Sciences, Computational Life Sciences*, 2(3):213–220, 2010.
58. B. E. Perrin, L. Ralaivola, A. Mazurie, S. Bottani, J. Mallet, and F. d'Alche Buc. Gene networks inference using dynamic Bayesian networks. *Bioinformatics*, 19(2):138–148, 2003.
59. L. Petzold. Automatic selection of methods for solving stiff and nonstiff systems of ordinary differential equations. *SIAM Journal of Scientific and Statistical Computing*, 4:136–148, 1983.
60. J. W. Pinney, D. R. Westhead, and G. A. McConkey. Petri Net representations in systems biology. *Biochemical Society Transactions*, 31:1513–1515, 2003.
61. T. Quarles, A. R. Newton, D. O. Pederson, and A. Sangiovanni-Vincentelli. *SPICE 3 Version 3F5 User's Manual*. Departement Electrical Engineering Computer Sciences, University of California, 1994.
62. H. Salis and Y. Kaznessis. Accurate hybrid stochastic simulation of a system of coupled chemical or biochemical reactions. *The Journal of Chemical Physics*, 122(5):054103, 2005.
63. M. Santillán. On the use of the Hill functions in mathematical models of gene regulatory networks. *Mathematical Modelling of Natural Phenomena*, 3(2):85–97, 2008.
64. M. A. Savageau. Comparison of classical and autogenous systems of regulation in inducible operons. *Nature*, 252:546–549, 1974.
65. M. A. Savageau. *Biochemical Systems Analysis. A Study of Function and Design in Molecular Biology*. Addison-Wesley, 1976.

66. R. Storn and K. Price. Differential evolution–a simple and efficient heuristic for global optimization over continuous spaces. *Journal of Global Optimization*, 11(4):341–359, 1997.

67. M. Sugimoto, S. Kikuchi, and M. Tomita. Reverse engineering of biochemical equations from time-course data by means of genetic programming. *BioSystems*, 80:155–164, 2005.

68. G. Szederkenyi, J. R. Banga, and A. A. Alonso. Inference of complex biological networks: distinguishability issues and optimization-based solutions. *BMC Systems Biology*, 5(177), 2011.

69. M. Vigelius, A. Lane, and B. Meyer. Accelerating reaction-diffusion simulations with general-purpose graphics processing units. *Bioinformatics*, 27(2):288–290, 2011.

70. J. N. Weiss. The Hill equation revisited: uses and misuses. *The FASEB Journal*, 11(11): 835–841, 1997.

71. O. Wolkenhauer, M. Ullah, W. Kolch, and C. Kwang-Hyun. Modeling and simulation of intracellular dynamics: choosing an appropriate framework. *IEEE Transactions on Nanobiosciences*, 3(3):200–7, 2004.

72. S. Xu and Y. Rahmat-Samii. Boundary conditions in particle swarm optimization revisited. *IEEE Transactions on Antennas and Propagation*, 55(3):760–765, 2007.

73. Y. Zhou, J. Liepe, X. Sheng, M. P. H. Stumpf, and C. Barnes. GPU accelerated biochemical network simulation. *Bioinformatics*, 27(6):874–876, 2011.

7

MODELING DYNAMIC GENE EXPRESSION IN *STREPTOMYCES COELICOLOR*: COMPARING SINGLE AND MULTI-OBJECTIVE SETUPS

Spencer Angus Thomas and Yaochu Jin
Department of Computing, University of Surrey, Guildford, UK
Emma Laing and Colin Smith
Department of Microbial and Cellular Sciences, University of Surrey, Guildford, UK

7.1 INTRODUCTION

Even the simplest of organisms can have an extremely complex network [1] spanning many levels of interactions from cellular to gene, to protein, and beyond. It is possible to model biological networks using gene regulatory networks (GRNs), which are groups of genes that interact through the production of their proteins. How these GRNs are constructed for large biological networks and how transcription factors are able to regulate the expression of thousands of genes in response to environmental changes is a fundamental problem in biology [2]. Furthermore, the reverse engineering of biological networks from expression data and the inference of the complexity in

Evolutionary Computation in Gene Regulatory Network Research, First Edition.
Edited by Hitoshi Iba and Nasimul Noman.

the networks is a current problem in computational and biological sciences [3]. The challenges of network reconstruction increase with the size of the network and suffer "underdeterminism," where there are insufficient data to build statistical models for the number of genes in the network [4]. Hase et al. [5] observed that in order to infer large-scale networks, it is vital to know *a priori* information of high performance algorithms so methods can be selected based on the details of the problem. However, systematic tests are difficult for reconstruction algorithms as the availability of real gene expression data is limited. As a result, many methods use synthetic data to benchmark their algorithms by comparing results to the known network. Artificial data sets, however, lack the level of complexity and uncertainty that real experimental data sets contain. Moreover, when using experimental data, in general, one may not know what the "real" network is. A greater emphasis is needed on the inference of unknown networks and the use of experimental data as opposed to benchmarking algorithms with artificially generated gene expression profiles.

Because of the lack of available data and the curse of dimensionality, the complexity of network reconstruction increases significantly with the number of genes. For larger networks, simplifications may be required as the number of genes means that a complete model capturing all biological interactions is computationally infeasible. When considering all possible interactions, even a small number of genes can lead to complex network dynamics [6] and an interaction topology that spans many levels, as illustrated in Figure 7.1. Because of this high complexity, collaborations between mathematicians, computer scientists, and biologist attempt to model the biological networks of large organisms such as some yeast (2000 genes) [7,8], *Escherichia coli* (5000 genes) [9], *Streptomyces coelicolor* (8000 genes) [10,11], flies (13,525 genes) [12], and humans (22,287 genes) [12].

7.1.1 Modeling Gene Expression

A common modeling technique for biological networks is to use GRNs, where interactions are modeled on the gene level only and are mediated by their protein products. Despite this simplification, GRNs are able to exhibit a range of network dynamics such as oscillations [13,14], bi-stability (toggle switches) [15–17], a combination of these [18,19], and many others. Analyzing GRNs is an important research area in systems biology [20,21], and understanding the relationship between the structure and dynamics of GRNs is vital in the understanding of natural evolution [22]. Ordinary differential equations can model rich dynamics and are able to combine detailed parametrized models with regulatory connectivity.

Topological models are of particular importance to our understanding of the behavior of GRNs due to the modularity of biological systems. The functions of these modules are important biological processes, such as AND and OR gates for time delays and robustness [17,20,23,24]. Swain et al. [25] illustrated the importance of topology by using a caterpillar and butterfly analogy, where the two insects contain the same genes, the connectivity of which is changed during the chrysalis phase resulting in the physical difference between the two. In some cases the topology of the network is more important than the parametrization, as the structure can determine the dynamic

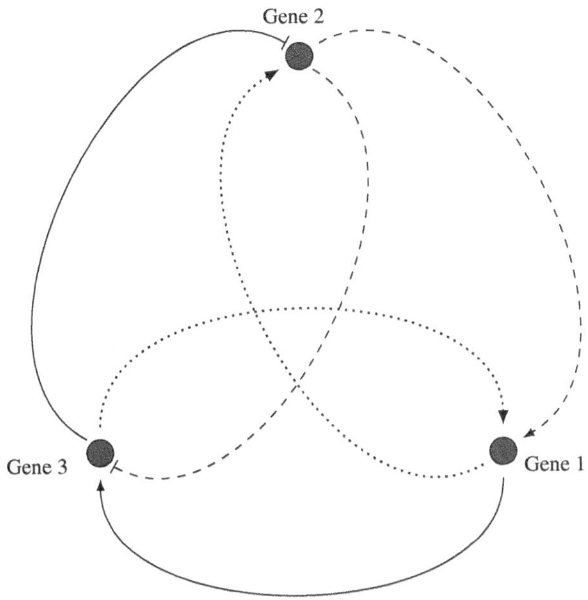

Figure 7.1 A gene regulatory network of a three-gene system interacting over several levels. This demonstrates how the complexity of modeling increases rapidly when considering multiple levels of interactions. Illustrated here are gene-level (solid line), protein-level (dashed line), and metabolite-level (dotted line) interactions. Lines with arrow heads and bar ends represent activating and repressive regulations, respectively.

behavior of the network [25, 26]. The importance of topology over parametrization is also present in more complex models, such as the *Drosophila* segment polarity network. This model contains 48 free parameters, which when randomized, each had a 90% chance of being compatible with the desired behavior regardless of parameter magnitude or range [27]. The authors in Ref. [27] observed the desired dynamics approximately 1 in every 200 runs, much more frequent than at random. The main issue with topological models, however, is the lack of a quantitative metric for comparison between models [28], though some metrics do exist [2]. One can use measures such as specificity, sensitivity, precision, and recall for benchmarking, though for the practical case of an unknown topology these measures are useless. For competing models of the same unknown system, the fit to the experimental data can be used as a measure of model quality. For models of similar fits to the data, the simpler model, that is, fewer nodes and/or connections, is preferred as it is easier to understand and less prone to over fitting through Occam's Razor [4].

The parametrization of a network is also important as it allows us to investigate the modeling of the connections within a GRN. The difference between a simple linear connection and a more complex nonlinear connection between genes could significantly affect their dynamical behavior. In Ref. [29] Ingram et al. found that even for a relatively simple motif, a bi-fan, network dynamics varies greatly for different connection types and parameter sets. The authors demonstrated that this simple topology can be tuned to give a desired output and, therefore, a general statement about a network's functionality cannot be based solely on structure. Gonze [18] observed regions in the parameter space that determined the dynamic behavior of a fixed network structure and the bifurcation values for changing the network dynamics.

Collectively these studies indicate that both topology and parametrization are important in general and may determine the functionality of a network together.

7.1.2 Reverse Engineering Biological Networks from Expression Data

To some degree, it is possible to infer regulatory networks from expression profiles of the genes in an organism [30–33]. This is an important challenge in systems biology [34]. Gene regulatory networks are the most important organization level within a cell [35] and the study of them is a growing area of research [26]. Reconstruction from time-series data is particularly important as it provides insight into the dynamic interactions between genes and can serve as an intermediate step between systems biology and bioinformatics [36]. Reverse engineering networks from expression data using computational techniques are common place [36–40], and can provide insight into network structures, regulation type, and strength, as well as the prediction of regulatory targets. However, the inference of biological GRNs from expression data is one of the most complex tasks in bioinformatics [41] due to the complexity in molecular interactions [1]. As GRNs are not fully understood [25], reconstruction from expression data still remains an open issue in biology [3] and is the focus of many computational techniques [42].

Despite experimental advances in data collection techniques, significant costs lead to limited availability for fine grain time-series data. In 2011, Penfold and Wild [43] noted that for microarray time-series data for three replicates each with 25 time points costs in the range of £30,000 (over $45,000). This has led to the underdeterminism of such problems, where there are insufficient time-series data available to statistically reconstruct large networks [4, 26, 38, 44–47]. Furthermore, specific growth conditions for many organisms mean that much of the data is heterogeneous and cannot necessarily be used together [45]. Presently, the inference of small networks remains a challenge as available methods cannot identify all connections and often yield false regulations [48]. Currently, there are a few algorithms that use large time-series data sets and typically consist of around 5–10 time points [49, 50]. Moreover, many techniques additionally use artificially generated expression profiles [51] that do not exhibit the levels of noise and complexity as experimental measurements. It should be noted, however, that this is mainly due to the unavailability of time-series data, but the task is also hindered by the complexity of biological networks [52]. In addition to short time-series data, most reverse engineering algorithms have been tested on small networks only. For artificial data sets, this usually consists of up to 30 genes and very much less for experimental data sets. In Ref. [53], the authors conducted an extensive comparison of inference algorithms and observed that for synthetic data performance was good up to networks of 30 genes, and for real data sets 24 genes are considered a large network.

A recent paper by Lee et al. [54], however, infers networks based on 31 time points, which is considerably larger than most data sets. In the same paper, the authors infer a large-scale network of up to 125 genes, much larger than most studies. Lee et al. demonstrate good reproduction of the data over the profiles; however, the data used are artificially generated and the optimization took almost 10 h for the largest network when computed in parallel over 25 CPUs. This does mark an impressive leap forward in terms of large-scale reconstruction, particularly, as the data, although artificially generated, are not continuous as in the majority of studies. Nevertheless, progress is still required as higher level organisms contain thousands of genes. At an increase rate of computation time in Ref. [54], a network of 1000 genes would require around 175 h of computation time when inferred in parallel across 25 CPUs. With such high computational costs, more investigation is essential for large-scale inference to improve the performance and efficiency of reconstruction algorithms.

7.1.3 The Life Cycle of *Streptomyces coelicolor*

Streptomyces are soil dwelling gram-positive bacteria that have genomes of nearly 8000 protein-coding genes [10]. An important aspect of these bacteria, other than their unusually large genome, is their high *G+C* content, of 72.4%, and their multi-cellular life cycle [10]. The growth of *Streptomyces* progresses from a vegetative mycelium (normal growth) to the formation of aerial mycelium (tall structures) and eventually releasing exospores in the propagative phase. Vegetative mycelium is growth related and involves all the normal cellular activities associated with cell growth and division. *Streptomyces* undergo a morphological switch from primary metabolism to secondary metabolism, which is non-growth linked and is non-essential but many important activities occur during this phase that help the bacterium to survive. After this morphological change, *Streptomyces* produce "secondary metabolites" including a range of antibiotics [55–58] that are the majority of those used in human and veterinary medicines [59, 60]. Furthermore, *Streptomyces* also produce many other medicinal compounds including anti-parasitic agents, anticancer and anti-tumor drugs, anti-fungals, anti-hypertensive, herbicides, and immunosuppressants [11,61].

Depletion of environmental nutrients is widely believed to contribute to the production of antibiotics as the onset of secondary metabolites coincides with the morphological differentiation from mycelium growth to developing aerial hyphae, indicating that these two processes are governed by overlapping regulatory networks [55,60,62–64]. The onset of secondary metabolism in *Streptomyces* is due to phosphate starvation as illustrated in Figure 7.2. Here, we can see that a decrease in phosphate (purple) coincides with an increase in biomass (brown) resulting in the formation of aerial mycelium between 20 and 35 h. The point at which *Streptomyces* begin producing secondary metabolites is generally referred to as the *stationary phase* [55] because the biomass remains constant. This begins at 35 h as shown in Figure. 7.2 and is the point where antibiotics begin to be produced (thick red, blue, and dark blue lines). The switch to secondary metabolism and the production of antibiotics is a tightly

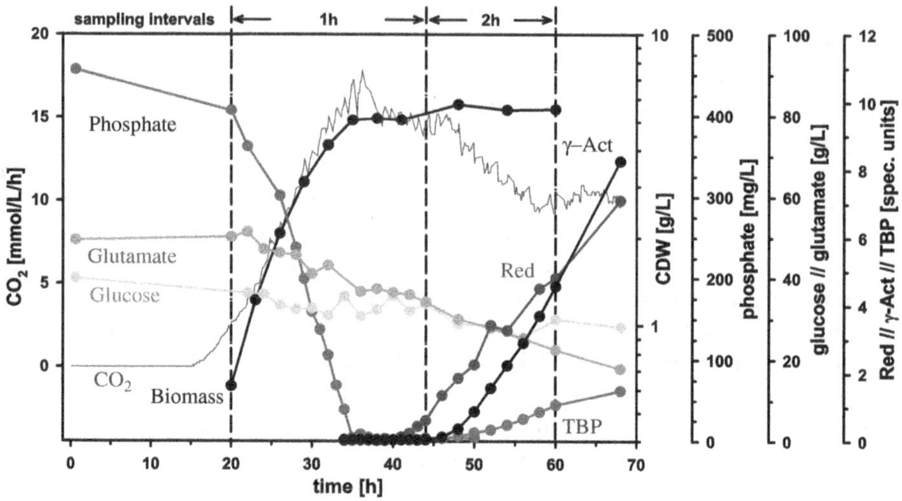

Figure 7.2 Dependence of antibiotic production on phosphate levels. The figure is reproduced with the addition of curve labels from Ref. [55] © 2010 Nieselt et al. under the Creative Commons Attribution License (http://creativecommons.org/licenses/by/2.0). Each line on the graph has a label of the same color that indicates the measure variable. Antibiotics produced here are undecylprodigiosin (Red), actinorhodin (γ-Act), and total blue pigment (TBP) are labeled on the right side of the figure.

regulated process triggered by nutrition depletion [65] and is designed to kill off competing microorganisms [60]. The bacteria themselves have co-evolved systems to aid resistance to the antibiotics that they produce [59]; however, they can also be lethal to their producers [65]. The production of antibiotics provides protection for the bacteria during phosphate starvation, so it can harvest phosphate from within the bacteria [60] such as from cellular material [66] and the cell walls [67].

7.1.4 The PhoP Sub-Network

Phosphate starvation leads to the onset of the production of antibiotics to provide protection against competing organisms so that spores can be produced from recycled mycelial components [60]. The gene responsible for phosphate harvesting when environmental sources are depleted is *phoP* (SCO4230), which acts as part of a two-component stress response system with *phoR* (SCO4229) [58]. These genes are part of the pho regulon, which is switched on in the event of phosphate starvation [60]. The *phoRP* system is present in other organisms and has been studied in *Bacillus subtilis* and *E. coli* [68]. Recent experiments conducted by the SysMO consortium [69] have provided the most extensive time series for *S. coelicolor* currently available. This experiment measured the genome-wide expression for three replicates of the wild-type strain and a *phoP* mutant strain, which all vary in the number of time points. The three replicate experiments consist of 32, 8, and 16 time points and a *phoP*-mutant strain (where *phoP* has been removed from the genome) containing 36 time points. In addition to this extensive time-series data set, which focuses on *phoP*, another experiment [66] has provided ChIP–chip data for *S. coelicolor* that can be used with the time-series data to model the PhoP (the protein produced by the *phoP* gene) network.

Although secondary metabolism can be controlled by the *phoRP* system, it is unclear if this is direct or indirect control as the mutant strains are unable to assimilate low phosphate levels to maintain growth [60]. Previously, the pho regulon in *B. subtilis* has been artificially stimulated without phosphate starvation by overproducing *phoP* in the absence of functional *phoR* [70]. If conducted in *Streptomyces*, this may be able to determine the *phoRP* systems' control over secondary metabolism [60]. The extent of *phoP* control goes beyond phosphate as it directly represses the nitrogen metabolism regulator GlnR, in addition to binding directly to its promoter, some of its targets, and other nitrogen metabolism genes [56]. The repression of nitrogen assimilating genes by the PhoP system during phosphate starvation is due to the slowing of growth reducing the need for nitrogen; therefore, expression of these genes is wasteful [56]. In addition to controlling the metabolism of nitrogen, PhoP also inhibits genes that assimilate ammonium for primary metabolism [71] and post-translational regulator genes [72] through GlnR [56]. The complexity of the *phoRP* system is furthered by its two-way cross regulation between it and the *afsKRS* system, activation of which is thought to be blocked by PhoP that represses the promoter of *afsS* [57]. The *afsKRS* system is known to influence antibiotic production and an overexpression of *afsR* and *afsB* leads to an increase in production of some antibiotics [73, 74]. Because of the number of the PhoP binding sites [66, Suppl. Table 1] and complexity of its interactions with genes, little is known about the extent of PhoP's control over genes associated with secondary metabolism and antibiotic production [60]. However, it is known that the phosphate control system is not a simple regulatory circuit and is part of a complex network, with strong links with other stress responses and bacterial virulence [68].

7.1.5 Computational Approach

Because of its obvious medical implications, the link between PhoP and antibiotic production is an important problem in biological modeling, and has been studied for several decades. Many of the recent investigations are collaborations between biologists, mathematicians, and computer scientists as a large number of genes involved, combined with limited and noisy data, make the modeling and reconstruction of these networks non-trivial. Currently, there are many methods to reverse engineer networks from biological data, such as Bayesian and statistical techniques [75–77], and informatics and correlation techniques [78–80]. These methods, however, struggle when the problem is underdetermined [4, 26, 38] where there are insufficient data to statistically infer the network. For the PhoP network, there are 387 direct regulatory targets [66] and it would require much more data than are financially feasible at present for statistical-based inference methods.

Reverse engineering networks from experimentally captured data using computational techniques are common places [36–40] and can provide insight into network structures, regulation type, as well as predict regulatory targets. These predictions can be experimentally verified and compared to models that can be updated with new data sets. Modeling biological systems through all their significant interacts would result in a high level of complexity [1]; thus, often simplifications are made. When making assumptions and simplifications to modeling biological processes, even for a small network, reconstruction can be computationally expensive due to the complexity of interactions, as indicated in Figure 7.1. The computational cost of network inference increases significantly with network size, which is demonstrated in Figure 7 of Ref. [75]. Because of these high computational costs, some investigations are using data integration methods [39, 45, 81–86] and utilizing mathematical and computational techniques such as decoupling [87, 88] and parallel computation [54, 89] to improve efficiencies and runtimes.

Here, we investigate several computational configurations for modeling the dynamic gene expression of networks of increasing size within the PhoP sub-network using the SysMO data set that contains 32 time points for every gene [69]. We investigate various optimization-based modeling methods to reproduce the dynamic gene expression profiles of the genes in these networks. Here we compare a single objective setup (SOS) for full network optimization with a comparable multi-objective setup (MOS), and also investigate these methods in a decoupled optimization arrangement. Furthermore, we conduct the above configurations for unprocessed gene expression (raw) data, which vary in expression level, and for normalized gene expression data, where all profiles are on the same range. The PhoP network combines the challenges of reconstruction from time-series data with the computation difficulties of large-scale network inference and additional complexity from experimental (as opposed to artificial) data.

7.2 REGULATORY NETWORKS AND GENE EXPRESSION DATA

Our method uses real biological data from microarray experiments, which give genome-wide expression profiles for all genes in *S. coelicolor*. Using these data we are able to build a model of the gene regulations that can reproduce the experimental expression profiles and predict regulatory interactions, that is, activating or repressive. Here we use the time-series data from the SysMO consortium [69] containing 32 time points for each gene in varying time intervals from 20 to 60 h with phosphate starvation occurring at 35 h. This is a much larger time-series data set compared to many data sets available [90] and, thus, can show trends in gene expression giving more insight into the regulatory interaction than steady-state data.

7.2.1 Bacterial Sub-Networks

Here we investigate several networks of various sizes targeted by PhoP involved in pyrimidine production (PyrR), antibiotic synthesis (RedD and CdaR), or formation of the aerial mycelium and sporulation (WhiB). In addition to these functional sub-networks, we also model the 55 of the 387 direct targets of PhoP that encode deoxyribonucleic acid (DNA)-binding proteins as well as the other 332 direct targets identified in Ref. [66]. These larger sub-networks are involved in a number of different biological functions, and are considered as we are interested in the number of genes rather than any biological processes. For the largest gene sub-set, we model 100, 200, and all 332 genes as separate sub-networks in order to gain more information regarding the performance with increasing network size. The sub-sets of 100 and 200 genes are derived from the full set of 332 by selecting genes in sequential order of SCO-numbers (identification numbers). In this study, we model eight sub-networks varying in size from 7 to 332 genes. Note that names in italics refer to gene names, otherwise they refer to the protein produced by that gene and its regulatory network.

PyrR Network Although PyrR has no links with antibiotic production, it is a direct target of PhoP [66] and is a simple sub-network containing only eight genes. In addition, PyrR is one of the most variable transcriptional regulators in the PhoP network [55]. PyrR has seven direct targets, all of which cannot regulate other genes on the gene level, and any regulatory interactions from these genes are beyond the scope of this model. The PyrR network contains *pyrR* (SCO1488) and targets SCO1481-87, which are involved in pyrimidine production [96].

WhiB Network Unlike the other sub-networks, the WhiB sub-network contains genes of non-sequential SCO-numbers that are summarized in Table 7.1. These genes are involved in sporulation from the aerial mycelium [96]. Deletion of some of these genes (*whiA,-B,-G-H*) in *Streptomyces* leads to the development of aerial mycelium but failure to produce spores as they prevent subdivision of the hyphae into compartments that become spores and development of spore walls [91, 92]. Despite the fact that they can regulate on the gene level, the transcription of WhiB is not severely dependent on *whiG* or *whiH* [93]; therefore, to simplify the model we do not consider feedback loops in this network, that is, all connections are from *whiB* only.

Table 7.1 WhiB sub-network genes identified from Ref. [94]

SCO No.	Gene name	Function	References
3034	*whiB*	Sporulation regulatory protein	[96]
4543	*whiJ*	Hypothetical protein	[94, 96]
4767	*whiD*	Putative regulatory protein	[96]
5621	*whiG*	RNA polymerase σ factor	[96]
6029	*whiI*	Two-component regulator	[94, 96]
1950	*whiA*	Related to sporulation septation	[96]
5315	*whiE*[a]	Polyketide cyclase	[96]
5316	*whiE*[a]	Acyl carrier protein	[94, 96]
5317	*whiE*[a]	Polyketide β-ketoacyl synthase β	[94, 96]
5318	*whiE*[a]	Polyketide β-ketoacyl synthase α	[94, 96]
5321	*whiE*[a]	Polyketide hydroxylase	[94, 96]
5819	*whiH*	Sporulation transcription factor	[96]

WhiB sub-network genes and their functions.

[a] *whiE* is a locus of eight genes [95].

RedD Network The RedD cluster is responsible for producing the antibiotic *undecylprodigiosin*, known as red due to its color. This gene has been identified as an indirect target of PhoP through differential gene expression analysis in Ref. [66]. This is similar to PyrR in that it is a small sub-network of 14 genes where only 1 gene, *redD*, encodes DNA-binding proteins. Therefore, as with PyrR, we have one gene that can regulate the cluster on the gene level an there is no feedback. This sub-network is slightly larger than the PyrR network and consists of greater levels of nonlinearity in the gene expression profiles. The Red Cluster consists of genes SCO5877 (*redD*), SCO5878, and SCO5886-98 [55, 66].

CdaR Network Calcium-dependent antibiotics are produced from a cluster of genes (SCO3210-3249) where only two genes, *cdaR* (SCO3217) and *absA2* (SCO3225), produce putative DNA-binding regulatory proteins, that is, act as regulator genes [66]. The latter gene, *absA2,* is part of a two-component response regulatory system, where AbsA1 represses the phosphorylation of AbsA2 [97] and it is the phos-AbsA2 that is believed to regulate *cdaR*, though it is not known if this is direct or indirect [98]. The method we use here only considers interactions at the gene level; therefore, we do not model the regulation from the phosphorylated AbsA2. For the CdaR sub-network mirrors the others with 1 regulatory gene and 40 target genes making this sub-network considerably larger than the others.

Direct PhoP Targets For both the 55 and 332 sub-sets of the PhoP direct targets we model only connections from PhoP. The former group (55) is able to encode gene regulation via proteins [66]; however, this is omitted for consistency with the other sub-networks. Therefore, we have no feedback loops and outward connections from the regulator gene *phoP* only. The function of these genes ranges across many biological processes as PhoP is a global regulator. These sub-sets were selected due to their size unlike the others, which are selected based their functionality. In these larger sub-set, there may be an overlap with other smaller networks, that is, PyrR, WhiB, and CdaR are all in the group of 55. The larger sub-set of 332 genes is also sub-divided to give a larger range of network sizes. Here we model three variations of this network, the first 100 genes, the first 200 genes (in order of ascending SCO numbers), and the full 332 genes in the sub-network. This will provide a greater understanding of the effect of network size on the performance of the inference process. Due to their ability to regulate other genes, the subset of 55 are referred to as 'Master' genes, conversely, the subset of 332 can not regulate others and are referred to as 'Slave' genes.

7.2.2 Data Normalization

The level of gene expression can vary in scale from gene to gene as demonstrated in Figure 7.3. When modeling networks of many genes this can require large range for the parameters in the connection equations to fit the expression profiles. For an optimization problem this can lead to enormous search spaces for all genes, which may result in sub-optimal solutions. In order to examine this potential problem we

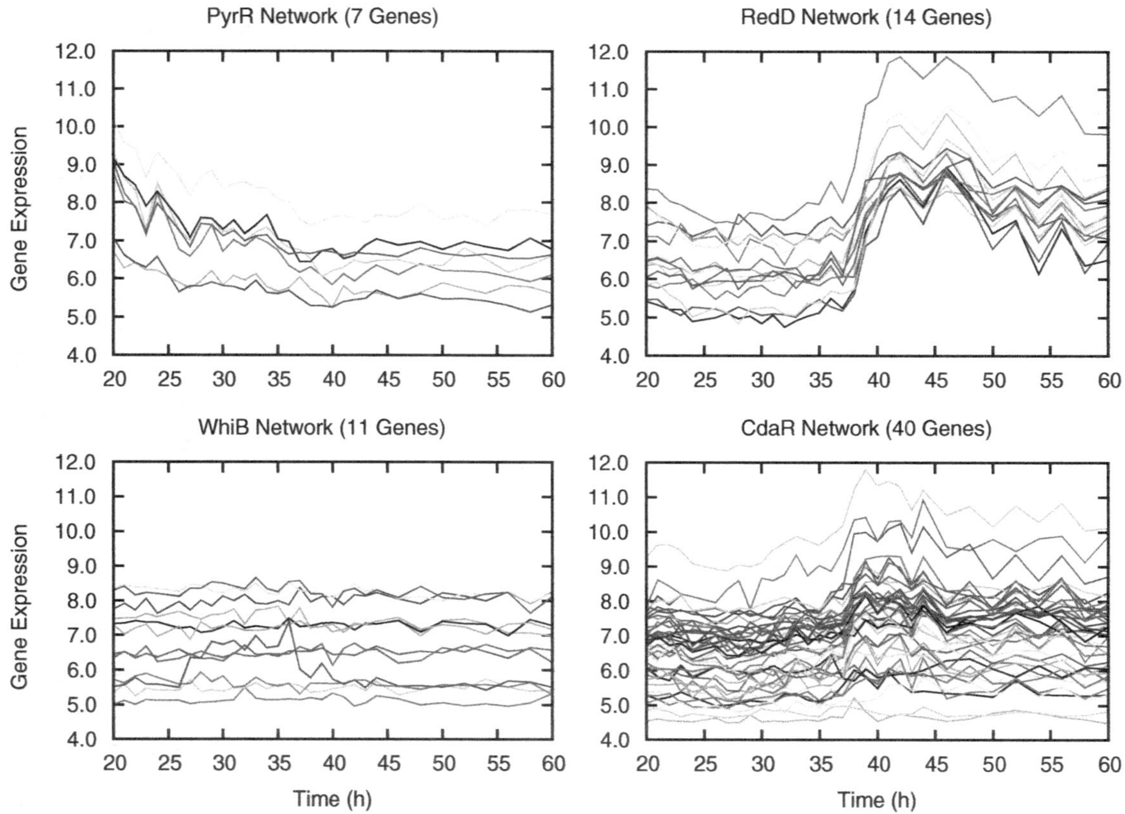

Figure 7.3 Unprocessed (raw) gene expression for the PyrR, WhiB, RedD, and CdaR networks [69]. The number of regulatory targets in each network is given in parentheses.

investigate the use of unprocessed "raw" gene expression data compared to gene expression levels that are normalized. Note that by unprocessed here we mean the data published in Ref. [69]; data processing and normalization in this work refers to our setup of the data and has no reference to the experimentation and analysis conducted by Lamarche et al. [69]. Here we normalize the data so that the area of each expression profile is unity:

$$\int \tilde{x}(t)dt = 1, \tag{7.1}$$

thus all expression profiles are on the same scale and a universal parameter range exists for all connections. This allows comparisons between parameters across the genes, for example, comparing regulation strength of one gene with another. To normalize the expression profiles, each of the gene expression measurements is divided by the area of the expression profile,

$$\tilde{x}(t) = \frac{x(t)}{N}, \tag{7.2}$$

where

$$N = \int_{t_0}^{t_f} x(t)dt. \tag{7.3}$$

This leads to all gene expression profiles within the sub-network, being on the same scale as illustrated in Figure 7.4. By providing a universal parameter range for all genes we have localized the search space compared to the raw data method.

7.3 OPTIMIZATION USING EVOLUTIONARY ALGORITHMS

As biological networks are often large [26], particularly for more complex organisms, sophisticated reconstruction techniques are required. Because of their success in a large range of optimization problems, evolutionary algorithms (EAs) have been used to reconstruct GRNs from time-series data [25, 53, 75, 99]. In addition, EAs have the advantage of not requiring detailed prior knowledge of the system, but also have the flexibility to add biological information during inference [53]. As optimization algorithms require only an evaluation of candidate solutions, they are a powerful tool for modeling complex problems in biology [100]. In order to fully reconstruct a GRN, one must identify both the topology and the parameterization of the connections. This results in a vast search space even for small networks, which can contain a complex structure and set of dynamics [6]. Their ability to deal with very complex and underdetermined problems and their high levels of flexibility make optimization algorithms an attractive method for network inference [38, 101].

Evolutionary algorithms have been widely used to evolve the parameters and structure of models [22, 53, 102, 103] including GRNs [13, 15, 19] and have been

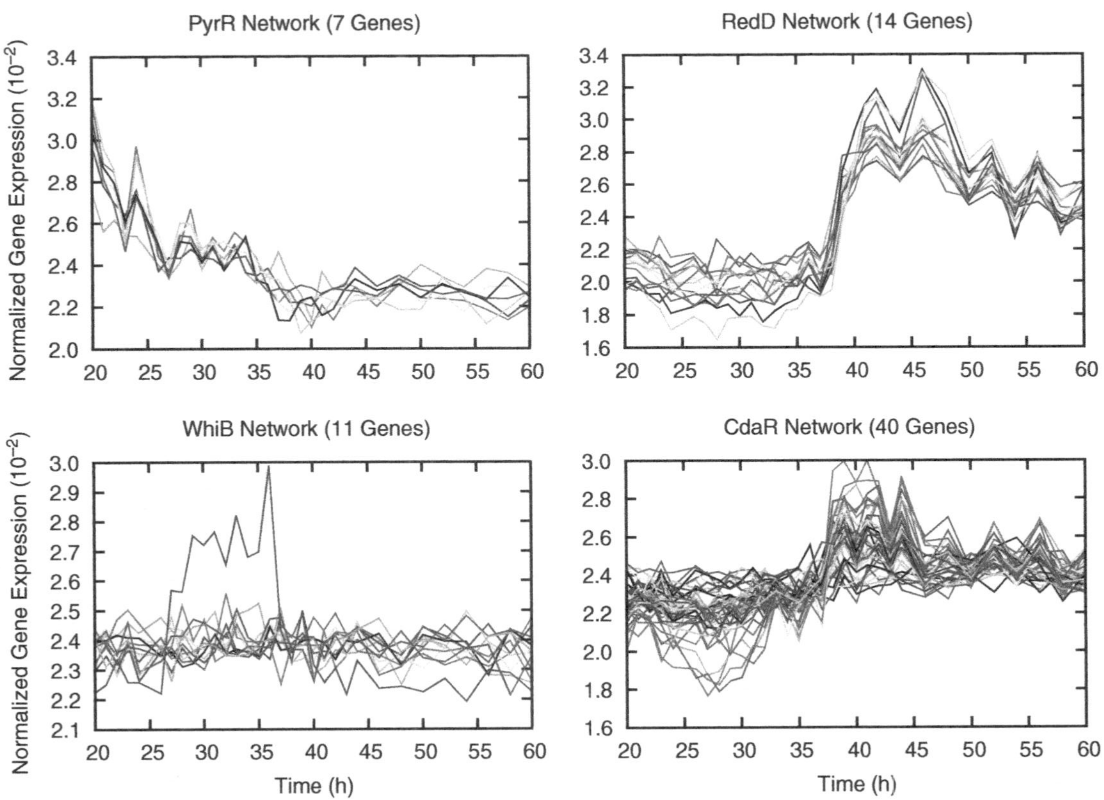

Figure 7.4 Normalized gene expression for the PyrR, WhiB, RedD, and CdaR network, raw data from Ref. [69]. The number of regulatory targets in each network is given in parentheses.

used by the authors previously for the PyrR network in *S. ceolicolor* [104]. Here, we use non-dominant sorting genetic algorithm (NSGA)-II [105] for our optimization as it is one of the most widely used multi-objective optimization algorithms [106]. Additionally, NSGA-II can be setup for a single objective problem [13] providing consistency between our single objective and multi-objective representations of the problem. This real-coded multi-objective genetic algorithm is able to handle continuous optimization problems through the use of simulated binary cross-over and polynomial mutation [107]. It is also an elitist algorithm that is suited to network reconstruction as it retains good solutions and has been demonstrated to aid convergence for continuous optimization problems [13, 15, 102, 105, 108, 109].

7.4 MODELING GENE EXPRESSION

One common method of modeling GRNs dynamics is to use differential equations [75, 110] as they can provide detailed temporal information about the system. A common representation of gene regulation is through the use of Hill functions [38, 111], which are nonlinear equations that are derived from Michaelis–Menten enzymatic kinetics [21, Appendix A]. Hill functions modeling the regulation of gene x_i by gene x_j can be activating, where protein production is promoted in the target gene, or repressive, where production is suppressed. The mathematical form for activating and repressive Hill functions is given by

$$H_{ij}^{a}(x_j) = \frac{(x_j)^{n_{ij}}}{(\theta_{ij})^{n_{ij}} + (x_j)^{n_{ij}}}, \qquad H_{ij}^{r}(x_j) = \frac{1}{1 + (x_j/\theta_{ij})^{n_{ij}}}, \tag{7.4}$$

respectively. Here θ is the reaction threshold and n is the Hill coefficient that determines how steep the interaction curve is. The state of gene x_i, regulated by gene x_j, at time t is then given by

$$\dot{x}_i = \omega_{ij} H_{ij}(x_j) - \gamma_i x_i, \tag{7.5}$$

where $\dot{x}$ is the time derivative of x, ω is the regulation weight, and γ is the rate of degradation of the protein produced by x. We can use an optimization algorithm to determine the parameters in Equations (7.4) and (7.5) based on gene expression time series to determine the connection parameters and characteristics such as the regulation type.

7.4.1 Single Objective Setup

In order to optimize the network parameters based on the state of the genes, we define an objective function as an error minimization problem. Here we calculate the sum squared error (SSE) between the model prediction and the experimental gene expression profile. As our objective function is to be minimized, the EA will vary the network parameters to provide the best fit to the expression profiles. The SSE is calculated over all N time points for all M genes and is rewritten, for convenience, as

$$F = \sum_{i=0}^{M} \sum_{t=0}^{N} \left(g_i(t) - x_i(t)\right)^2, \tag{7.6}$$

where $g_i(t)$ and $x_i(t)$ are the values of the experimental gene expression level and the model gene expression level, respectively. The SSE given in Equation (7.6) determines the fit of the entire network to the expression data of all genes. We use Equation (7.5) to model all connections. Here, we have four free regulation parameters and one Boolean value to determine the regulation type, giving a total of five parameters for each connection in the network. The regulation parameter is used to determine if the connection is activating or repressive, H^a_{ij} or H^r_{ij} in Equation (7.4). This is a common objective for biological network reconstruction [75, 77, 112] and can be applied to both real and Boolean networks [26].

7.4.2 Multi-Objective Setup

In Section 7.4.1 we detailed a single objective optimization representation of the problem of reverse engineering regulatory networks from expression data. However, the use of multiple objectives can provide a better exploration of the parameter space in GRN reconstruction [113], which is vital for large-scale inference problems. Additionally, the benefits of multiobjectivization [114], where a single objective problem is represented with multiple objectives, may enhance the optimization process. This is achieved either by using an additional objective to guide the search though the parameter space or by decomposing the original objective into multiple objectives that are sub-sets of the original. Therefore, this representation does not necessarily yield conflicting objectives unlike multi-objective problems in general. This MOS may solve the problem more efficiently than the single objective configuration. A local optimal solution in the single objective setup corresponds to one of the two objectives in the MOS, and as it is a local optimum; by definition, the other objective value must be worse. This solution will therefore be located in the top right region of the objective space and, as we have a minimization problem, correspond to a poor candidate solution. Therefore, this solution will be dominated by other solutions during the non-dominated sorting in NSGA-II, and is unlikely to remain in the population over successive generations. The MOS may be less prone to local optima than the single objective equivalent based on this feature. The ability to escape local optima becomes increasingly important as the number of genes, and problem dimension, increases. Based on this, here we also investigate several MOSs to compare with the typical SOS commonly used.

As in the single objective case, we optimize the four network parameters, however we define two separate, but not necessarily conflicting, objectives to be optimized simultaneously. Both objectives are to minimize Equation (7.6) using Equation (7.5); however, the first objective models all connections using an activating regulation and the second objective models them all as repressive regulations, Equation (7.4). Thus, we have a multi-objective problem with the following objectives:

$$F_1 = \sum_{i}^{M} f_i^a \quad , \quad f_i^a = \sum_{t}^{N} \left[g_i(t) - x_i^a(t) \right]^2 , \tag{7.7}$$

$$F_2 = \sum_{i}^{M} f_i^r \quad , \quad f_i^r = \sum_{t}^{N} \left[g_i(t) - x_i^r(t) \right]^2 , \tag{7.8}$$

where x_i^a and x_i^r are the connections to gene x_i using H_{ij}^a and H_{ij}^r, respectively, in Equation (7.4). Here, the SSE per gene is summed for M genes in the network for only activating regulations, F_1, and only repressive regulations, F_2. The SSE for a gene in the network is given as the sum of the squared difference between the gene expression profile $g(t)$ and the model $x(t)$ over N time points. In this setup, a given set of network parameters, ω, θ, n, and γ, are evaluated using both objectives (7.7) and (7.8) producing a Pareto front of possible solutions for the problem; see Figure 7.5. In general, for multi-objective optimization problems the objectives are conflicting resulting in tradeoff solutions that cannot be compared without user preference or system constraints [115]. In this case, however, the problem is actually a single objective problem with the goal to minimize the SSE between the model and the data for all network genes. Therefore, we can select the best solution from the Pareto front based on the objective values and do not have to add additional preferences or constraints. More specifically, our "preference" for best fit to the data for each gene

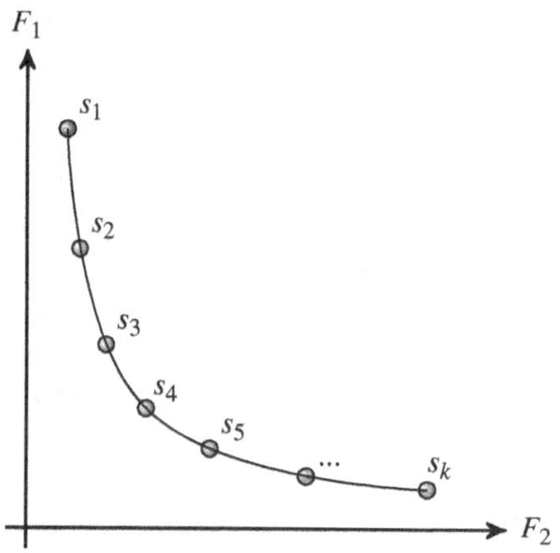

$$s_1 = \begin{cases} {}^1F_1 = {}^1f_1^a + {}^1f_2^a + \cdots + {}^1f_M^a \\ {}^1F_2 = {}^1f_1^r + {}^1f_2^r + \cdots + {}^1f_M^r \end{cases}$$

$$s_2 = \begin{cases} {}^2F_1 = {}^2f_1^a + {}^2f_2^a + \cdots + {}^2f_M^a \\ {}^2F_2 = {}^2f_1^r + {}^2f_2^r + \cdots + {}^2f_M^r \end{cases}$$

$$\cdots\cdots\cdots\cdots\cdots\cdots$$

$$s_k = \begin{cases} {}^kF_1 = {}^kf_1^a + {}^kf_2^a + \cdots + {}^kf_M^a \\ {}^kF_2 = {}^kf_1^r + {}^kf_2^r + \cdots + {}^kf_M^r \end{cases}$$

$$s^b = \sum_{i=1}^{M} f_i^b \; ; \; f_i^b = \min \left\{ \min \{ f_i^a \} ; \min \{ f_i^r \} \right\}$$

(a) (b)

Figure 7.5 (a) The Pareto front of non-dominated solutions to the multi-objective problem in Equations (7.7) and (7.8). Illustrated here is a front of k solutions for a network of M genes. (b) Pareto solutions and the corresponding sum squared errors for individual genes (see the text for details). The overall best solution, s^b, is obtained by selected the lowest f value for each gene in turn yielding, in general, a network of mixed regulation types.

is implicit in this setup; thus, solutions can be selected from the front automatically. By arranging this problem as a multi-objective problem, we do not have to optimize the regulation parameter in the system directly. The final network produced by this method is not restricted to one regulation type and can be a mix of regulations.

To select the best solution, we search the Pareto front to obtain the best parameter values for each connection. As defined in Equations (7.7) and (7.8), the two objectives are determined though the sum of the SSE for each gene in the network. Thus, by retaining the individual gene SSEs, we are able to search the Pareto front to obtain the solution (parameter set) that corresponds to the lowest SSE for each gene. Each solution on the Pareto front has two objective values, each of which corresponds to summed sequence of individual gene SSEs (produced from a specific parameter set), Equations (7.7) and (7.8). Once the optimization process has finished, for each gene, the Pareto front is searched for the parameter set that yields to best objective. Figure 7.5(a) illustrates the Pareto front and Figure 7.5(b) details how each solution corresponds to a summation of individual SSEs for each connection. For each gene, in turn, the front is searched to obtain the solution corresponding to the minimum SSE from either F_1 (activating connection, f_i^a) or F_2 (repressive connection, f_i^r). That is for each gene we evaluate $f_i^a < f_i^r$ and select the minimum f_i that indirectly determines the regulation type. The result is a network containing mixed regulation types that are each selected based on the lowest SSE.

This method has the potential to conduct a better search of the parameter space through its representation as a multi-objective problem. The overhead of searching the Pareto front, performed after the optimization, even for a large number of individuals and genes, is negligible compared to the optimization time. This multi-objective optimization process removes the need for the direct optimization of the regulation type and, therefore, reduces the number of parameters in the problem. Reducing the number of parameters from five to four for each of the n connections results in a large reduction in the search space and could therefore enhance convergence to the global optimum. Here we use 100 individuals and 1000 generations in the optimization stage.

7.4.3 Decoupled Approach

Biological systems are known to be sparse [30] therefore we assume that the connections within the network are independent and treat them as separate, a process known as decoupling [87,88]. This reduces the dimensionality of the single objective optimization problem from one $5n$ problem, where n is the number of connections, to n problems each with a dimension of five in the SOS. Here the number five corresponds to the four network parameters and the one Boolean regulation type parameter. Similarly for the MOS, a $4n$ problem becomes n lots of four dimension problems. Furthermore, due to the decoupling, this technique can also benefit from the use of parallel computation, where each connection can be optimized on a separate CPU. This process can significantly reducing the computational runtime for large networks, though is not implemented here. The objective function here remains the same as for the full network optimization case with the exception of the summation over the number of genes. In this setup Equation (7.6) becomes

$$f_i = \sum_{t=0}^{N} (g_i(t) - x_i(t))^2. \tag{7.9}$$

Here the total network SSE, F_{dec}, is calculated after the optimization process is completed by summing the individual connections' objective value from Equation (7.9),

$$F_{\text{dec}} = \sum_{i=0}^{M} f_i, \tag{7.10}$$

for all connections.

In this single objective case $x_i(t)$ is based on one of the equations in Equation (7.4) depending on the connection parameter determined by the optimizer as in the full network case. For the multi-objective optimization setup, however, the decoupled method differs slightly. In the multi-objective decoupled arrangement, each connection is optimized in a similar way to the full network case:

$$f_i^a = \sum_{t}^{N} \left(g_i(t) - x_i^a(t)\right)^2 \tag{7.11}$$

$$f_i^r = \sum_{t}^{N} \left(g_i(t) - x_i^r(t)\right)^2. \tag{7.12}$$

However, the selection of solution from the Pareto front is much simpler. In this method, we have a Pareto front for each connection due to the problem being decoupled. Therefore, as we have a minimization problem, the lowest objective value for each connection is always one of the extreme solutions. This is clearly illustrated in Figure 7.6. For each connection we simply evaluate $f_i^a < f_i^r$ to obtain the optimal parameter set and regulation type, see Figure 7.6. Here we use 100 individuals and 100 generations as this setup converges much faster than the full network setup thus requiring fewer generations in the optimization.

7.5 RESULTS

7.5.1 Comparing Objectives from Un-Normalized Data

For the case of the un-normalized (raw) data, each of the gene expression profiles within a single network potentially lies on a different scale; thus, determining a network error is not so straightforward. Comparing the SSE between two genes within the same network may be meaningless if the expression profiles vary in scale,

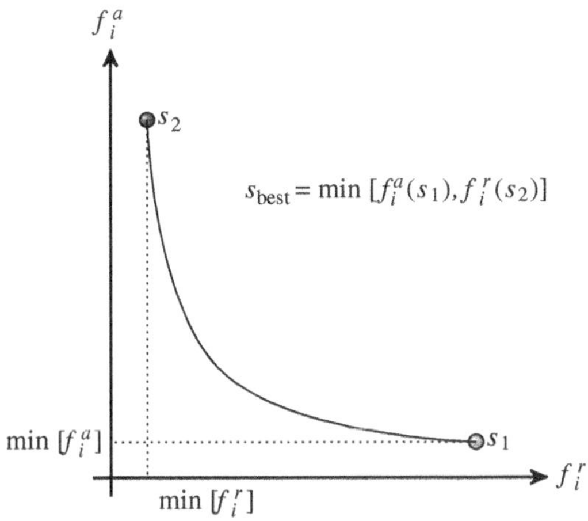

Figure 7.6 Pareto front of non-dominated solutions to the multi-objective problem, Equations (7.11) and (7.12), with a decoupled arrangement. Solution s_1 corresponds to the best activating connection, and s_2 corresponds to the best repressive solution, thus evaluating the minimum of these two yields the optimal parameters for each connection.

with larger scale expression values tending to have larger SSEs. Therefore, if we take the ratio of SSE (δx) and the area of the expression profile ($\int_{t_i}^{t_f} x(t)dt$) for each gene, we are left with a dimensionless quantity, $\delta x_i'$, that has removed the difference in expression scale:

$$\delta x_i' = \frac{\delta x_i}{\int_{t_i}^{t_f} x(t)dt}. \tag{7.13}$$

For each gene, the dimensionless SSEs ($\delta x_i'$) can be summed to give the total error for the network, ΔX:

$$\Delta X = \sum_{i=1}^{M} \delta x'_i, \tag{7.14}$$

where i is the ith network gene and M is the total number of genes within the network. This normalization of the objective functions not only provides a meaningful network error measurement, but it also enables comparison between the different networks and arrangements. We can now compare the raw data setup with the normalized data arrangement directly. No further processing is required for the normalized data case, as each expression profile has the same area; see Equation (7.1). Therefore, Equation (7.13) reduces to $\delta x_i' = \delta x_i$ and the total network error would remain the sum of the errors in the individual genes.

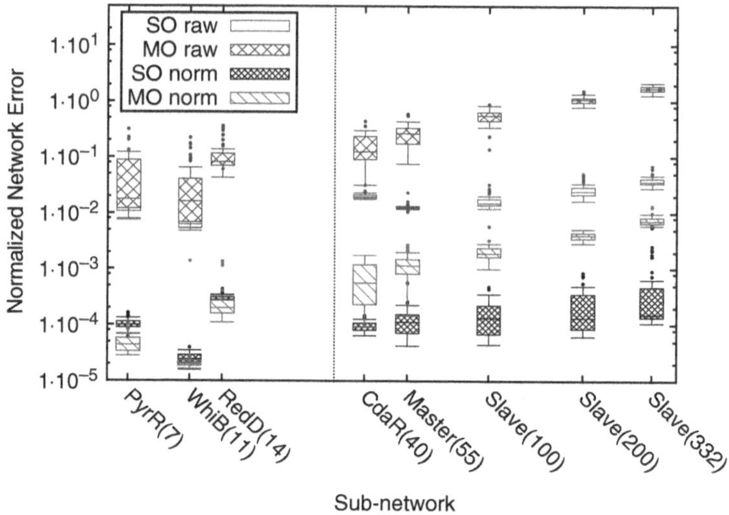

Figure 7.7 Box plots for the sub-networks for the single and multi-objective setups for both normalized and un-normalized (raw) gene expression data. Box plots are the distribution of sum squared error over 50 independent simulations of each setup. Labels on the *x*-axis give the sub-network name and the number of target genes within the network. Outliers are given as circles and averages are median values. The dashed line indicates when the problem becomes underdetermined, that is, when the number of genes exceeds the number of data points. SO, single objective; MO, multi-objective.

7.5.2 Full Network Optimization

Figure 7.7 shows the box plots of the total network SSE for 50 randomly initialized optimizations of each of the networks. All raw data here are post-processed using Equation (7.14), thus enabling a direct comparison between raw and normalized setup performances.

Raw versus Normalized Data For the SOS, the use of normalized data yields much lower network errors compared with the raw data. The difference is two orders of magnitude across the entire range of network sizes. This illustrates the benefits of using normalized data as it has produced much better objective values. The SOS normalized distributions are bigger compared to the raw SOS cases for the larger networks; however, the difference between the distributions is still large. For the MOS, the same outcome is seen as for the SOS case. In addition, it also reduces the distribution range for the smaller sub-networks. Figure 7.7 clearly shows that the use of normalized data is beneficial and improves both the SOS and the MOS.

Single versus Multiple Objective Setup For the raw data, the MOS case produces a large spread of network errors over the 50 simulations indicated by the long box plots, which overlap with the SOS distributions for the three smallest sub-networks. The MOS performs worse than SOS over the whole range of network sizes and appears to diverge with increasing network size. Similar behavior is seen when using the normalized data and comparing the SOS and MOS cases. Both the raw and normalized data SOS results appear to be approximately constant with increase network size, whereas both the MOS experiments increase with network size. The MOS normalized case, however, does not exhibit the larger distributions of network errors as in the SOS case.

The results here indicate that the use of normalized data can improve performance on modeling gene expression for large-scale networks compared to raw data. This improvement is due to the localization of the search space for all genes. Localization enhances convergence of the optimizer and results in lower errors in each gene and, therefore, the whole network. These results also show that an SOS is better than an MOS for both raw and normalized data over the network sizes tested. The MOS with normalized data does outperform the SOS raw arrangement; however, trends indicate that the SOS raw arrangement would perform better than the MOS normalized case for larger networks that tested here. The normalized data set with the SOS performs the best of all the methods tested. Despite the increase in distribution of errors with increasing network size, the trend indicates that increasing network size results in a small increase in the average network error.

7.5.3 Decoupled Network Optimization

Figure 7.8 shows the results from the decoupled approach for the sub-networks modeled. The use of the decoupled methodology leads to much narrower distributions

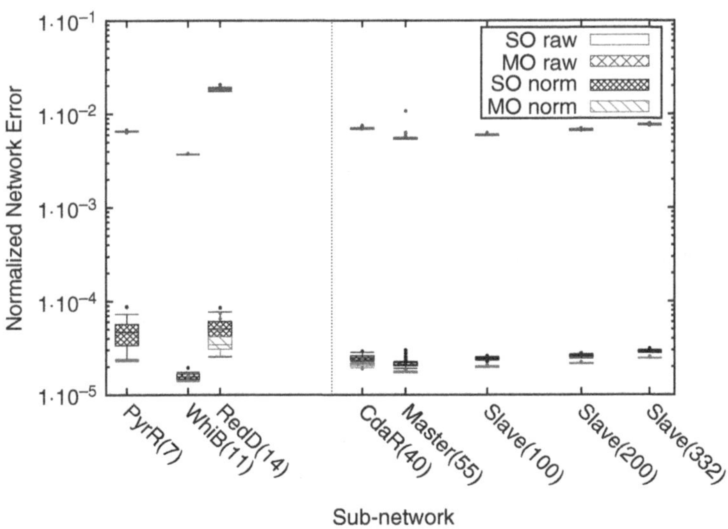

Figure 7.8 Box plots of sum squared error distributions over 50 simulations for the decoupled arrangement, plots as shown in Figure 7.7. A significant difference between the normalized and raw data is clear here with a separation between distributions of around 2.5 orders of magnitude. SO, single objective; MO; multi-objective.

compared to the full network arrangement for both raw and normalized data sets and for both SOS and MOS. In the full network case, some distributions span an order of magnitude, where the largest ones in the decoupled approach are approximately half an order of magnitude, and are the minority. Almost all distributions are so narrow that they are not fully observable in Figure 7.8.

Raw versus Normalized Data A significant difference between using raw and normalized data is evident in Figure 7.8 where distributions based on data sets are separated by approximately 2.5 orders of magnitude. The use of normalized data has led to a reduction in network error as in the full optimization case. Additionally, the large difference between the RedD network and the other sub-network has greatly been reduced, though it has also led to a slightly wider distribution (taking the scale into account) over the 50 simulations. Also for the SOS, the normalized data have also produced a wider distribution of the PyrR network (relative to the other sub-networks) compared to the raw data. However this distributions is still around 2 orders of magnitude lower than the raw data case.

Single versus Multiple Objective Setup For the raw data set, the difference between the SOS and MOS is small. For clarity, the distributions shown in Figure 7.8 have been plotted in difference scale in Figures 7.9(a) and 7.9(b) for the raw and normalized data, respectively. The largest ones for sub-networks show that the MOS results in lower total network errors than the SOS, though this difference is small; see Figure 7.9(a). Over the range of sub-network sizes, the decoupled method has removed the poor performance of the MOS and, thus, has led to comparable results to the SOS. For the normalized data, we can see that the MOS produces narrower distributions than the SOS with lower average values in all sub-networks. Moreover, the average value of the MOS distributions is lower than the minimum SOS distribution value for all sub-networks except for RedD. This indicates significant differences between these methodologies, and even in the exception RedD, the median of MOS is below the lower quartile of the SOS. Furthermore, the maximum (excluding outliers) of the MOS is approximately equal to the average of the SOS. This is shown in Figure 7.9(b). From this figure it is clear the MOS with normalized data produces the best performance as it results in the lowest total network error. Moreover, for the four largest sub-networks, the MOS produces narrower distributions that are approximately 15% lower in total network error. This appears to be a stable trend and may indicate that the MOS produces lower network errors for larger sub-networks than SOS in general.

7.6 DISCUSSION

Using benchmark synthetic data sets is useful in comparing an algorithm's performance [40, 43, 53]; however, it may not be a true representation of the algorithm's performance on real data sets. Currently, there are very few studies that use real data and models are not often based on time-series data. Topological models are useful in

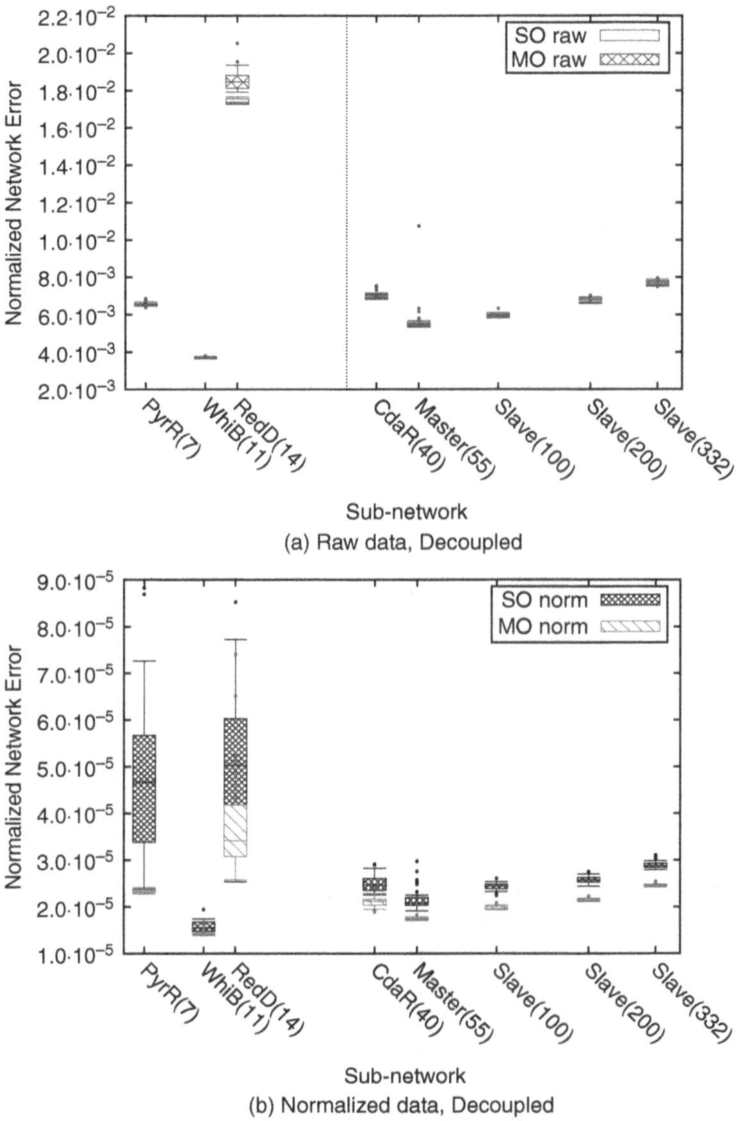

Figure 7.9 Box plots from Figure 7.8 separated for clarity: the performance of both the single objective setup and multi-objective setup (a) on the raw gene expression data and (b) on the normalized gene expression data. SO, single objective; MO; multi-objective.

identifying motifs and functionalities of networks as well as comparing similarities between different networks and organisms. However, dynamic models are vital in our understanding of gene regulation and interactions within a network. Here, we have investigated several techniques and setups for modeling dynamic GRN interactions based on gene expression data. We have investigated the effect of an MOS and observe a performance hindrance when compared to an SOS for a full network optimization technique. This hindrance has less effect when using normalized data, but is still present. When using a decoupled optimization technique, the MOS is a form of multiobjectivization and is comparable in performance to the SOS when using raw data, and performs significantly better when using normalized data. These results show that the use of multiobjectivization can enhance the optimization search through objective decomposition, whereas a multi-objective representation (that is not a form of multiobjectivization) may hinder performance.

We have demonstrated that the use of normalized data leads to reductions in network error compared with raw gene expression data. This reduction is seen in all arrangements, after correcting for gene expression scale to compare setups and network sizes. The improvement through the use of normalized data is due to the localization of the objective search space. By having a universal parameter range, due to all expression profiles being on the same scale, the optimizer can search the objective space much more efficiently than for the raw data case. For the normalized data setup, the optimal solutions for all connections in a network lie within a local region of the parameter space, whereas this is not the case when using raw data. Moreover, the parameter space when using raw data may be vast with optimal connection values far apart for different genes within a network due to expression scale. The normalization of the gene expression data is a straightforward task that can be applied to data sets prior to optimization and has a significant effect on the results as indicated here.

It is clear that in the full network cases the MOS is not providing optimal solutions to the problem. This is a consequence of the methodology, as solutions on the front represent a collection of individual gene SSEs, which are searched to provide the parameter values for each connection. The selection during the optimization is based on the lowest total SSE for a network, that is, the sum of the individual connection SSEs, yielding solutions with low SSE for all connections rather than for individuals. As a result, the optimal parameter values for each connection are not necessarily selected as the overall network SSE may be high. Although, in principle, this sounds analogous to the SOS, the difference lies in the selection during optimization, where solutions are evaluated based on their total SSE and then post-optimization individual connection parameters are selected. This problem of sub-optimality increases with sub-network size leading to a higher rate of increase in SSEs for larger networks in the MOS compared to the SOS.

The performance enhancement using the decoupled case is due to the reduction in the parameter search space. This method improved performance in terms of lower network error and reduced variance between simulations compared to the full network case. The reduced variance is evident by the narrowing of the distributions over the 50 simulations as shown in Figure 7.8 compared to that shown in Figure 7.7. Additionally, we observe a reduction in the number of outliers (simulation results greater/less than $1.5 \times$ median) over the 50 simulations. This is a result of the reduced variance of the distribution and can be attributed to the increased convergence power of the decoupled arrangement to optimal solutions. For the raw data, the SOS and MOS perform comparably over the range of network sizes; however, for the normalized data we observe a clear improvement of the MOS over the SOS. The best combination of objective setup and data type observed here is the MOS decoupled arrangement with normalized data. This produces the lowest network errors for all of the sub-networks investigated here and demonstrated high reproducibility of the results through low variance. Moreover, this methodology is significantly better than the next best combination, SOS decoupled configuration with normalized data, over all network size.

For the largest network used here (332 genes) the computational runtimes for the full network optimization are approximately 20 and 30 min for the SOS and MOS, respectively (for both the raw and normalized data). For the decoupled arrangements, the SOS runtime was approximately 15 min (for both raw and normalized data) and for the MOS, simulations took 20 and 10 min for the raw and normalized data, respectively. Therefore, the MOS method with normalized data has the shortest runtime in addition to providing the best solutions to the problem.

In this work, we have only considered networks where genes are regulated by only one gene. In general, genes may be targeted by more than one regulator that will add an additional layer of complexity to the problem by increasing the search space. However, we have shown here that this simplification can be effective for reconstructing large networks (up to 332 genes) from time-series data. As biological networks are known to be sparse [30], this is a good first approximation and may be sufficient for the majority of networks. Large networks, such as those used here, are likely to contain genes that have additional regulators, thus an extension of this method could be to add additional regulatory connections for genes that the initial method is unable to fit. This can easily be implemented after the method used here and applied to genes where the SSE between the model and data expression profiles is above a threshold. Extending the method this way would only increase the problem complexity when necessary and avoid overfitting by maintaining biological sparsity.

7.7 CONCLUSIONS

The results here suggest that normalized data enhance convergence of this optimization problem and result in consistent solutions, that is, small variance, being obtained over numerous runs. This effect is independent of objective methodology and observed in SOS and MOS for both the full and decoupled optimization arrangements. Furthermore, the use of a decoupled approach also improves consistency between optimization runs; thus, the results of each simulation can be taken with a high degree of confidence as the variance is low in the majority of cases. Additionally, we observe a reduction in outlier solutions when using normalized data. We examine all comparable methods for optimization and conclude that the best performance is obtained by using a decoupled approach with normalized expression data in conjunction with the novel multi-objective technique developed here. This configuration performs better than the single objective alternative for all network sizes, and exhibits a significant reduction in network SSE compared to the full network optimization setup. In Ref. [53], the authors noted that the application for multiobjectivization to larger networks still requires investigation. Here we have demonstrated the effectiveness of such an approach to enhance optimization across various networks that appears independent of increasing network sizes.

We have developed a novel MOS for modeling the dynamic behavior of genes in a given network and demonstrated that a decoupled arrangement multi-objective can outperform a standard single objective approach. Our novel configuration of a decoupled multi-objective optimizer setup performs comparably to a single objective case when using raw gene expression data; however, when using normalized gene expression data, it consistently performs better with increasing network size. The improved performance with increasing network size occurs from a slight reduction in dimensionality for each connection, reduced from five to four variables. Here, the small improvements over each connection lead to greater performance over larger networks. This novel setup shows little dependence on increasing network size and thus can be applied to large-scale network reconstruction problems. In addition, this methodology, in general, exhibits little variation over numerous independent simulations and thus provides consistently good solutions to this problem. Our method is able to escape local optima as, due to the representation of the problem, these solutions are dominated during the optimizations stage, see Section 7.4.2, resulting in narrow distributions of optimized solutions over a number of runs. Furthermore, this multi-objective approach has the smallest computational runtime of the methodologies tested here and can be applied to other networks and data sets.

REFERENCES

1. Mustafa Khammash. Reverse engineering: the architecture of biological networks. *BioTechniques*, 44(3):323–329, 2008.
2. Haiyuan Yu and Mark Gerstein. Genomic analysis of the hierarchical structure of regulatory networks. *Proceedings of the National Academy of Sciences of the United States of America*, 103(40):14724–14731, 2006.
3. Patrizia F. Stifanelli, Teresa M. Creanza, Roberto Anglani, Vania C. Liuzzi, Sayan Mukherjee, Francesco P. Schena, and Nicola Ancona. A comparative study of covariance selection models for the inference of gene regulatory networks. *Journal of Biomedical Informatics*, 46(5):894–904, 2013.
4. Guy Karlebach and Ron Shamir. Modelling and analysis of gene regulatory networks. *Nature Reviews. Molecular Cell Biology*, 9(10):770–780, 2008.
5. Takeshi Hase, Samik Ghosh, Ryota Yamanaka, and Hiroaki Kitano. Harnessing diversity towards the reconstructing of large scale gene regulatory networks. *PLoS Computational Biology*, 9(11):e1003361, 2013.
6. Hana El-Samad, Stephen Prajna, Antonis Papachristodoulou, John Doyle, and Mustafa Khammash. Advanced methods and algorithms for biological networks analysis. *Proceedings of the IEEE*, 94(4):832–853, 2006.
7. Mark Schena, Dari Shalon, Ronald W. Davis, and Patrick O. Brown. Quantitative monitoring of gene expression patterns with a complementary DNA microarray. *Science*, 270(5235):467–470, 1995.

8. Eberhard O. Voit and Tomas Radivoyevitch. Biochemical systems analysis of genome-wide expression data. *Bioinformatics*, 16(11):1023–1037, 2000.
9. Oksana Lukjancenko, Trudy M. Wassenaar, and David W. Ussery. Comparison of 61 sequenced *Escherichia coli* genomes. *Microbial Ecology*, 60(4):708–720, 2010.
10. Giselda Bucca, Emma Laing, Vassilis Mersinias, Nicholas Allenby, Douglas Hurd, Jolyon Holdstock, Volker Brenner, Marcus Harrison, and Colin Smith. Development and application of versatile high density microarrays for genome-wide analysis of *Streptomyces coelicolor*: characterization of the HspR regulon. *Genome Biology*, 10(1):R5, 2009.
11. Stephen D. Bentley, Keith F. Chater, Ana-M. Cerdeo-Trraga, Greg L. Challis, Nicholas R. Thompson, Keith D. James, David E. Harris, Michael A. Quail, Helen M. Kieser, David Harper, Alex Bateman, Stephanie Brown, Govind Chandra, Carton W. Chen, Mark Collins, Ann Cronin, Andrew Fraser, Arlette Goble, Juan Hidalgo, Tony Hornsby, Simon Howarth, Hsuan-Chung Huang, Tobias Kieser, Natasha L. Larke, Lee Murphy, Karen Oliver, Susan O'Neil, Ester Rabbinowitsch, Marie-Adéle Rajandream, Kim Rutherford, Simon Rutter, Kath Seeger, David Saunders, Sarah Sharp, Robert Squares, Steven Squares, Kate Taylor, Tim Warren, Andreas Wietzorrek, John Woodward, Bart G. Barrell, Julian Parkhill, and David A. Hopwood. Complete genome sequence of the model actinomycete *Streptomyces coelicolor* a3(2). *Nature*, 417(6885):141–147, 2002.
12. Thomas Schlitt and Alvis Brazma. Current approaches to gene regulatory network modelling. *BMC Bioinformatics*, 8(Suppl 6):S9, 2007.
13. Spencer Angus Thomas and Yaochu Jin. Single and multi-objective *in silico* evolution of tunable genetic oscillators. In: Robin C. Purshouse, Peter J. Fleming, Carlos M. Fonseca, Salvatore Greco, and Jane Shaw (eds.). *Evolutionary Multi-Criterion Optimization*, Volume 7811, *Lecture Notes in Computer Science*, pp. 696–709. Springer, Berlin, 2013.
14. Jesse Stricker, Scott Cookson, Matthew R. Bennett, William H. Mather, Lev S. Tsimring, and Jeff Hasty. A fast, robust and tunable synthetic gene oscillator. *Nature*, 456(7221):516–519, 2008.
15. Spencer A. Thomas and Yaochu Jin. Combining genetic oscillators and switches using evolutionary algorithms. In: Yaochu Jin (ed.). *Computational Intelligence in Bioinformatics and Computational Biology (CIBCB), 2012 IEEE Symposium on*, pp. 28–34, IEEE, New York, 2012.
16. David Angeli, James E. Ferrell, and Eduardo D. Sontag. Detection of multistability, bifurcations, and hysteresis in a large class of biological positive-feedback systems. *Proceedings of the National Academy of Sciences of the United States of America*, 101(7):1822–1827, 2004.
17. Chunguang Li, Luonan Chen, and Kazuyuki Aihara. A systems biology perspective on signal processing in genetic network motifs [life sciences]. *IEEE Signal Processing Magazine*, 24(2):136–147, 2007.
18. Didier Gonze. Coupling oscillations and switches in genetic networks. *Biosystems*, 99(1):60–69, 2010.
19. Spencer A. Thomas and Yaochu Jin. Evolving connectivity between genetic oscillators and switches using evolutionary algorithms. *Journal of Bioinformatics and Computational Biology*, 11(3):1341001-1–1341001-15, 2013.
20. Uri Alon. Network motifs: theory and experimental approaches. *Nature Reviews Genetics*, 8(6):450–461, 2007.

21. Uri Alon. *An Introduction to Systems Biology: Design Principles of Biological Circuits.* CRC Press, Taylor and Francis Group, Boca Raton, FL, 2006.
22. Yaochu Jin and Bernhard Sendhoff. Evolving *in silico* bistable and oscillatory dynamics for gene regulatory network motifs. In: Jun Wang (ed.). *Evolutionary Computation, 2008. CEC 2008 (IEEE World Congress on Computational Intelligence). IEEE Congress on*, pp. 386–391, IEEE, New York, 2008.
23. Ricard V. Solé and Sergi Valverde. Are network motifs the spandrels of cellular complexity? *Trends in Ecology and Evolution*, 21(8):419–422, 2006.
24. Ron Milo, Shai Shen-Orr, Shalev Itzkovitz, Nadav Kashtan, Dmitri B. Chklovskii, and Uri Alon. Network motifs: simple building blocks of complex networks. *Science*, 298(5594):824–827, 2002.
25. Martin Swain, Thomas Hunniford, Johannes Mandel, Niall Palfreyman, and Werner Dubizky. Modeling gene-regulatory networks using evolutionary algorithms and distributed computing. In: Tony Hey, David W. Walker, and Carl Kesselman (eds.). *Cluster Computing and the Grid, 2005. CCGrid 2005, IEEE International Symposium on*, Volume 1, pp. 512–519, IEEE, New York, 2005.
26. Jennifer Hallinan. Gene networks and evolutionary computation. In: Gary B. Fogel, David W. Corne, and Yi Pan (eds.). *Computational Intelligence in Bioinformatics*, pp. 67–96. John Wiley & Sons, Inc., Hoboken, NJ, USA, 2007.
27. George von Dassow, Eli Meir, Edwin M. Munro, and Garrett M. Odell. The segment polarity network is a robust developmental module. *Nature*, 406:188–192, 2000.
28. Dion J. Whitehead, Andre Skusa, and Paul J. Kennedy. Evaluating an evolutionary approach for reconstructing gene regulatory networks. In: Jordan Pollack, Mark Bedau, Phil Husbands, Takashi Ikegami, and Richard A. Watson (eds.). *Ninth International Conference on the Simulation and Synthesis of Living Systems (ALIFE9)*. MIT Press, Cambridge, MA, 2004.
29. Piers J. Ingram, Michael P. H. Stumpf, and Jaroslav Stark. Network motifs: structure does not determine function. *BMC Genomics*, 7:108, 2006.
30. Robert D. Leclerc. Survival of the sparsest: robust gene networks are parsimonious. *Molecular Systems Biology*, 4(213):1–6, 2008.
31. Hailong Zhu, R. Shyama Prasad Rao, Tao Zeng, and Luonan Chen. Reconstructing dynamic gene regulatory networks from sample-based transcriptional data. *Nucleic Acids Research*, 40(21):10657–10667, 2012.
32. Irene M. Ong, Jeremy D. Glasner, and David Page. Modelling regulatory pathways in *E. coli* from time series expression profiles. *Bioinformatics*, 18(Suppl 1):S241–S248, 2002.
33. David Simcha, Laurent Younes, Martin Aryee, and Donald Geman. Identification of direction in gene networks from expression and methylation. *BMC Systems Biology*, 7:118, 2013.
34. Wei-Po Lee and Yu-Ting Hsiao. Inferring gene regulatory networks by incremental evolution and network decomposition. In: Haiban Duan (ed.). *Optimization and Systems Biology*, Volume 9, *Lecture Notes in Operations Research*, pp. 311–324, IEEE, New York, 2008.
35. Anton Crombach and Paulien Hogeweg. Evolution of evolvability in gene regulatory networks. *PLoS Computational Biology*, 4(7):e1000112, 2008.
36. Hendrik Hache, Hans Lehrach, and Ralf Herwig. Reverse engineering of gene regulatory networks: a comparative study. *EURASIP Journal on Bioinformatics and Systems Biology*, 8:1–12, 2009.

37. Hidde de Jong. Modeling and simulation of genetic regulatory systems: a literature review. *Journal of Computational Biology*, 9(1):67–103, 2002.
38. Alina Sîrbu, Heather J. Ruskin, and Martin Crane. Stages of gene regulatory network inference: the evolutionary algorithm role. In: Eisuke Kita (ed.). *Evolutionary Algorithms*. InTech, Rijeka, Croatia, 2011.
39. Michael Hecker, Sandro Lambeck, Susanne Toepfer, Eugene van Someren, and Reinhard Guthke. Gene regulatory network inference: data integration in dynamic models—a review. *Biosystems*, 96(1):86–103, 2009.
40. Mukesh Bansal, Vincenzo Belcastro, Alberto Ambesi-Impiombato, and Diego di Bernardo. How to infer gene networks from expression profiles. *Molecular Systems Biology*, 3(78):1–10, 2007.
41. Mariana Recamonde Mendoza and Ana Lúcia C. Bazzan. Evolving random Boolean networks with genetic algorithms for regulatory networks. In: Natalio Krasnogor (ed.). *Proceedings of the 13th Annual Conference on Genetic and Evolutionary Computation*, GECCO'11, pp. 291–298, ACM, New York, 2011.
42. Chien-Hua Peng, Yi-Zhi Jiang, An-Shun Tai, Chun-Bin Liu, Shih-Chi Peng, Chun-Ta Liao, Tzu-Chen Yen, and Wen-Ping Hsieh. Causal inference of gene regulation with subnetwork assembly from genetical genomics data. *Nucleic Acids Research*, 42(5):2803–2819, 2013.
43. Christopher A. Penfold and David L. Wild. How to infer gene networks from expression profiles, revisited. *Interface Focus*, 1(6):857–870, 2011.
44. Edward Keedwell and Ajit Narayanan. Discovering gene networks with a neural-genetic hybrid. *IEEE/ACM Transactions on Computational Biology and Bioinformatics*, 2(3):231–242, 2005.
45. Alina Sîrbu, Heather J. Ruskin, and Martin Crane. Integrating heterogeneous gene expression data for gene regulatory network modelling. *Theory in Biosciences*, 131(2):95–102, 2012.
46. Briti Sundar Mondal, Arup Kumar Sarkar, Md. Mahmudul Hasan, and Nasimul Noman. Reconstruction of gene regulatory networks using differential evolution. In: A. M. Shadullah (ed.). *Computer and Information Technology (ICCIT), 2010 13th International Conference on*, pp. 440–445, IEEE, New York, 2010.
47. Andrea Rau, Florence Jaffrzic, Jean-Louis Foulley, and Rebecca W. Doerge. Reverse engineering gene regulatory networks using approximate Bayesian computation. *Statistics and Computing*, 22(6):1257–1271, 2012.
48. Leon Palafox, Nasimul Noman, and Hitoshi Iba. Reverse engineering of gene regulatory networks using dissipative particle swarm optimization. *IEEE Transactions on Evolutionary Computation*, 17(4):577–587, 2013.
49. Michael Weber, Sebastian Henkel, Sebastian Vlaic, Reinhard Guthke, Everardus van Zoelen, and Dominik Driesch. Inference of dynamical gene-regulatory networks based on time-resolved multi-stimuli multi-experiment data applying netgenerator v2.0. *BMC Systems Biology*, 7:1, 2013.
50. Neal S. Holter, Amos Maritan, Marek Cieplak, Nina V. Fedoroff, and Jayanth R. Banavar. Dynamic modeling of gene expression data. *Proceedings of the National Academy of Sciences of the United States of America*, 98(4):1693–1698, 2001.
51. Shinichi Kikuchi, Daisuke Tominaga, Masanori Arita, Katsutoshi Takahashi, and Masaru Tomita. Dynamic modeling of genetic networks using genetic algorithm and S-system. *Bioinformatics*, 19(5):643–650, 2003.

52. Timothy S. Gardner, Diego di Bernardo, David Lorenz, and James J. Collins. Inferring genetic networks and identifying compound mode of action via expression profiling. *Science*, 301(5629):102–105, 2003.

53. Alina Sirbu, Heather J. Ruskin, and Martin Crane. Comparison of evolutionary algorithms in gene regulatory network model inference. *BMC Bioinformatics*, 11:59, 2010.

54. Wei-Po Lee, Yu-Ting Hsiao, and Wei-Che Hwang. Designing a parallel evolutionary algorithm for inferring gene networks on the cloud computing environment. *BMC Systems Biology*, 8:5, 2014.

55. Kay Nieselt, Florian Battke, Alexander Herbig, Per Bruheim, Alexander Wentzel, Øyvind M Jakobsen, Håvard Sletta, Mohammad T. Alam, Maria E. Merlo, Jonathan Moore, Walid A. M. Omara, Edward R. Morrissey, Miguel A. Juarez-Hermosillo, Antonio Rodrguez-Garca, Merle Nentwich, Louise Thomas, Mudassar Iqbal, Roxane Legaie, William H. Gaze, Gregory L. Challis, Ritsert C. Jansen, Lubbert Dijkhuizen, David A. Rand, David L. Wild, Michael Bonin, Jens Reuther, Wolfgang Wohlleben, Margaret C. M. Smith, Nigel J. Burroughs, Juan F. Martn, David A. Hodgson, Eriko Takano, Rainer Breitling, Trond E. Ellingsen, and Elizabeth M. H. Wellington. The dynamic architecture of the metabolic switch in *Streptomyces coelicolor*. *BMC Genomics*, 11:10, 2010.

56. Antonio Rodríguez-García, Alberto Sola-Landa, Kristian Apel, Fernando Santos-Beneit, and Juan F. Martín. Phosphate control over nitrogen metabolism in *Streptomyces coelicolor*: direct and indirect negative control of glnr, glna, glnii and amtb expression by the response regulator phop. *Nucleic Acids Research*, 37(10):3230–3242, 2009.

57. Fernando Santos-Beneit, Antonio Rodrguez-Garca, Alberto Sola-Landa, and Juan F. Martín. Cross-talk between two global regulators in *Streptomyces*: Phop and afsr interact in the control of afsS, pstS and phoRP transcription. *Molecular Microbiology*, 72(1):53–68, 2009.

58. Antonio Rodríguez-García, Carlos Barreiro, Fernando Santos-Beneit, Alberto Sola-Landa, and Juan F. Martín. Genome-wide transcriptomic and proteomic analysis of the primary response to phosphate limitation in *Streptomyces coelicolor* m145 and in a δphoP mutant. *Proteomics*, 7(14):2410–2429, 2007.

59. Andy Hesketh, Chris Hill, Jehan Mokhtar, Gabriela Novotna, Ngat Tran, Mervyn Bibb, and Hee-Jeon Hong. Genome-wide dynamics of a bacterial response to antibiotics that target the cell envelope. *BMC Genomics*, 12:226, 2011.

60. Gilles P. van Wezel and Kenneth J. McDowall. The regulation of the secondary metabolism of *Streptomyces*: new links and experimental advances. *Natural Products Reports*, 28(7):1311–1333, 2011.

61. Mamoru Komatsu, Takuma Uchiyama, Satoshi mura, David E. Cane, and Haruo Ikeda. Genome-minimized *Streptomyces* host for the heterologous expression of secondary metabolism. *Proceedings of the National Academy of Sciences of the United States of America*, 107(6):2646–2651, 2010.

62. Tatsuichiro Higashi, Yuko Iwasaki, Yasuo Ohnisi, and Sueharu Horinouchi. A-factor and phosphate depletion signals are transmitted to the grixazone biosynthesis genes via the pathway-specific transcription activator grir. *Journal of Bacteriology*, 189(9):861–880, 2007.

63. Kenneth J. McDowall, Arinthip Thamchaipenet, and Iain S. Hunter. Phosphate control of oxytetracycline production by *Streptomyces rimosus* is at the level of transcription from promoters overlapped by tandem repeats similar to those of the DNA-binding sites of the OmpR family. *Journal of Bacteriology*, 181(10):3025–3032, 1999.

64. Mohammad T. Alam, Maria E. Merlo, The STREAM Consortium, David A. Hodgson, Elizabeth M.H. Wellington, Eriko Takano, and Rainer Breitling. Metabolic modeling and analysis of the metabolic switch in *Streptomyces coelicolor*. *BMC Genomics*, 11:202, 2010.
65. Anushree Chatterjee, Laurie Drews, Sarika Mehra, Eriko Takano, Yiannis N. Kaznessis, and Wei-Shou Hu. Convergent transcription in the butyrolactone regulon in *Streptomyces coelicolor* confers a bistable genetic switch for antibiotic biosynthesis. *PLoS One*, 6(7):e21974, 2011.
66. Nicholas E. E. Allenby, Emma Laing, Giselda Bucca, Andrzej M. Kierzek, and Colin P. Smith. Diverse control of metabolism and other cellular processes in *Streptomyces coelicolor* by the phoP transcription factor: genome-wide identification of in vivo targets. *Nucleic Acids Research*, 40(19):9543–9556, 2012.
67. Franois Voelker and Stphane Altaba. Nitrogen source governs the patterns of growth and pristinamycin production in streptomyces pristinaespiralis. *Microbiology*, 147(9):2447–2459, 2001.
68. Martin G. Lamarche, Barry L. Wanner, Sébastien Crépin, and Josée Harel. The phosphate regulon and bacterial virulence: a regulatory network connecting phosphate homeostasis and pathogenesis. *FEMS Microbiology Reviews*, 32(3):461–473, 2008.
69. STREAM Consortium. SysMO project 10: STREAM—global metabolic switching in *Streptomyces colicolor*. Available from: http://www.sysmo.net/index.php?index=62. Last accessed date 11 November, 2015.
70. Mitsuo Ogura, Hirotake Yamaguchi, Ken ichi Yoshida, Yasutaro Fujita, and Teruo Tanaka. DNA microarray analysis of *Bacillus subtilis* degu, coma and phop regulons; an approach to comprehensive analysis of *B. subtilis* 2-component regulatory systems. *Nucleic Acids Research*, 29(18):3804–3813, 2001.
71. Yvonne Tiffert, Petra Supra, Reinhild Wurm, Wolfgang Wohlleben, Rolf Wagner, and Jens Reuther. The *Streptomyces coelicolor* glnr regulon: identification of new glnr targets and evidence for a central role of glnr in nitrogen metabolism in actinomycetes. *Molecular Microbiology*, 67(4):861–880, 2008.
72. Jens Reuther and Wolfgang Wohlleben. Nitrogen metabolism in *Streptomyces coelicolor*: transcriptional and post-translational regulation. *Journal of Molecular Microbiology and Biotechnology*, 12(1–2):139–146, 2007.
73. Sueharu Horinouchi, Morikazu Kito, Makoto Nishiyama, Kaoru Furuya, Soon-Kwang Hong, Katsuhide Miyake, and Teruhiko Beppu. Primary structure of afsr, a global regulatory protein for secondary metabolite formation in streptomyces coelicolor a3(2). *Gene*, 95(1):49–56, 1990.
74. Sueharu Horinouchi and Teruhiko Beppu. Production in large quantities of actinorhodin and undecylprodigiosin induced by afsB in streptomyces lividans. *Agricultural and Biological Chemistry*, 48(8):2131–2133, 1984.
75. Spencer Angus Thomas and Yaochu Jin. Reconstructing biological gene regulatory networks: where optimization meets big data. *Evolutionary Intelligence*, 7(1):29–47, 2014.
76. Juliane Schäfer and Korbinian Strimmer. An empirical Bayes approach to inferring large-scale gene association networks. *Bioinformatics*, 21(6):754–764, 2005.
77. Matthieu Vignes, Jimmy Vandel, David Allouche, Nidal Ramadan-Alban, Christine Cierco-Ayrolles, Thomas Schiex, Brigitte Mangin, and Simon de Givry. Gene regulatory network reconstruction using Bayesian networks, the dantzig selector, the lasso and their meta-analysis. *PLoS One*, 6(12):e29165, 2011.

78. Marlene Castro-Melchor, Salim Charaniya, George Karypis, Eriko Takano, and Wei-Shou Hu. Genome-wide inference of regulatory networks in *Streptomyces coelicolor*. *BMC Genomics*, 11:578, 2010.

79. Xiujun Zhang, Xing-Ming Zhao, Kun He, Le Lu, Yongwei Cao, Jingdong Liu, Jin-Kao Hao, Zhi-Ping Liu, and Luonan Chen. Inferring gene regulatory networks from gene expression data by path consistency algorithm based on conditional mutual information. *Bioinformatics*, 28(1):98–104, 2012.

80. Adam A. Margolin, Ilya Nemenman, Katia Basso, Chris Wiggins, Gustavo Stolovitzky, Riccardo D. Favera, and Andrea Califano. ARACNE: an algorithm for the reconstruction of gene regulatory networks in a mammalian cellular context. *BMC Bioinformatics*, 7(Suppl 1):S7, 2006.

81. Jesper Tegnèr, M. K. Stephen Yeung, Jeff Hasty, and James J. Collins. Reverse engineering gene networks: integrating genetic perturbations with dynamical modeling. *Proceedings of the National Academy of Sciences of the United States of America*, 100(10):5944–5949, 2003.

82. Florian Geier, Jens Timmer, and Christian Fleck. Reconstructing gene-regulatory networks from time series, knock-out data, and prior knowledge. *BMC Systems Biology*, 1:11, 2007.

83. Mudassar Iqbal, Yvonne Mast, Rafat Amin, David A. Hodgson, The STREAM Consortium, Wolfgang Wohlleben, and Nigel J. Burroughs. Extracting regulator activity profiles by integration of de novo motifs and expression data: characterizing key regulators of nutrient depletion responses in *Streptomyces coelicolor*. *Nucleic Acids Research*, 40(12):5227–5239, 2012.

84. Guy Karlebach and Ron Shami. Constructing logical models of gene regulatory networks by integrating transcription factor–DNA interactions with expression data: an entropy-based approach. *Journal of Computational Biology*, 19(1):30–41, 2012.

85. Kevin Y. Yip, Roger P. Alexander, Koon-Kiu Yan, and Mark Gerstein. Improved reconstruction of *in silico* gene regulatory networks by integrating knockout and perturbation data. *PLoS One*, 5(1):e8121, 2010.

86. Christoph Kaleta, Anna Gohler, Stefan Schuster, Knut Jahreis, Reinhard Guthke, and Swetlana Nikolajewa. Integrative inference of gene-regulatory networks in *Escherichia coli* using information theoretic concepts and sequence analysis. *BMC Systems Biology*, 4:116, 2010.

87. Nasimul Noman and Hitoshi Iba. Inferring gene regulatory networks using differential evolution with local search heuristics. *IEEE/ACM Transactions on Computational Biology and Bioinformatics*, 4(4):634–647, 2007.

88. Eberhard O. Voit and Jonas Almeida. Decoupling dynamical systems for pathway identification from metabolic profiles. *Bioinformatics*, 20(11):1670–1681, 2004.

89. Mei Xiao, Luwen Zhang, Bing He, Jiang Xie, and Wu Zhang. A parallel algorithm of constructing gene regulatory networks. In: Luonan Chen, Xiang-Sun Zhang, Ling-Yun Wu, and Yong Wang (eds.). *Optimization and Systems Biology*, Volume 11, *Lecture Notes in Operations Research*, pp. 184–188, World Publishing Corporation, Beijing, 2009.

90. Chao Sima, Jianping Hua, and Sungwon Jung. Inference of gene regulatory networks using time-series data: a survey. *Current Genomics*, 10(6):416–429, 2009.

91. Anne McVittie. Ultrastructural studies on sporulation in wild-type and white colony mutants of *Streptomyces coelicolor*. *Journal of General Microbiology*, 81(2):291–302, 1974.

92. David A. Hopwood, Hansrudolf Wildermuth, and Helen M. Palmer. Mutants of *Streptomyces coelicolor* defective in sporulation. *Journal of General Microbiology*, 61(3):397–408, 1970.
93. Juan Soliveri, Krystal L. Brown, Mark J. Buttner, and Keith F. Chater. Two promoters for the whib sporulation gene of streptomyces coelicolor a3(2) and their activities in relation to development. *Journal of Bacteriology*, 174(19):6215–6220, 1992.
94. Strepdb, the streptomyces annotation server. Available from: http://strepdb.streptomyces.org.uk/. Last accessed date 11 November, 2015.
95. Klas Flardh and Mark J. Buttner. *Streptomyces morphogenetics*: dissecting differentiation in a filamentous bacterium. *Nature Reviews Microbiology*, 7(1):36–49, 2009.
96. Andrea Franceschini, Damian Szklarczyk, Sune Frankild, Michael Kuhn, Milan Simonovic, Alexander Roth, Jianyi Lin, Pablo Minguez, Peer Bork, Christian von Mering, and Lars J. Jensen. String v9.1: protein–protein interaction networks, with increased coverage and integration. *Nucleic Acids Research*, 41(D1):D808–D815, 2013.
97. Todd Anderson, Paul Brian, Perry Riggle, Renqiu Kong, and Wendy Champness. Genetic suppression analysis of non-antibiotic-producing mutants of the *Streptomyces coelicolor* absa locus. *Microbiology*, 145(9):2343–2353, 1999.
98. Todd B. Anderson, Paul Brian, and Wendy C. Champness. Genetic and transcriptional analysis of absa, an antibiotic gene cluster-linked two-component system that regulates multiple antibiotics in streptomyces coelicolor. *Molecular Microbiology*, 39(3):553–566, 2001.
99. Shin Ando and Hitoshi Iba. Inference of gene regulatory model by genetic algorithms. In: Jong-Hwan Kim (ed.). *Evolutionary Computation, 2001. Proceedings of the 2001 Congress on*, Volume 1, pp. 712–719, IEEE, New York, 2001.
100. Yong Wang, Xiang-Sun Zhang, and Luonan Chen. Optimization meets systems biology. *BMC Systems Biology*, 4(Suppl 2):S1, 2010.
101. Melanie Mitchell. *An Introduction to Genetic Algorithms*. MIT Press, Cambridge, MA, 1999.
102. Yaochu Jin and Yan Meng. Emergence of robust regulatory motifs from *in silico* evolution of sustained oscillation. *Biosystems*, 103(1):38–44, 2011.
103. Dominique Chu. Evolving genetic regulatory networks for systems biology. In: Kay Chen Tan and Jian Xin Xu (eds.). *Evolutionary Computation, 2007. CEC 2007, IEEE Congress on*, pp. 875–882, IEEE, New York, 2007.
104. Spencer A. Thomas, Yaochu Jin, Emma Laing, and Colin P. Smith. Reconstructing regulatory networks in streptomyces using evolutionary algorithms. In: Yaochu Jin and Spencer A. Thomas (eds.). *Computational Intelligence (UKCI), 2013 13th UK Workshop on*, pp. 24–30, IEEE, New York, 2013.
105. Kalyanmoy Deb, Amrit Pratap, Sameer Agarwal, and T. Meyarivan. A fast and elitist multiobjective genetic algorithm: NSGA-II. *Evolutionary Computation, IEEE Transactions on*, 6(2):182–197, 2002.
106. Hans J. F. Moen, Nikolai B. Hansen, Harald Hovland, and Jim Tørresen. Many-objective optimization using taxi-cab surface evolutionary algorithm. In: Robin C. Purshouse, Peter J. Fleming, Carlos M. Fonseca, Salvatore Greco, and Jane Shaw (eds.). *Evolutionary Multi-Criterion Optimization*, Volume 7811, *Lecture Notes in Computer Science*, pp. 128–142, Springer, Berlin, 2013.

107. Kalyanmoy Deb and Ram Bhushan Agrawal. Simulated binary crossover for continuous search space. *Complex Systems*, 9:115–148, 1995.

108. Yaochu Jin, Yan Meng, and Bernhard Sendhoff. Influence of regulation logic on the easiness of evolving sustained oscillation for gene regulatory networks. In: Chrystopher L. Nehaniv (ed.). *Artificial Life, 2009. ALife'09, IEEE Symposium on*, pp. 61–68, IEEE, New York, 2009.

109. Mariana Recamonde Mendoza and Ana Lúcia C. Bazzan. Evolving random Boolean networks with genetic algorithms for regulatory networks reconstruction. In: Natalio Krasnogor (ed.). *Proceedings of the 13th Annual Conference on Genetic and Evolutionary Computation*, GECCO'11, pp. 291–298, ACM, New York, 2011.

110. Sohei Ito, Naoko Izumi, Shigeki Hagihara, and Naoki Yonezaki. Qualitative analysis of gene regulatory networks by satisfiability checking of linear temporal logic. In: Isidore Rigoutsos, Christodoulos A. Floudas, Junhyong Kim, and Harel Weinstein (eds.). *Proceedings of the 2010 IEEE International Conference on Bioinformatics and Bioengineering*, BIBE'10, pp. 232–237, IEEE, New York, 2010.

111. Hidde De Jong and Johannes Geiselmann. Modeling and simulation of genetic regulatory networks by ordinary differential equations. In: Edward R. Dougherty, Ilya Shmulevich, Jie Chen, and Z. Jane Wang (eds.). *Genomic Signal Processing and Statistics*, Hindawi Publishing Corporation, New York, pp. 201–239, 2005.

112. Jie Xiong and Tong Zhou. Gene regulatory network inference from multifactorial perturbation data using both regression and correlation analyses. *PLoS One*, 7(9):e43819, 2012.

113. Tim Hohm and Eckart Zitzler. Multiobjectivization for parameter estimation: a case-study on the segment polarity network of *Drosophila*. In: Günther Raidl (ed.). *GECCO'09: Genetic and Evolutionary Computation Conference (GECCO 2009)*, pp. 209–216, ACM, New York, 2009.

114. Julia Handl, Douglas B. Kell, and Joshua Knowles. Multiobjective optimization in bioinformatics and computational biology. *IEEE/ACM Transactions on Computational Biology and Bioinformatics*, 4(2):279–292, 2007.

115. Kalyanmoy Deb. *Multi-Objective Optimisation using Evolutionary Algorithms*, 1st ed. Wiley, Kanpur, India, 2001.

8

RECONSTRUCTION OF LARGE-SCALE GENE REGULATORY NETWORK USING S-SYSTEM MODEL

Ahsan Raja Chowdhury

Faculty of Information Technology, Monash University, Australia

Madhu Chetty

Faculty of Science and Technology, Federation University Australia, Australia

8.1 INTRODUCTION

Systems biology is an evolving field that allows holistic analysis by uncovering the system-level principles of a biological system. According to Brazhink et al. [6], a biological system can be visualized as a multi-layered network of different processes. An abstract view of such a system, shown in Figure 8.1, considers the processes as involving variables such as metabolites, proteins, and genes in separate layers. However, in reality, the variables in each layer interact not only with the variables in the same layer but also with the variables in other layers. In most of these underlying interactions, genes play an important role in carrying out the complex biochemical interactions. Thus, a gene regulatory network (GRN) acquires significance as it can reveal the underlying biological processes of living organisms, and provide new insights into the causes of complex diseases or for designing new drugs [16]. As a functional circuitry of a living organism, GRN exhibits the regulatory relationships

Evolutionary Computation in Gene Regulatory Network Research, First Edition.
Edited by Hitoshi Iba and Nasimul Noman.

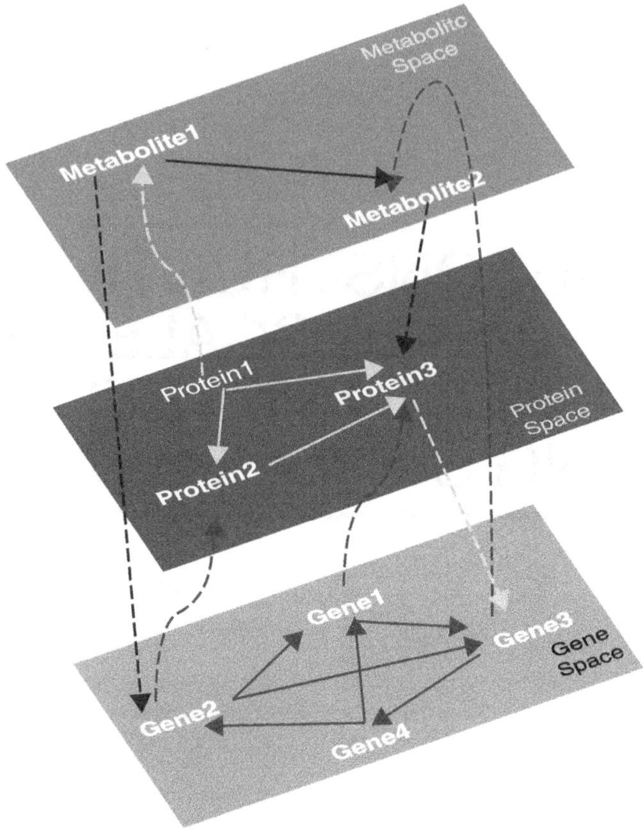

Figure 8.1 Different layers of processes in a typical biological system (adapted from Ref. [6]).

among genes of a cellular system. The responses of genes are represented by their expression profiles, which are captured during the deoxyribonucleic acid (DNA) microarray experiments [43]. A GRN (also referred to as genetic network or gene network) represents the relationship among genes of a genome. In an equivalent graph representation of an N-gene GRN, a gene is shown to be regulated by all N genes, including the self-regulation. However, investigations by Bolouri and Davidson [5] show that, on average, a gene can be regulated typically by four to eight other genes. Figure 8.2(a) shows a GRN of 10 genes having 11 regulations, while Figure 8.2(b) shows a large-scale GRN of 100 genes where, as we can easily realize, it is extremely difficult to count the number of regulations (= 193) from this form of graphical representation.

8.1.1 Significance of Inferring Large-Scale Gene Regulatory Networks

A GRN plays a crucial role in controlling various mechanisms inside a cell. Understanding the interactions among genes helps in understanding the inner details of

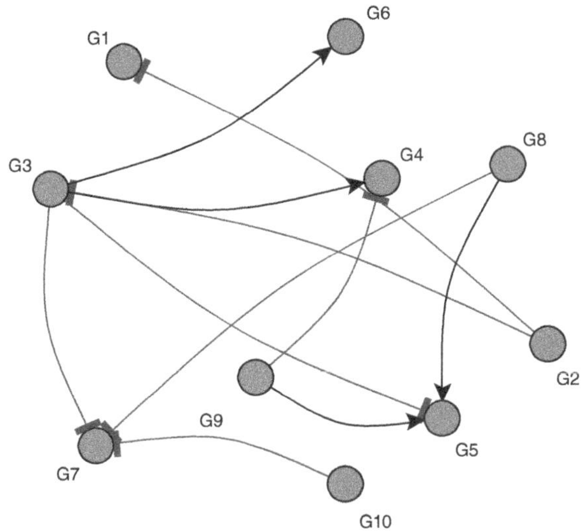

(a) A GRN of 10 genes

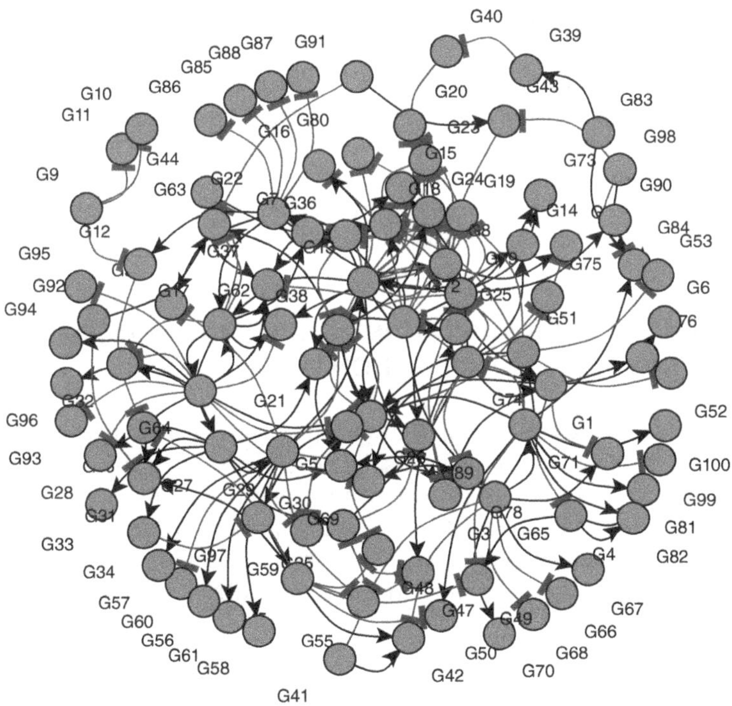

(b) A large-scale GRN

Figure 8.2 (a) A sample GRN of 10 genes with 11 interactions. (b) A large-scale GRN of 100 genes with 193 interactions. The networks are generated using GeneNetWeaver tool [42], where panels (a) and (b) are taken from Dream_Challenges → Dream3_In-Silico_Size_10 → InsilicoSize10-Ecoli1 and Dream_Challenges → Dream4_In-Silico_Size_100 → InsilicoSize100_5, respectively.

cells, and the ability to accurately reconstruct a GRN serves this important objective. Among various available computational techniques, evolutionary algorithms (EAs) are commonly applied for optimizing GRNs during reverse engineering, that means, to optimally learn the model parameters used to represent a network. Since, real-life GRNs consist of thousands of genes, effective and efficient techniques are required to reconstruct the GRN accurately in a reasonable computational time frame.

Reverse engineering GRN is the process of identifying genetic interactions using time-series data with an appropriate mathematical model. These gene-expression data go beyond a generic view of the genome and differentiate between genes within (i) different tissues of the same organism and (ii) different states of the cells in the same tissue. With the advent of cutting-edge technologies for gene profiling, massive amounts of biological data are now available to researchers enabling them to unravel the underlying transcriptional regulations in gene circuits using model-based identification methods.

Thus, reconstructing the genome-wide GRN is a crucial step in uncovering the complete biochemical networks of cells. A GRN helps in understanding interactions at the cellular level and has immense potential for application in genetic engineering. Moreover, knowledge about GRNs provides valuable evidence for the therapeutic studies of complex diseases [3, 38]. Since a real-life GRN consists of thousands of genes, both the model and the method should be robust enough to cope with that amount of genes. However, the model of our particular interest, that is, the S-system model, is currently suitable for modeling small-scale GRN of 5–20 genes and medium-scale GRN having up to 40–50 genes. Hence, inferring a GRN with the S-system model having a large number of genes (say, more than 50) can be designated as large-scale GRN modeling. Such large-scale modeling requires suitable technique from both modeling and optimization methodology perspectives.

8.2 REVERSE ENGINEERING GRN WITH S-SYSTEM MODEL AND EVOLUTIONARY COMPUTATION

Reverse engineering a GRN as an optimization problem depends not only on selecting an appropriate model, but also on the use or development of a suitable optimization method. Furthermore, the performance of the optimization method is also significantly reliant on the nature of the data, noise level present in the data, sampling time, and the type of regulations. This section is devoted to describing the details of different variants of the S-system model for representing GRNs and efficient optimization methods.

8.2.1 S-System Model

The interaction among the components (e.g., metabolites, genes, and proteins) is nonlinear in nature. Therefore, a nonlinear model is more appropriate for modeling nonlinear processes of GRNs [4, 8, 18, 50]. S-system model, a particularly well-established nonlinear model [41], is capable of capturing various dynamics of the complex regulations. Moreover, the S-system is able to represent the regulations of both production and degradation phases, while the other models can only represent the interactions of the production phase. The S-system has been considered as an excellent balance between model complexity and mathematical tractability; it is complex enough to represent a wide range of dynamics, yet is simple enough to allow certain analytical studies.

The S-system, with a set of tightly coupled nonlinear differential equations, is among the best known nonlinear differential-equation-based models for GRNs. These advantages have led to the successful application of this model to the analysis of biochemical networks [11, 12, 23, 26, 40, 53].

8.2.1.1 Canonical S-System Model The S-system model, proposed by Savageau [41], is a well-known system for biochemical networks and has mainly attracted attention for GRN inference in the last decade [23, 26, 30]. The S-system approach is found to be both promising and challenging for GRN modeling. For an N-gene network, the component processes are characterized by the following power-law-based S-system equation:

$$\frac{dX_i}{dt} = \alpha_i \prod_{j=1}^{N} X_j^{g_{ij}} - \beta_i \prod_{j=1}^{N} X_j^{h_{ij}}, \quad i = 1, \dots, N, \tag{8.1}$$

where X_i is the expression level of ith gene. Non-negative parameters α_i and β_i are called rate constants, and real-valued exponents g_{ij} and h_{ij} are referred to as kinetic orders. If $g_{ij} = 0$, it implies that there is no activation or inhibition from Gene-j to Gene-i. If $g_{ij} > 0$, Gene-j activates Gene-i and if $g_{ij} < 0$, Gene-j inhibits Gene-i. Compared to g_{ij}, the term h_{ij} has an opposite effect on the genes i and j, that is, for $h_{ij} > 0$, Gene-j inhibits Gene-i and if $h_{ij} < 0$, Gene-j activates Gene-i. The term $\alpha_i \prod X_j^{g_{ij}}$ models the process of ribonucleic acid (RNA) *synthesis*, while the term $\beta_i \prod X_j^{h_{ij}}$ models the process of RNA *degradation*. The set of parameters that define the S-system model is given as $\theta = \{\alpha, \beta, g, h\}$. To infer a GRN of N genes using the S-system model, $2 \times N(N + 1)$ parameters must be estimated. Thus, to reverse engineer a small network of 5 genes ($N = 5$), we need to estimate $2 \times 5(5 + 1) = 60$ parameters in total, with 12 parameters representing each gene. An investigation of the computational time demonstrates that, in the optimization task for solving tightly coupled S-system model, 95% of the total optimization time is consumed by the numerical integration operation [54]. In order to deal with the problem of high dimensionality and computation complexity for solving S-system equations, different decoupling approaches have been applied to decompose the canonical system into smaller problems.

8.2.1.2 Decoupled S-System Model As mentioned in Section 8.2.1.1 regarding the computational complexity of canonical S-system modeling, to overcome these limitations, a decoupled system has been proposed by Maki et al. [30]. The decoupled system divides the given problem into N sub-problems, each having $2 \times (N + 1)$ number of parameters. In the ith sub-problem, corresponding to ith gene, the parameter set $\theta_i = \{\alpha_i, \beta_i, g_{ij}, h_{ij}\}$ is estimated by solving the decoupled S-system equation:

$$\frac{dX_i}{dt} = \alpha_i \prod_{j=1}^{N} Y_j^{g_{ij}} - \beta_i \prod_{j=1}^{N} Y_j^{h_{ij}}. \tag{8.2}$$

For solving Equation (8.2), $Y_{i=j}$ is obtained by numerical integration, whereas $Y_{i\neq j}$ is obtained by pre-calculations directly via observed times-series data. With the help of direct estimation rather than numerical calculation, decoupling greatly reduces the computational burden. For direct estimation of time-series data, researchers commonly use the technique of linear spline interpolation [36].

Other than the above decoupling approach by problem decomposition, another form of decoupling with the use of linear programming (LP) is possible [49]. According to that method [49], assuming $\frac{dX_i(t)}{dt} > 0$ at time t in Equation (8.1), the following inequality is obtained (after taking logarithm of each side):

$$\log\alpha_i + \sum_{j=1}^{N} g_{ij}\log X_j(t) > \log\beta_i + \sum_{j=1}^{N} h_{ij}\log X_j(t). \tag{8.3}$$

Here, $X_j(t)$ being known from the observed data, Equation (8.3) is a linear inequality, if $\log\alpha_i$ and $\log\beta_i$ are considered as parameters. In the case of $\frac{dX_i(t)}{dt} < 0$, a similar inequality is obtained and the relative ratios of the parameters are obtained by solving these inequalities using LP. Since the method is very fast in solving the canonical problem, it is suitable for solving larger genetic networks. However, the approach is confined to determining unique parameters and is susceptible to noise.

Another approach proposed by Voit and Almeida [54] decouples the canonical equation to a set of algebraic equations. This method substitutes the left-hand side of Equation (8.1) with estimated slopes, obtained from the observed time-series data. Then the new form of the equations is solved either in a parallel or in a serial manner. Although the method reported satisfactory results only for a small genetic network and it showed great improvement in computation time, the reconstruction process failed to infer the parameters precisely.

8.2.2 An Evolutionary Framework: Differential Evolution

Various optimization algorithms are available in the literature that can be used to solve complex problem such as GRN reconstruction. Simulated annealing, LP, and EAs are a few examples of popular optimization techniques that have been used frequently in inferring parameters for GRNs. Among these, evolutionary optimization is becoming prevalent in solving critical and real-world problems in industry, medicine, and defense [45]. Among the EAs that aim to learn the parameters of a GRN, the differential evolution (DE) is of particular interest as it is suitable for handling real-valued, multi-dimensional, and multi-modal problems. Differential evolution is a versatile and robust EA for continuous function optimization [47].

Differential evolution, first proposed by Price and Storn [47] in 1997, is a simple, yet powerful, population-based stochastic search algorithm for global optimization. From the very beginning, DE has been used for many optimization problems due to its robustness and effectiveness [46, 47, 56] and its capability of handling problems having nonlinear and multi-modal objective functions. Let S be the search space of the problem under consideration so that $S \subset \mathbb{R}^N$. Let D be the optimization parameters (=$2N(N+1)$ for the N-gene network while using an S-system model) and P defines the number of instances (also known as individuals) of the population. According to the working principle of DE, a population with P individuals can be represented as

$$X_G = \{X_{1,G}, X_{2,G}, \ldots, X_{P,G}\}, \tag{8.4}$$

where

$$X_{i,G} = \{X_{i,1,G}, X_{i,2,G}, \ldots, X_{i,D,G}\} \quad i = 1, 2, \ldots, P. \tag{8.5}$$

Here G in the above two equations implies the generation number of the optimization, while the subsequent generation is represented as G+1. This iterative algorithm seeks a better solution in every generation by applying evolutionary operations (described later), and the fit solutions among the current and new solutions are propagated in the successive generation (denoted as G+1).

At first, the parameter vectors are initialized with an initial population generation technique. Usually, the initial population is chosen randomly between the lower ($X_{i,\mathrm{low}}$) and upper ($X_{i,\mathrm{upper}}$) bounds defined for each X_ith parameter. Although the initial population sometimes produces an unrealistic candidate solution, the random initialization is performed to cover the search space.

Following the initialization phase, DE enters the iterative phase, where several vector operations are performed. There are three main operations [47] performed in each evolution: *mutation*, *crossover*, and *selection*. During each evolution, DE employs both mutation and crossover to produce a *trial vector* $U_{i,G}$ for each target vector $X_{i,G}$. Then a selection operation is invoked that chooses the better of both the trial and target vectors based on the problem's fitness criteria, and the better vector is placed in the next-generation individual. Each of these operations is elaborated below.

8.2.2.1 *Mutation*

In a genetic algorithm (GA), mutation is traditionally understood as changing the value of a parameter/variable to another value [19]. In the case of binary parameters, 0 (zero) is flipped to 1 (one), and vice versa, during the mutation operation (MO). However, DE generates a new parameter vector by adding the vector of an individual to the weighted difference between the vectors of two other individuals. For each target vector $X_{i,G}$, with $i = 1, 2, \ldots, P$, a mutant vector $V_{i,G+1}$ is generated accordingly:

$$V_{i,G+1} = X_{r_1,G} + F(X_{r_2,G} - X_{r_3,G}). \tag{8.6}$$

Here, $0 < \{r_1, r_2, r_3\} \leq P$ represents three random indices where $r_1 \neq r_2 \neq r_3 \neq i$, and i is the index of the current individual. $F \in [0, 2]$ is a real and constant mutation scale factor used to control the amplification of $X_{r_2,G} - X_{r_3,G}$. Various techniques are also proposed in the literature to select the individuals (r_1, r_2, and r_3) other than random selection [19].

8.2.2.2 *Crossover* A crossover is defined as producing a new individual by taking one part from one individual and the remaining part from another individual [19]. However, in DE, the parameters of the mutant vector $V_{i,G+1}$ are mixed with those of the target vector $X_{i,G}$ to generate the trial vector, $U_{i,G+1}$. This parameter mixing is referred to as a crossover operation in DE. Among the available strategies for performing the crossover, /DE/rand/1/bin [47] is described below:

$$\mathbf{U}_{i,j,G+1} = \begin{cases} V_{i,j,G+1}, & \text{if Rand}(j) \leq \text{CR or } j = j_{\text{rand}}, \\ X_{i,j,G}, & \text{if Rand}(j) > \text{CR and } j \neq j_{\text{rand}}. \end{cases} \tag{8.7}$$

Here i and j are the indices of individuals and parameter vectors, respectively, and CR implies the crossover factor, typically CR $\in [0, 1]$. Rand(j) $\in [0, 1]$ is the jth evaluation of a uniform random number generator, where the index j_{rand} is another randomly chosen integer within the range [1, P]. However, when the control parameters from the trial vector go out of bounds, Ronkkonen et al. [39] have suggested the following equation to reflect these parameters back from the bound by the amount of violation:

$$\mathbf{U}_{i,j,G+1} = \begin{cases} 2 * X_{i,j,\text{low}} - U_{i,j,G+1} & \text{if } U_{i,j,G} < X_{i,j,\text{low}}, \\ 2 * X_{i,j,\text{upp}} - U_{i,j,G+1} & \text{if } U_{i,j,G} > X_{i,j,\text{upp}}. \end{cases} \tag{8.8}$$

8.2.2.3 *Selection* While the mutation and crossover operations are performed in DE, the selection procedure compares the vector $X_{i,G}$ and its corresponding trial vector $U_{i,G+1}$, and selects a vector according to the survival selection mechanism [19, 47]. The following equation is used when the minimization of fitness function (f) is used as the selection criteria:

$$\mathbf{X}_{i,G+1} = \begin{cases} U_{i,G+1}, & \text{if } f(U_{i,G+1}) \leq f(X_{i,G}), \\ X_{i,G}, & \text{otherwise.} \end{cases} \tag{8.9}$$

The above equation ensures the survival of the better individuals for the next generation.

Although DE exhibited excellent performance for solving various multi-modal and complex problems, Fan and Lampinen have proposed the trigonometric mutation operation (TMO) [17] for DE in order to achieve faster convergence and ensure robustness. Incorporating the TMO in the DE algorithm, the authors have proposed the *trigonometric differential evolution (TDE)* [17]. Similar to DE, the TDE is a population-based search algorithm and includes a certain number of candidate solutions (known as individuals), applies evolutionary operators (i.e., mutation, crossover, and selection) over the current generation and produces the individuals for the subsequent generation.

The TMO can be defined using the following equation:

$$\begin{aligned} y_i = \tfrac{1}{3}(x_{r_1} + x_{r_2} + x_{r_3}) + (p_{r_2} - p_{r_1})(x_{r_1} - x_{r_2}) + (p_{r_3} - p_{r_2})(x_{r_2} - x_{r_3}) \\ + (p_{r_1} - p_{r_3})(x_{r_3} - x_{r_1}), \end{aligned} \tag{8.10}$$

where

$$\begin{aligned} p_{r_1} = |f(x_{r_1})|/p', \quad p_{r_2} = |f(x_{r_2})|/p', \quad p_{r_3} = |f(x_{r_3})|/p', \\ p' = |f(x_{r_1})| + |f(x_{r_2})| + |f(x_{r_3})|. \end{aligned} \tag{8.11}$$

The selection of r_1, r_2, and r_3 follows the same rule as conventional MO. Noman and Iba used the traditional mutation and trigonometric mutation based on a probability F_t and $(1 - F_t)$, respectively. This MO is then followed by crossover and selection operation to form the individuals for the next generation. The TDE was reported as having better convergence properties [34] than the well-known DE [47] and also efficient in genetic network inference [31,33]. Hence, in this research, we have also used the TDE as the optimization algorithm for reverse engineering genetic networks.

8.2.3 Model Evaluation Criteria

All GAs use a fitness function as a measure of the "goodness" of a solution. Although numerous fitness functions are reported in the literature, this section describes the most popular and widely used fitness functions for inferring S-system parameters.

8.2.3.1 Mean Squared Error Based Fitness Function For GRN reconstruction using the S-system model, the most commonly used evaluation criterion (known as mean squared error (MSE)) is the difference between the target time-series data and the computed data. However, because any meaningful conclusion about complex dynamics cannot be derived using a single set of time-course data, multiple sets of time-course data are often considered [48]. The fitness function for the canonical system, considering multiple sequence data, becomes

$$f^{\mathrm{MSE}} = \sum_{i=1}^{N} \sum_{k=1}^{M} \sum_{t=1}^{T} \left\{ \frac{X_{k,i}^{\mathrm{cal}}(t) - X_{k,i}^{\mathrm{exp}}(t)}{X_{k,i}^{\mathrm{exp}}(t)} \right\}^2. \tag{8.12}$$

Here, M is the total number of datasets, N implies the number of genes, T is the number of sampling points, $X_{k,i}^{\text{cal}}(t)$ represents the numerically calculated expression level of Gene-i at time t in kth dataset, and $X_{k,i}^{\text{exp}}(t)$ represents the experimentally observed expression level of Gene-i at time t in kth dataset. The search algorithm determines optimal network parameters by minimizing the fitness function f^{MSE}. The decoupled form of Equation (8.12) is written as follows by dropping the first summation term:

$$f_i^{\text{MSE}} = \sum_{k=1}^{M} \sum_{t=1}^{T} \left\{ \frac{X_{k,i}^{\text{cal}}(t) - X_{k,i}^{\text{exp}}(t)}{X_{k,i}^{\text{exp}}(t)} \right\}^2. \tag{8.13}$$

While inferring the decoupled S-system, the ith sub-problem estimates the optimal parameter set $\Omega_i = \{\alpha_i, \beta_i, g_{ij}, h_{ij}\}$ for which f_i^{MSE} is the minimum.

8.2.3.2 Mean Squared Error Based Fitness Function with Penalty Term

A major problem in the S-system-based GRN inference method is to determine the non-regulation (zero-valued) parameters. If all the zero-valued parameters can be found, the values of remaining non-zero parameters can easily be determined. However, finding zero-valued parameters is not straightforward. Kimura et al. [26] added an effective penalty term to the fitness function given in Equation (8.13) for penalizing non-zero regulations and obtaining a sparse network [23]. The modified fitness function for solving the decoupled S-system equation becomes

$$f_i^{\text{MSE}} = \sum_{k=1}^{M} \sum_{t=1}^{T} \left\{ \frac{X_{k,i}^{\text{cal}}(t) - X_{k,i}^{\text{exp}}(t)}{X_{k,i}^{\text{exp}}(t)} \right\}^2 + c \sum_{j=1}^{N-I} (|G_{ij}| + |H_{ij}|). \tag{8.14}$$

Here G_{ij} and H_{ij} are formed by sorting the absolute values of g_{ij} and h_{ij}, respectively, in a non-decreasing order by their absolute values. I is the maximum allowed in-degree of the network and c is the balance factor. Noman and Iba [33] modified it further by combining both the kinetic orders in the following way:

$$f_i^{\text{MSE}} = \sum_{k=1}^{M} \sum_{t=1}^{T} \left\{ \frac{X_{k,i}^{\text{cal}}(t) - X_{k,i}^{\text{exp}}(t)}{X_{k,i}^{\text{exp}}(t)} \right\}^2 + c \sum_{j=1}^{2N-I} (|K_{ij}|). \tag{8.15}$$

Here, K_{ij} are the absolute values of kinetic orders of Gene-i sorted in non-descending order. To address various limitations of the regularized squared relative error of the above fitness functions, Chowdhury et al. [13] proposed a novel fitness function referred to as *adaptive squared relative error* (ASRE), which is given below:

$$\text{ASRE} = \sum_{t=1}^{T} \left(\frac{X_i^{\text{cal}}(t) - X_i^{\text{exp}}(t)}{X_i^{\text{exp}}(t)} \right)^2 + B_i \times C_i \frac{2N}{2N - r_i}. \tag{8.16}$$

Here r_i is the total number of actual regulators. B_i is a balancing factor used to maintain desired balance between the two terms of ASRE. C_i is the penalty factor for the ith gene, defined as

$$C_i = \begin{cases} 1, & \text{if } J < r_i < I, \\ 1 + (J - r_i)^2, & \text{if } r_i \le J, \\ 1 + (r_i - I)^2, & \text{if } r_i \ge I, \end{cases} \tag{8.17}$$

with I and J being the maximum and minimum in-degrees, respectively. In the ASRE criterion, in contrast to a *fixed* weighting factor c as in Equation (8.15), the penalty factor C_i takes the form of an inverse power law. This is motivated by the fact that biological networks often have a scale-free structure, in which the node connectivity degree x distributes according to a power-law distribution, $P(x) \propto x^{-\gamma}$, with the scaling parameter $\gamma \in [2, 3]$ for various networks in nature, society, and technology [44]. Gene regulatory networks generally have low in-degrees, with the number of genes having high in-degree diminishing according to a power-law form. Note that in our formulation, we also enforce a minimum in-degree J; thus, genes with the number of in-degrees falling in-between the minimum and maximum number of in-degrees $[J, I]$ are not penalized ($C_i = 1$), while genes falling out of this region are penalized according to an inverse power-law term ($C_i = 1 + d^\gamma$, where $\gamma = 2$ and d is the number of missing or violated regulations). This fitness function reported very good results for small-scale GRNs and is not directly suitable for inferring large-scale GRNs [13].

8.2.3.3 *Information Criteria-Based Fitness Function* There also exist other fitness functions based on information criteria, such as Akaike information criteria (AIC), Bayesian information criteria, and generalized cross-validation. Akaike information criteria [1] is most commonly used in statistical modeling to show the discrepancy between the target and the estimated model. Let $\epsilon_i(t)$ be the difference between $X_i^{cal}(t)$ and $X_i^{exp}(t)$ for Gene-i. If $\epsilon_i(t)$ is assumed to be the normal distribution with mean $\mu_i = 0$ and standard deviation σ_i, which are constant over time for Gene-i, then the log-likelihood Λ_i of the expression data for this gene and for a set of parameters Ω_i of the same is given by

$$\Lambda_i(\Omega_i, \sigma_i) = -\frac{1}{\sigma_i^2} \sum_{t=1}^{T} (\epsilon_i(t))^2 - \frac{T}{2} \ln(2\pi\sigma_i^2). \tag{8.18}$$

The maximum likelihood estimate of σ_i^2 is obtained accordingly as follows:

$$\sigma_i^2(\max) = \frac{1}{T} \sum_{t=1}^{T} (\epsilon_i(t))^2. \tag{8.19}$$

Substituting Equation (8.19) in Equation (8.18), the log likelihood of the estimated model is obtained, from which AIC is defined as [1]

$$\text{AIC} = -2\Lambda + 2D. \tag{8.20}$$

Here, D is the number of parameters included in the model. This AIC-based fitness value has been further modified to Equation (8.21) [32] by combining Equation (8.15):

$$f_i^{\text{AIC}} = -2\Lambda + 2D + c \sum_{j=1}^{2N-I} (|K_{ij}|). \tag{8.21}$$

The additional penalty term becomes useful when the number of regulations in the skeletal network is higher than the maximum in-degree (denoted as I). Noman and Iba [21, 32–34] studied TDE with both MSE- and AIC-based fitness functions on S-system-based reconstruction.

8.2.4 Limitations of S-System Modeling in Inferring Large-Scale GRN

Since EAs are efficient and robust in solving complex problems, the inference of GRNs using the S-system model uses EAs extensively as a learning technique. This section reviews the limitations of algorithms in reverse engineering GRN that uses an EA as the inference method and the S-system as the modeling framework.

Tominaga et al. [51] initiated the inverse problem of parameter estimation of the S-system model using a classical GA. The method obtained reasonable accuracy in learning all 60 parameters of a 5-gene network from noise-free data. Later, Kikuchi et al. [23] enhanced this technique by introducing the concept of "real-coded genetic algorithm." Their proposed method, called PEACE1, demonstrated superior performance over the work of Tominaga et al. [51] for the same network and input data. Kikuchi et al. [23] also introduced a "pruning term" and proposed a new fitness function. Although the parameter values of the inferred network were very close to the target values, PEACE1 was found unsuitable for inferring a large network due to its computational complexity.

Kimura et al. [25] proposed a memetic algorithm for reconstructing GRNs using the decomposed S-system formalism. This was a successful attempt to infer network parameters from data that contain noise. Later, they extended their work with a cooperative co-evolutionary algorithm and including an improved fitness function [26]. Kimura et al. [24] also used the LP technique with their previously proposed method, genetic local search with distance independent diversity control (GLSDC) [25]. This method exhibited great promise in computational efficiency; however, the accuracy of the inferred network was not satisfactory.

Iba et al. accomplished a systematic and specific improvement toward reverse engineering a GRN with the S-system model and EA. First of all, Sakamoto and Iba [40] proposed a genetic-programming-based reconstruction method introducing the least mean square technique for inferring GRNs. This method successfully inferred a small-scale network from both noise-free data and data with noise. Later, Noman and Iba performed extensive analysis [31–34] on the S-system model for GRN inference using versatile and robust DE. For the first time, they used TDE, a variant of DE, in the inference procedure and proposed a memetic algorithm. The Hill-climbing local search algorithm and the information-criteria-based new fitness function, proposed by Noman et al., proved highly effective in finding good solutions for small- and medium-scale networks. On the other hand, Ando and Iba [2] used the hybrid evolutionary method of genetic programming and statistical analysis to reconstruct a genetic network. Although this method possesses strong theoretical underpinnings, it was limited to showing the performance evaluation for small-scale networks.

Similar to the method of Ando and Iba [2], Tsai et al. [52] used hybrid differential evolution for identifying the parameters of the GRNs. They have shown superior performance of their proposed method over the methods of Kikuchi et al. [23] (PEACE1) and Kimura et al. [26] (GLSDC) for reverse engineering a small-scale network. However, Tsai et al. used the canonical form of S-system equations (Equation 8.1), and the method was not scalable to large network reconstruction. Further, the method reported poor results when noise was present in the data. Ko et al. [27] also proposed a method using hybrid differential evolution that includes a modified collocation approximation method to avoid numerical integration. This method reported excellent performance in running time complexity; however, it was limited to exhibiting the results from a single small-scale network.

Liu et al. [28, 29] proposed an inference method using the separable estimation method and GA. To improve the accuracy of the method, (i) a new objective function based on L_1 regularization was proposed, and (ii) an approximation for the decoupled S-system equations was performed using a five-point numerical derivative method. However, the performance evaluation of the method was shown for a very small 5-gene network and the method was limited to working with a single dataset.

Recently, Chowdhury et al. [13–15] have proposed a time-delayed S-system model, which is capable of simultaneously inferring instantaneous and time-delayed regulations present in the GRN. In accordance with the new modeling approach, the authors have also proposed a new inference mechanism by adapting our cardinality-based fitness criteria [12]. Although the proposed method obtained excellent results for all the GRN considered in Ref. [13], the proposed method is also limited to inferring small- and medium-scale networks.

All the methods discussed so far exhibit the limitation of various approaches to inferring small- and medium-scale genetic networks, that is, 10–20-gene network and 40–50-gene network, respectively. As mentioned earlier, S-system model-based GRN inference methods are susceptible to computational complexity due to the rigorous numerical integration involved. Moreover, inferring a large number of parameters for the S-system model also leads to astronomically high computational time. Although the decoupled S-system model can reduce the computation burden by approximation, applications are still limited to GRNs of 10–20 genes, or at most 50 genes.

8.3 THE PROPOSED FRAMEWORK FOR INFERRING LARGE-SCALE GRN

We propose a three-stage computational framework for inferring large-scale GRNs with the S-system model. Stage 1 takes the microarray time-series data of the considered organism, along with the prior biological knowledge of that organism, that is, a list of regulatory genes (RGs) and target genes (TGs). For the well-studied organisms, the biological knowledge is available in the literature; however, for less-studied ones, comparative genomic techniques can be employed to map RGs from the source to the target organism. Next, separating the expression profiles for RGs and TGs, we create two sub-networks: SubNet-1 that consists of RGs and the interactions among RGs only and SubNet-2, consisting of all RGs, single TGs, and all the interactions on that TG. While performing the optimization, we iteratively consider every TG in the SubNet-2 and estimate the parameters of that TG. Then we solve both the networks using an EA (TDE, in particular) and the newly proposed fitness criteria. From the biological knowledge of genetic network [9, 22], we safely assume that a TG cannot be regulated by more than 10–15 RGs. Hence, we allow a TG to have at most $I_{\max}$ regulations, and penalize the solution of Gene-i that has regulations more than I_i (initially $I_i = I_{\max}/2$). Other than random initialization, the proposed method initializes every individual of the population with a self-degradation ($|h_{i,i}| > 0$) and allows at most $I_{\max}$ parameters to have non-zero values. This is an attempt to incorporate prior biological knowledge into the initial population, which, in turn, helps the optimization to converge more quickly than random initialization. We also include a new and efficient multiple-cardinality-based diversification (MCD) procedure (described later), which is especially designed to work effectively with the proposed multiple-cardinality-based fitness criteria. In the final stage of the proposed framework, the results are combined to form the entire network. All the existing S-system-based methods [12, 23, 26] consider the entire network of N genes for optimization, whereas the proposed method considers two networks and further applies decoupling technique for parameter estimation.

8.3.1 Adapted S-System Model

Among the N genes in a GRN, we state that N_{RG} genes are RGs and N_{TG} are TGs, where $0 \leq \{N_{\mathrm{RG}}, N_{\mathrm{TG}}\} \leq N$. According to the type of regulations, we rewrite the S-system equation (8.2) as follows:

$$\frac{dX_i}{dt} = \alpha_i \prod_{j=1}^{N_{\mathrm{RG}}+N_{\mathrm{TG}}} X_j^{g_{ij}} - \beta_i \prod_{j=1}^{N_{\mathrm{RG}}+N_{\mathrm{TG}}} X_j^{h_{ij}}, \quad i = 1, \ldots, N_{\mathrm{RG}} + N_{\mathrm{TG}}. \tag{8.22}$$

Since we know that the TGs do not regulate RGs, neither in the production phase nor the in degradation phase, we represent SubNet-1 (consisting of RGs only and the regulations among the RGs) and SubNet-2 (consisting of RGs and TGs, and the regulations from RGs to TGs), respectively, with the following equations:

$$\frac{dX_i}{dt} = \alpha_i \prod_{j=1}^{N_{\mathrm{RG}}} X_j^{g_{ij}} - \beta_i \prod_{j=1}^{N_{\mathrm{RG}}} X_j^{h_{ij}}, \quad i = 1, \ldots, N_{\mathrm{RG}} \tag{8.23}$$

$$\frac{dX_i}{dt} = \alpha_i \prod_{j=1}^{N_{\mathrm{RG}}+1} X_j^{g_{ij}} - \beta_i \prod_{j=1}^{N_{\mathrm{RG}}+1} X_j^{h_{ij}}, \quad i = N_{\mathrm{RG}} + 1, \ldots, N_{\mathrm{RG}} + N_{\mathrm{TG}}. \tag{8.24}$$

However, due to the decoupled S-system formulation, parameter learning of single Gene-i is performed one at a time. Thus, in the decoupled form for SubNet-1, we estimate $\Omega_i^1 = \{\alpha_i, \beta_i, \{g_{ij}, h_{ii}\}_{j=1,\ldots,N_{\mathrm{RG}}}\}$ parameters for each Gene-i. On the other hand, we estimate $\Omega_i^2 = \{\alpha_i, \beta_i, \{g_{ij}, h_{ii}\}_{j=1,\ldots,N_{\mathrm{RG}}+1}\}$ parameters by considering the SubNet-2 as an (N_{RG}+1)-gene network, and iteratively learn the parameters for all N_{TG} genes. However, we also consider the assumption of Ref. [10] regarding the presence of self-degradation for each genes (i.e., $|h_{i,i}| > 0$) and allow at most $I_{\max}$ regulations for any gene, that is, $\hat{C}(g_{i,j}) < I_{\max}$, where $\hat{C}(g_{i,j})$ is the number of regulation on Gene-i in the production phase. On the other hand, SubNet-1 is an N_{RG}-gene network that allows all possible interactions among RGs in the production but only self-inhibition in the degradation. The in-degrees of each gene are updated according to the adaptive regulatory genes cardinality (ARGC) algorithm [12].

8.3.2 New Fitness Function

Chowdhury et al. [12, 13] proposed two multiple-cardinality-based fitness functions, where each of the fitness functions split the entire search space into three regions; gene with regulations (i) less than minimum in-degree (region-1) (ii) greater than maximum in-degree (region-3), and (iii) in between maximum and minimum in-degrees (region-2). Genes in region-1 and region-3 are penalized with an exponential penalty term, while the genes in region-2 are allowed to stay in the region without being penalized. In order to cope with large-scale GRN inference, we modify the fitness function of Ref. [13] by introducing a fourth region. We consider that, while penalizing the solutions of region-3 ($r_i > I_i$), we throw out a solution from the optimization with an infinite penalty value if it takes the regulations more than $I_{\max}$. This penalization will essentially remove the unfeasible solutions from the competition and allows the remaining individuals to survive during the optimization process:

$$\mathrm{RSRE} = \sum_{t=1}^{T} \left(\frac{X_i^{\mathrm{cal}}(t) - X_i^{\mathrm{exp}}(t)}{X_i^{\mathrm{exp}}(t)} \right)^2 + B_i \times C_i \frac{r_{\max}}{r_{\max} - r_i}. \tag{8.25}$$

Here, for ith gene, r_{max} is the maximum number of parameter, r_i is the number of total regulations, and C_i is the scaling factor defined as follows:

$$C_i = \begin{cases} 1, & \text{if } I_i \geq r_i \geq J_i, \\ \infty, & \text{if } r_i > I_{max}, \\ 1 + (J - r_i)^2, & \text{if } r_i \leq J, \\ 1 + (r_i - I)^2, & \text{if } r_i \geq I. \end{cases} \tag{8.26}$$

Here, r_{max} is the number of parameters for a single gene (N_{RG} and $N_{RG} + 1$ for SubNet-1 and SubNet-2, respectively). I_i and J_i are max in-degree and min in-degree, respectively, for Gene-i.

8.3.3 Multiple-Cardinality-Based Diversification

For evolutionary optimization, especially when dealing with large-scale GRN modeling, getting stuck in local minima is considered to be one of the prime problems. Further, the number of sub-optimal solutions increases over the generations due to different selection pressures, which, in turn, can spread over the entire population. Being trapped in the local minima can also result in the loss of diversity in the population [7], consequently leading to premature convergence. Hence, it is necessary to trigger a mechanism within the optimization framework to increase the level of diversity. This can be done when the diversity drops below a certain threshold value or when the evolution is stagnant for a specified number of generations. A conventional diversification strategy is to replace a certain percentage of individuals from the population with new, randomly generated individuals, either in every generation or whenever current diversity goes below a pre-defined threshold value [20,55]. In most cases, randomly generated individuals replace the existing individuals with lower fitness values that are selected randomly.

Although we have developed a new cardinality-based fitness function, especially designed for inferring large-scale GRNs, where the in-degree values are updated with an adaptive algorithm, namely ARGC, still the optimization may get trapped into local minima and the population may lose diversity. In order to maintain the diversity, we propose a MCD technique that works according to the steps of Algorithm 1.

The MCD algorithm creates $R_1 + R_2$ new individuals to replace the existing individuals. For every R_1 individuals, we select a random number R_R such that $J_i \leq R_R \leq I_i$, randomly select R_R kinetic order values, and assign random values within the range [−3.00, 3.00]. The remaining $2N - R_R$ kinetic order values are set to 0 and two rate constants (α, β) are initialized randomly within their corresponding limits. We also create R_2 individuals in a similar way, with the exception of assigning random values to all the kinetic order values. While replacing the individuals, we select R_1+R_2 individuals randomly from the current population so that none of the selected individuals are within the best R_3 fitness values, and replace them with newly created individuals. The "*PositionToReplace()*" procedure in the above algorithm actually finds a single position k, where the individual with the kth index is not among the best R_3 individuals and that kth position has not been selected before for replacement. For this proposed cardinality-based diversification, we set R_1, R_2, and R_3 as 20%, 10%, and 50%, respectively, of total individuals P. The value of these three parameters (R_1, R_2, and R_3) indicates that 30% of the individuals other than the best 50% individuals are replaced by another new 30% of individuals, where 20% are initialized with cardinality constraints and 10% are entirely initialized with random values. It should be noted that we apply the proposed diversification procedure when the best individual is unchanged for 30 consecutive generations.

Algorithm 1: *MCD (Ind,* I_i, J_i, R_1, R_2*)*

Input: P= Population size, *Ind*= N individuals, I_i/J_i= maximum and minimum in-degrees, respectively, of current individual
Output: Updated *Ind* after diversification

```
For a=1 to R_1 Do
    Flag_j ← 0, ∀_{j=1,...,R_1}
    R_R ←RandomInt(J_i, I_i)
    For b=1 to R_R Do
        k ←RandomInt(1, 2N) & Flag_k ≠ 1
        Flag_k ← 1
        NewInd_a.Kinetics_k ←RandomFloat(-3.00, 3.00)
    End For
    For b=1 to 2N Do
        If Flag_k=0 Then
            NewInd_a.Kinetics_k ←0
        End If
    End For
    NewInd_a.Alpha ←RandomFloat(0, 20.00)
    NewInd_a.Beta ←RandomFloat(0, 20.00)
End For
For a=1 to R_2 Do
    NewInd_{R_1+a}.Kinetics_k ←RandomInt(-3.00, 3.00), ∀_{k=1,...,2N}
    NewInd_{R_1+a}.Alpha ← RandomFloat(0, 20.00)
    NewInd_{R_1+a}.Beta ← RandomFloat(0, 20.00)
End For
For a=1 to R_1 + R_2 Do
    k ←PositionToReplace()
    Ind_k ← NewInd_a
End For
```

8.4 EXPERIMENTAL RESULTS

The framework proposed in the previous section is evaluated experimentally using three *in silico* networks: Net-1, Net-2, and Net-3. Data for Net-1 (20-gene network

[31]) are generated using the S-system equation and the well-known fourth-order Runge–Kutta method for numerical integration. Net-2 ($N = 50$, $|N_{RG}| = 9$, $|N_{TG}| = 41$, and total regulations = 91) and Net-3 (N=100, $|N_{RG}| = 36$, $|N_{TG}| = 64$, and total regulations = 249) are generated using the GeneNetWeaver tool [37], which is used to generate *in silico* benchmarks in the DREAM challenge initiative [37]. This tool generates biologically plausible network topologies and dynamics of any given size by extracting random sub-networks of *Saccharomyces cerevisiae* and *Escherichia coli* [37]. We used the tool to generate two networks and the corresponding time-series data as in the DREAM4 challenges with 10 different perturbations for each experiment. In addition to noise-free data, all three networks are tested with four different levels of Gaussian noise (5%, 10%, 15%, and 20%).

The proposed algorithm is implemented in C++ using a 2.16 GHz Dual-core CPU PC with 3 GB of RAM. This code and data for all three networks can be made available on request. The parameter values for the TDE algorithm were set as mutation factor $F_o = 0.5$, TMO $F_t = 0.05$, crossover factor CF = 0.8, population size $P = 100$. The maximum in-degrees (I_{max}) were set to three for Net-1 and eight for the remaining two networks based on the knowledge of the genetic networks [9,22,31]. The in-degrees I_i and J_i are, respectively, set to $I_{max}/2$ and 1 for each gene, and updated with ARGC algorithm [12] in every $l = 50$ generations. We have executed the proposed optimization method for 5 trials with 500 generations in each trial.

We plot the sensitivity (S_n) and specificity (S_p) to obtain the receiver operating characteristic (ROC) graphs for all the three networks in different noise conditions. The ROC graphs have been used as Leon et al. have shown in Ref. [35]. Two more well-known performance measures, that is, precision (P_r) and F-score (F), have been applied for further evaluation of the networks and the comparisons of the proposed method with other methods. Net-1 (both SubNet-1 and SubNet-2) is evaluated with two existing S-system-based methods (ALG [31] and REGARD [12]) and a non-S-system-based method BANJO [57], the widely used dynamic Bayesian network-based GRN reconstruction algorithm. For the two large networks (Net-2 and Net-3), both ALG [31] and REGARD [12] failed to converge and produce any result for a single gene even with the decoupled S-system model within 10 h on the same computer.

8.5 DISCUSSIONS

The ROC graphs for Net-1, shown in Figures 8.3(a)–8.3(e), illustrate that the proposed method not only infers the true positives correctly but also the false positives with equal accuracy; hence the ROC plots for the proposed method for all the graphs are either at or near the optimal point (1, 1) or close to that point. On the other hand, the ROC points for two S-system-based existing methods [12,31] are away from the optimal point, while this is even inferior for other existing method BANJO [57]. The results from ROC graphs demonstrate the effectiveness of the proposed method in comparison to the two existing S-system-based methods ALG [31] and REGARD [12], and a non-S-system-based method BANJO (Bayesian network) [57].

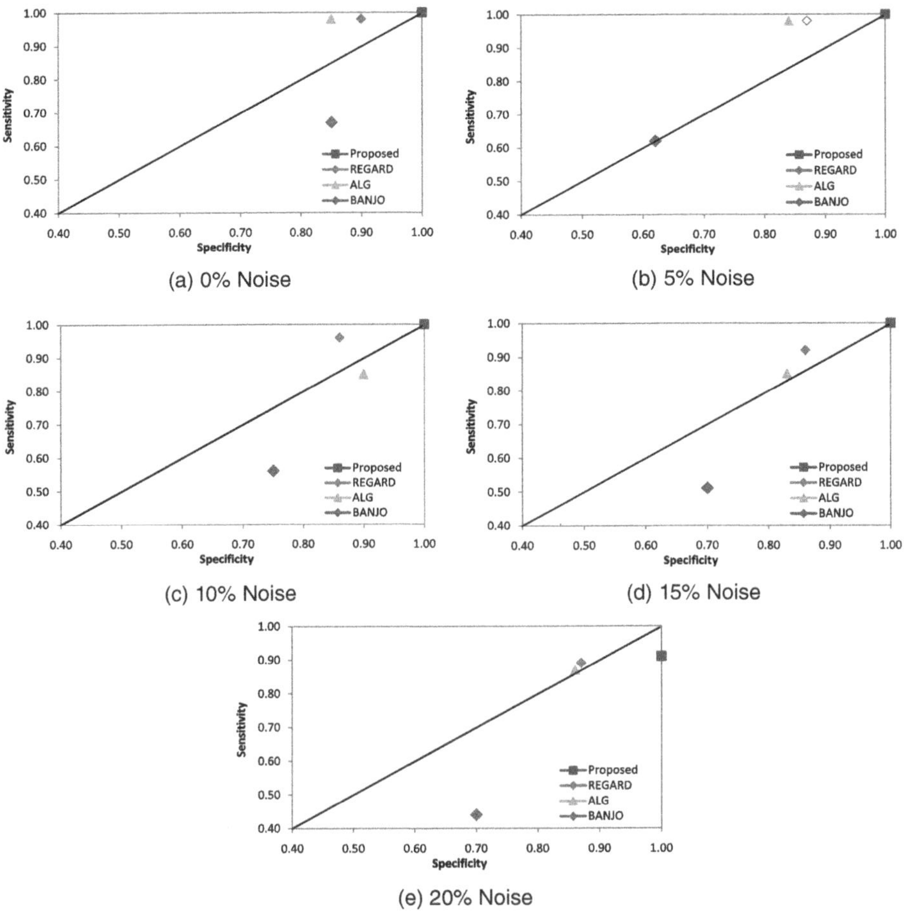

Figure 8.3 Receiver operating characteristic (ROC) points shown in ROC graphs for different methods (proposed, REGARD [12], ALG [31], and BANJO [57]) in various noise conditions.

Since, the larger networks Net-2 and Net-3 are sub-networks extracted from real networks, the microarray data have noise in it and hence the reverse engineering process may end up by (i) not inferring all the true regulations, and/or (ii) inferring some false regulations. Thus, the performance of the proposed method is not at par with the performance of Net-1. However, for both Net-2 and Net-3, the proposed method outperforms the existing state-of-the-art method BANJO [57], as shown in Figures 8.4(a) and 8.4(b). Figures 8.5(a)–8.5(d) show that, while evaluating the performance in terms of precision and *F*-score, the proposed method exhibits superiority over BANJO for the networks in all the different noise conditions. Regarding the time responses for four TGs of Net-1 in all noise conditions, as shown in Figure 8.6, we observe that the time responses for these four TGs closely follow the trends of the target expression patterns. From Figures 8.4 and 8.5 it is clear that the proposed

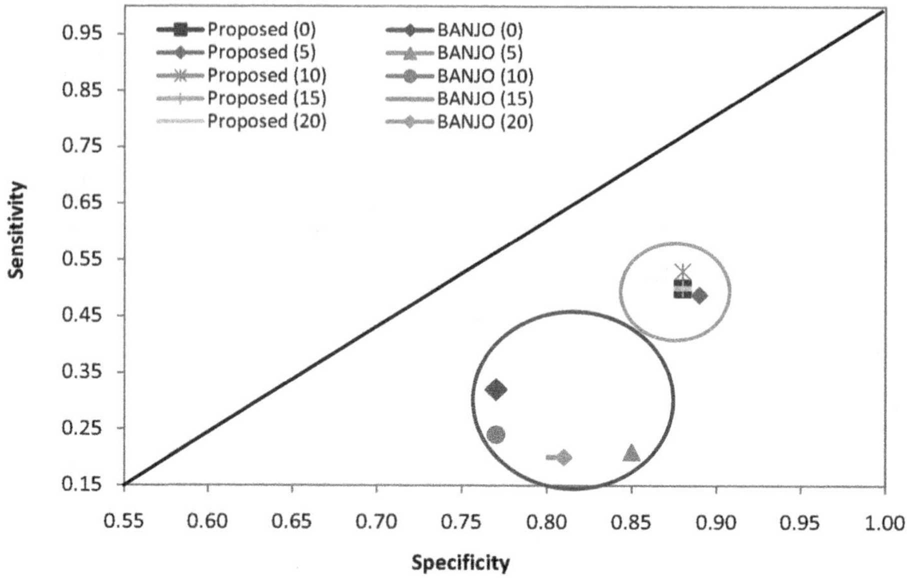

(a) ROC graph for Net-2

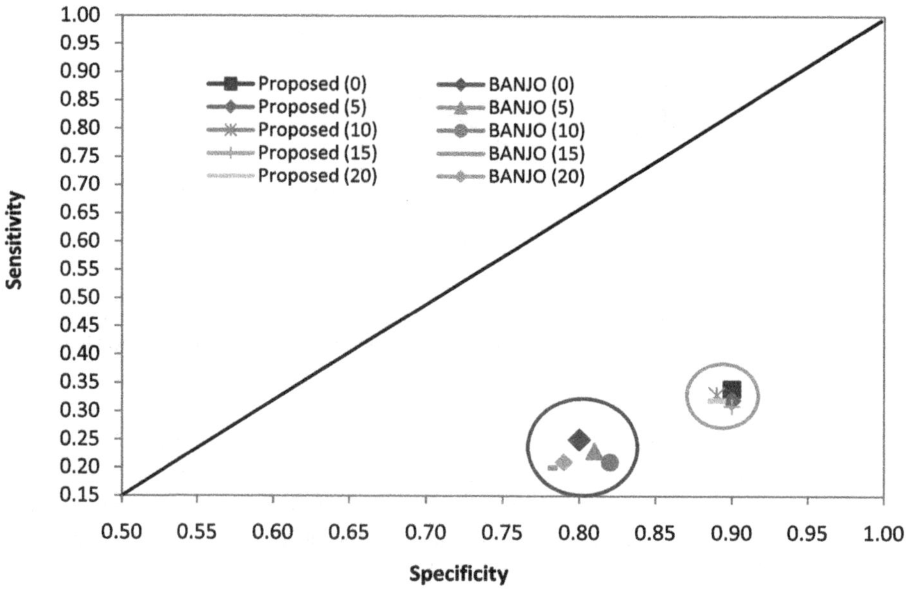

(b) ROC graph for Net-3

Figure 8.4 Receiver operating characteristic (ROC) points shown in ROC graphs for Net-2 and Net-3 for proposed method and BANJO [57] in various noise conditions. The bigger circle and the smaller circle inside the graph show the positions of the ROC points for the proposed method and BANJO, respectively.

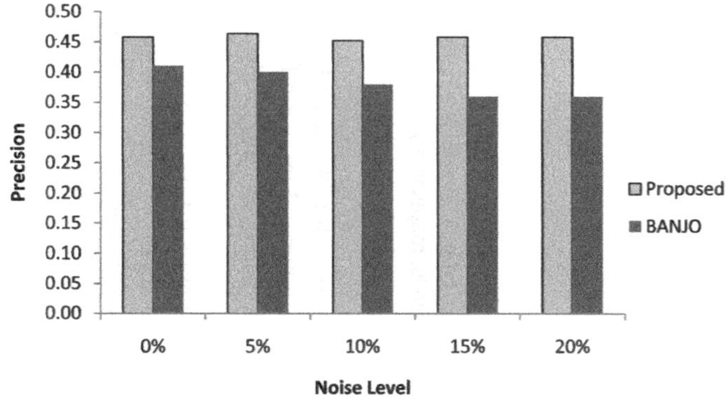

(a) Precision(Net-2)

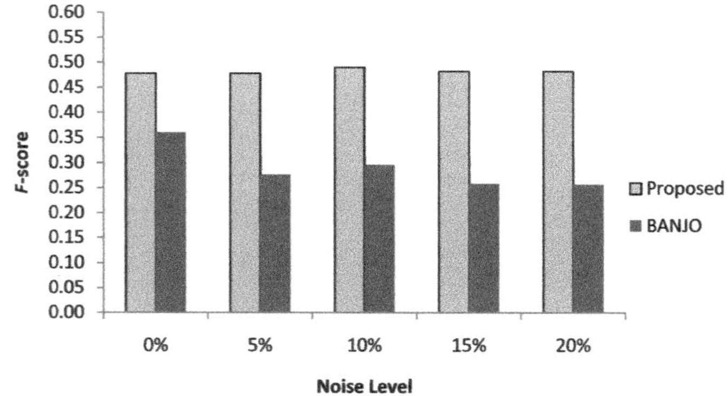

(b) *F*-score(Net-2)

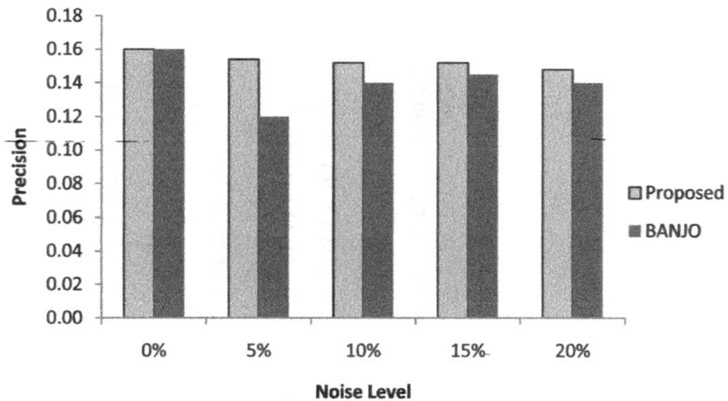

(c) Precision(Net-3)

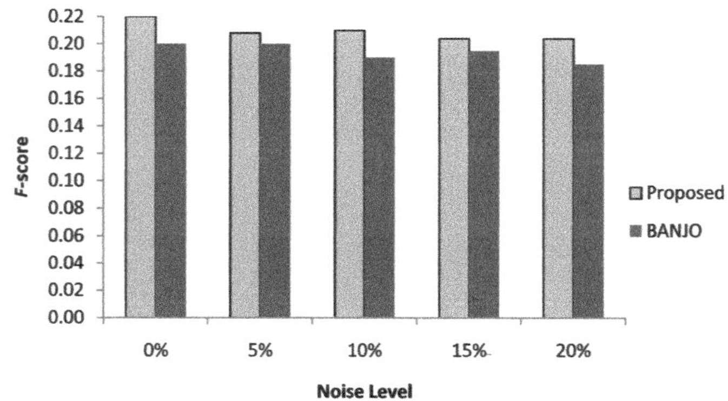

(d) *F*-score(Net-3)

Figure 8.5 Evaluation of the proposed method with BANJO [57] in terms of precision and *F*-score.

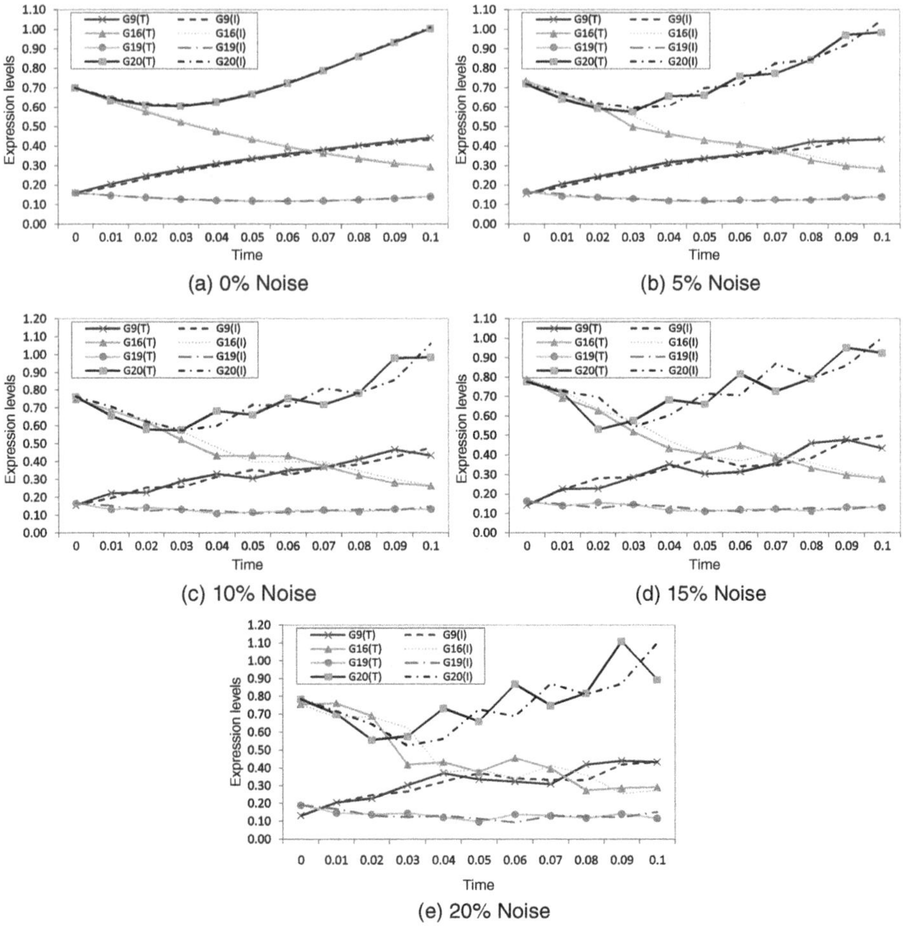

Figure 8.6 Target and inferred expression profiles for four target genes of Net-2 with five different levels of noise. T/I after the gene names in the graphs implies target/inferred. The horizontal and vertical axes in the graphs represent time and expression levels, respectively.

method outperforms BANJO [57] at all five levels of noise for Net-2 and Net-3. Although the results of TGs of Net-1 are more accurate (shown in Figure 8.6 for four TGs of Net-1), the time expressions of TGs of Net-2 and Net-3 certainly follow the trend of the target expressions.

Furthermore, let us also consider the issue of computational time to reconstruct the network. The S-system-based methods require large execution time not only due to learning of a large number of parameters but also due to the numerical integration. The proposed method resolves this issue by decomposing the network into subnetworks and by efficient partitioning of the search space by designing a new fitness function. On an average, the proposed method required around 3 min to estimate the parameters of each gene for Net-1 using the decoupled equation. On the other hand, the average times for ALG [31] and REGARD [12] were close to 8 h and 4 h, respectively. For larger networks, the proposed method required around 2 min and 13 min with Net-2 and Net-3, respectively. However, both the existing methods failed to converge within the 10 h of execution. The computation time for a single gene in the decoupled S-system largely depends on the number of RGs for the proposed method, which is the total number of genes (both RGs and TGs) for both ALG [31] and REGARD [12].

8.6 CONCLUSION

Revealing genetic regulations from high-throughput data using computational techniques is always considered as a challenging task. While there have been some recent efforts using various modeling approaches to reverse engineering large genetic networks, the current state-of-the-art S-system modeling techniques are limited to small-, and at most medium-scale networks due to the large number of model parameters. In this chapter, we first discussed the existing state-of-the-art methods for inferring GRN using the S-system model. Later, we have presented the three-stage computational framework to reconstruct large-scale genetic networks with the S-system model by decomposing a network into two sub-networks based on prior biological knowledge and independently inferring the regulations of the decomposed networks. For optimization, we have proposed a regulatory genes-cardinality-based fitness function that effectively narrows down the search based on population statistics resulting in faster convergence. Investigations carried on two large-scale genetic networks and a medium-scale network, with varying noise level in each network, show excellent performance over both well-known S-system and non-S-system-based models.

ACKNOWLEDGMENTS

This work has been supported by a Post Publication Award of Monash University, Australia. The authors acknowledge the useful discussions with Dr. Nguyen Xuan Vinh, Research Fellow, University of Melbourne, during initial stages of the research work.

REFERENCES

1. H. Akaike. Information theory and an extension of the maximum likelihood principle. In: *International Symposium on Information Theory*, pp 267–281, 1973.
2. S. Ando and H. Iba. Construction of genetic network using evolutionary algorithm and combined fitness function. *Genome Informatics*, 14:94–103, 2003.
3. J. E. Bailey. Lessons from metabolic engineering for functional genomics and drug discovery. *Nature Biotechnology*, 17(7):616–618, 1999.

4. M. Bansal, G. D. Gatta, and D. di Bernardo. Inference of gene regulatory networks and compound mode of action from time course gene expression profiles. *Bioinformatics*, 22(7):815–822, 2006.
5. H. Bolouri and E. H. Davidson. Modeling transcriptional regulatory networks. *Bioessays*, 24(12):1118–1129, 2002.
6. P. Brazhnik, A. de la Fuente, and P. Mendes. Gene networks: how to put the function in genomics. *Trends in Biotechnology*, 20(11):467–472, 2002.
7. E. Burke, S. Gustafson, and K. G. Diversity in genetic programming: an analysis of measures and correlation with fitness. *IEEE Transactions on Evolutionary Computation*, 8(1):47–62, 2004.
8. I. Cantone, L. Marucci, F. Iorio, M. A. Ricci, V. Belcastro, M. Bansal, S. Santini, M. di Bernardo, D. di Bernardo, and M. P. Cosma. A yeast synthetic network for in vivo assessment of reverse-engineering and modeling approaches. *Cell*, 137:172–181, 2009.
9. K. Chen and N. Rajapaksy. The evolution of gene regulation by transcription factor and micrornas. *Nature Reviews Genetics*, 8:93–103, 2007.
10. D. Y. Cho, K. H. Cho, and B. T. Zhang. Identification of biochemical networks by s-tree based genetic programming. *Bioinformatics*, 22:1631–1640, 2006.
11. A. R. Chowdhury and M. Chetty. An improved method to infer gene regulatory network using S-system. In: *IEEE Congress on Evolutionary Computation*, pp. 1012–1019, 2011.
12. A. R. Chowdhury, M. Chetty, and N. X. Vinh. Adaptive regulatory genes cardinality for reconstructing genetic networks. In: *IEEE Congress on Evolutionary Computation*, pp. 1–8, 2012.
13. A. R. Chowdhury, M. Chetty, and N. X. Vinh. Incorporating time-delays in S-system model for reverse engineering genetic networks. *BMC Bioinformatics*, 14:196, 2013.
14. A. R. Chowdhury, M. Chetty, and N. X. Vinh. On the analysis of time-delayed interactions in genetic network using S-system model. In *International Conference on Neural Information Processing (2)*, pp. 616–623, 2013.
15. A. R. Chowdhury, M. Chetty, and N. X. Vinh. Reverse engineering genetic networks with time-delayed S-system model and Pearson correlation coefficient. In *International Conference on Neural Information Processing (2)*, pp. 624–631, 2013.
16. P. Csermely, T. Korcsmáros, H. J. Kiss, G. London, and R. Nussinov. Structure and dynamics of molecular networks: a novel paradigm of drug discovery: A comprehensive review. *Pharmacology and Therapeutics*, 138:333–408, 2013.
17. H. Y. Fan and J. Lampinen. A trigonometric mutation operation to differential evolution. *Journal of Global Optimization*, 27:105–129, 2003.
18. T. S. Gardner, D. di Bernardo, D. Lorenz, and J. J. Collins. Inferring genetic networks and identifying compound mode of action via expression profiling. *Science*, 301(5629):102–105, 2003.
19. D. E. Goldberg. *The Design of Innovation: Lessons from and for Competent Genetic Algorithm*. Norwell, MA, USA, Kluwer Academic, 2002.
20. J. J. Grefenstette. Genetic algorithms for changing environments. In: *Parallel Problem Solving From Nature II*, Elsevier, Amsterdam, pp. 137–144, 1992.
21. M. M. Hasan, N. Noman, and H. Iba. A prior knowledge based approach to infer gene regulatory networks. In: *International Symposium on Biocomputing*, pp. 15–17, 2010.
22. L. He and G. J. Hannon. Micrornas: small RNAs with a big role in gene regulation. *Nature Reviews Genetics*, 5:522–531, 2004.

23. S. Kikuchi, D. Tominaga, M. Arita, K. Takahashi, and M. Tomita. Dynamic modeling of genetic networks using genetic algorithm and S-system. *Bioinformatics*, 19(5):643–650, 2003.

24. S. Kimura, Y. Amano, K. Matsumura, and M. Okada-Hatakeyama. Effective parameter estimation for S-system models using lpms and evolutionary algorithms. In: *IEEE Congress on Evolutionary Computation*, pp. 1–8, 2010.

25. S. Kimura, M. Hatakeyama, and A. Konagaya. Inference of S-system models of genetic networks from noisy time-series data. *Chem-Bio Informatics Journal*, 4(1):1–14, 2004.

26. S. Kimura, K. Ide, A. Kashihara, M. Kano, M. Hatakeyama, R. Masui, N. Nakagawa, S. Yokoyama, S. Kuramitsu, and A. Konagaya. Inference of S-system models of genetic networks using a cooperative coevolutionary algorithm. *Bioinformatics*, 21(7):1154–1163, 2005.

27. C.-L. Ko, F.-S. Wang, Y.-P. Chaob, and T.-W. Chena. S-system approach to modeling recombinant *Escherichia coli* growth by hybrid differential evolution with data collocation. *Biochemical Engineering Journal*, 28:1016, 2008.

28. L.-Z. Liu, F.-X. Wu, and W.-J. Zhang. Alternating weighted least squares parameter estimation for biological S-systems. In: *IEEE International Conference on Systems Biology*, pp. 6–11, 2012.

29. L.-Z. Liu, F.-X. Wu, and W.-J. Zhang. Inference of biological S-system using the separable estimation method and the genetic algorithm. *IEEE/ACM Transactions on Computational Biology and Bioinformatics*, 9(4):955–965, 2012.

30. Y. Maki, T. Ueda, M. Okamoto, N. Uematsu, K. Inamura, K. Uchida, Y. Takahashi, and Y. Eguchi. Inference of genetic network using the expression profile time course data of mouse p19 cells. *Genome Informatics*, 13:382–383, 2002.

31. N. Noman. A memetic algorithm for reconstructing gene regulatory networks from expression profile. PhD thesis, Graduate School of Frontier Sciences at the University of Tokyo, 2007.

32. N. Noman and H. Iba. On the reconstruction of gene regulatory networks from noisy expression profiles. In: *IEEE Congress on Evolutionary Computation*, pp. 2543–2550, 2006.

33. N. Noman and H. Iba. Reverse engineering genetic networks using evolutionary computation. *Genome Informatics*, 16(2):205–214, 2006.

34. N. Noman and H. Iba. Inferring gene regulatory networks using differential evolution with local search heuristics. *IEEE/ACM Transactions on Computational Biology and Bioinformatics*, 4:634–647, 2007.

35. L. Palafox, N. Noman, and H. Iba. Reverse engineering of gene regulatory networks using dissipative particle swarm optimization. *IEEE Transactions on Evolutionary Computation*, 17(4):577–587, 2013.

36. W. Press, S. Teukolsky, W. Vetterling, and B. Flannery. *Numerical Recipies in C, 2nd ed.* Cambridge University Press, Cambridge, 1995.

37. R. Prill, J. Saez-Rodriguez, L. alexopoulos, P. Sorger, and G. Stolovitzky. Crowdsourcing network inference: the dream predictive signaling network challenges. *Science Signalling*, 4(189):mr7, 2011.

38. R. Ram and M. Chetty. A Markov-blanket-based model for gene regulatory network inference. *IEEE/ACM Transactions on Computational Biology and Bioinformatics*, 8(2):353–367, 2011.

39. J. Ronkkonen, S. Kukkonen, and K. Price. Real-parameter optimization with differential evolution. In: *IEEE Congress on Evolutionary Computation*, volume 1, pp. 506–513, 2005.
40. E. Sakamoto and H. Iba. Inferring a system of differential equations for a gene regulatory network by using genetic programming. In: *IEEE Congress on Evolutionary Computation*, pp. 720–726, 2001.
41. M. Savageau. *Biochemical Systems Analysis. A Study of Function and Design in Molecular Biology*. Addison-Wesley Publishing Company, Massachusetts, 1976.
42. T. Schaffter, D. Marbach, and D. Floreano. Genenetweaver: *in silico* benchmark generation and performance profiling of network inference methods. *Bioinformatics*, 27(16):2263–2270, 2011.
43. M. Schena. *DNA Microarrays: A Practical Approach*. Oxford University Press, Oxford, 1993.
44. P. Sheridan, T. Kamimura, and H. Shimodaira. A scale-free structure prior for graphical models with applications in functional genomics. *PLoS One*, 5(11):e13580, 2010.
45. W. M. Spears, K. A. D. Jong, T. Bäck, D. B. Fogel, and H. de Garis. An overview of evolutionary computation. In: *European Conference on Machine Learning*, pp. 442–459, 1993.
46. R. Storn. System design by constraint adaptation and differential evolution. *IEEE Transactions on Evolutionary Computation*, 3(1):22–34, 1999.
47. R. Storn and K. V. Price. Differential evolution—a simple and efficient heuristic for global optimization over continuous spaces. *Journal of Global Optimization*, 11:341–359, 1997.
48. F. Streichert, H. Planatscher, C. Spieth, H. Ulmer, and A. Zell. Comparing genetic programming and evolution strategies on inferring gene regulatory networks. In: *Genetic and Evolutionary Computation Conference*, p. 2004, 2004.
49. S. M. T. Akutsu and S. Kuhara. Inferring qualitative relations in genetic networks and metabolic pathways. *Bioinformatics*, 18(8):727–734, 2000.
50. J. Tegner, M. K. S. Yeung, J. Hasty, and J. J. Collins. Reverse engineering gene networks: integrating genetic perturbations with dynamical modeling. *National Academy of Sciences*, 100(10):5944–5949, 2003.
51. D. Tominaga, N. Koga, and M. Okamoto. Efficient numerical optimization algorithm based on genetic algorithm for inverse problem. In: *Genetic and Evolutionary Computation Conference*, pp. 251–258, 2000.
52. K.-Y. Tsai and F.-S. Wang. Evolutionary optimization with data collocation for reverse engineering of biological networks. *Bioinformatics*, 21(7):1180–1188, 2005.
53. E. O. Voit. Biochemical systems theory: a review. *ISRN Biomathematics*, 2013:1–53, 2013.
54. E. O. Voit and J. Almeida. Decoupling dynamical systems for pathway identification from metabolic profiles. *Bioinformatics*, 20:1670–1681, 2004.
55. H. Wang, D. Wang, and Y. Shengxiang. A memetic algorithm with adaptive hill climbing strategy for dynamic optimization problems. *Soft Computing*, 13(8–9):763–780, 2009.
56. L. Wang, H. Ni, R. Yang, V. Pappu, M. B. Fenn, and P. M. Pardalos. Feature selection based on meta-heuristics for biomedicine. *Optimization Methods and Software*, 29(4):1–18, 2013.
57. J. Yu, V. A. Smith, P. P. Wang, A. J. Hartemink, and E. D. Jarvis. Advances to Bayesian network inference for generating causal networks from observational biological data. *Bioinformatics*, 20:3594–3603, 2004.

III

EAs FOR EVOLVING GRNs AND REACTION NETWORKS

9

DESIGN AUTOMATION OF NUCLEIC ACID REACTION SYSTEM SIMULATED BY CHEMICAL KINETICS BASED ON GRAPH REWRITING MODEL

Ibuki Kawamata

Department of Bioengineering and Robotics, Graduate School of Engineering, Tohoku University, Miyagi, Japan

Masami Hagiya

Department of Computer Science, Graduate School of Information Science and Technology, The University of Tokyo, Tokoyo, Japan

9.1 INTRODUCTION

Deoxyribonucleic acid (DNA) and ribonucleic acid (RNA) are used as a genetic information carrier and a functional module in nature. Gene regulatory network (GRN) focuses on the topology of how genes coded in DNA interact each other. In this chapter, we rather focus on direct interactions among nucleic acids than reactions through genes.

Evolutionary Computation in Gene Regulatory Network Research, First Edition.
Edited by Hitoshi Iba and Nasimul Noman.

Specifically, we regard nucleic acids as a programmable material to directly implement artificial dynamic systems. Basic reactions employed in a system are simple base-pairing and enzymatic reactions. Functional systems, which we call *nucleic acid reaction systems*, can be programmed and driven by cascading such basic reactions.

Typical modules of nucleic acid reaction systems are combinatorial circuit [1–3], sensing switch [4,5], and autonomous actuators [6,7]. By combining such modules, applications such as molecular sensors [8–10], drug delivery system [11], cell diagnosis system [12], and implementation of neural network [13] have been demonstrated. Such systems are designed to contribute to subjects which can be categorized as follows: engineering nano-scale machines, reconstructing living organism, and the treatment of diseases.

To explore further complex systems, techniques to analyze and strategies to design a new module with the aid of a computer play an important role in the field. Straightforward way to represent nucleic acids and simulate their dynamic behaviour is employing atomic [14, 15] and coarse-grained [16, 17] models. Simulations based on such molecular dynamics, however, are computationally expensive to predict the behavior of large dynamic nucleic acid reaction systems. To overcome the problem, relatively abstract modeling, programming language, and frameworks for composable enzyme-free nucleic acid reaction systems have been proposed [2,18,19]. Computer-aided design tool is also developed for scaling up an enzymatic reaction network [20].

With such modeling and simulating techniques, however, designing nucleic acid reaction system that has desired functionality still requires human trial and error to find out a topology of interest. The design process becomes difficult when the number of molecular species in a system increases. It is because the whole network of the system is extremely large by cross-talks among nucleic acids, which makes it unintelligible to prevent undesired reactions.

To eliminate the human trial and error in the designing process, it is possible to apply an evolutionary computation using heuristic algorithms that search for a semi-optimal system. In the case of GRN, evolutionary computation has been successfully applied to various dynamic systems such as oscillatory behavior [21,22]. Although the evolutionary computations had significant impact because of the design of nontrivial topologies, the designed GRNs were not so easy to implement as actual systems. One of the reasons for the problem comes from unrealistic models employed in the design of GRN. For instance, the models are based on pseudo-gene [23], simple network [24,25], and a combination of standard parts library [26]. Those models are in the higher level of abstraction and assume unrealizable properties of genes and their regulation. For actual implementation, a system composed of nucleic acids has more advantage than GRN because they are programmable materials and their reactions are realizable as simple chemical reactions [27].

We introduce our original approaches to model, simulate, and automatically design nucleic acid reaction systems with desired functions. This chapter is composed of five sections, which are arranged in the following order. In Section 9.1, we have overviewed the potential applications of nucleic acid reaction systems and pointed out the problems of conventional modeling, simulation, and designing methods. In

Section 9.2, our graph-based modeling of nucleic acids is illustrated using simple examples. In Section 9.3, we explain a technique to simulate a nucleic acid reaction system represented by our graph-based modeling. Section 9.4 describes our automatic design algorithm capable of searching for a system with desired function. Finally we discuss the limitations of our approaches in Section 9.5.

9.2 NUCLEIC ACID REACTION SYSTEM

9.2.1 Domain-Level Modeling

Basic properties and a standard representation of nucleic acid reaction systems are first explained. From a chemical point of view, a strand of DNA is a polymer of deoxyribonucleotide monomers. Each nucleotide contains one of the four bases adenine, thymine, guanine, and cytosine (written as "A", "T", "G", and "C", respectively). On the other hand, RNA is a polymer of ribonucleotide monomers, which employs "U" (uracil) instead of "T." Due to the chemical structure of each base, nucleotides are capable of forming hydrogen bonds between complementary pairs ("A"-"T" and "G"-"C" for DNA and "A"-"U" and "G"-"C" for RNA). As a consequence, two complementary strands hybridize into a well known double stranded structure [28]. To illustrate dynamic systems, a simple representation of nucleic acids has been exploited (Figure 9.1), which focuses on the topological information of DNA [29, 30]. In the representation, single stranded DNA (ssDNA) is drawn as a straight arrow representing the direction from $5'$ to $3'$ ends of DNA. Only complementary strands in anti-parallel directions can hybridize. A sequence of nucleotides is divided into subsequences in accordance with reaction units (called *domain* or *abstract base*), where each unit is denoted by a letter. We use uppercase and lowercase letters to represent complementary domains and connect them by a line that corresponds to hydrogen bonds.

Nucleic acid reaction systems using RNA can be also represented by the domain level representation. To distinguish RNA from DNA, zigzag arrow is assigned for RNA [32]. Since DNA is more stable and less reactive compared with RNA, one can select DNA or RNA (or both of them) as a material for nucleic acid reaction system depending on interest. For example, DNA is favorable for constructing a dynamic nano-scale device that requires geometric structure [33, 34]. In contrast, it is possible to take advantage of RNA to moderate gene expressions by a mechanism called RNA interference (RNAi) [35–37].

9.2.2 Hydrogen Bond Reactions

Dynamics of reaction systems can be achieved by repeating basic reactions of nucleic acids. We exploit two types of reactions—hydrogen bond and enzymatic reactions. The reactions of hydrogen bond are classified into three basic reactions, which are hybridization, denaturation, and branch migration. In contrast, the variation of enzymes can cause a wide variety of basic reactions.

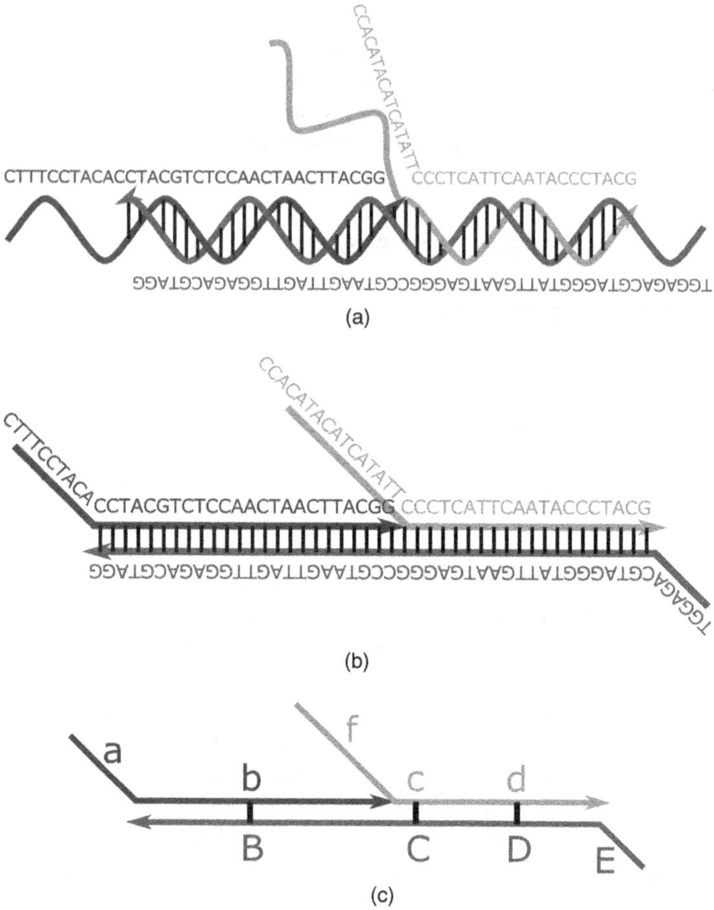

Figure 9.1 Domain level representation is explained in a step-by-step manner using DNA catalytic gate [31] as an example. (a) All sequences of nucleotides and hydrogen bonds are shown using double helical arrows and lines. (b) Single stranded DNA is represented as a straight arrow. (c) Letters are allocated to domains that are units of reactions.

In the domain level representation, hybridization and denaturation are association and dissociation between complementary domains, respectively. By branch migration, unhybridized domain displaces neighboring hybridized domain that has the same sequence.

Only the three hydrogen bond reactions (simply called hybridization reactions) take place in the example of catalytic reaction (Figure 9.2). Without a catalyst, the substrate and the fuel cannot react because domains of the substrate are not accessible to the fuel strand. In the presence of the input, however, they hybridize together and release the output and the signal strands. Since the input serves as catalyst that can promote multiple cycles of reactions, the concentration of output can be larger than that of input.

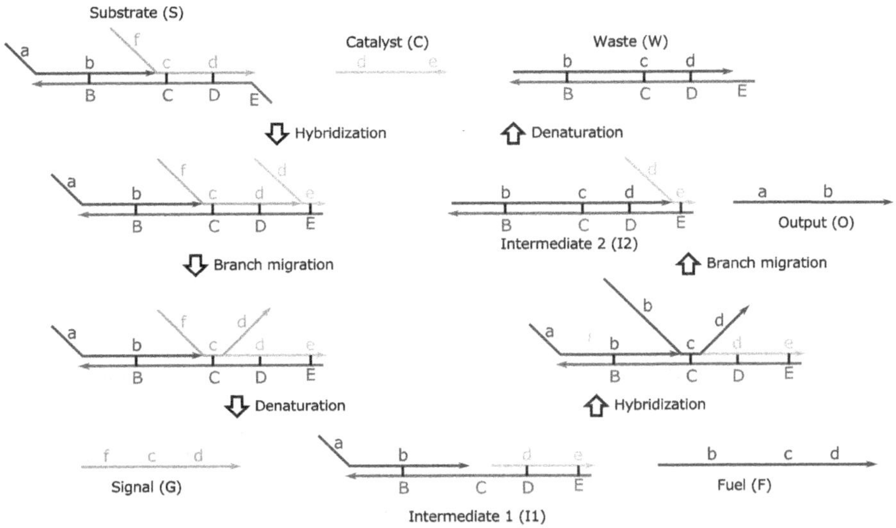

Figure 9.2 Hydrogen bond reactions of catalytic gate [31]. One cycle of reaction starts from the top and continue in anticlockwise direction. The gate consists of the substrate (S) and the fuel (F), while the catalyst (C) serves as an input of the gate. S and C first hybridize together and result in an intermediate structure. By branch migration and denaturation, the structure separates into the signal (G) and the intermediate 1 (I1). The I1 and the F turn into the intermediate 2 (I2) and the output (O) by another sequential hybridization and branch migration. Finally, I2 releases C and becomes the waste product (W) by denaturation.

9.2.3 Enzymatic Reactions

Further functionalities can be implemented by employing enzymatic reactions. Enzymes that react with nucleic acids include, but not limited to, polymerase, nuclease, and ligase. Polymerase synthesizes a new sequence that is complementary to template nucleic acids. The polymerase is called DNA-directed or RNA-directed polymerase depending on the nucleic acids of the template. Furthermore, depending on nucleic acids of the newly synthesized strand, it is called DNA or RNA polymerase. For example, RNA-directed DNA polymerase synthesizes a new DNA strand from RNA template.

DNA polymerase is fundamentally primer-dependent, in which the enzyme always extends the 3′ end of primer strands. This type of polymerase has been exploited to amplify the amount of DNA by polymerase chain reaction (PCR) to detect small amount of DNA. On the other hand, some of RNA polymerases employ primer-independent polymerization. If a new RNA strand is polymerized from double stranded DNA (dsDNA) template starting from a special promoter region, the reaction is called transcription.

Nucleases cleave phosphate backbone of nucleic acids in various manners. For instance, DNA and RNA are decomposed by deoxyribo- and ribo-nuclease, respectively. In contrast, ligase join 3′ and 5′ ends of two DNA strands, which plays an important role in replicating DNA [38].

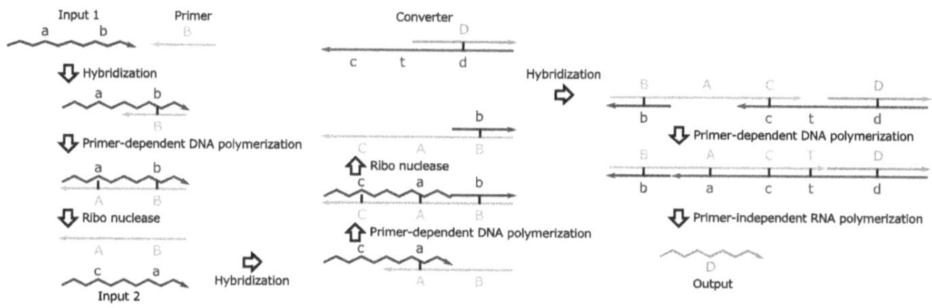

Figure 9.3 Enzymatic reactions of the AND gate of RTRACS [32]. The gate consists of DNA primer and converter, while the inputs and output are ssRNA. After the primer and the input 1 hybridize together, the primer is extended by an RNA-directed primer-dependent DNA polymerase. Ribo-nuclease then decomposes only the input 1. The remained ssDNA hybridizes to the input 2 for another polymerization and decomposition, which results in a partially double stranded DNA. The structure hybridizes to the converter and forms fully double stranded DNA by polymerization. Since promoter region (indicated by "T" in the figure) becomes double stranded, downstream of the dsDNA is transcribed to the output by a primer-independent RNA polymerization.

Reverse-transcription-and-transcription-based autonomous computing system (RTRACS) [32, 39, 40] makes use of polymerization reactions to implement a logic operation (Figure 9.3). The primer ssDNA of RTRCS is extended by a collaboration of two ssRNA inputs and polymerization. The extended structure and the converter then form a full duplex DNA structure by hybridization and polymerization. Promoter region in the double stranded structure is recognized by a primer-independent polymerase to initiate transcription. The system can be regarded as an AND gate because the output ssRNA is synthesized if and only if both inputs are given.

9.2.4 Graph-Based Model

To systematically predict the dynamic behavior of a nucleic acid reaction system, we propose a model based on a graph data structure that is derived from the domain level representation. As an example, the graph-based model of three examples including the DNA catalytic gate are shown in Figure 9.4. We represent domains, hydrogen bonds, and phosphate backbones as nodes, undirected edges, and directed edges of graph data structure, respectively. The advantage of the model compared with other conventional models is the flexibility to represent various secondary structures of nucleic acids. A flexible representation is useful not only for the semi-automated simulation of nucleic acid reaction systems but also for the design of modules that take advantage of complex structures of nucleic acids.

One parameter is allocated to each domain in order to tell the differences among domains. The parameter corresponds to the length of nucleotides sequence in a domain. As shown in the domain level representation of the DNA catalytic gate (Figure 9.1), 10, 24, 4, 16, 6, and 16 are allocated to the domains "a", "b", "c", "d", "e", and "f" in Figure 9.4(a), respectively. The parameter determines the kinetic constant of denaturation reaction, which is explained in detail in the next simulation section.

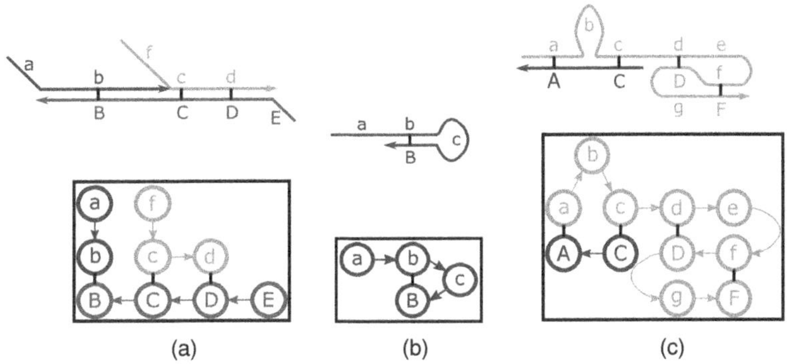

Figure 9.4 Graph-based models of the (a) catalytic gate [31], (b) hairpin loop, (c) bulge loop, and pseudoknot. Domain level representations and corresponding graph-based models are shown in the top and the bottom of the figure, respectively. Domain "c" of (b), "b" of (c), and from "d" to "F" of (c) are hairpin loop, bulge loop, and pseudoknot, respectively. We surround a connected graph with a rectangle shape because a connected graph corresponds to one molecule.

We further model basic chemical reactions to systematically illustrate the topology of nucleic acid reaction systems. Since a DNA molecule is represented by a graph based model, a chemical reaction is defined as a transformation of graph, or graph rewriting. A graph rewriting rule is in a form $L \rightarrow R$ that is similar to normal chemical equation. L and R correspond to patterns of graph data structures before and after the reaction of the rewriting rule, respectively. By replacing a subgraph that matches L into R, a new graph is derived from a graph. The topology of the catalytic gate is shown in Figure 9.5 as an example. All basic rewriting rules are summarized in the following section.

9.3 SIMULATION BY CHEMICAL KINETICS

9.3.1 Enumeration of Structure

The graph rewriting model provides a simulation method based on chemical kinetics. From an initial configuration, ordinary differential equations (ODEs) can be systematically built. The initial configuration is a set of structures and their concentrations. By numerically solving the ODEs, the dynamic behavior of a system can be predicted. Our simulator has two parts of calculations: enumerating structures that are producible from initial configuration and analyzing the concentration changes by a numerical method.

At the beginning of a simulation, the simulator enumerates whole structures in a system to determine the number of variables. By iteratively applying rewriting rules from the initial configuration, all the producible structures are derived. During the enumeration, the simulator prohibits a structure that contains multiple identical strands because it is necessary to prevent an explosion of structure space.

After the enumeration process, the simulator assigns variables, which represent concentrations of molecules, to all the structures. Based on chemical kinetics, the

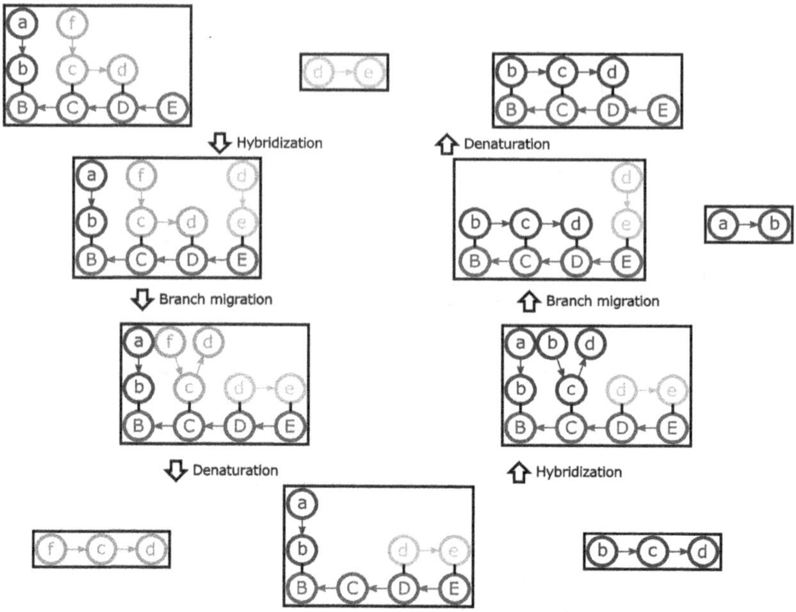

Figure 9.5 Transition of graph-based model of the catalytic gate [31] by applying graph rewriting rules. The figure is arranged in the same way as Figure 9.2. Although each reaction corresponds to a transition of graphs, the transition of branch migration reaction is applied twice for the reaction on the right to displace both "b" and "d" at once. A graph can be separated into multiple connected graphs as a result of graph transitions such as branch migration and denaturation.

simulator formalizes ODE for each variable. The ODEs are numerically solved to predict the time evolution of the concentrations of molecules. We actually carry a standard numerical calculation called Runge–Kutta–Fehlberg-4,5 method to solve the ODEs [41]. The rate constants of basic reactions are defined by changing the order of magnitude while the units of concentration and time are arbitrary.

Transitions of basic rewriting rules for hybridization reactions are summarized in Table 9.1. Although the rate constants of hybridization and branch migration are fixed, that of denaturation depends on the length parameter. Suppose the length parameter of a domain to dissociate is l, the rate constant k_d is calculated by $8.0 \times (-2)^l$. The function is chosen because we assume that the free energy of double helix decreases proportional to the length of nucleotide sequence [42]. Note that time-derivation term of each structure becomes a summation of all expressions of possible reactions. For example, if both bimolecular hybridization and denaturation reactions in the Table 9.1 take place, the term of C_2 is formalized as $\frac{d}{dt}C_2 = k_d C_4 - k_{h1} C_1 C_2$.

Similarly, basic rewriting rules of enzymatic reactions are summarized in Table 9.2. Unlike the hybridization reactions, rules of enzymatic reactions are capable of operating directed arrows that represent phosphate backbones of nucleic acids. Any enzymatic reactions that produce a structure with more than two identical strands are also prohibited as in the case of hybridization reactions.

Even the simulator prohibits some reactions, combinatorial explosion of structure space is inevitable. To ignore unimportant structures that may not be the main

Table 9.1 Transition by basic hybridization reactions. Basic rewriting rules of hybridization reactions are summarized

Reaction name	Schematic example	Equations	Rate constant
Bimolecular hybridization	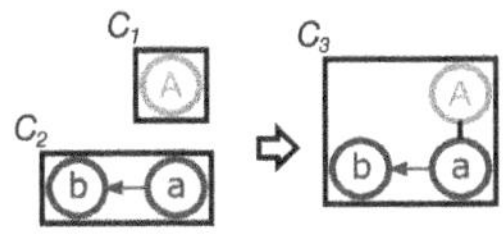	$\frac{d}{dt}C_1 = -k_{h1}C_1C_2$ $\frac{d}{dt}C_2 = -k_{h1}C_1C_2$ $\frac{d}{dt}C_3 = k_{h1}C_1C_2$	$k_{h1} = 0.1$
Unimolecular hybridization by extending	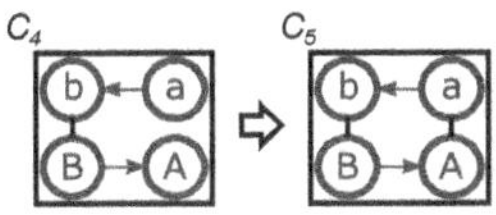	$\frac{d}{dt}C_4 = -k_{h2}C_4$ $\frac{d}{dt}C_5 = k_{h2}C_4$	$k_{h2} = 1.0$
Unimolecular hybridization with interval	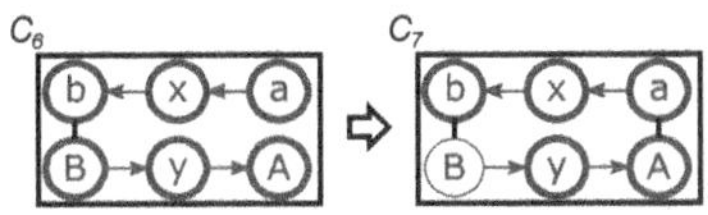	$\frac{d}{dt}C_6 = -k_{h3}C_6$ $\frac{d}{dt}C_7 = k_{h3}C_6$	$k_{h3} = 0.1$
Denaturation	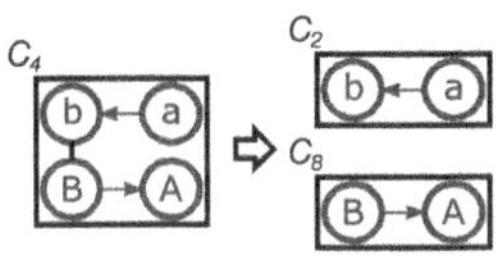	$\frac{d}{dt}C_4 = -k_dC_4$ $\frac{d}{dt}C_2 = k_dC_4$ $\frac{d}{dt}C_8 = k_dC_4$	k_d is defined as a function
Branch migration	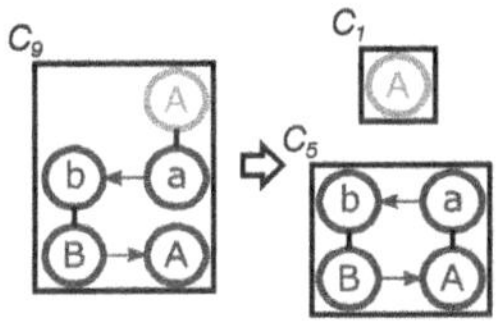	$\frac{d}{dt}C_9 = -k_bC_9$ $\frac{d}{dt}C_1 = k_bC_9$ $\frac{d}{dt}C_5 = k_bC_9$	$k_b = 0.01$

First column describes the name of reaction rule. Second column illustrates a typical example of graph rewriting by applying the reaction rule. Concentration of a structure is denoted by a variable written in the upper left of the rectangle that is surrounding the structure. ODEs and rate constant for the reaction rule are listed in the third and fourth column, respectively.

products of a system, we introduce a threshold. Structures that have a concentration lower than 10^{-5} are disregarded. By the threshold, the rewriting rules are applied to produce new structures only when the enumeration process becomes necessary. The simulator produces structures dynamically and checks whether the concentration of each structure exceeds the threshold. To be more precise, the duration of a simulation is divided into intervals so that the simulator is able to check the concentration at the beginning of each interval.

9.3.2 Time Evolution of Catalytic Gate and RTRACS

To show that graph-based model is capable of predicting the behavior of nucleic acid reaction systems, the catalytic gate [31] and the AND gate of RTRACS [32] are selected as benchmarks. The result of kinetic simulations of the catalytic gate is shown in Figure 9.6. Catalytic amplification was verified because the concentration

Table 9.2 Transition by basic enzymatic reactions. Meaning of columns are the same as in the previous Table

Reaction name	Schematic example	Equations	Rate constant
Primer-dependent DNA-directed DNA polymerization	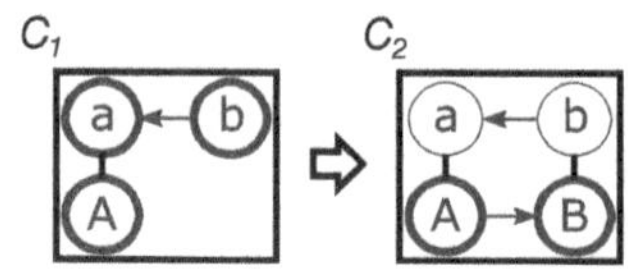	$\frac{d}{dt}C_1 = -k_{ddd}C_1$ $\frac{d}{dt}C_2 = k_{ddd}C_1$	$k_{ddd} = 0.005$
Primer-dependent RNA-directed DNA polymerization	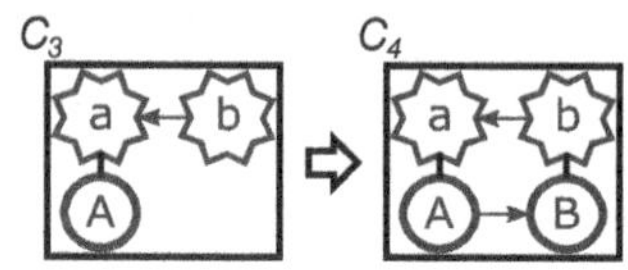	$\frac{d}{dt}C_3 = -k_{drd}C_3$ $\frac{d}{dt}C_4 = k_{drd}C_3$	$k_{drd} = 0.005$
Primer-dependent RNA-directed RNA polymerization	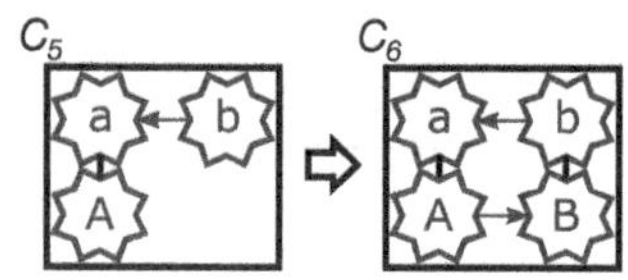	$\frac{d}{dt}C_5 = -k_{drr}C_5$ $\frac{d}{dt}C_6 = k_{drr}C_5$	$k_{drr} = 0.005$
Primer-independent DNA-directed RNA polymerization	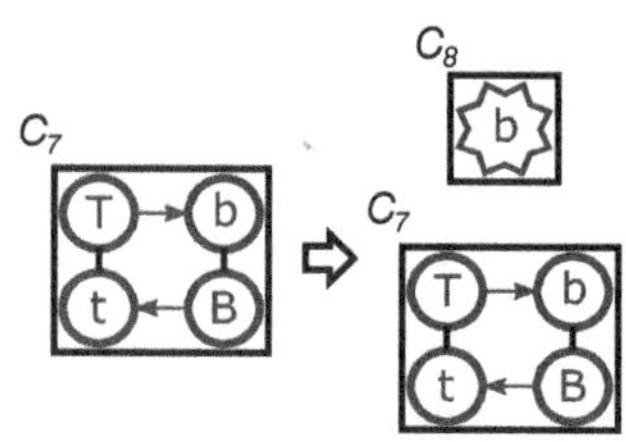	$\frac{d}{dt}C_8 = k_{idr}C_7$	$k_{idr} = 0.005$
Decay	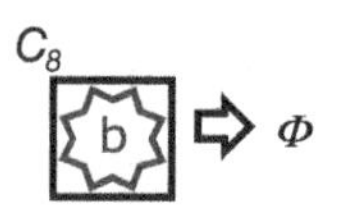	$\frac{d}{dt}C_8 = -k_{decay}C_8$	$k_{decay} = 0.001$

Nodes of RNA are represented by zigzag shapes, which is consistent with the domain level representation of RTRCS. In the case of primer-independent DNA-directed RNA polymerization, downstream of promoter double stranded domain 'T' is transcribed to an mRNA. By a decay reaction, all the RNA nodes disappear from a graph. When the result of graph rewriting by a decay reaction is an empty graph, it is denoted by Φ.

of output increased higher than that of input in most of the cases. The results were in good agreement with the experimental results shown in the original paper. We then simulated the AND gate of RTRACS [32] to examine the effect of enzymatic reactions. As in the result (Figure 9.7), the concentration of RNA output increased if and only if both RNA inputs were added. Since the output is synthesized by transcription, the concentration of output showed no increase for the false combinations of inputs where promoter region cannot form dsDNA. This result indicates that the simulator can predict the behavior of a nucleic acid reaction system driven by enzymatic reactions. It is possible to adapt the simulator for further design automation by using the simulator as a tool to evaluate nucleic acid reaction systems.

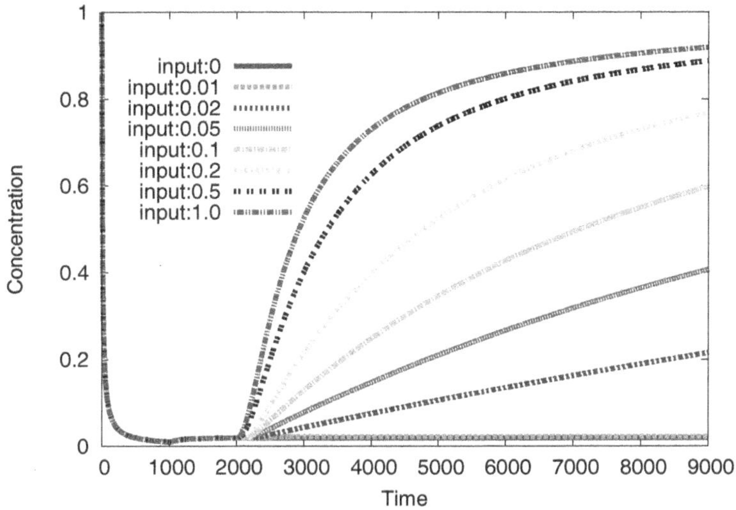

Figure 9.6 Simulation result of the catalytic gate [31]. *x* and *y* axes are time and concentration of the output strand, respectively. First changes in first 2000 time units were caused by making a gate. Input was added at 2000 time units, which leads to the increase of the concentration of output. The graph legends indicate the concentration of input. When the concentration of input was 0.01, that of output did not increase because the concentration of input was too small so that an intermediate structure could not exceed the threshold.

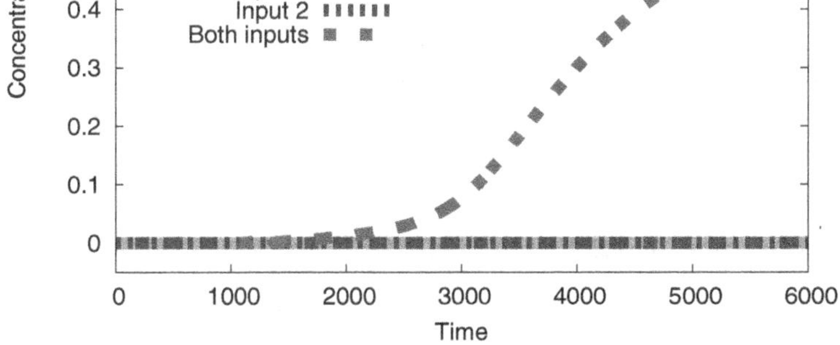

Figure 9.7 Simulation result of the AND gate of RTRACS [32]. *x* and *y* axes are time and concentration of the output strand, respectively. The inputs were added at 1000 time units. The concentration of the output began to increase from about 2000 time units when both of the inputs were added. The graph legends indicate the combination of inputs.

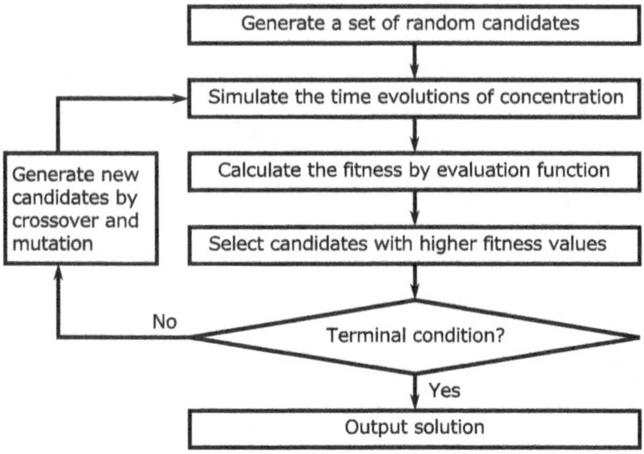

Figure 9.8 Flowchart of automatic design algorithm. After generating a set of random initial candidates, the algorithm iterates the process of simulation, evaluation, selection, and generation until the terminal condition is satisfied.

9.4 AUTOMATIC DESIGN OF NUCLEIC ACID REACTION SYSTEM

9.4.1 Algorithm of Evolutionary Computation

In this section, our strategy and the results of automatic design of nucleic acid reaction systems are explained. First, we briefly overview genetic algorithm (Figure 9.8). The algorithm tries to maximize a fitness value, which is a typical approach for combinatorial optimization problem.

Starting from a set of randomly generated candidates, the algorithm searches for a topology with bigger fitness value by iterating from one generation to another. Fitness value of each candidate is first calculated by predicting the dynamical behavior of the candidate. A new set of candidates is then generated by using two operations, which are crossover and mutation. After that, fitness values are again calculated for the new candidates. From old and new sets, the algorithm selects candidates with higher fitness values and keeps them as a current set. After iterating the processes, the method outputs a system with the biggest fitness value among produced candidates as a semi-optimal solution.

9.4.2 Genotype of Nucleic Acid Reaction System

The target of our evolutionary algorithm is a gate made of nucleic acid reaction system. To carry the algorithm, a model to represent genotype, a method to predict its phenotype, an evaluation function to calculate the fitness, and operations of crossover and mutation over the genotype are necessary. As a genotype of a single candidate, we adopted sequences of letters. The genotype defines the search space of the algorithm.

We chose 13 letters as an alphabet, which consists of letters from "a" to "f", from "A" to "F," and null letter. A genotype of a gate is represented as a sequence of 16 letters, which are divided into four four-letter sequences. The alphabet is regarded as a set of domains for nucleic acid reactions while each four-letter sequence corresponds to a single stranded nucleic acids. Once a candidate is represented by a sequence of letters, it is possible to chemically implement the system by applying sequence design method to each domain.

The null letter is included in the alphabet to represent a strand with less than four letters and a gate with less than four strands. These parameters such as 13 and 16 that restrict the search space are imposed to eliminate an ineffective search. The parameters may be adjusted depending on the function of interest in order to optimize the search space of the automatic design.

For example, a genotype in the following sequence of letters

a, b, null, null; f, c, b, null; null, null, null, null; E, D, C, B

represents the substrate of the catalytic gate whose phenotype is shown by a graph-based model in Figure 9.4(a). Thanks to the flexibility of the graph-based model, the search space comprises a system that utilizes various secondary structures such as hairpin loop, bulge loop, and pseudoknot. To simplify the notation, the genotype is described by ignoring null letters. The example can be written as "ab", "fcb", "EDCB".

To represent inputs to the gate, additional sequences are introduced to the genotype depending on the target. We target two types of gates, one of which is a combinatorial circuit (logic gate) and the other is a state transition machine (automaton). In the case of logic gate that utilizes single stranded nucleic acids as inputs, two four-letter sequences are added to the genotype to represent the input strands. In contrast, two letters are added for the automaton to represent domains that are stimulated by inputs.

By those definitions, a random candidate can be generated by randomly choosing 24 (for logic gate) and 18 (for automaton) letters from the alphabet. The size of the set of candidates is set to be 256. The length parameter of each domain is also randomly chosen between 3 and 34 for each trial of evolutionary computation.

The operations of crossover and mutation are defined in a straightforward manner. For the crossover, one positions of the letters is randomly selected. From two candidates, a new candidate is generated by swapping letters after the position (known as one-point crossover). A mutation is defined as an operation to flip a randomly chosen letter into another random letter. More precisely, the mutated candidate is produced by repeating such flip for random times (five times at most).

9.4.3 Simulation of Phenotype, Generation, and Selection

The graph-based model is capable of simulating the dynamic behavior of a given genotype. A gate is constituted by hierarchical simulations, in which four strands are simulated separately in the beginning. Then, the results of the first two simulations are merged for another round of simulations, while the same simulation is carried for the last two. Finally, both results are further merged and simulated to constitute a gate, to which inputs can cause successive reactions.

From the result of the simulation, a fitness value is calculated by an evaluation function. The evaluation functions differ from target to target depending on the function of interest. Concrete definitions of the evaluation functions of logic gate and automaton are explained in Sections 9.4.4 and 9.4.5, respectively.

For a new set, 256 new candidates are generated from the current set one by one. One candidate is generated by either crossover or mutation. The probability to choose crossover is 50% and to choose mutation is also 50%. We set relatively high probability for the mutation because the effect of mutation was large in our problem. In the case of crossover, two candidates are chosen from the current sets with the probability proportional to the fitness value (known as roulette wheel selection). The same probability is also applied for mutation though only one candidate is chosen.

When a new set is generated, the current set becomes an old set. After evaluating the new set, top 50% of the candidates among the old and the new sets are selected for the next current set. As a terminal condition, we run the algorithm for 100 and 200 generations for the designs of logic gate and automaton, respectively.

9.4.4 Evaluation Function of Logic Gate

To calculate a fitness value that determines how well a system behaves as a desired system, an evaluation function is required. We define the evaluation function of a logic gate by three criteria. The first criterion v_1 is calculated as

$$v_1 = \max\left\{\min(T(s)) - \max(F(s)) | s \in S\right\},$$

where S is a set of single stranded structures in a system, and $T(s)$ and $F(s)$ are sets of concentrations of single strand s after simulations by adding specific combinations of inputs to the gate. $T(s)$ contains only the results from the combinations of inputs that satisfy the logic, while $F(s)$ contains only the results from the combinations of inputs that do not satisfy the logic.

In an AND gate for example, $T(s)$ has only one element that is the concentration of s after simulating with both inputs. In contrast, $F(s)$ has three elements that are the concentrations of s after simulations with remaining three combinations. This value indicates the smallest difference between true and false states, which means how well the system works as desired logic gate.

The second criterion v_2 is given as

$$v_2 = \max\left\{average(T(s)) - average(F(s)) | s \in S\right\},$$

where S, $T(s)$ and $F(s)$ are the same sets in the first criterion, and *average* calculates an average value from the elements of a given set. This criterion is introduced to make the evaluation function smoother.

The last criterion v_3 is calculated as

$$v_3 = \frac{1}{|S|}\sum_{s \in S}(\max(C(s)) - \min(C(s))),$$

where $C(s) = T(s) \cup F(s)$. This value indicates how well each strand is involved in the reaction, which is an important factor to minimize unnecessary strands.

Finally, the fitness value of a candidate is calculated as a weighted average of these criteria by a function

$$0.8v_1 + 0.1v_2 + 0.1v_3.$$

The output of the gate is dynamically selected in the process of evaluation by

$$output = \underset{s \in S}{argmax}(\min(T(s)) - \max(F(s))),$$

which is a strand that decides v_1.

9.4.5 Evaluation Function of Automaton

Another target is an automaton in which the order of inputs can control the output. We assume that the automaton is a gate composed of DNA that has the following features.

1. The gate has two inputs and one output.
2. Inputs to the gate are a small molecule and light that block the hydrogen bond between complementary domains.
3. Other state transitions of the gate are driven by hybridization reactions.
4. The final states of the automaton can be different if the orders to add inputs are different.
5. The gate emits the output when the state changes to one of the final states.

To apply the automatic design method to the automaton, we extended the simulator to support the second feature.

Extended version of the simulator has an operation to change the rate constants of hybridization and denaturation of specific domain at specific time units. If the operation is executed, the rate constants of hybridization and denaturation turn to 0.0 and 1.0, respectively, which leads to the separation of the domains. For chemical implementation of such feature, the strategies of ATP [43] and UV [44] sensors can be adopted.

To evaluate an automaton, we defined an evaluation function by the following expression,

$$v = \max\left\{|C_{1,2}(s) - C_{2,1}(s)|\,|s \in S\right\},$$

where $C_{1,2}(s)$ and $C_{2,1}(s)$ are the concentrations of the structure s after simulations that adds inputs in the orders denoted by the indexes. This fitness value indicates the difference between the two ordering of inputs, and we can distinguish the two states when this fitness value is large. Similar to the logic gate, the output of the gate is dynamically selected during the evaluation process by

$$output = \underset{s \in S}{argmax}(|C_{1,2}(s) - C_{2,1}(s)|).$$

9.4.6 Automatically Designed Logic Gates Driven by Hybridization Reaction

As a result of executing the automatic design method, logic gates with novel topologies were successfully designed. Examples of OR and AND gates utilizing only hybridization reaction are shown in Figures 9.9 and 9.10, respectively. Note that letters and its order of the genotype are rearranged to make them intuitively understandable. Because unused domains do not effect the evaluation function, such domains can be preserved during the evolution process.

In the designed OR gate, both input 1 and input 2 can release the output by sequential hybridization and branch migration. The design of OR gate was too simple so that all trials of evolutionary computation converged to a system that had the same topology.

In the designed AND gate, the gate and input 1 first hybridize to temporarily release the output which will hybridize to the remaining gate immediately. Eventually, input 2 can completely release the output. The evolutions of the highest fitness value among candidates in respect to the number of generation are shown in Figure 9.11. The evolutions were successively achieved for all trials because fitness values converged to relatively high values (bigger than 0.75). Interestingly, the designs of AND gate did not converge to a single topology, which indicates that AND operation can be achieved by multiple topologies.

9.4.7 Automatically Designed AND Gate Driven by Enzymatic Reaction

To demonstrate the ability of the kinetic simulator and the automatic design method, a logic gate that utilizes enzymatic reaction was also designed. The gate consists of DNA while inputs and output of the gate are RNA, which is inspired from RTRACS. The enzymatic reactions involved in the system are also the same to RTRACS, that are decay, primer-dependent DNA polymerization, and transcription.

The designed AND gate is shown in Figure 9.12, in which the gate first hybridizes to input 1 and then, the 3′ end of DNA strand is extend by polymerization. After a decay reaction of input 1, input 2 hybridizes to the extended domain. By extending hybridized input 2, the promoter region denoted by "T" become double stranded DNA. The downstream of the promoter is transcribed to produce the output RNA.

Although the mechanism of the behavior of the gate was similar to that of RTRACS, the topology of the system was not the same. The result indicates that our methodology can apply to design a system using not only DNA but also RNA and enzymatic reactions.

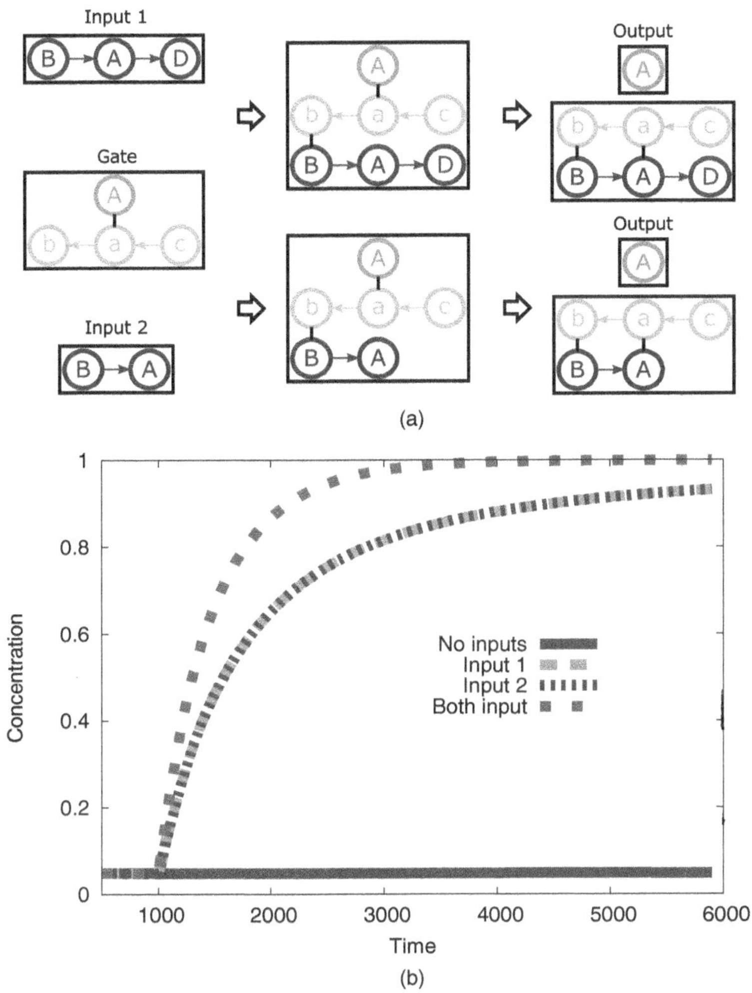

Figure 9.9 Designed enzyme-free OR gate. Obtained genotypes of the gate were "cab," and "A," while those of inputs were "BA" and "BAD." (a) Two transitions to produce the output with either of inputs are illustrated. Domains "c" and "D" were unused for the reaction. (b) The simulation result of the OR gate is shown. *x* and *y* axes are time and concentration of the output, respectively. The concentration of output increased if at least one input was added at 1000 time units.

9.4.8 Automatically Designed Automaton Sensing the Stimuli from Environment

The design of logic gates was rather regarded as a proof of concept, because important criteria to cascade such gates were not included. For example, one of the inputs of the OR gate or enzymatic AND gate was a subsequence of the other input. Such properties are undesirable to implement a logic circuit by integrating multiple modules in a real system.

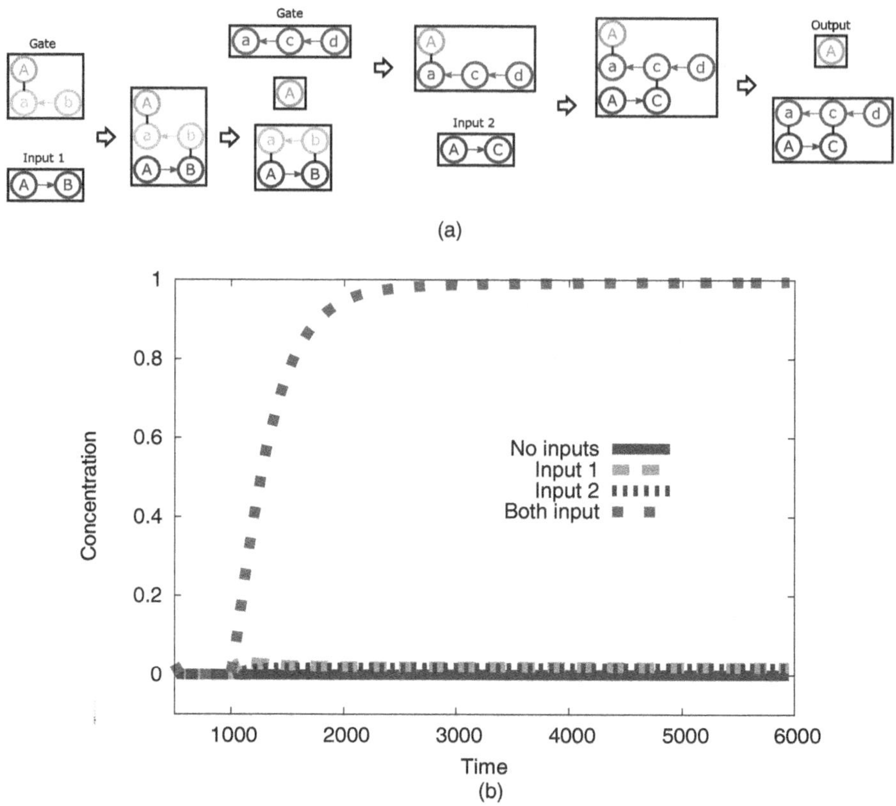

Figure 9.10 Designed enzyme-free AND gate. Obtained genotypes of the gate were "A," "ba," and "dca," while those of inputs were "AB" and "AC." (a) Transitions to produce the output with both inputs are illustrated. Domain "d" was unused for the reaction. (b) The simulation result of the AND gate is shown. *x* and *y* axes are time and concentration of the output, respectively. The concentration of the output increases if and only if both inputs are added at 1000 time units.

The design of the automaton does not need to care such undesirable phenomenon. This is because the inputs of the automaton are assumed to be irrelevant to each other. Furthermore, the output of the automaton is rather a dynamic control of nucleic acids than a cascadable single strand.

One of the nontrivial designed automaton by the algorithm is illustrated in Figure 9.13. In the initial state, the gate is in one of the two states which can switch back and forth by branch migration reactions. When either of inputs is added, the systems promotes the reaction irreversibly.

By input 1, hybridization between domains "a" and "A" is blocked. The reaction keeps the output connected to the gate. Subsequent input 2 blocks the hybridization between "b" and "B," which has nothing to do with the output. In contrast, when the connection between "b" and "B" is first blocked by input 2, the output is released by a branch migration reaction. Subsequent input 1 does not affect the output. The reaction mechanism of this topology matches the features of desired automaton.

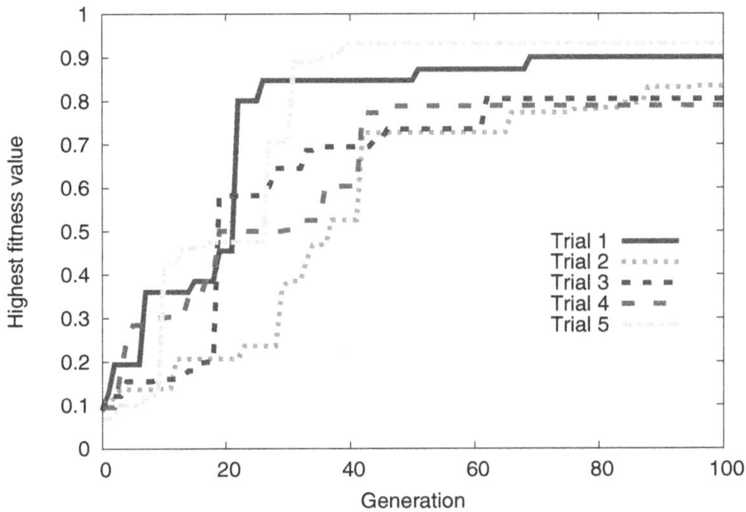

Figure 9.11 Evolutions of fitness values of the AND gate. *x* and *y* axes are the generation and fitness value, respectively. Each line connects a sequence of plots, which reflects the transition of the highest fitness value in candidates during the evolutionary computation. All the trials obtained relatively high fitness values.

The evolutions of fitness values among five trials are shown in Figure 9.14. Unlike the design of enzyme-free AND gate, some of the trials to design the automaton did not reach high fitness values. Even in the successive trials, the number of generations required to obtain the solution was longer than the design of AND gate. This may be due to the complexity of the function.

9.5 DISCUSSION AND CONCLUSION

9.5.1 Discussion

In this section, we discuss the advantages and limitations of our methods by comparing with related researches.

One of the strongest points of the proposed graph-based model was the ability to represent various structures of nucleic acids. Flexible representation of nucleic acid was important because conventional modules have taken the advantage of various secondary structures such as hairpin loop [45], nonlinear structure [46], and kissing structure [47]. The model was capable of analyzing and designing such modules due to a flexible representation by graph. In contrast, conventional techniques to analyze and design reaction systems limited the structure of nucleic acids to hairpin loop [18], linear structure [19], and simple gate [2].

The flexibility of graph-based model allowed the representation of various chemical reactions as rewriting rules. In general, graph rewriting has been commonly applied in chemistry [48], molecular biology [49], and biological pathway [50]. In

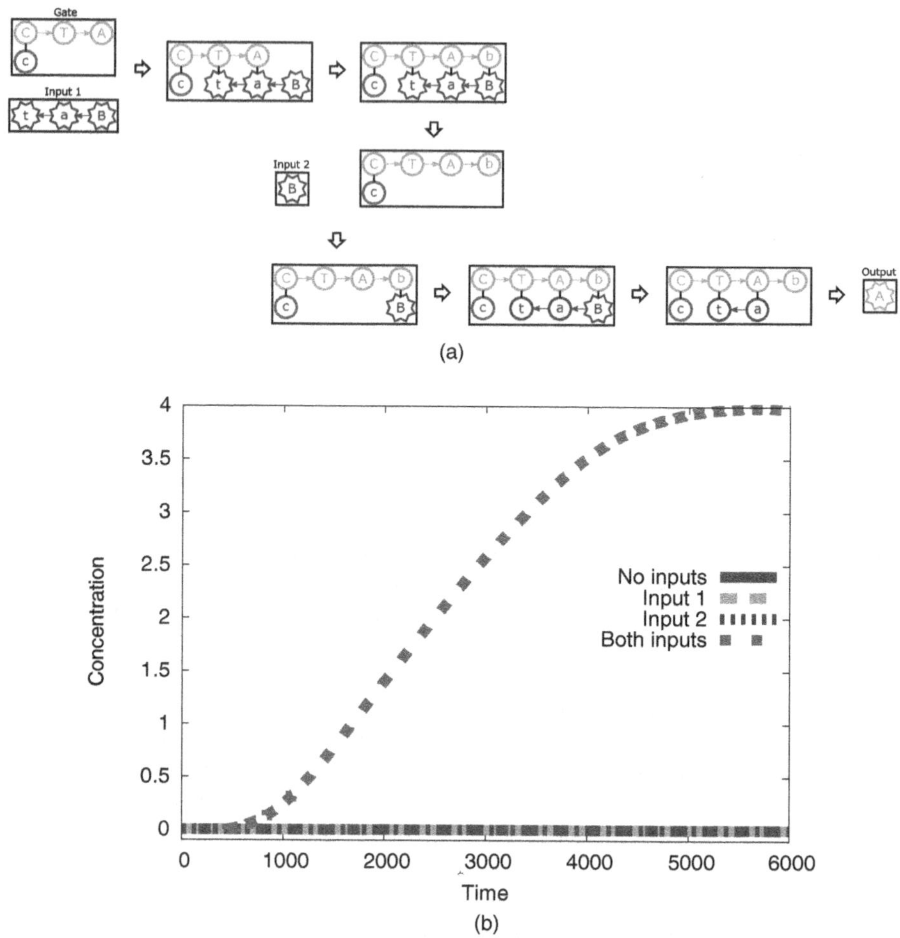

Figure 9.12 Designed AND gate that utilizes enzyamtic reactions. Obtained genotypes of the gate were "CTA," and "c", while those of inputs were "Bat" and "B". (a) The gate and its transitions to produce the output with both inputs are illustrated. (b) The simulation result of the AND gate is shown. *x* and *y* axes are time and concentration of the output, respectively. The concentration of output increased if and only if both inputs were added.

this chapter, we proposed a set of transition rules for nucleic acid reactions, which were directed to analyze and design nucleic acid reaction systems with hybridization and enzymatic reactions.

The target structures of the graph-based model were nucleic acids with Watson–Crick complementary hydrogen bonds and normal phosphate backbone. Nodes of the graph-based model have at most one for each undirected, incoming directed, and outgoing directed edges. Obviously, the model does not cover other structures such as triplex with Hoogsteen base pairs [51], G-quadruplex [52], and so on. Hydroxy group

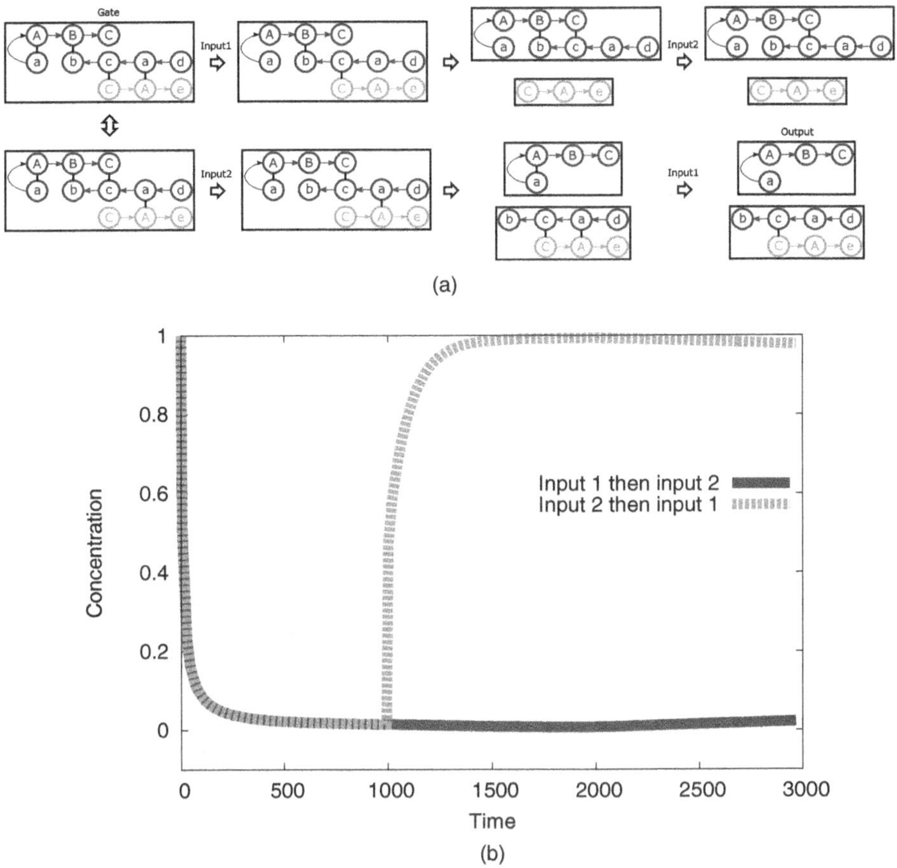

Figure 9.13 Designed automaton. Obtained genotypes of the gate were “aABC,” “dacb,” and “CAe,” while the domains to be blocked by inputs were “a” and “b.” (a) Transitions to produce the output with two ordering of inputs are illustrated. Domains “d” and “e” were unused for the reaction. (b) The simulation result of the automaton is shown. *x* and *y* axes are time and concentration of the output, respectively. First and second inputs are added in 1000 and 2000 time units. The concentration of output increased if input 2 was added before input 1.

modification of the end of phosphate backbone was also not supported, although the information is important for some enzymatic reactions such as ligation.

We applied the graph-based model to predict the dynamic behavior of a system by solving ODE-based on chemical kinetics. Conventional programming languages and simulation techniques for nucleic acid reaction system, however, were further applied to a model checking of chemical system [53], integrating large scale circuit [54], and the implementation of arbitrary chemical reactions [27] with the aid of limitations of those models. In contrast, our simulation technique cannot be directly applied to such purposes since the combinatorial explosion became a crucial problem.

Our simulation method did not thoroughly apply all possible rewriting rules due to the threshold and the restriction on graph imposed in order to minimize the number

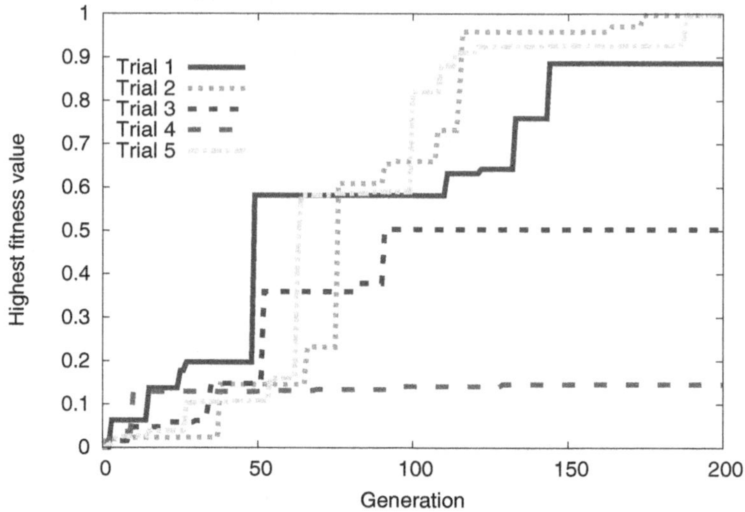

Figure 9.14 Evolutions of fitness values of the automaton. *x* and *y* axes are the generation and fitness value, respectively. Each line connects a sequence of plots, which reflects the transition of the highest fitness value in candidates during the evolutionary computation. Although trials 1, 2, and 5 obtained relatively high fitness values, trials 3 and 4 did not.

of structures. The threshold was introduced to eliminate unimportant structures, which leads to an efficient simulation. Simulation becomes incorrect, however, when the concentration of important structure cannot exceed the threshold. Without the restriction on graph, the simulation can spontaneously produce unbounded number of structures, which made the simulation intractable.

We focused on continuous simulations because target nucleic acid reaction systems follow the law of mass action. Our graph rewriting model, however, can be directly applied to stochastic simulation [55]. In that case, fitness value of nucleic acid reaction system always differs for each run, which may require further refinement for evolutionary computation.

Applying simulation technique to design a new topology of nucleic acid reaction by evolutionary computation is our original idea. In the area of genetic network, such evolutionary computations were commonly applied to design an interesting topology (e.g., oscillatory behavior). Some of them were based on graph models [24, 25]. In those researches, however, the implementation of the designed topology into a real gene regulatory network seems extremely difficult because inverse problem to construct genes from given topology was a challenging task to solve. On the other hand, nucleic acid reaction system is ready for implementation because sequence design methods using other types of evolutionary computations are well researched [56, 57].

In fact, we implemented the automatically designed enzyme-free AND gate (Figure 9.12) in *in vitro* chemical experiment as a demonstration. The result of the chemical experiment is shown in Figure 9.15. The fluorescent intensity of output

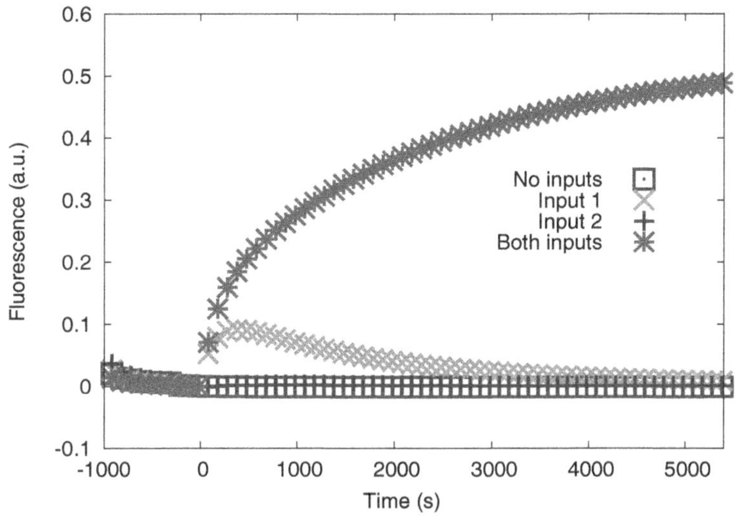

Figure 9.15 Chemical experimental result of automatically designed enzyme-free AND gate. *x* and *y* axes are time in second and fluorescent intensity in arbitrary unit, respectively. To read out the output, fluorophore and quencher molecules were attached to the output and gate. Fluorescent intensity was normalized by the maximum intensity recorded in advance. Inputs were added at 0 s. For technical reason, unnecessary domains in the system were omitted for the experiment.

strand increased if and only if both inputs were added. Because input 1 can temporally release the output, the intensity slightly increased when only input 1 was added. Although we observed such undesired increase, the experimental result has an agreement with the simulation result. It may be possible to further optimize our simulation method by employing realistic rate constants that are derived by chemical experiment [42, 58].

The number of single strands is a simple scale to measure the complexity of nucleic acid modules. For example, the modules of hairpin pathway [18], see-saw gate [2], and enzyme-free logic gate [1] consist of 1, 5, and 3 strands, respectively. It may be possible to state that the ability of the automatic design method catches up with a design by human trial and error in some functions, because there are not so many differences between the numbers of molecular species. The limits to design much complex systems by evolutionary computation lied in the search space of the topology. The combinatorial optimization method cannot escape from the trade-off between the broadness of the search space and the efficiency of the heuristic algorithm.

The design of DNA automaton was a challenging task because molecular machine that detects the ordering of inputs was nontrivial for human. The state transitions of the designed automaton were not so simple to understand intuitively even though there were only four steps to release the output. Since the search spaces were the same among evolutionary computations, difference of the target function affected the efficiency of the design. It suggests that the evaluation function to calculate the fitness value requires further optimization to design much optimal solution.

9.5.2 Conclusion

A novel graph-based model to predict the dynamics of nucleic acid reaction systems is proposed since analyzing and designing the systems become rapidly important from biological and engineering aspect. Employing graph brings about a flexible representation of nucleic acid compared with other conventional methods to design or program a system driven by hybridization reactions. The graph-based model includes not only hybridization reactions but also enzymatic reactions as graph rewriting rules. As a result, the behavior of nucleic acid reaction systems are successfully explained in terms of the time evolution of concentration.

As an application of kinetic simulation, design automation of the topology of nucleic acid reaction system by evolutionary computation is addressed. By defining appropriate search space and evaluation function, logic gate and state transition machine are successfully designed. Implementation of automatically designed system is possible, which is verified by chemical experiment of one of the logic gates. Expected applications of this research is the development of a nontrivial nucleic acid reaction systems that can serve as modules of molecular device.

REFERENCES

1. Georg Seelig, David Soloveichik, David Yu Zhang, and Erik Winfree. Enzyme-free nucleic acid logic circuits. *Science*, 314(5805):1585–1588, 2006.
2. Lulu Qian and Erik Winfree. A simple DNA gate motif for synthesizing large-scale circuits. *Journal of the Royal Society Interface*, 8(62):1281–1297, 2011.
3. Johann Elbaz, Oleg Lioubashevski, Fuan Wang, Françoise Remacle, Raphael D. Levine, and Itamar Willner. DNA computing circuits using libraries of DNAzyme subunits. *Nature Nanotechnology*, 5(6):417–422, 2010.
4. David E. Huizenga and Jack W. Szostak. A DNA aptamer that binds adenosine and ATP. *Biochemistry*, 34(2):656–665, 1995.
5. Louis C. Bock, Linda C. Griffin, John A. Latham, Eric H. Vermaas, and John J. Toole. Selection of single-stranded DNA molecules that bind and inhibit human thrombin. *Nature*, 355(6360):564–566, 1992.
6. Tosan Omabegho, Ruojie Sha, and Nadrian C. Seeman. A Bipedal DNA Brownian motor with coordinated legs. *Science*, 324(5923):67–71, 2009.
7. Jonathan Bath, Simon J. Green, and Andrew J. Turberfield. A free-running DNA motor powered by a nicking enzyme. *Angewandte Chemie International Edition*, 44(28):4358–4361, 2005.
8. Sara Tombelli, Maria Minunni, and Marco Mascini. Analytical applications of aptamers. *Biosensors and Bioelectronics*, 20(12):2424–2434, 2005.
9. Teresa Mairal, Veli Cengiz Özalp, Pablo Lozano Sánchez, Mònica Mir, Ioanis Katakis, and Ciara K. O'Sullivan. Aptamers: molecular tools for analytical applications. *Analytical and Bioanalytical Chemistry*, 390(4):989–1007, 2008.
10. Eun Jeong Cho, Joo-Woon Lee, and Andrew D. Ellington. Applications of aptamers as sensors. *Annual Review of Analytical Chemistry*, 2:241–264, 2009.

11. Shawn M. Douglas, Ido Bachelet, and George M. Church. A logic-gated nanorobot for targeted transport of molecular payloads. *Science*, 335(6070):831–834, 2012.

12. Suvir Venkataraman, Robert M. Dirks, Christine T. Ueda, and Niles A. Pierce. Selective cell death mediated by small conditional RNAs. *Proceedings of the National Academy of Sciences*, 107(39):16777–16782, 2010.

13. Lulu Qian, Erik Winfree, and Jehoshua Bruck. Neural network computation with DNA strand displacement cascades. *Nature*, 475(7356):368–372, 2011.

14. Prabal K. Maiti, Tod A. Pascal, Nagarajan Vaidehi, Jiyoung Heo, and William A. Goddard. Atomic-level simulations of seeman DNA nanostructures: The paranemic crossover in salt solution. *Biophysical Journal*, 90(5):1463–1479, 2006.

15. Mahmoud Moradi, Volodymyr Babin, Christopher Roland, and Celeste Sagui. Reaction path ensemble of the B-Z-DNA transition: a comprehensive atomistic study. *Nucleic Acids Research*, 41(1):33–43, 2012.

16. Thomas E. Ouldridge, Ard A. Louis, and Jonathan P. K. Doye. DNA nanotweezers studied with a coarse-grained model of DNA. *Physical Review Letters*, 104(17):178101, 2010.

17. Jonathan P. K. Doye, Thomas E. Ouldridge, Ard A. Louis, Flavio Romano, Petr Šulc, Christian Matek, Benedict E. K. Snodin, Lorenzo Rovigatti, John S. Schreck, Ryan M. Harrison, and William P. J. Smith. Coarse-graining DNA for simulations of DNA nanotechnology. *Physical Chemistry Chemical Physics*, 15(47):20395–20414, 2013.

18. Peng Yin, Harry M. T. Choi, Colby R. Calvert, and Niles A. Pierce. Programming biomolecular self-assembly pathways. *Nature*, 451(7176):318–322, 2008.

19. Andrew Phillips and Luca Cardelli. A programming language for composable DNA circuits. *Journal of the Royal Society Interface*, 6(Suppl 4):S419–S436, 2009.

20. Nathanaël Aubert, Clément Mosca, Teruo Fujii, Masami Hagiya, and Yannick Rondelez. Computer-assisted design for scaling up systems based on DNA reaction networks. *Journal of The Royal Society Interface*, 11(93):20131167, 2014.

21. Hongqing Cao, Francisco J Romero-Campero, Stephan Heeb, Miguel Cámara, and Natalio Krasnogor. Evolving cell models for systems and synthetic biology. *Systems and Synthetic Biology*, 4(1):55–84, 2010.

22. Yasuaki Kobayashi, Tatsuo Shibata, Yoshiki Kuramoto, and Alexander S. Mikhailov. Robust network clocks: Design of genetic oscillators as a complex combinatorial optimization problem. *Physical Review*, 83(6):060901, 2011.

23. Barry Drennan and Randall D. Beer. Evolution of repressilators using a biologically-motivated model of gene expression. *Proceedings of the Tenth International Conference on the Simulation and Synthesis of Living Systems*, 10:22–27, 2006.

24. Dominique Chu. Evolving genetic regulatory networks for systems biology. In *Proceedings of IEEE Congress on Evolutionary Computation 2007 (CEC2007)*, pages 875–882, 2007.

25. Yasuaki Kobayashi, Tatsuo Shibata, Yoshiki Kuramoto, and Alexander S. Mikhailov. Evolutionary design of oscillatory genetic networks. *The European Physical Journal B*, 76(1):167–178, 2010.

26. Mario A. Marchisio and Jörg Stelling. Automatic design of digital synthetic gene circuits. *PLoS Computational Biology*, 7(2):e1001083, 2011.

27. David Soloveichik, Georg Seelig, and Erik Winfree. DNA as a universal substrate for chemical kinetics. *Proceedings of the National Academy of Sciences*, 107(12):5393–5398, 2010.

28. James D. Watson and Francis H. C. Crick. Molecular structure of nucleic acids. *Nature*, 171(4356):737–738, 1953.

29. Akio Nishikawa, Masayuki Yamamura, and Masami Hagiya. DNA computation simulator based on abstract bases. *Soft Computing—A Fusion of Foundations, Methodologies and Applications*, 5(1):25–38, 2001.

30. Luca Cardelli. Strand algebras for DNA computing. *Natural Computing*, 10(1):407–428, 2011.

31. David Yu Zhang, Andrew J. Turberfield, Bernard Yurke, and Erik Winfree. Engineering entropy-driven reactions and networks catalyzed by DNA. *Science*, 318(5853):1121–1125, 2007.

32. Masahiro Takinoue, Daisuke Kiga, Koh-Ichiroh Shohda, and Akira Suyama. Experiments and simulation models of a basic computation element of an autonomous molecular computing system. *Physical Review E*, 78(4):041921, 2008.

33. Hongzhou Gu, Jie Chao, Shou-Jun Xiao, and Nadrian C. Seeman. A proximity-based programmable DNA nanoscale assembly line. *Nature*, 465(7295):202–205, 2010.

34. Kyle Lund, Anthony J. Manzo, Nadine Dabby, Nicole Michelotti, Alexander Johnson-Buck, Jeanette Nangreave, Steven Taylor, Renjun Pei, Milan N. Stojanovic, Nils G. Walter, Erik Winfree, and Hao Yan. Molecular robots guided by prescriptive landscapes. *Nature*, 465(7295):206–210, 2010.

35. Peter Brodersen and Olivier Voinnet. The diversity of RNA silencing pathways in plants. *TRENDS in Genettics*, 22(5):268–280, 2006.

36. David C. Baulcombe. Amplified silencing. *Science*, 315(5809):199–200, 2007.

37. Julia Pak and Andrew Fire. Distinct populations of primary and secondary effectors during RNAi in C. elegans. *Science*, 315(5809):241–244, 2007.

38. Reiji Okazaki, Tuneko Okazaki, Kiwako Sakabe, Kazunori Sugimoto, and Akio Sugino. Mechanism of DNA chain growth, I. Possible discontinuity and unusual secondary structure of newly synthesized chains. *Proceedings of the National Academy of Sciences*, 59(2):598–605, 1968.

39. Masahiro Takinoue, Daisuke Kiga, Koh-Ichiroh Shohda, and Akira Suyama. RNA oscillator: limit cycle oscillations based on artificial biomolecular reactions. *New Generation Computing*, 27(2):107–127, 2009.

40. Anton Kan, Koh-Ichiroh Shohda, and Akira Suyama. A DNA based molecular logic gate capable of a variety of logical operations. *Lecture Notes in Computer Science*, 7433:86–97, 2012.

41. E. Fehlberg. Klassische Runge-Kutta-Formeln vierter und niedrigerer Ordnung mit Schrittweiten-Kontrolle und ihre Anwendung auf Wärmeleitungsprobleme. *Computing*, 6(1):61–71, 1970.

42. David Yu Zhang and Erik Winfree. Control of DNA strand displacement kinetics using toehold exchange. *Journal of the American Chemical Society*, 131(47):17303–17314, 2009.

43. Razvan Nutiu and Yingfu Li. Structure-switching signaling aptamers. *Journal of the American Chemical Society*, 125(16):4771–4778, 2003.

44. Xingguo Liang, Hidenori Nishioka, Nobutaka Takenaka, and Hiroyuki Asanuma. A DNA nanomachine powered by light irradiation. *ChemBioChem*, 9(5):702–705, 2008.

45. Kensaku Sakamoto, Hidetaka Gouzu, Ken Komiya, Daisuke Kiga, Shigeyuki Yokoyama, Takashi Yokomori, and Masami Hagiya. Molecular computation by DNA hairpin formation. *Science*, 288(5469):1223–1226, 2000.
46. Wei Li, Yang Yang, Hao Yan, and Yan Liu. Three-input majority logic gate and multiple input logic circuit based on DNA strand displacement. *Nano Letters*, 13(6):2980–2988, 2013.
47. Georg Seelig, Bernard Yurke, and Erik Winfree. Catalyzed relaxation of a metastable DNA fuel. *Journal of the American Chemical Society*, 128(37):12211–12220, 2006.
48. Martin Mann, Heinz Ekker, and Christoph Flamm. The Graph Grammar Library - a generic framework for chemical graph rewrite systems. *Lecture Notes in Computer Science*, 7909:52–53, 2013.
49. Francesc Rosselló and Gabriel Valiente. Graph transformation in molecular biology. *Lecture Notes in Computer Science*, 3393:116–133, 2005.
50. Vincent Danos, Jérôme Feret, Walter Fontana, and Jean Krivine. Scalable simulation of cellular signaling networks. *Lecture Notes in Computer Science*, 4807:139–157, 2007.
51. Marco Brucale, Giampaolo Zuccheri, and Bruno Samori. The dynamic properties of an intramolecular transition from DNA duplex to cytosine-thymine motif triplex. *Organic & Biomolecular Chemistry*, 3(4):575–577, 2005.
52. Satoru Nagatoishi, Takahiko Nojima, Elzbieta Galezowska, Bernard Juskowiak, and Shigeori Takenaka. G quadruplex-based FRET probes with the thrombin-binding aptamer (TBA) sequence designed for the efficient fluorometric detection of the potassium ion. *ChemBioChem*, 7(11):1730–1737, 2006.
53. Matthew R. Lakin, David Parker, Luca Cardelli, Marta Kwiatkowska, and Andrew Phillips. Design and analysis of DNA strand displacement devices using probabilistic model checking. *Journal of the Royal Society Interface*, 9(72):1470–1485, 2012.
54. Lulu Qian and Erik Winfree. Scaling up digital circuit computation with DNA strand displacement cascades. *Science*, 332(6034):1196–1201, 2011.
55. Daniel T. Gillespie. Exact stochastic simulation of coupled chemical reactions. *The Journal of Physical Chemistry*, 81(25):2340–2361, 1977.
56. Fumiaki Tanaka, Masashi Nakatsugawa, Masahito Yamamoto, Toshikazu Shiba, and Azuma Ohuchi. Towards a general-purpose sequence design system in DNA computing. In *Proceedings of the 2002 World on Congress on Computational Intelligence*. IEEE, 2002.
57. Soo-Yong Shin, In-Hee Lee, Dongmin Kim, and Byoung-Tak Zhang. Multiobjective evolutionary optimization of DNA sequences for reliable DNA computing. *IEEE Transactions on Evolutionary Computation*, 9(2):143–158, 2005.
58. Kevin Montagne, Raphael Plasson, Yasuyuki Sakai, Teruo Fujii, and Yannick Rondelez. Programming an in vitro DNA oscillator using a molecular networking strategy. *Molecular System Biology*, 7:466, 2011.

10

USING EVOLUTIONARY ALGORITHMS TO STUDY THE EVOLUTION OF GENE REGULATORY NETWORKS CONTROLLING BIOLOGICAL DEVELOPMENT

Alexander Spirov

Computer Science and CEWIT, SUNY Stony Brook, Stony Brook, NY, USA; and the Sechenov Institute of Evolutionary Physiology and Biochemistry, St.-Petersburg, Russia

David Holloway

Mathematics Department, British Columbia Institute of Technology, Burnaby, B.C., Canada

10.1 INTRODUCTION

Developmental biologists approach the embryological formation of organisms in terms of the temporal and spatial regulation of gene expression. How does gene regulation precisely specify in what cells and at what time a gene is expressed (to a given level)? Much of the work in the field is on the gene–gene interactions, frequently in extended regulatory networks, which control given events in development.

Evolutionary Computation in Gene Regulatory Network Research, First Edition.
Edited by Hitoshi Iba and Nasimul Noman.

Computational work on developmental gene regulatory networks (GRNs), to construct the "wiring diagrams" of gene connections and characterize network dynamics, has been going for many decades. More recently, there has been increasing focus on the evolutionary dynamics of how these GRNs arose. Since developmental GRNs are fundamental to creating the body plan of a species (and the functions associated with the form), understanding the evolution of GRNs is fundamental to understanding the dynamics of speciation: how is an organism's form (phenotype) maintained, despite environmental variation and genetic mutation, and how is such stability balanced against evolvability, or the potential to respond with fit phenotypic changes to environmental or genetic changes? Computational simulations of GRN evolution have become very powerful tools for addressing such fundamental issues in evolutionary theory. At the same time, evolutionary algorithms (EAs) have become a major approach for optimization problems. EA has been inspired by the general principles of biological evolution, such as mutation, selection, and reproduction. We feel there is a large potential for developing new EA techniques based on more detailed mechanisms of biological evolution, for instance, using sexual reproduction or retroviral mechanisms of gene crossover. At the same time, the techniques used by computational biologists simulating evolution can be computationally intensive, and could benefit from improvements from computer science EA approaches (e.g., Ref. [1]). For a recent methodological guide to the computational biology of evolutionary simulations, see also Ref. [2].

In general, we will be discussing evolutionary computations with the general cycle (Figure 10.1):

(a) An initial population is chosen. In simple cases, individuals can be different parameter sets for a given GRN; more complex cases include individuals with different connectivities or member genes.
(b) Individuals are tested for fitness against given criteria. For example, for spatial expression problems, individuals are scored by how well they recreate experimental patterns (e.g., an individual's parameters are used in a differential equation model of the patterning process, and the simulated pattern is scored against experimental data).
(c) Low-scoring individuals are selected out of the population.
(d) New individuals are introduced into the population to replace those just selected out. Generation of new individuals can be specified by inheritance rules from parent individuals.
(e) Parameters are mutated. This can, for example, alter gene–gene interaction strengths, including adding or eliminating interactions. More detailed approaches can mutate at a gene's sequence level, for example, altering cis-regulatory sequences (to alter regulator binding strengths), or distinguishing between point mutation and crossover operations, which involve entire regions of a sequence.
(f) Steps (b)–(e) are repeated for some number of generations.

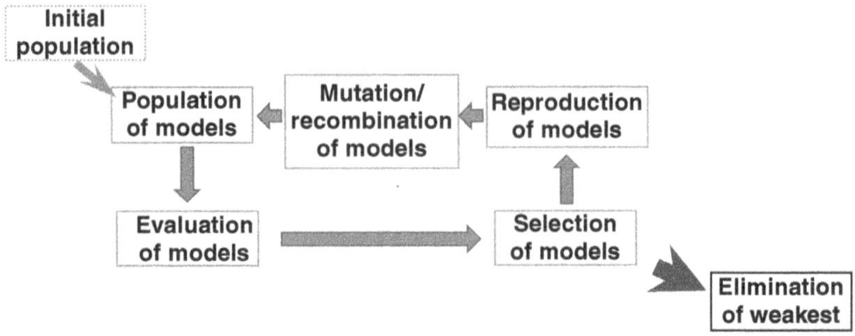

(a)

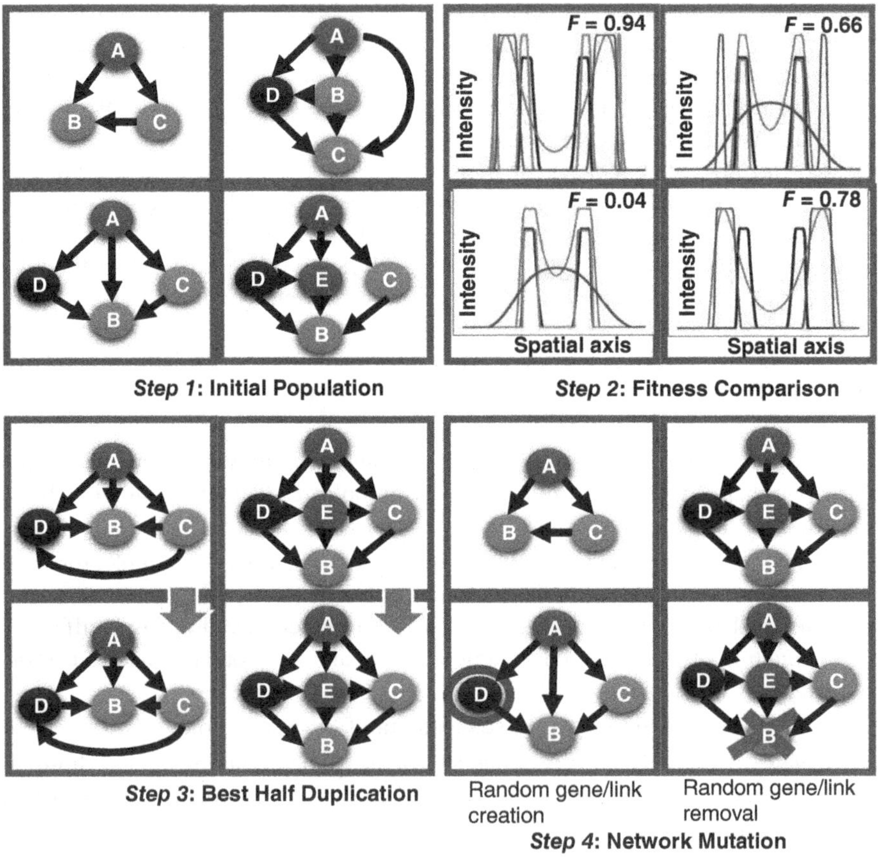

(b)

Figure 10.1 Overview of evolutionary computation approach. (a) General cycle of evaluation, selection, reproduction, and mutation to create spatial gene expression patterns (see the text). *F*-values represent the relative fitness for each spatial pattern. (b) A more detailed schematic of this process, for an initial population of GRNs (Step 1), evaluation for fitness to produce a particular spatial gene expression pattern (Step 2), selection and reproduction of the fittest GRNs (Step 3), and GRN mutation (Step 4), changing parameters (kinetics) or the network connections themselves, as shown here (after Ref. [3]).

In this chapter, we will outline the general approaches being used by computational biologists to simulate evolution, and review some of the major impacts this work is having on biological theory (especially, understanding the dynamics of evolution). With this survey of evolutionary simulation, we aim to inspire computer scientist practitioners of EA with the biology, as well as pointing to areas in the biological modeling, which could most benefit from increased computational efficiencies.

10.2 COMPUTATIONAL APPROACHES FOR THE EVOLUTION OF DEVELOPMENTAL GRNs

We distinguish two broad approaches for simulating GRN evolution in developmental problems. The first, the "coarse-grained" approach, treats the genes as nodes and their interactions as edges in a network (as depicted in Figure 10.1). The interactions and their strengths can be coded as a regulatory matrix. This approach has been used very extensively for studying evolutionary principles over the past two decades. It is, however, coarse grained in the sense that real gene regulation and biological evolution both operate at the deoxyribonucleic acid (DNA) sequence level. Fine-grained approaches have been increasingly developed in recent years, modeling at the level of sequence information (or at least a region of sequence, or regulatory module). With the corresponding increase in genome sequencing and comparative genomics, this provides powerful predictive value for evolutionary dynamics at the molecular genetic level.

10.2.1 Coarse-Grained Approaches

Coarse-grained simulations of GRN evolution were pioneered by Andreas Wagner et al. [4,5]. In their approach, genes are nodes of the network, which interact (network edges) only one on one (co-factors are not modeled), leading to Boolean (on–off) expression levels. These assumptions are too simple for a number of developmental phenomena (such as spatially patterned gene expression in response to a concentration gradient of a regulator), but the models are very fast to solve, and have had significant impact on the understanding of evolutionary principles (see Ref. [6] for further perspective).

10.2.1.1 W Matrix In Wagner's core model, a GRN is represented by the state of the network genes $1-N$:

$$\vec{S}(t) := (S_1(t), \ldots, S_N(t)). \tag{10.1}$$

$S_i(t)$ is binary—the gene is either expressed or not. Cross- and autoregulatory interactions causing the state to change are modeled by difference equations:

$$S_i(t+\tau) = \sigma\left[\sum_{j=1}^{N} w_{ij}S_j(t)\right] = \sigma[h_i(t)], \tag{10.2}$$

where the expression state of gene G_i at time $t + \tau$, $S_i(t + \tau)$, is a function of a weighted sum, $h_i(t)$, of the expression state of all network genes at time t. $\sigma(x)$ is the sign function ($\sigma(x) = -1$ for $x < 0, \sigma(x) = +1$ for $x > 0$ and $\sigma(0) = 0$), and τ is a time constant whose value depends on biochemical parameters such as the rate of transcription or the time necessary to export mRNA into the cytoplasm for translation. The real constants w_{ij} represent the strength of regulatory interaction of the product of G_j with G_i, (activation, $w_{ij} > 0$; repression, $w_{ij} < 0$). w_{ij} are the elements of the W matrix representing regulation in the GRN. The approach is shown schematically in Figure 10.2; see Ref. [7] for a recent review.

10.2.1.2 Extensions of the *W* Matrix Approach

The W matrix approach has been used extensively over the past two decades. To address particular topics, it has been refined and extended in a number of ways.

Interaction function Siegal and Bergman [8, 9] introduced a sigmoidal function for σ in Equation (10.2) that allowed for dynamic switching of a repressive state into an activating state. This was then extended to allow for recombination (allele r or R, with equal probability) to modify σ [10]. Masel [11] set $\sigma(x) = 1$ (expressed) if $x \geq 0$ and $\sigma(x) = 0$ (not expressed) if $x < 0$, a more realistic mapping, biologically, than the negative values allowed in Equation (10.2). The balance between activating and inhibiting connections in W matrix models, including oscillations, was explored by McDonald et al. [12].

Co-factors Biological gene regulation in eukaryotes commonly involves co-factors, in which the effect of a transcription factor (TF) is amplified or diminished by the presence of a different TF. Extension to a triple-index representation, W_{ijk}, allows for TFs j and k to simultaneously affect the expression of gene i [13].

Spatially distributed gene expression A critical aspect of developmental gene expression is its spatial dependence; genes must be expressed in distinct domains in order for tissues to differentiate in the correct locations. Although simulations of spatially dependent GRNs have a long tradition in developmental biology, it is more recently that the evolution of these GRNs has been studied. Anterior–posterior segmentation of the insect body plan is extremely well characterized through comparative biology, as well as through developmental biology of the fruit fly *Drosophila melanogaster*, and has become a test case for evolutionary simulations of spatially dependent gene expression. While Boolean gene states have been used with spatial patterning, researchers are generally turning toward continuous (DE) representations of gene states, to address the concentration-dependent aspects of developmental patterning. Evolutionary coarse-grained segmentation models include those of Salazar-Ciudad et al. [130, 131]; Sole et al. [14], Francois et al. [15, 16], Fujimoto et al. [17], and ten Tusscher and Hogeweg [18, 19] (overview of their approach shown in Figure 10.3). For reasons of computational simplicity, these models assume a one-dimensional embryo consisting of a row of cells (Figure 10.3b), ignoring morphogenetic processes and patterning in higher dimensions. Some of these network

\	**TF_1**	**TF_2**	**TF_3**	**...**	**TF_n**
G_1	$W_{g1 \leftarrow tf1}$	$W_{g1 \leftarrow tf2}$	$W_{g1 \leftarrow tf3}$	...	$W_{g1 \leftarrow tfn}$
G_2	$W_{g2 \leftarrow tf1}$	$W_{g2 \leftarrow tf2}$	$W_{g2 \leftarrow tf3}$	...	$W_{g2 \leftarrow tfn}$
G_3	$W_{g3 \leftarrow tf1}$	$W_{g3 \leftarrow tf2}$	$W_{g3 \leftarrow tf3}$	...	$W_{g3 \leftarrow tfn}$
...	...	...	...	...	...
G_n	$W_{gn \leftarrow tf1}$	$W_{gn \leftarrow tf2}$	$W_{gn \leftarrow tf3}$	...	$W_{gn \leftarrow tfn}$

(a)

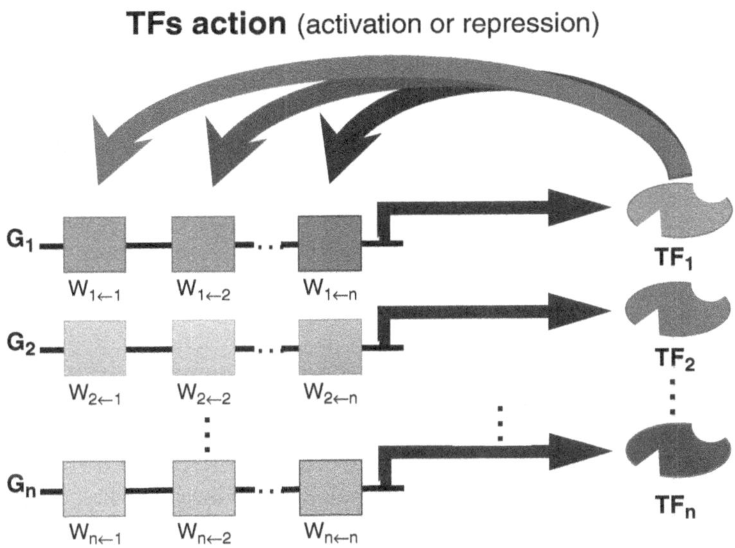

(b)

Figure 10.2 Representation of a GRN via the connectivity or regulation matrix $W = W_{i \leftarrow j}$ (the W element represents the action of the jth gene on the ith gene; can also be written as W_{ij}). (a) Matrix representation of gene regulation by transcription factors (TFs) encoded by the genes. In biological terms, the w_{ij}s represent both the TF binding strength and the transcriptional activation (or repression) ability of the jth factor on the ith gene. w_{ij} values are relative. The ith row of W corresponds to the entire cis-regulatory region of gene G_i. The matrix $W = (w_{ij})$ corresponds to all regulatory DNA elements relevant to regulatory interactions among network genes. The more zero entries W has, the fewer regulatory interactions exist among network genes. (b) Each gene (horizontal arrow) is regulated by the products of the other genes (boxes). Strength and direction of regulation (depicted as different color saturation levels) are a function of both the regulatory element and the abundance of its corresponding gene product. Genotype is represented as the matrix W, and phenotype is the vector **TF**.

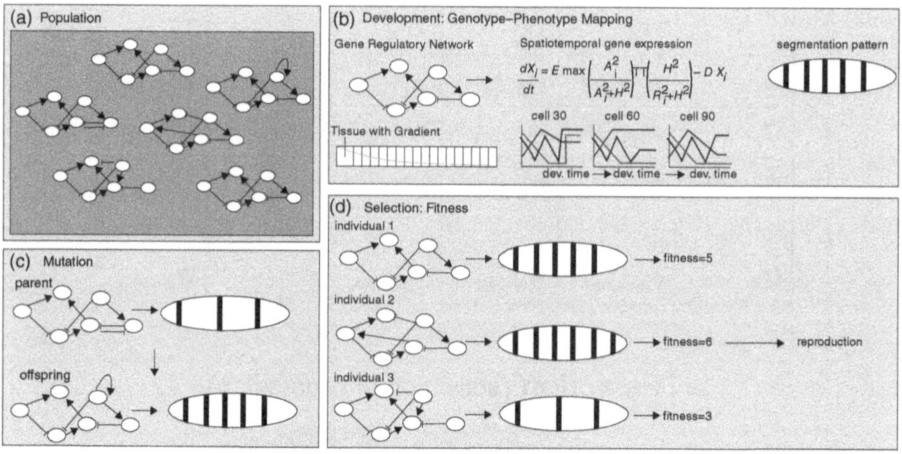

Figure 10.3 An example of modeling evolution of a GRN controlling spatially dependent gene expression, in this case anterior–posterior segmentation of the fly body plan. (a) The population of individuals, each with a GRN. (b) GRN topology specifies the gene expression response to a maternal signaling gradient (modeled continuously with DEs). This produces variations on a striped expression pattern. (c) GRNs are mutated. (d) The reproduction chances of individuals are fitness based; fitness increases with the segment number. Reprinted with permission from Ref. [19], copyright 2013 Springer Science and Business Media.

models, (Refs. [15, 16, 18, 19]) are still Boolean and do not incorporate diffusion or transport for modeling spatial pattern formation, while others (Refs. [17, 130, 131]) have developed continuous evolutionary models for insect segmentation, with Hill-equation kinetics. With the exception of the work in Refs. [18, 19], which used a genome encoding the regulatory network, studies used the GRN directly as the genotype, with mutations occurring directly on the network (as depicted in Figure 10.3c) rather than on the genome sequence. The approaches differed, in details, on how the genotype is mapped to the phenotype (Figure 10.3b and Ref. [19]); please see Ref. [19] for a recent review on how these evolutionary developmental simulation studies shed light on some classic biological problems.

10.2.1.3 Increased Efficiency in Solving Coarse-Grained Models

Though an advantage of coarse-grained approaches is their speed, several groups have worked on improving computational efficiency, in order to access larger numbers of generations (simulation cycles), particularly for spatially dependent models. Fomekong-Nanfack et al. [20] developed an evolutionary strategy (ES) for finding parameters in segmentation models, some 5–140 times faster than an earlier parallel simulated annealing approach [21]. Their "island" ES algorithm operates on N populations of individuals, each with a population size of i. Simulation cycles are generally run within islands, with the best individuals copied between islands every m generations. Jostins and Jaeger [22] developed an asynchronous communication between parallel islands that offered further speed-up. A different parallel method,

DEEP (Differential Evolution Entirely Parallel), has been developed by Kozlov and Samsonov [23].

10.2.1.4 Digital Organisms At a more general level, evolution has been studied within the framework of artificial life. Digital organisms can replicate, mutate, and have fitness criteria. This approach does not tend to address the biological details of gene regulation, but it can offer a platform to study general principles of evolving systems. For instance, Lenski et al. [24] used two classes of organism; one simple (rapidly replicating) and one complex (able to accelerate replication according to "metabolic" rewards). They found, after millions of mutations were introduced, that the complex organisms were more robust to single mutations than the simple organisms. Ofria et al. [25] found that while traits may appear suddenly in an evolutionary process, the information content of the genome is gradually varying. They argued that "... it is nearly impossible for a significantly complex trait to appear without reusing existing information." Clune et al. [26] addressed the classic question of whether ontogeny recapitulates phylogeny, finding, in digital organisms, that traits that evolved earlier tended to occur earlier in an individual's development. By adding a type of epistasis (gene background effect), Valverde et al. [27] observed that modular control of expression (clustered gene effects) was spontaneously generated (also see Section 10.2.2.2). Covert et al. [28], using digital organisms, described how deleterious mutations could be beneficial, opening access to new regions of the fitness landscape. Batut et al. [29] applied digital organisms to the evolution of marine cyanobacteria; matching observations that reduced selection strength led to genome loss, with a majority of the loss in non-coding sequences.

10.2.2 Fine-Grained Approaches

To address the biology more realistically, a number of groups have incorporated DNA structure into evolutionary models. The finer level of detail brings significant computing costs, but can answer evolutionary questions on the molecular genetic level (see, e.g., the review in Ref. [30]). Different levels of abstraction can be made within the fine-grained approaches. It is rare to model at base-pair resolution, but a number of groups include TF binding sites (TFBSs; including cooperative or competitive effects between TFs), as well as addressing the modular structure of cis-regulatory regions (reflecting that many genes have multiple enhancers [31,32]).

10.2.2.1 Evolution of TFBSs ten Tusscher and Hogeweg [33] extended the basic *W* matrix approach to allow for multiple TFBSs. This allowed them to model different types of mutations: in TFBSs; in the coding part of the gene; as well as those affecting segments of the genome (see also Ref. [34]). At slightly more detail, Pujato et al. [35] included structural models of TF–DNA complexes to find binding strengths. Mutation altered TFBS strengths and created or destroyed binding sites (BSs). The Sinha group [36, 37] has developed thermodynamic models for binding individual TFBSs and for entire cis-regulatory modules (CRMs). Simulations of TFBS evolution for 37 well-characterized CRMs in *Drosophila* segmentation corroborated the BS conservation at short times and BS loss at longer times seen across 12 *Drosophila* genomes.

10.2.2.2 Evolution of Modular Control Many genes have multiple CRMs, with distinct sequences of TFBSs. For spatial patterning, different CRMs can control different regions of gene expression. We have developed a model for the evolution of multiple CRM structure in one of the major *Drosophila* segmentation genes, *hunchback* (*hb*) [38,39]. Figure 10.4(a) shows some of the experimentally determined structure, in terms of BSs for specific *hb* TFs, separated into two of the three known enhancers active during segmentation. Figure 10.4(b) shows how we represent this structure in symbolic strings. The expression levels of *hb* depend on the arrangement of activating and repressing TFBSs, modeling co-activation and co-repression. In each generation of an evolutionary computation, *hb* spatial patterns are selected against experimental *hb* patterns. TFBSs mutate and evolve to optimize fits to the data. We have shown that the multiple CRMs observed in *hb* could have readily evolved.

10.3 USING EVOLUTIONARY COMPUTATIONS TO INVESTIGATE BIOLOGICAL EVOLUTION

Evolutionary computations have been applied in a number of areas, allowing researchers to test the limits of theoretical concepts. This has led to a number of refinements of classic evolutionary theory in recent years.

10.3.1 Evolvability and Robustness

Several related concepts from the 1950s have had a major impact on the evolutionary theory of development. These include *canalization*, whereby species show low phenotypic variation, despite ample genetic and environmental variation (also termed as "robustness" [40–42]; and *genetic assimilation*, in which a phenotypic change induced by an environmental perturbation becomes stabilized in the genotype [43,44].

10.3.1.1 Canalization The understanding of how canalization operates has been greatly advanced in recent years through evolutionary modeling (reviewed in Ref. [45]). A number of different computational approaches have exhibited canalization, highlighting different aspects of the evolutionary mechanisms involved.

Stabilizing selection (in which individuals are removed from both ends of a phenotypic distribution, maintaining the mean) has been shown, by modeling at several levels, to contribute to canalization. In simulations where phenotypes had different degrees of sensitivity to microenvironmental variation (developmental noise [46]), stabilizing selection produced canalization. Explicit modeling of individual development and evolution also showed this [4]. Modeling at the population-genetic level, Wagner et al. [47] distinguished that canalization against environmental variation can increase with stabilizing selection, but canalization against mutation increases only

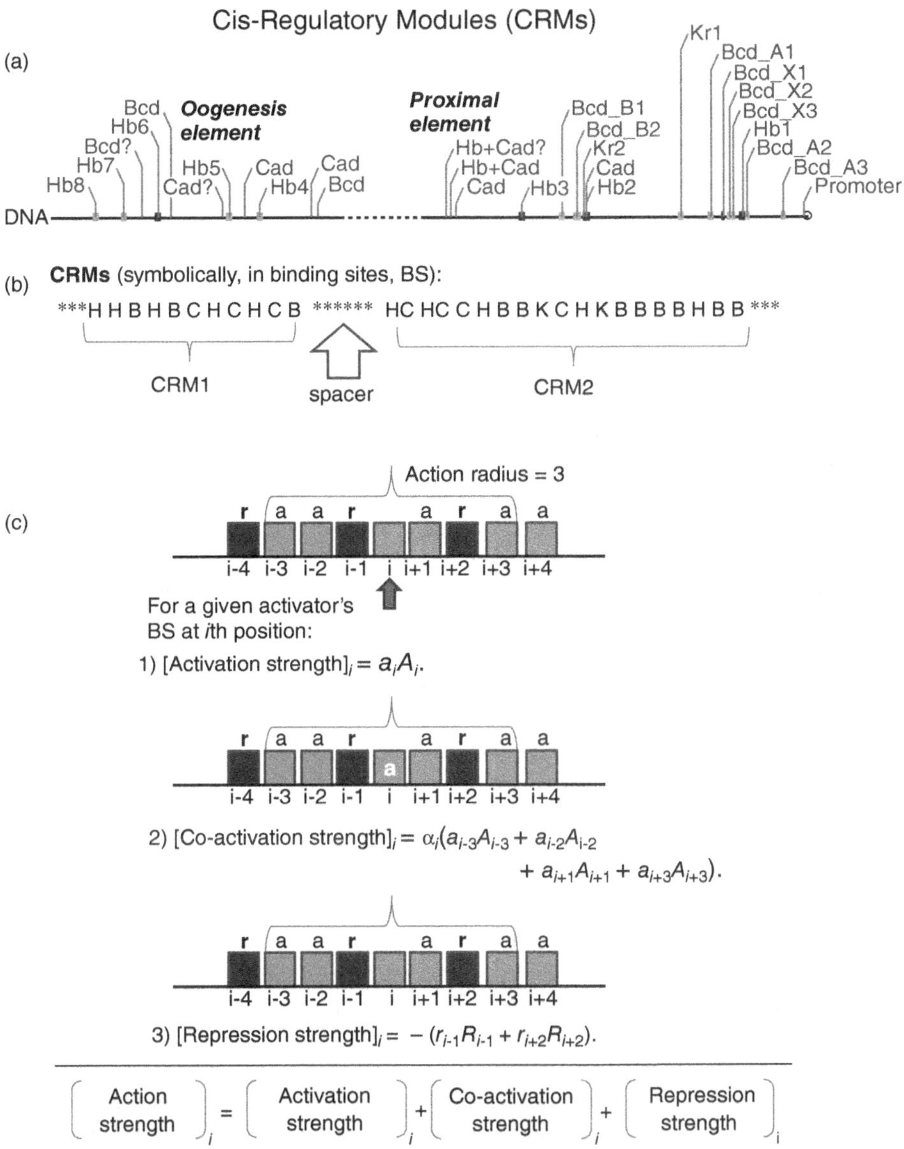

Figure 10.4 Representation of *hb* regulatory region (after Ref. [39]). (a) Schematic of two of the *hb* cis-regulatory modules (CRMs), showing binding sites (BSs) for specific TFs (visualized with *Genamics Expression* software). (b) Representation of the BS for *hb* TFs (Bcd—B, Cad—C, Ems—E, Gt—G, Hb—H, Kr—K, Kni—N, Tll—T, etc.) as strings of characters, neglecting distance along the DNA between BSs. Asterisks denote spacers (non-BS DNA) separating CRMs. (c) Strength of an activating BS is calculated by a three-step algorithm which sums activation (including co-activation) and repression strengths dependent on neighboring TFs (a short action radius of three BSs is used): (1) local activation strength is tallied; (2) neighboring activation is added; (3) neighboring repression is added. The summed action strength is shown at the bottom of the figure.

up to a point. Above a threshold, stabilizing selection purges the genetic variation from a population, and genetic canalization no longer applies.

Indeed, further work indicates that selection for environmental canalization may produce genetic canalization as a by-product [8, 9], that more densely connected networks evolve to be more robust to perturbation than sparser networks under stabilizing selection, and that phenotypic robustness to genetic perturbation does not directly indicate adaptive canalization (reviewed in Ref. [48]). Siegal and Bergman's [8] simulations indicate that increased GRN complexity is associated with robustness. Canalization is not due to direct selection for reduced variability, but because GRNs evolve complexity to buffer against the effects of lethal gene deletions. "We argue that canalization may be an inevitable consequence of complex developmental-genetic processes and thus requires no explanation in terms of evolution to suppress phenotypic variation" [8] (see also Refs. [5, 49]). Systematic computational tests of how an arbitrary null (knockout) mutation affects the expression of other genes reveal that arbitrary genes in complex gene networks have the property of buffering genetic variation and, therefore, have the potential to act as evolutionary capacitors. In a complex GRN, there are many gene products which could appear as "phenotypic" capacitors, such that their removal increased phenotypic variability [9]. This is consistent with experimental observations that the buffering of other genes' expression levels is disrupted when a single gene is knocked out. An extensively studied example is the molecular chaperone Hsp90 [50, 51], but GRN dynamics suggest there should be many more. Experiments in yeast have indicated more than 300 gene products whose removal increased variation [52]. Leclerc [104] has noted that real gene networks appear to be robust, but tend to be sparsely, not densely, connected. He showed that when the costs of complexity are taken into account, that robustness implies a parsimonious network structure that is sparsely connected; that selection will favor sparse networks when topology is free to evolve.

Gombar et al. [53] simulated epigenetic regulation (e.g., chromatin remodeling), and found that this not only facilitated decoupling of mutational and environmental robustness, but it also increased GRN sensitivity to noise (without the loss of robustness). The authors speculated that this epigenetic effect was a necessary condition for multicellularity to arise. Furusawa and Kaneko [54] found that positive feedback through epigenetic regulation aids fast and precise evolutionary responses to environmental change.

Canalization as a by-product of selection against environmental variation appears to be quite general, being seen in simulations of ribonucleic acid (RNA) evolution [55] and metabolic network evolution [56] (where it was suggested that environmental noise is needed for metabolic networks to be robust to gene loss). In general, "It may be that any developmental system that greatly reduces the dimensionality of phenotypes relative to the dimensionality of genotypes will produce the appearance of canalization by releasing hidden genetic variation when perturbed" [48].

10.3.1.2 Genetic Assimilation As Waddington first stated the genetic assimilation concept, "... if selection was practiced for the readiness of a strain of organisms to respond to an environmental stimulus in a particular manner, genotypes might eventually be produced which would develop into the favoured phenotype even in the absence of the environmental stimulus. A character which had originally been an 'acquired' one might be said to have become genetically assimilated" [43] (see also Refs. [40, 44]). Computations in the past decade have refined this idea. For instance, Masel [11], using a variant of the Wagner model (Section 10.2.1.2), showed that genetic assimilation can occur in the absence of selection for the trait.

Genetic assimilation requires an underlying phenotypic plasticity—a capacity for the genotype to produce multiple phenotypes in response to non-genetic perturbations [57–59]. A factor in this response is the release of cryptic genetic variation, which occurs due to neutral drift, but may be invisible until a large enough perturbation occurs [60]. Mathematical analysis of Wagner's canalization model [61] indicates that evolution toward a particular phenotype stabilizes other (cryptic) phenotypes, making GRNs more robust to genetic perturbations (loss, mutation, etc.). Computations by Lande [62] quantify the relation between plasticity dynamics and assimilation, wherein plasticity must increase in order to allow evolution to a perturbed environment, but then decrease in order to maintain the new optimum. Espinosa-Soto et al. [63] also characterized the plasticity–assimilation relation, using classic W-matrix simulations [64].

10.3.1.3 Noise Lehner and Kaneko [65] have used an approach from statistical physics to derive a "fluctuation–response" relation for evolution akin to the "fluctuation–dissipation" theorem relating Brownian noise and motion in fluids. This describes a fundamental relation between environmental noise and evolvability. They found an optimal range of noise, which could contribute to evolution: above a critical noise level "error catastrophe" resulted; below this, GRNs can get stuck in fitness landscapes, and not evolve due to perturbations (i.e., low plasticity) [66–68]. They also found that evolution of robustness to noise could result in robustness to genetic mutation (similar to Ref. [8]), due to a proportionality between the environmental (noise) and genetic (mutation) contributions to phenotypic variance [69–72].

A number of these projects used numerical simulations to explore the theoretical principles. In Ref. [67], it was found that networks acquired both mutational and noise robustness if the phenotypic variance induced by mutations was smaller than that observed in an isogenic population (where variance is only due to gene expression noise). They found adaptive states, characterized by high production and decay terms, were also less affected by gene expression noise. Developmental noise could aid both adaptability and robustness, with the highest adaptability at high noise levels that nearly destroy robustness.

In Ref. [71], two interacting and evolving networks were modeled: one a GRN controlling protein expression and the other a metabolic network depending on the GRN proteins and environmentally supplied nutrients. Growth of the "cell" containing these networks is limited by metabolite influx. They found that in any cell with stochastic gene expression, the cell's state was selected by noise—it did not need to be driven by a specific signalling network.

Kuwahara and Soyer [73] used an intrinsic method (Gillespie algorithm) to model noise in a gene with autofeedback, and corroborated that noise (in the resulting bistable switch) aids evolvability. Extending the approach to consider protein–gene and protein–protein interactions, van Dijk et al. [74] have predicted that GRNs with monomeric TFs, rather than dimeric, are more robust to mutations affecting regulatory interactions.

Quantifying Waddington's concept, Kaneko [75] identified genetic assimilation as the proportionality between genetic and environmental contributions to phenotypic variability. He found that the number of genes with environmental contribution greater than genetic contribution increased as robustness of fitness increased, with a concomitant decrease in the degrees of freedom for gene mutation [76].

10.3.1.4 Adaptation Biologically, stable end states might not be the critical feature for speciation or survival. How fast a species approaches a stable state, its rate of adaptation, might be more important. Genes in a GRN will generally change expression in response to an environmental perturbation, with some genes returning rapidly to prior levels and some maintaining altered expression to adapt [77]. Neyfakh et al. [78], using a simple model of a cell with a GRN controlling basic metabolism of an environmental nutrient, found a rapid rise in fitness in response to an environmental perturbation, which then plateaued. The resulting GRNs were quite diverse, many with what appeared to be functionally non-essential genes (see also Section 10.3.3.1 on gene co-option). Cuypers and Hogeweg [79] (reviewed in Ref. [80]) further worked with this model, and found that the plateauing after the initial rise in fitness could be characterized by a period of streamlining gene loss.

Does this dynamic merely reflect the plasticity needed for genetic assimilation, or does plasticity itself affect adaptation rate (i.e., what is the relation between plasticity and evolvability)? Fierst [81], using the Wagner model [4], showed that "a history of phenotypic plasticity may ... shorten the waiting time for the generation of phenotypic variance from new mutations and recombination. Rather than acting as a short-term alternative, phenotypic plasticity may facilitate future adaptation and genetic evolution." On the other hand, plasticity can contribute to liability in developmental GRNs. Draghi and Whitlock [82], with a continuous extension of Wagner's model [5], quantified how phenotypic plasticity increased variance, in mutations and in standing genomes, which could be amplified for correlated genes.

Francois and Siggia [16] (see also Refs. [83,84]) used small GRNs to characterize two finer aspects of the response to perturbations: responsiveness, the speed at which genes altered expression, and quality of adaptation, or how well the GRN maintained fitness to the new environmental condition (see also Ref. [85] for an application to temperature compensation in circadian clocks). Ma et al. [86] did a complete analysis of all 3-node GRNs, and found only two topologies with an adaptive response, a negative feedback loop with a buffering node and an incoherent feed-forward loop (FFL). However, this does not appear to generalize to large GRNs: simulations with large GRNs [77] did not find few gene motifs underlying adaptation; rather, adaptation was cooperative across the network (and contributed to robustness).

Modeling has made the original concepts of Waddington much more precise; the quantitative questions raised by the modeling, such as what degree of plasticity is optimal for the development of particular organisms, may now be at the stage to allow for experimental testing.

10.3.2 Crossover

Genetic material is altered in mutation and in reproduction. Both have profound impacts on evolutionary dynamics. There is great potential to refine EA by drawing from the details of biological reproduction mechanisms. In particular, point mutation tends to destroy meaningful "words" (genes) in the process of evolution; crossover of longer segments of genetic material can maintain these building blocks [87, 88]. For instance, Martin and Wagner [7] added recombination and point mutation steps to the *W* matrix approach and reported that recombination may be a much stronger factor in GRN evolution than point mutation. When both steps are active, they found genotype diversity to be increased, the deleterious effect of point mutations to be decreased, and the emergence of cis-regulatory complexes (combinatorial regulatory interactions). In this section, we discuss several approaches to simulating reproduction mechanisms and how this reflects on the role of reproduction in evolutionary dynamics.

10.3.2.1 Sexual Reproduction The role of sexual reproduction in evolution has been intensively studied, and has long been understood to increase population fitness. Azevedo et al. [89] quantified this effect using an extension of the *W* matrix approach (see also Ref. [33] for sexual reproduction in *W* matrices). At closer detail, Livnat et al. [90, 91] argued that sex does not maximize fitness generally, but specifically improves mixability, the capacity of a particular genome to function with a wide variety of genetic partners. Sex also has the potential to destroy favorable combinations of genes. Misevic et al. [92, 93] used a digital organisms' approach to compare sexual reproduction and asexual reproduction. They found clustering of genes (modularity) for similar traits with sexual reproduction, which may contribute to maintaining desirable characteristics.

Sexual reproduction frequently involves sexual selection, which can drive sexual dimorphism. Rapid evolution of dimorphism suggests that the underlying GRNs may favor evolvability over phenotypic stability. However, Fierst [81, 94] found, with a *W* matrix approach, that GRNs could produce dimorphic characteristics with both high robustness and high evolvability; coarse-grained dynamics may have the capacity to do both things.

Heterosis, or "hybrid vigor," is the tendency for crosses of individuals from genetically distinct populations to improve particular traits (e.g., growth rate). Emmrich et al. [95, 96] have recently developed a GRN evolutionary simulation approach with hybridization to quantify this process.

10.3.2.2 Retroviral Crossover Sexual reproduction is not the only mechanism used in biology for crossover of large segments of intact genetic information. We have developed a genetic algorithm (RetroGA) with crossover inspired by the mechanism by which retroviruses create DNA strands from RNA [87, 88] (which is then copied into a host). Biologically, the DNA strand is generally produced from two parent RNA strands, with the read-off switching between strands depending on jump sequences (regions of homology). We have tested this mechanism on Royal Road and Royal Staircase functions [97], designed to evaluate the retention of building blocks (e.g., genes), as well as on biological test cases in bacterial promoter evolution and *Drosophila* segmentation. These tests are characterized by basin-subportal architecture of the fitness landscape, with long periods of neutral evolution followed by quick jumps to higher fitness (upon creation of a new building block). In all cases, the RetroGA algorithm significantly sped up evaluation in comparison to point mutation.

10.3.3 GRN Outgrowth

GRNs can change size in the process of evolution, by gaining or losing genes. Genes can be newly created or destroyed within a GRN. But it is likely much more common for pre-existing genes to have small mutations in their CRMs, which allows them to be regulated by TFs from other GRNs, effectively recruiting them into that GRN [98, 99]. A number of groups have simulated ways in which simpler ancestral GRNs can evolve into more complex derived GRNs.

10.3.3.1 Gene Co-Option Computations indicate that co-option (recruitment) of genes and outgrowth of GRNs (increase in network nodes) occur spontaneously, without particular selective pressure for network complexity. Incorporating gene addition and gene withdrawal operators into a standard GA, we found a net increase of genes in GRNs (Figure 10.5) [100, 101]. We tested the algorithm on a coarse-grained (continuous, partial DE) GRN model for *Drosophila* segmentation, and found that the derived networks showed increased developmental robustness to fluctuations in maternal signalling factors. Co-opted genes frequently matched experimentally observed patterns, and became essential nodes of the derived GRN. Similar facility of network outgrowth has been observed in Ref. [102]. In simulations of protein networks, these authors found that adding mechanisms for protein addition, duplication, or deletion, as well as creation or destruction of interactions produced network growth, in the absence of any specific selection pressure for larger networks. Tsuda and Kawata [103] found similar results with a fine-grained GRN model (with TF binding in individual CRMs): complex GRNs tended to evolve if beneficial gene duplications were fixed in the genome. In the very fine-scaled work of He et al. [36], which was corroborated against *Drosophila* sequence data and expression patterns, it was observed that complex networks were favored by the evolutionary process (characterized by large numbers of TFBSs in the enhancers). With relevance also to Section 10.3.1, Pujato et al. [35] found that in simpler GRNs (with less connections) robustness (to environmental perturbations) depended on local sequences decreasing the creation or destruction of TFBSs. For more complex GRNs (with more connections), robustness depends on regulating TFBSs with high-level network wide effects. Iwasaki et al. [60] found a strong correlation between the amount of cryptic variation and the network size, leading them to propose that network expansion could be a source of evolvability. Work by Leclerc [104] places some caution on universal GRN outgrowth: if a cost is entered for GRN complexity, evolution can favor sparser networks.

\	TF_1	TF_2	...	TF_n
G_1	$W_{g1 \leftarrow tf1}$	$W_{g1 \leftarrow tf2}$	...	$W_{g1 \leftarrow tfn}$
G_2	$W_{g2 \leftarrow tf1}$	$W_{g2 \leftarrow tf2}$	...	$W_{g2 \leftarrow tfn}$
...	...	...	...	...
G_n	$W_{gn \leftarrow tf1}$	$W_{gn \leftarrow tf2}$	...	$W_{gn \leftarrow tfn}$

(a)

\	TF_1	TF_2	...	TF_n	TF_{C1}
G_1	$W_{g1 \leftarrow tf1}$	$W_{g1 \leftarrow tf2}$	...	$W_{g1 \leftarrow tfn}$	$W_{g1 \leftarrow tfc1}$
G_2	$W_{g2 \leftarrow tf1}$	$W_{g2 \leftarrow tf2}$	...	$W_{g2 \leftarrow tfn}$	$W_{g2 \leftarrow tfc1}$
...	...	...	...	...	...
G_n	$W_{gn \leftarrow tf1}$	$W_{gn \leftarrow tf2}$	...	$W_{gn \leftarrow tfn}$	$W_{gn \leftarrow tfc1}$
G_{C1}	$W_{gc1 \leftarrow tf1}$	$W_{gc1 \leftarrow tf2}$	...	$W_{gc1 \leftarrow tfn}$	$W_{gc1 \leftarrow tfc1}$

(b)

Figure 10.5 Modeling co-option of a new gene into a network during evolution. (a) Initial n-gene network ($G_1, \dots, G_n$), producing n TFs ($TF_1, \dots, TF_n$). (b) New gene (G_{C1}), producing a new factor TF_{C1}, can be incorporated by chance and retained if it maintains or improves fitness (gray colored row and column). The initial n-gene network can become completely rewired by co-opting G_{C1}.

Computations suggest that gene co-option can provide robustness to a GRN from parasite attack. In studies of co-evolution of a host and a parasite, Salathe and Soyer [105] found that parasite selective pressure led to recruitment and retention of new nodes to a host GRN, which buffered the host to loss of particular genes. When the parasites were removed, the host GRN could shrink again without the loss of fitness.

In several studies [106–108], we have characterized how gene co-option can maintain GRN functionality to form segmentation patterns, despite the attack by genetic parasites (transposons, e.g., Refs. [109–114]). With a coarse-grained model [107, 108], we simulated the attack of a transposon on the ability of a GRN to read a maternal positional information gradient critical to segmentation. Computations showed an "arms race"—if the GRN could co-opt new genes rapidly enough, these could adapt to a new role of reading the maternal gradient before transposon infection spreads. More recently [108], we have observed similar host–parasite dynamics in a fine-grained model, in which transposons insert at particular base-pair sequences. These studies illustrate how gene co-option can lead to robustness, not against environmental perturbations, but against a particular type of gene loss due to the ubiquitous selective pressure of transposons on the genome. The transposon mechanism also sped evolutionary searches, which may be applicable to EA optimization in general (see also Ref. [115]).

10.3.3.2 *Modularity of Cis-Regulation* Fine-scale modeling (at the level of TFBS evolution) can be used to investigate TFBS clustering, or the origin of CRMs. In *Drosophila*, the well-studied gene *hb* is known to have three CRMs actively driving expression during segmentation. The CRMs control expression in distinct spatial regions, though there is some overlap and redundancy. To study the evolvability of this genomic structure, we constructed a representation of an ancestral *hb* cis-regulatory region as a string of 104 characters [38, 39]. Each character represents a TFBS for known regulators of *hb*, or a neutral spacer (Figure 10.4). TFBSs are either activating or repressing. Total regulatory strength from the string was used to solve pattern forming (DE) equations, with the results scored (fitness) against experimental *hb* patterns. Evolution on populations of thousands of individuals readily produced clustering or modularity, such that the cis-regulatory region formed three distinct CRMs. Depending on overall regulation strength (R_a), the spatial domains controlled by each CRM ranged from entirely redundant to fully distinct (Figure 10.6). For $R_a \geq 79$, we observed the emergence of a distinct "stripe element" expression, and a close association between the model output and the actual expression domains seen for the three biological CRMs. We conclude from these simulations that modularity is readily evolvable. It can produce redundant expression, a basal form of redundancy against mutations (in which the loss of a modules' function is not catastrophic). Modularity also allows for a mix of redundancy and distinct expression, as seen biologically; this allows for some redundancy, but also the capacity to differentially regulate the temporal and spatial aspects of particular domains (biologically, the anterior *hb* domains are expressed before the mid and posterior domains).

The fine-scale modeling of Duque et al. [37] corroborates a tendency to clustering, and, in addition, calibrates the rate of maintenance of TFBS clusters against divergence in evolutionary time: across 37 well-studied CRMs in segmentation, they could fit divergence times between 12 different *Drosophila* species. Saito et al. [116], in evolutionary computations of simple and complex GRNs, found that complex GRNs tend to evolve redundancy. Soyer's work with protein networks shows similar tendency toward modularity [117].

10.3.4 Characterization of GRN Space

A very powerful use of modeling is to map out the range of possibilities for a given dynamic mechanism. Many authors have done this with evolutionary simulations, characterizing parameter spaces or fitness functions for particular evolutionary mechanisms.

With a *W* matrix approach to represent the genotype-to-phenotype mapping, Borenstein and Krakauer [118] studied how the developmental plan maps several genotypes to the same phenotype, and consequently generates a much smaller number of distinct phenotypes (which depend on the specific mapping). They found that gene background (epistasis) created a clustering of phenotypes in the space of possibilities (morphospace), with the result that low-interaction ancestral GRNs spanned a larger volume in the space than the highly connected derived GRNs. Payne et al. [119] looked at this in terms of input (coded in a gene's cis-regulatory region) to

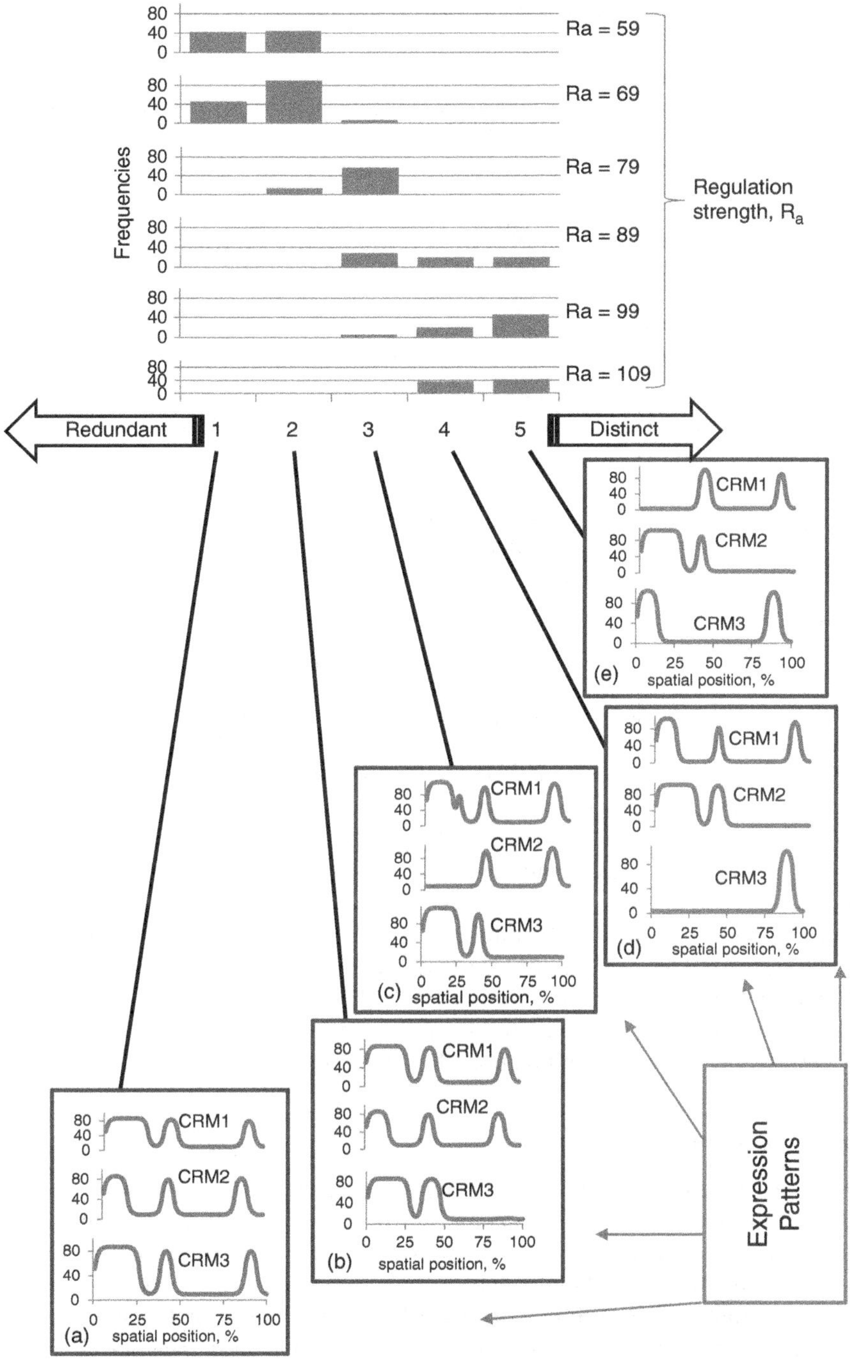

Figure 10.6 *hb* gene expression patterns driven from each of three evolved CRMs. As overall regulation strength, R_a, is increased, the expression domains go from fully redundant (a) to fully distinct (e). Patterns 3–5 (c)–(e) show anterior–posterior separation of domains, as seen biologically. Pattern 3 (c) is closest to the biological domain–CRM correspondence, especially the middle domain, with no anterior expression.

output (gene expression) functions. Clustering can result from the fact that many GRNs can produce the same output; innovation can be measured in terms of an increasing distance from ancestral forms in the phenotype space.

Focusing on small GRNs, Nochomovitz and Li [120] surveyed the phenotype space for 3- and 4-node networks. They found that certain phenotypes can be coded by a very broad range of GRNs, allowing for flexible design. These tended to have a conserved set of connections for phenotypic stability, as well as a more evolvable set of connections. If a fourth gene was added to a 3-node net, it was frequently involved in the conservative suppression of phenotypic variation. Feed-forward loops are very common small GRN motifs in biological networks. Widder et al. [121, 122] used an analytical approach (made feasible by the small GRN size) to do a complete survey of the GRN-function landscape FFLs. They found the abundance of the FFL motif could be predicted from its high evolvability.

Large-scale surveys of phenotype space can provide broad insights into the dynamics of evolution. While the first assumption might be that species evolve toward steady states, Pinho et al. [123] found that GRNs (in the W matrix approach) more often showed cyclic behavior (periodic orbits) than monotonic evolution toward fixed points. They reported that stable evolution toward fixed points was more common in sparser networks.

Multifunctional GRNs can produce different expressions at different stages of development. Using a genotype space characterizing millions of GRNs and their functions, Payne and Wagner [124] found that the number of multifunctional GRNs declines exponentially depending on the number of functions. The total number of multifunction GRNs can remain high, but they become increasingly non-clustered as the number of functions increases. This implies that the historical trajectory of a GRN is important for acquiring new functions and that multifunctional GRNs are particularly susceptible to mutation. This is, however, a sequential view of the acquisition of multifunctionality; Warmflash et al. [125] considered that GRNs are under selective pressure from multiple constraints simultaneously. Frequently, these cannot all be equally optimized. The authors of Ref. [125] found that coding a Pareto optimization into their GRN evolutionary mechanism produced a more efficient search for fit GRNs than ad hoc combination of design criteria.

10.3.5 Epistasis

GRN computations can be used to study epistatic effects, by comparing a general null hypothesis of additive gene action (where genetic background has no effect) to epistatic models with background effects. Carter et al. [126] found that epistasis can be a strong modifier of evolution: if positive (cooperative), it can enhance evolvability; if negative (antagonistic), it can favor canalization. Sanjuan and Nebot [127] used computations to test whether simpler genomes tended to have antagonistic epistasis. Simulations indicated that multifunctionality and a lack of redundancy in small GRNs was associated with negative epistasis, and the converse of these in large networks was associated with positive epistasis (which was associated with increased robustness to mutations). Draghi and Plotkin [128] found epistasis to have low prevalence early in adaptation but to be more common later in adaptation. Accompanying shifts from antagonistic to synergistic effects paralleled the small versus large GRN results in Refs. [35, 118]. This may be useful for comparing bacterial versus eukaryotic evolutionary strategies.

Cotterell and Sharpe [129] tested epistatic effects in spatial gradient-reading GRNs (e.g., that convert graded input into discrete spatial domains in segmentation). They found that such GRNs were connected, in terms of gene additions and deletions, through unstable, non-functional GRNs. However, epistatic gene-background effects could stabilize the non-functional GRNs, so that evolution between stable networks occurred by these pathways.

10.3.6 Body Segmentation

In addition to the studies on spatially distributed gene expression discussed in Section 10.2.1.2, several lines of investigation have used GRN computations to elucidate probable paths of evolution for body segmentation mechanisms in insects. Our work [106–108] has focused on evolution of GRNs that produce segmentation patterns simultaneously, characteristic of long germ band insects such as *Drosophila*. Many insects have short germ band segmentation, in which only two or three segments are formed at a time (compared to seven in *Drosophila*). These short patterns are propagated down the body to form higher numbers of final segments. Salazar-Ciudad et al. [130, 131] simulated stripe forming GRNs, and found that small numbers of stripes (up to three) tended to be formed by hierarchical networks, while higher numbers of stripes tended to be formed simultaneously. These simultaneous networks could become hierarchical with further evolution. ten Tusscher and Hogeweg [18, 19] have computed GRN evolution on large populations, and found that the sequential (short germ band) mechanism evolves readily from an ancestral unsegmented state (also see the work in Ref. [15–17]). They argue that the ease with which this mechanism evolves supports independent origins in vertebrates, annelids, and arthropods. They suggest that the simultaneous (long germ) mechanism reflects particular historical constraints during evolution.

10.4 CONCLUSIONS

The increasing use of GRN simulations in recent years has had a significant impact on the understanding of a number of evolutionary principles in developmental biology. Starting from discrete, Boolean representations (*W* matrices), refinements have included modeling of continuous gene expression levels, consideration of co-factor effects, modeling at TFBS resolution, and extension to spatially distributed gene expression (a fundamental aspect of body plan formation in development). This chapter has outlined the techniques developed in this area of computational biology, and the main biological impacts these have had. Our hope is to provide some inspiration for computer scientists in EA that the types of evolutionary dynamics from the detailed biology can enrich development of new and more efficient search algorithms. In turn, we hope an increased interaction between computational biology and computer science can lead to improvements in the efficiency and scope of the developmental simulations (for instance, in Section 10.3.4, the use of multi- rather than single-objective optimization). Several recent computer science reviews [115, 132–135] have called for development of a field of computational evolution. A number of the topics outlined (what does evolution act on; how does evolution happen; and what are the practical applications of this knowledge) are, as discussed in this chapter, being approached by computational biologists. Some topics, however, such as modeling evolution of nested, multi-level organisms, deserve much more attention (see Refs. [136, 137] for other recent reviews.) Increased interaction will benefit both computational biologists and computer scientists. To summarize, some of the areas we feel will be especially fruitful for interdisciplinary exchanges:

1. Evolvability and robustness: one of the major findings described above is that environmental noise can contribute to genetic robustness. For EA searches in general, noise addition may offer a means for increasing the reliability of evolved networks to solve a given problem. Noise also supports evolvability, speeding evolution in response to changing constraints. The Kaneko group's relation between noise and evolvability may have general applicability to noise in EA. Feedback mechanisms such as epigenetic remodeling can also increase reliability and speed, and could be built into EA approaches.
2. Crossover: there are numerous ways in which genetic material is altered biologically. Point mutation, the most common means in EA, is one of these. Others include transposition, inversion, deletion, and gene duplication [138]. As described in Section 10.3.2, a number of groups are modeling sexual reproduction, with child GRNs or sequences created from parent GRNs, creating large scale or modular inheritance of genetic material. We have found, with retroviral [87, 88] and transposon [106–108] crossover mechanisms (with tags in the genetic string for insertion sites), that modular genetic alteration can be highly effective for searches in "basin-subportal" architectures characterized by long periods of neutral evolution interspersed with discrete jumps in fitness. Modular inheritance is critical for maintaining meaningful "words" or genes during evolution, and has the potential for large improvements in EA speed.
3. Network outgrowth: simulations indicate that network growth is natural, if genes are available for co-option. (Biologically, this is commonly through modification of cis-regulatory regions, creating BSs for new TFs.) Computations indicate that larger, more evolved networks may have more cryptic variation, and can more readily respond to changes in constraints. This could be considered as a general strategy for EA. We have studied co-option as a means for a host to deal with parasitic attacks. This suggests two lines of future research: co-option could be used in EA to make optimization robust to attacks; and co-evolution (i.e., of host–parasite) could be a means for multi-objective optimization. We also found modularity of cis-regulatory regions can evolve

readily, and this could be a means in EA to form both redundant and differential controls of different functions of a target solution (cf. the CRM to spatial domain correspondence in Figure 10.6).

4. Co-factors/epistasis: multi-factor regulation of genes greatly increases GRN connectivity, with far richer dynamics, than simple one-to-one gene interactions. Simulations have shown that epistasis can tune evolutionary searches between canalization and evolvability—this could be a tool for guiding EA searches in general.

ACKNOWLEDGEMENTS

The authors thank the U.S. NIH for financial support, grant R01-GM072022. A.V.S. thanks The Russian Foundation for Basic Research, grant 13-04-02137.

REFERENCES

1. Noman, N., Palafox, L. and Iba, H. (2013). Evolving genetic networks for synthetic biology. *New Generation Computing* 31, 71–88.
2. Spirov, A. and Holloway, D. (2013). Using evolutionary computations to understand the design and evolution of gene and cell regulatory networks. *Methods* 62, 39–55.
3. Francois, P. and Siggia E. (2010). Predicting embryonic patterning using mutual entropy fitness and in silico evolution. *Development* 137, 2385–2395.
4. Wagner, A. (1994). Evolution of gene networks by gene duplications: a mathematical model and its implications on genome organization. *Proceedings of the National Academy of Sciences of the United States of America* 91, 4387–4391.
5. Wagner, A. (1996). Does evolutionary plasticity evolve? *Evolution* 50, 1008–1023.
6. Sirbu, A., Ruskin, H.J. and Crane, M. (2010). Comparison of evolutionary algorithms in gene regulatory network model inference. *BMC Bioinformatics* 11, 59.
7. Martin, O.C. and Wagner, A. (2009). Effects of recombination on complex regulatory circuits. *Genetics* 183, 673–684.
8. Siegal, M.L. and Bergman, A. (2002). Waddington's canalization revisited: developmental stability and evolution. *Proceedings of the National Academy of Sciences of the United States of America* 99, 10528–10532.
9. Bergman, A. and Siegal, M.L. (2003). Evolutionary capacitance as a general feature of complex gene networks. *Nature* 424, 549–552.
10. MacCarthy, T. and Bergman, A. (2007). Co-evolution of epistasis and recombination favors asexual reproduction. *Proceedings of the National Academy of Sciences of the United States of America* 104, 12801–12806.
11. Masel, J. (2004). Genetic assimilation can occur in the absence of selection for the assimilating phenotype, suggesting a role for the canalization heuristic. *Journal of Evolutionary Biology* 17, 1106–1110.
12. McDonald, D., Waterbury, L., Knight, R. and Betterton, M.D. (2008). Activating and inhibiting connections in biological network dynamics. *Biology Direct* 3, 49.

13. Jaeger, J., Sharp, D.H. and Reinitz, J. (2007). Known maternal gradients are not sufficient for the establishment of gap domains in *Drosophila melanogaster. Mechanisms of Development* 124, 108–28.
14. Sole, R.V., Fernandez, P. and Kauffman, S.A. (2003). Adaptive walks in a gene network model of morphogenesis: insights into the Cambrian explosion. *International Journal of Developmental Biology* 47, 685–693.
15. Francois, P., Hakim, V. and Siggia, E.D. (2007). Deriving structure from evolution: metazoan segmentation. *Molecular Systems Biology* 3, 154.
16. Francois, P. and Siggia, E. (2008). A case study of evolutionary computation of biochemical adaptation. *Physical Biology* 5, 026009.
17. Fujimoto, K., Ishihara, S. and Kaneko, K. (2008). Network evolution of body plans. *PLoS One* 3, e2772.
18. ten Tusscher, K.H. and Hogeweg, P. (2011). Evolution of networks for body plan patterning; interplay of modularity, robustness and evolvability. *PLoS Computational Biology* 7, e1002208.
19. ten Tusscher, K.H.W.J. (2013). Mechanisms and constraints shaping the evolution of body plan segmentation. *The European Physical Journal E: Soft Matter* 36, 54.
20. Fomekong-Nanfack, Y., Kaandorp, J.A. and Blom, J.G. (2007). Efficient parameter estimation for spatio-temporal models of pattern formation: case study of *Drosophila melanogaster. Bioinformatics* 23, 3356–3363.
21. Chu, K., Deng, Y. and Reinitz, J. (1999). Parallel simulated annealing by mixing of states. *Journal of Comparative Physiology* 148, 646–662.
22. Jostins, L. and Jaeger, J. (2010). Reverse engineering a gene network using an asynchronous parallel evolution strategy. *BMC Systems Biology* 4, 17.
23. Kozlov, K. and Samsonov, A. (2011). DEEP—differential evolution entirely parallel method for gene regulatory networks. *Journal of Supercomputing* 57, 172–178.
24. Lenski, R.E., Ofria, C., Collier, T.C. and Adami, C. (1999). Genome complexity, robustness and genetic interactions in digital organisms. *Nature* 400, 661–664.
25. Ofria, C., Huang, W. and Torng, E. (2008). On the gradual evolution of complexity and the sudden emergence of complex features. *Artificial Life* 14, 255–263.
26. Clune, J., Pennock, R.T., Ofria, C. and Lenski, R.E. (2012). Ontogeny tends to recapitulate phylogeny in digital organisms. *The American Naturalist* 180, E54–E63.
27. Valverde, S., Sol, R.V. and Elena, S. (2012). Evolved modular epistasis in artificial organisms. *Artificial Life* 13, 111–115.
28. Covert, A.W., Lenski, R.E., Wilke, C.O. and Ofria, C. (2013). Experiments on the role of deleterious mutations as stepping stones in adaptive evolution. *Proceedings of the National Academy of Sciences of the United States of America* 110, E3171–E3178.
29. Batut, B., Parsons, D.P., Fischer, S., Beslon, G. and Knibbe, C. (2013). *In silico* experimental evolution: a tool to test evolutionary scenarios. *BMC Bioinformatics* 14, S11.
30. Gutierrez, J. and Maere, S. (2014). Modeling the evolution of molecular systems from a mechanistic perspective. *Trends in Plant Science* 19, 292–303.
31. Hong, J.W., Hendrix, D.A. and Levine, M.S. (2008). Shadow enhancers as a source of evolutionary novelty. *Science* 321, 1314.
32. Perry, M.W., Boettiger, A.N. and Levine, M. (2011). Multiple enhancers ensure precision of gap gene-expression patterns in the *Drosophila* embryo. *Proceedings of the National Academy of Science of the United States of America* 108, 13570–13575.

33. ten Tusscher, K.H. and Hogeweg, P. (2009). The role of genome and gene regulatory network canalization in the evolution of multi-trait polymorphisms and sympatric speciation. *BMC Evolutionary Biology*, 9, 159.
34. Crombach, A. and Hogeweg, P. (2008). Evolution of evolvability in gene regulatory networks. *PLoS Computational Biology* 4, e1000112.
35. Pujato, M., MacCarthy, T., Fiser, A. and Bergman, A. (2013). The underlying molecular and network level mechanisms in the evolution of robustness in gene regulatory networks. *PLoS Computational Biology* 9, e1002865.
36. He, X., Duque, T.S.P.C. and Sinha, S. (2012). Evolutionary origins of transcription factor binding site clusters. *Molecular Biology and Evolution* 29, 1059–1070.
37. Duque, T., Samee, M.A.H., Kazemian, M., Pham, H.N., Brodsky, M.H. and Sinha, S. (2013). Simulations of enhancer evolution provide mechanistic insights into gene regulation. *Molecular Biology and Evolution* 31, 184–200.
38. Spirov, A.V. and Holloway, D.M. (2012). Evolution in silico of genes with multiple regulatory modules, on the example of the *Drosophila* segmentation gene hunchback. *Proceedings of the IEEE Symposium on Computational Intelligence in Bioinformatics and Computational Biology* 2012, 244–251.
39. Zagrijchuck, E.A., Sabirov, M.A., Holloway, D.M. and Spirov, A.V. (2014). In silico evolution of the *hunchback* gene indicates redundancy in cis-regulatory organization and spatial gene expression. *Journal of Bioinformatics and Computational Biology* 12, 1441009.
40. Waddington, C.H. (1942). Canalization of development and the inheritance of acquired characters. *Nature* 150, 563–565.
41. Schmalhausen, I. (1949). I. *Factors of Evolution*. Philadelphia, PA: Blakiston; II. *Factors of Evolution: The Theory of Stabilizing Selection*. Chicago: University of Chicago Press (reprinted 1986).
42. Rendel, J.M. (1959). Canalization of the acute phenotype of *Drosophila*. *Evolution* 13, 425–439.
43. Waddington, C.H. (1956). Genetic assimilation of the bithorax phenotype. *Evolution* 10, 1–13.
44. Waddington, C.H. (1953). Genetic assimilation of an acquired character. *Evolution* 7, 118–126.
45. Visser, J., Hermisson, J., Wagner, G.P., Meyers, L.A., Bagheri-Chaichian, H., et al. (2003). Perspective: evolution and detection of genetic robustness. *Evolution* 57, 1959–1972.
46. Gavrilets, S. and Hastings, A. (1994). A quantitative-genetic model for selection on developmental noise. *Evolution* 48, 1478–1486.
47. Wagner, G.P., Booth, G. and Bagheri-Chaichian, H. (1997). A population genetic theory of canalization. *Evolution* 51, 329–347.
48. Stearns, S.C. (2002). Progress on canalization. *Proceedings of the National Academy of Sciences of the United States of America* 99, 10229–10230.
49. Huerta-Sanchez, E. and Durrett, R. (2007). Wagner's canalization model. *Theoretical Population Biology* 71, 121–130.
50. Rutherford, S.L. and Lindquist, S. (1998). Hsp90 as a capacitor for morphological evolution. *Nature* 396, 336–342.
51. Yeyati, P.L., Bancewicz, R.M., Maule, J. and van Heyningen, V. (2007). Hsp90 selectively modulates phenotype in vertebrate development. *PLoS Genetics* 3, e43.

52. Levy, S.F. and Siegal, M.L. (2008). Network hubs buffer environmental variation in Saccharomyces cerevisiae. *PLoS Biology* 6, e264.
53. Gombar, S., MacCarthy, T. and Bergman, A. (2014). Epigenetics decouples mutational from environmental robustness. Did it also facilitate multicellularity? *PLoS Computational Biology* 10, e1003450.
54. Furusawa, C. and Kaneko, K. (2013). Epigenetic feedback regulation accelerates adaptation and evolution. *PLoS One* 8, e61251.
55. Ancel, L.W. and Fontana, W. (2000). Plasticity, evolvability and modularity in RNA. *Journal of Experimental Zoology* 288, 242–283.
56. Soyer, O.S. and Pfeiffer, T. (2010). Evolution under fluctuating environments explains observed robustness in metabolic networks. *PLoS Computational Biology* 6, e1000907.
57. West-Eberhard, M.J. (2003). *Developmental Plasticity and Evolution*. New York, NY: Oxford University Press.
58. Pigliucci, M. and Murren, C.J. (2003). Genetic assimilation and a possible evolutionary paradox: can macroevolution sometimes be so fast as to pass us by? *Evolution* 57, 1455–1464.
59. Gilbert, S.F. and Epel, D. (2008). *Ecological Developmental Biology: Integrating Epigenetics, Medicine, and Evolution*. Sunderland, MA: Sinauer Associates.
60. Iwasaki, W.M., Tsuda, M.E. and Kawata, M. (2013). Genetic and environmental factors affecting cryptic variations in gene regulatory networks. *BMC Evolutionary Biology* 13, 91.
61. Le Cunff, Y. and Pakdaman, K. (2012). Phenotype-genotype relation in Wagner's canalization model. *Journal of Theoretical Biology* 314, 69–83.
62. Lande, R. (2009). Adaptation to an extraordinary environment by evolution of phenotypic plasticity and genetic assimilation. *Journal of Evolutionary Biology* 22, 1435–1446.
63. Espinosa-Soto, C. Martin, O.C. and Wagner, A. (2011). Phenotypic plasticity can facilitate adaptive evolution in gene regulatory circuits. *BMC Evolutionary Biology* 11, 5.
64. Espinosa-Soto, C., Martin, O.C. and Wagner, A. (2011). Phenotypic plasticity can increase phenotypic variability after non-genetic perturbations in gene regulatory circuits. *Journal of Evolutionary Biology* 24, 1284–1297.
65. Lehner, B. and Kaneko, K. (2011). Fluctuation and response in biology. *Cellular and Molecular Life Sciences* 68, 1005–1010.
66. Kaneko, K. and Furusawa, C. (2006). An evolutionary relationship between genetic variation and phenotypic fluctuation. *Journal of Theoretical Biology* 240, 78–86.
67. Kaneko, K. (2008). Shaping robust system through evolution. *Chaos* 18, 026112.
68. Sakata, A., Hukushima, K. and Kaneko, K. (2009). Funnel landscape and mutational robustness as a result of evolution under thermal noise. *Physical Review Letters* 102, 148101.
69. Kaneko, K. and Furusawa, C. (2008). Relevance of phenotypic noise to adaptation and evolution. *IET Systems Biology* 2, 234–46.
70. Kaneko, K. (2012). Phenotypic plasticity and robustness: evolutionary stability theory, gene expression dynamics model, and laboratory experiments. *Advances in Experimental Medicine and Biology* 751, 249–78.
71. Kaneko, K. (2012). Evolution of robustness and plasticity under environmental fluctuation: formulation in terms of phenotypic variances. *Journal of Statistical Physics* 148, 686–704.

72. Kaneko, K. (2007). Evolution of robustness to noise and mutation in gene expression dynamics. *PLoS One* 2, e434.
73. Kuwahara, H. and Soyer, O.S. (2012). Bistability in feedback circuits as a byproduct of evolution of evolvability. *Molecular Systems Biology* 8, 564.
74. van Dijk, A.D.J., van Mourik, S. and van Ham, R.C.H.J. (2012). Mutational robustness of gene regulatory networks. *PLoS One* 7, e30591.
75. Kaneko, K. (2009). Relationship among phenotypic plasticity, phenotypic fluctuations, robustness, and evolvability; Waddington's legacy revisited under the spirit of Einstein. *Journal of Biosciences* 34, 529–42.
76. Kaneko, K. (2011). Proportionality between variances in gene expression induced by noise and mutation: consequence of evolutionary robustness. *BMC Evolutionary Biology* 11, 27.
77. Inoue, M. and Kaneko, K. (2013). Cooperative adaptive responses in gene regulatory networks with many degrees of freedom. *PLoS Computational Biology* 9, e1003001.
78. Neyfakh, A.A., Baranova, N.N. and Mizrokhi, L.J. (2006). A system for studying evolution of life-like virtual organisms. *Biology Direct* 1, 23.
79. Cuypers, T.D. and Hogeweg, P. (2012). Virtual genomes in flux: an interplay of neutrality and adaptability explains genome expansion and streamlining. *Biology and Evolution* 4, 212–229.
80. Hogeweg, P. (2012). Toward a theory of multilevel evolution: long-term information integration shapes the mutational landscape and enhances evolvability. In: Soyer O.S. (ed.), *Evolutionary Systems Biology*, Springer, Berlin, pp. 195–224.
81. Fierst, J.L. (2011). A history of phenotypic plasticity accelerates adaptation to a new environment. *Journal of Evolutionary Biology* 24, 1992–2001.
82. Draghi, J.A. and Whitlock, M.C. (2012). Phenotypic plasticity facilitates mutational variance, genetic variance, and evolvability along the major axis of environmental variation. *Evolution* 66, 2891–2902.
83. Francois, P. (2012). Evolution in silico: from network structure to bifurcation theory. In: Soyer O.S. (ed.), *Evolutionary Systems Biology*. Springer, Berlin, pp. 157–182.
84. Lalanne, J.B. and Francois, P. (2013). Principles of adaptive sorting revealed by in silico evolution. *Physical Review Letters* 110, 218102.
85. Francois, P., Despierre, N. and Siggia, E.D. (2012). Adaptive temperature compensation in circadian oscillations. *PLoS Computational Biology* 8, e1002585.
86. Ma, H.W., Kumar, B., Ditges, U., Gunzer, F., Buer, J., et al. (2004). An extended transcriptional regulatory network of Escherichia coli and analysis of its hierarchical structure and network motifs. *Nucleic Acids Research* 32, 6643–6649.
87. Spirov, A.V. and Holloway, D.M. (2011). Retroviral genetic algorithms: implementation with tags and validation against benchmark functions. In: Rosa, A., Kacprzyk, J., Filipe, J. (eds.), *International Conference on Evolutionary Computation Theory and Applications, Proceedings*. SciTePress, Setubal, Portugal, pp. 233–238. (DOI 10.5220/0003674102330238).
88. Spirov, A.V. and Holloway, D.M. (2012). New approaches to designing genes by evolution in the computer. In: Roeva O. (ed.), *Real-World Applications of Genetic Algorithms*. InTech, Rijeka, Croatia, pp. 235–260.
89. Azevedo, R.B.R., Lohaus, R., Srinivasan, S., Dang, K.K. and Burch, C.L. (2006). Sexual reproduction selects for robustness and negative epistasis in artificial gene networks. *Nature* 440, 87–90.

90. Livnat, A., Papadimitriou, C., Dushoff, J. and Feldman, M.W. (2008). A mixability theory for the role of sex in evolution. *Proceedings of the National Academy of Sciences of the United States of America* 105, 19803–19808.
91. Livnat, A., Papadimitriou, C., Pippenger, N. and Feldman, M.W. (2010). Sex, mixability, and modularity. *Proceedings of the National Academy of Sciences of the United States of America* 107, 1452–1457.
92. Misevic, D., Ofria, C. and Lenski, R.E. (2006). Sexual reproduction reshapes the genetic architecture of digital organisms. *Proceedings of the Royal Society of London. Series B: Biological Sciences* 273, 457–464.
93. Misevic, D., Ofria, C. and Lenski, R.E. (2010). Experiments with digital organisms on the origin and maintenance of sex in changing environments. *Journal of Heredity* 101, S46–S54.
94. Fierst, J.L. (2013). Female mating preferences determine system-level evolution in a gene network model. *Genetica* 141, 157–170.
95. Emmrich, P.M.F., Roberts, H.E. and Pancaldi, V. (2012). A gene regulatory network simulation of heterosis. In: Lones, M.A., Smith, S.L., Teichmann, S., Naef, F., Walker, J.A., and Trefzer, M.A. (eds.), *Information Processign in Cells and Tissues. Lecture Notes in Computer Science*, Volume 7223, Springer, Berlin, pp. 12–16.
96. Emmrich, P.M.F., Pancaldi, V., Roberts, H.E., Kelly, K.A. and Baulcombe, D.C. (2014). A gene regulatory model of heterosis and speciation. arXiv:1309.3772
97. van Nimwegen, E. and Crutchfield, J.P. (2000). Optimizing epochal evolutionary search population-size independent theory. *Computer Methods in Applied Mechanics and Engineering* 186, 171–194.
98. True, J.R. and Carroll, S.B. (2002). Gene co-option in physiological and morphological evolution. *Annual Review of Cell and Developmental Biology* 18, 53–80.
99. Carroll, S.B., Grenier, J.K. and Weatherbee, S.D. (2001). *From DNA to Diversity: Molecular Genetics and the Evolution of Animal Design*. Malden, MA: Blackwell Science.
100. Spirov, A.V. and Holloway, D.M. (2007). Recruiting new genes in evolving genetic networks: simulation by the genetic algorithms technique. In: AO, S.I., Douglas, C., Grundfest, W.S., Schruben, L., and Wu, X. (eds.), *Proceedings of the World Congress on Engineering and Computer Science*, Newswood Limited, Hong Kong, pp. 16–22.
101. Spirov, A.V. and Holloway, D.M. (2009). The effects of gene recruitment on the evolvability and robustness of gene networks. In: Ao, S.I., Rieger, B., and Chen, S.-S. (eds.), *Advances in Computational Algorithms and Data Analysis. Lecture Notes in Electrical Engineering*, Volume 14, Springer, Berlin, pp. 29–50.
102. Soyer, O.S. and Bonhoeffer, S. (2006). Evolution of complexity in signaling pathways. *Proceedings of the National Academy of Sciences of the United States of America* 103, 16337–16342.
103. Tsuda, M.E. and Kawata, M. (2010). Evolution of gene regulatory networks by fluctuating selection and intrinsic constraints. *PLoS Computational Biology* 6, e1000873.
104. Leclerc, R.D. (2008). Survival of the sparsest: robust gene networks are parsimonious. *Molecular Systems Biology* 4, 213.
105. Salathe, M. and Soyer, O.S. (2008). Parasites lead to evolution of robustness against gene loss in host signaling networks. *Molecular Systems Biology* 4, 202.
106. Spirov, A., Kazansky, A. and Holloway, D. (2012). Complexification of gene networks by co-evolution of genomes and genomic parasites. In: A. Rosa, A. Dourado, K. Madani,

J. Filipe, and J. Kacprzyk, *Proceedings of the 4th International Joint Conference on Computational Intelligence, Barcelona, Spain*, SciTePress, Setubal, Portugal, pp. 238–244.

107. Spirov, A., Sabirov, M. and Holloway, D.M. (2012). In silico evolution of gene co-option in pattern-forming gene networks. *The Scientific World Journal* (special issue on Computational Systems Biology) 2012, 560101.
108. Spirov, A., Zagriychuck, E. and Holloway, D. (2014). Evolutionary design of gene networks: forced evolution by genomic parasites. *Parallel Processing Letters* 24, 1440004.
109. Brosius, J. (1991). Retroposons—seeds of evolution. *Science* 251, 753.
110. Hurst, G.D.D. and Werren, J.H. (2001). The role of selfish genetic elements in eukaryotic evolution. *Nature Reviews Genetics* 2, 597–606.
111. Vansant, G. and Reynolds, W.F. (1995). The consensus sequence of a major Alu subfamily contains a functional retinoic acid response element. *Proceedings of the National Academy of Sciences of the United States of America* 92, 8229–8233.
112. Polak, P. and Domany, E. (2006). Alu elements contain many binding sites for transcription factors and may play a role in regulation of developmental processes. *BMC Genomics* 7, 133.
113. McGraw, J.E. and Brookfield, J.F.Y. (2006). The interaction between mobile DNAs and their hosts in a fluctuating environment. *Journal of Theoretical Biology* 243, 13–23.
114. Startek, M., Le Rouzic, A., Capy, P., Grzebelus, D. and Gambin, A. (2013). Genomic parasites or symbionts? Modeling the effects of environmental pressure on transposition activity in asexual populations. *Theoretical Population Biology* 90, 145–151.
115. Banzhaf, W., Beslon, G., Christensen, S., Foster, J.A., Képès, F., Lefort, V., Miller, J.F., Radman, M. and Ramsden J.J. (2006). Guidelines: From articial evolution to computational evolution: a research agenda. *Nature Reviews. Genetics* 7, 729–735.
116. Saito, N., Ishihara, S. and Kaneko, K (2014). Evolution of genetic redundancy: the relevance of complexity in genotype-phenotype mapping. *New Journal of Physics* 16, 063013.
117. Soyer, O.S. (2007). Emergence and maintenance of functional modules in signaling pathways. *BMC Evolutionary Biology* 7, 205.
118. Borenstein, E. and Krakauer, D.C. (2008). An end to endless forms: epistasis, phenotype distribution bias, and nonuniform evolution. *PLoS Computational Biology* 4, e1000202.
119. Payne, J.L., Moore, J.H. and Wagner, A. (2014). Robustness, evolvability, and the logic of genetic regulation. *Artificial Life* 20, 111–126.
120. Nochomovitz, Y. and Li, H. (2006). Highly designable phenotypes and mutational buffers emerge from a systematic mapping between network topology and dynamic output. *Proceedings of the National Academy of Sciences of the United States of America* 103, 4180–4185.
121. Widder, S., Sole, R.V. and Macia, J. (2012). Evolvability of feed-forward loop architecture biases its abundance in transcription networks. *BMC Systems Biology* 6, 7.
122. Widder, S., Sole, R.V. and Macia, J. (2013). Plasticity, evolvability and the abundance of feed-forward loops in transcription networks. In: Füllsack, M. (ed.), *Networking Networks, Origins, Applications, Experiments*. Turia + Kant, Vienna, pp. 81–100.
123. Pinho, R., Borenstein, E. and Feldman, M.W. (2012). Most networks in Wagner's model are cycling. *PLoS One* 7, e34285.

124. Payne, J.A. and Wagner, A. (2013). Constraint and contingency in multifunctional gene regulatory circuits. *PLoS Computational Biology* 9, e1003071.
125. Warmflash, A., Francois, P. and Siggia, E. (2012). Pareto evolution of gene networks: an algorithm to optimize multiple fitness objectives. *Physical Biology* 9, 056001.
126. Carter, A.J.R., Hermisson, J. and Hansen, T.F. (2005). The role of epistatic gene interactions in the response to selection and the evolution of evolvability. *Theoretical Population Biology* 68, 179–196.
127. Sanjuan, R. and Nebot, M.R. (2008). A network model for the correlation between epistasis and genomic complexity. *PLoS One* 3, e2663.
128. Draghi, J.A. and Plotkin, J.B. (2013). Selection biases the prevalence and type of epistasis among beneficial substitutions. *Evolution* 67, 3120–3131.
129. Cotterell, J. and Sharpe, J. (2013). Mechanistic explanations for restricted evolutionary paths that emerge from gene regulatory networks. *PLoS One* 8, e61178.
130. Salazar-Ciudad, I., Newman, S.A. and Sole, R.V. (2001). Phenotypic and dynamical transitions in model genetic networks. I. Emergence of patterns and genotype-phenotype relationships. *Evolution and Development* 3, 84–94.
131. Salazar-Ciudad, I., Newman, S.A. and Sole, R.V. (2001). Phenotypic and dynamical transitions in model genetic networks. II. Application to the evolution of segmentation mechanisms, Evolution and Development 3, 95–103.
132. Kuo, P.D., Leier, A. and Banzhaf, W. (2004). Evolving dynamics in an artificial regulatory network model. In: Yao, X., Burke, E., Lozano, J., Smith, J., Merelo-Guervos, J., Bullinaria, J., Rowe, J., Tino, R., Kaban, A., Schwefel, H.-P., *Parallel Problem Solving from Nature. Lecture Notes on Computer Science*, Volume 3242, Springer, Berlin, pp. 571–580.
133. Hu, T. and Banzhaf, W. (2010). Evolvability and speed of evolutionary algorithms in light of recent developments in biology. *Journal of Artificial Evolution and Applications* 2010, Article No.1.
134. Trefzer, M.A., Kuyucu, T., Miller, J.F. and Tyrrell, A.M. (2013). On the advantages of variable length GRNs for the evolution of multicellular developmental systems. *IEEE Transactions on Evolutionary Computation* 17, 100–121.
135. Huneman, P. (2012). Computer science meets evolutionary biology: pure possible processes, and the issue of gradualism. In: Pombo, O., Torres, J., Symons, J., Rahman, S., *Special Sciences and the Unity of Science*. Springer, Berlin, pp. 137–162.
136. Lim, W.A., Lee, C.M. and Tang, C. (2013). Design principles of regulatory networks: searching for the molecular algorithms of the cell. *Molecular Cell* 49, 202–212.
137. Taute, K.M., Gude, S., Nghe, P. and Tans, S.J. (2014). Evolutionary constraints in variable environments, from proteins to networks. *Trends in Genetics* 30, 192–198.
138. Watson, J., Geard, N. and Wiles, J. (2004). Towards more biological mutation operators in gene regulation studies. *BioSystems* 76, 239–248.

11

EVOLVING GRN-INSPIRED *IN VITRO* OSCILLATORY SYSTEMS

Quang Huy Dinh

Department of Information and Communication Engineering, Graduate School of Information Science and Technology, The University of Tokyo, Bunkyo, Tokyo, Japan

Nathanael Aubert

Department of Information Science, Ochanomizu University, Bunkyo, Tokyo, Japan

Nasimul Noman

School of Electrical Engineering and Computer Science, Faculty of Engineering and Built Environment, The University of Newcastle, Newcastle, New South Wales, Australia

Hitoshi Iba

Department of Information and Communication Engineering, Graduate School of Information Science and Technology, The University of Tokyo, Bunkyo, Tokyo, Japan

Yannic Rondelez

LIMMS/CNRSIIS, Institute of Industrial Science, The University of Tokyo, Meguro, Tokyo, Japan

11.1 INTRODUCTION

In a living organism, DNA works as the instruction manual to delineate its physical characteristics and behaviors. Genes and their products, proteins, form a unique framework to put the blueprint inscribed in DNA into effect. The interactions among

Evolutionary Computation in Gene Regulatory Network Research, First Edition.
Edited by Hitoshi Iba and Nasimul Noman.

genes in the form of proteins, commonly known as gene regulatory networks (GRN), provide precise and timely processing of information to ensure seamless progress in life. The specificity, versatility, and, above all, programmability of these molecular machines offer the potential to function as a powerful tool to synthesize new chemical systems that can have radical effects in different spheres of our life. Today, the programmability of these nano-machines is explored *in vivo*, *in vitro* and *in silico* for rational construction of molecular circuit for altering and/or controlling cell behavior. The focus of this chapter is to address the issue of generating target behaviors in *in vitro* chemical systems that utilizes the notions of GRN. Toward automatic construction of such systems we make use of a computational approach that works mimicking the natural evolution.

GRNs are dynamic systems with distinct control mechanism for regulating the concentration of different molecules over time in such a way that the developmental features and response behaviors emerge in the system as per requirement. Inside the cell, the concentrations of various compounds are essentially administered by dynamically creating and destructing those compounds' molecules with the assistance of stable DNA. The long-standing DNA is divided in to modules (e.g., genes and promoters) which synthesize the transitory molecules (e.g., RNAs and proteins). These transitory molecules in turn control the synthesis capability genes by their interactions in the form of activation or repression. Accumulating all these features GRNs assemble dynamic systems which can exhibit various dynamic functionalities. Replicating these ideas, we are expecting to design novel biochemical systems with enhanced information processing capacities. Today, synthetic biology is working just following the essential principles of the molecular systems in GRN and tweaking the molecular characteristics of genes and proteins [1]. Although the *in vivo* approach carries the potential to revolutionize biology, the practical implementation and integration of the novel circuits have been hindered by many key issues one of which is unexpected interference from neighboring components [2]. In contrast, the *in vitro* approach enjoys the benefit of being free of interference, hence more tractable from the perspective of quantitative modeling and feasibility for comprehensive studies [3, 4]. Consequently, the design and implementation of several *in vitro* circuits that can exhibit different types of information processing capabilities have been completed in recent years [3–12]. In particular, one molecular programming toolbox, known as Polymerase/Exonuclease/Nickase Dynamic Network Assembly toolbox (PEN DNA toolbox) that works utilizing three essential characteristics of GRN namely modularity, dynamism, and template control of the reactivity, has been very successful in building several interesting systems such as oscillator [3] and switchable memory [4].

Although the field of molecular programming opens doors to the rational building of chemical systems with the capability of complex information processing, construction of the reaction network targeting a precise dynamics remains a laboring task. As described in Ref.[13], identifying the network structure for a target behavior is one of the greatest challenges faced by the molecular programmers in their design efforts. Although for simple instances, different mathematical tools such as linear stability analysis (LSA) and chemical reaction network theory can be of great aid, for a complex system, integration of modules in a hierarchical manner does not guarantee the expected behavior because of many unanticipated interference. Therefore, in designing larger and complex system, especially when no prior knowledge about the circuit/function relationship is available, computational modeling, simulation, and optimization can be very useful in reverse engineering the circuit topology as well as parameters.

A wide spectrum of computational approaches have been used in reverse engineering GRN from various kinds of genomic data which ranges from clustering [14, 15], statistical methods [16, 17], neural network, and fuzzy logic [18, 19]. Besides, Evolutionary algorithm (EA) is a popular choice of algorithm for reverse engineering GRN—many of its applications have been presented in other chapters of this book. Recently, EAs have also found its applications in evolving biochemical network for a target function [20–29]. Many of these algorithms use a very high level representation of the networks which can give only an overall idea about the interaction among the biomolecules but lacks the details for wet-lab realization [30]. Although often the primary reason behind the choice of high-level mathematical representation is computational efficiency, there are other practical implementation issues involved in representing networks in evolutionary algorithms such as competing conventions, protecting innovation, and building block preserving mating operations. An EA called NeuroEvolution of Augmenting Topologies (NEAT) potentially solved many of these problems for evolving structure and parameters for neural networks [31]. NEAT uses unique historical marking for each gene in the genome representation that enables tracking of genes throughout the evolution process as well as designing of effective crossover operations. Besides, its innovation preservation mechanism and parsimonious complexification of network also guarantee the identification of the smallest network for a given task. The molecular components of PEN DNA toolbox such as compounds, regulation, and kinetic parameters have a direct correspondence with the nodes, edge, and weights of neural networks, respectively. We developed an evolutionary algorithm borrowing the philosophy of NEAT for automatic construction of biochemical systems. We call this framework ERNe (evolving reaction network) [32]. Given the target behavior, ERNe has been very successful in evolving different types of biochemical networks with necessary details for wetlab implementation. In this chapter, we present the working principle of ERNe and investigate its capability for evolving some novel interesting behaviors which have not been tried before.

The rest of this chapter is organized as follows. In Section 11.2, we describe the PEN DNA Toolbox, how it is modeled based on biochemistry, and its components. Section 11.3 provides the overview of the previous evolutionary approaches. Moving on, in Section 11.4, we describe the ERNe algorithm in detail. Section 11.5 shows the experimental results and finally discussion is given in Section 11.6.

11.2 PEN DNA TOOLBOX

In this Section, we will present various models of the PEN DNA toolbox with increasing level of details, their respective sets of equations, and show how big a minimal system from the toolbox, a simple autocatalytic template, becomes in each case. We would like to emphasize that all levels of modeling have their use, and are efficient when used in the right context. The most simple model (Section 11.2.2) should be used when scanning a large variety of topologies and parameters. Tuning specific systems requires a more precise, albeit more time-consuming approach (Section 11.2.3). Finally, when trying to check and predict the behavior of a single implementation, the highest amount of details available should be as predictive as possible (Section 11.2.4). Note that the way enzymes are modeled is a separate problem altogether. We argue that enzymatic saturation should be modeled whenever possible, as it has little performance cost for a large impact on the behavior of PEN systems (Section 11.2.5).

11.2.1 Overview

The PEN DNA toolbox was inspired by gene regulatory networks [3]. In such networks, regulation comes in two flavors: activatory or inhibitory, respectively increasing and decreasing the expression of a targeted gene. One of their goals was to reproduce such mechanisms while removing the complexity of using RNA and proteins in the network.

The toolbox mimics regulatory mechanisms through specific DNA compounds, activating or inhibiting the generation of other DNA compounds. Activators are often referred to as signal, while DNA compounds performing inhibition are simply called inhibitors. Activation is done through the help of a second class of DNA compounds called templates. At a coarse-grained level, templates are DNA compounds present in the solution that generates an output from an input, eventually releasing both. The output can itself be an activator, allowing the cascading of multiple reactions. In the PEN toolbox, the exact mechanism through which activation is done is based on a combination of polymerase and nickase enzymes [3]. However, other mechanisms, such as Qian and Winfree's seesaw gate [33] could implement templates.

From a biochemical point of view, activation is done as shown in Figure 11.1: a signal strand hybridizes with the first half (*input domain*) of a target template. Polymerase then extends the signal's 3′ end until the template is completely double-stranded. This structure is recognized by the nicking enzyme, which cuts (*nicks*) the backbone of the extended signal at a specified position. Both input and output are eventually released. Note that the output of the template can be another activator, an inhibitor or a DNA compound unrelated to the PEN toolbox. This last possibility

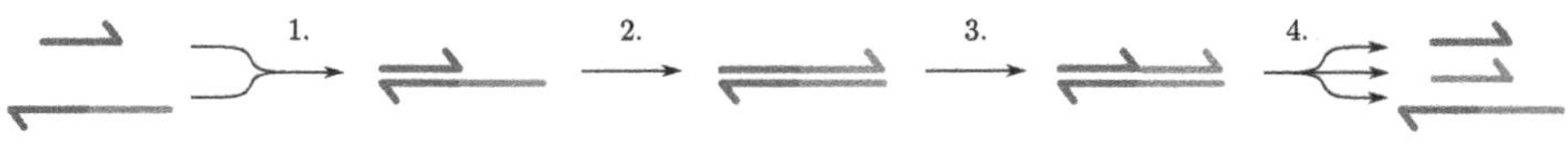

Figure 11.1 Idealized working of a template with its activator. **1.** Hybridization of the input with the template. **2.** Elongation by polymerase: the enzyme reads the template and extends the input using dNTP (monomeric DNA) present in the buffer. **3.** The newly formed duplex contains the recognition sequence of the nuclease, triggering the enzyme to cut right between the input and output. **4.** The nicked structure is not stable at the working temperature, eventually releasing both input and output.

allows the creation of hybrid systems with both the PEN toolbox and other computing paradigms to interact.

A third enzyme, called exonuclease, degrades DNA signal over time, as would happen to gene expression in a cell. Templates, which represent the reaction network instead of signal, are modified to prevent degradation. We can thus consider that the overall concentration of a given template compound is constant over time.

Despite having been introduced only a few years ago, the DNA PEN toolbox can already boast a variety of models [4,34,35], with a broad range of applications based on each model's respective ease to simulate and predictability.

Those approaches have in common that instead of modeling the DNA compounds all the way down to the actual nucleotide sequences (the actual ATGCs), they abstract these compounds to the domain level. A domain is a specific sequence of nucleotides to which we give a meaning. As such, activator DNA strands are considered as one domain, which gives templates two domains (input and output) and inhibitors three (partial input and output of the target template as well as terminal mismatch). More general models can take into account actual atoms [36], spacial configuration of DNA [37], or even specifically care about secondary structures (such as bulges, or hairpins) [38], depending on the targeted application.

The main advantage of domain level modeling is that it keeps the systems simple, reduces considerably the number of potential reactions. In particular, it removes partial interactions between unrelated strands, which would otherwise generate a number of possible reactions exponential in the number of compounds. This is also the downside of this type of modeling: it makes it too rough to take into account sequence dependent phenomenons, such as stacking (when two DNA strands at a nick "pile" on each other to increase the overall duplex stability [39]). When such reactions are known to be important, it is always possible to find a workaround to add them to the model (Section 11.4). Nevertheless, unexpected reactions, such as infamous parasitic reaction that can hijack whole systems [40], are beyond the scope of such type of models, which advocates for the use of sequence level modeling.

A schematic representation of both activation and inhibition, as well as the special case of autocatalytic activation, as implemented in the PEN toolbox is shown in Figure 11.2.

11.2.2 Simplified Model

At the lowest level of detail, we consider template compounds as black boxes (Figure 11.3). Those boxes have a simple transfer function inspired by Michaelis–Menten reaction rates. The rationale behind this model is to assume that hybridization and denaturation (two complementary DNA strands attaching and detaching, respectively) reach equilibrium much faster than enzyme-based reactions. This yields the transfer function in Figure 11.3(b) for each module. The complete set of equations can thus be easily built from the graph representation of a system by summing over templates generating a given compound, with the addition of a first order approximation of the exonuclease activity $-\text{exo}_s[s](t)$.

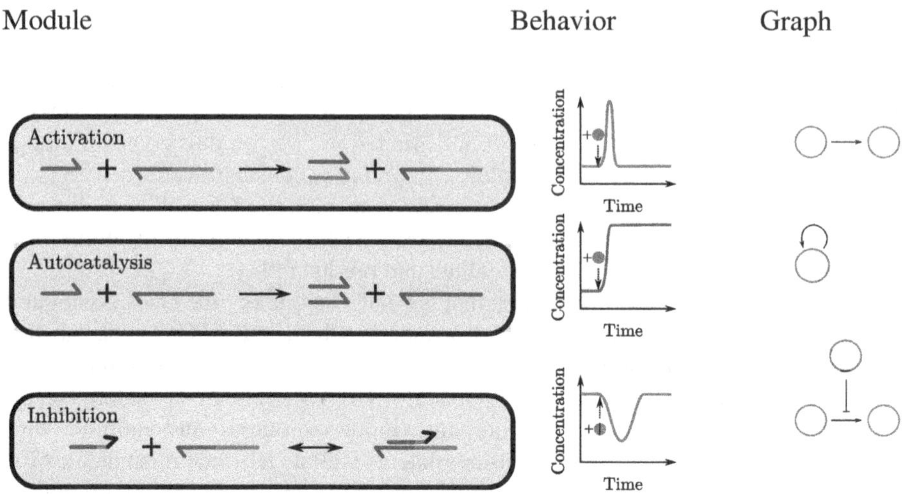

Figure 11.2 The modules of the PEN toolbox: activation, autocatalysis and inhibition.

This model of the DNA PEN toolbox was first introduced in Padirac *et al.*'s work on the bistable switch [4]. Because of the simplicity of the equations, large systems can be simulated extremely fast. This propriety was leveraged to evolve complex PEN toolbox systems, requiring thousands of separate evaluations. In this process, interesting patterns were discovered by the algorithm [41] (Figure 11.4).

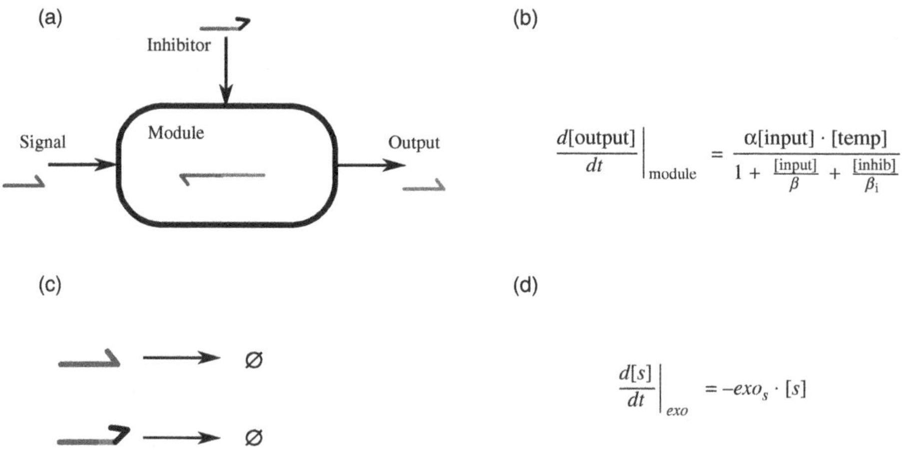

Figure 11.3 (a) Black box representation of a template. (b) The general form of the transfer function for this box, with α the optimal speed of the module, [.] the concentration of DNA compounds, β and β_i the Michaelis parameters of the input and inhibitor, respectively. (c) The reactions involving the exonuclease: each unprotected compound (anything but templates) is degraded over time. (d) The contribution of the exonuclease to the equation describing the evolution of the compounds s over time. exo_s, the activity of the exonuclease with respect to s, usually depends on the length of s. In general, this means that two values are possible, one for signals and one for inhibitors (slightly longer).

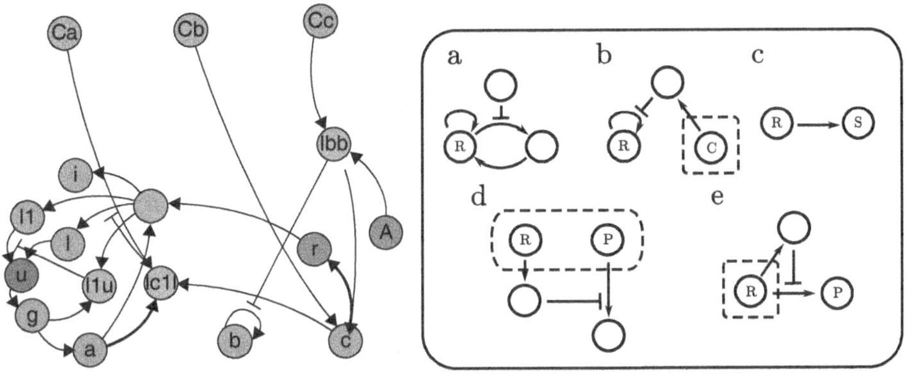

Figure 11.4 Left: a large evolved system sensing its environment (three nodes at the top) and computing a new internal state. Right: various recurring patterns generated during various runs. One may consider them "standard" subroutines that are combined together to make larger systems.

Applying the principles of this model, we can write the behavior of an autocatalytic template as a single equation:

$$\frac{d[s]}{dt}(t) = \frac{\alpha \cdot [temp][s](t)}{1 + \frac{[s](t)}{\beta}} - exo_s \cdot [s](t) \qquad (11.1)$$

11.2.3 Internal State of the Templates

The main problem with the previous model is that it does not keep track of the various steps necessary to generate an output. This removes a lot of delays that are actually characteristic of real systems. In fact, the lack of delay makes it harder to design oscillators [3], and probably other systems. Additionally, the previous model makes signal and inhibitor compounds much easier to degrade, as they are never considered to be in a double-stranded state, which would prevent the action of the exonuclease.

For these reasons, we can make a more complex, domain-level model that takes into account all the potential template configurations. The main assumption of such model is that a given signal strand is either completely free or completely double-stranded to its complementary domain. This allows us to abstract away the potential complexity of partially complementary structures as well as the actual transitions from single-stranded to double-stranded into the stability of the sequence, a parameter of the model.

While it is possible to keep the "black box" abstraction for the behavior of templates, the content of the box is much more complex. First of all, we need six variables per template to keep track of the current concentration of all configurations: free template, template with input, template with output, template with both, double-stranded template and inhibited template. The list of all reactions happening with a given template is depicted in Figure 11.5.

The expression for exonuclease does not change; however, the meaning of s in Figure 11.3(d) is slightly different, as it represents now the concentration of s that is not

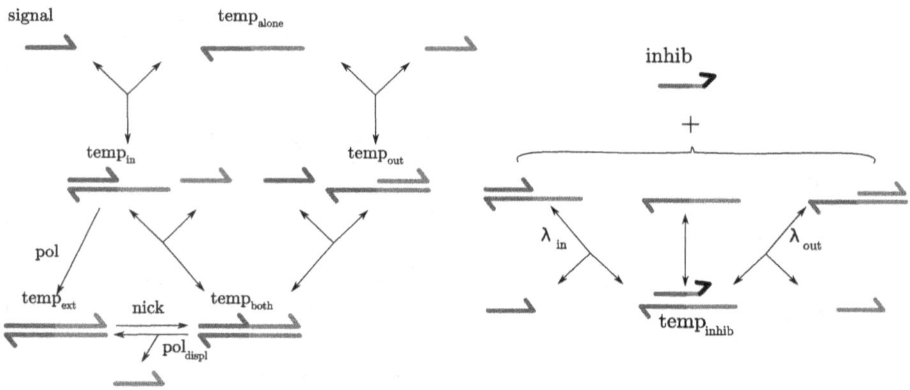

Figure 11.5 Left: Reactions related to activation. Non-enzymatic reactions (hybridization and denaturation) are reversible and as such are represented by a double-headed arrow. Enzymatic reactions (polymerization and nicking) are not reversible. Right: Reactions related to inhibition. Because an inhibitor always leave a short toehold on the left and on the right of the template, it is possible for both the input and the output to invade and remove the inhibitor. Note that this is done at a slower speed than hybridization, with a slowdown dependent on the number of free bases [42].

attached to a template. Note that it is still possible to get back the total concentration of any signal or inhibitor by summing over all configurations that contain it. Note that for an autocatalytic template, the $temp_{both}$ configuration has to be counted twice, as it contains two copies of the compound.

The set of equation describing an autocatalytic template thus becomes:

$$
\begin{aligned}
\frac{d[s]}{dt}(t) &= k_{duplex} \cdot K_s \cdot \left([temp_{in}](t) + [temp_{out}](t) + 2 \cdot [temp_{both}](t)\right) \\
&\quad - k_{duplex} \cdot [s](t) \cdot \left(2\,[temp_{alone}](t) + [temp_{in}](t) + [temp_{out}](t)\right) \\
&\quad + pol \cdot [temp_{both}](t) - exo_s(t) \cdot [s](t) \\
\frac{d[temp_{alone}]}{dt}(t) &= k_{duplex} \cdot \left([temp_{in}](t) + [temp_{out}](t)\right) \\
&\quad - 2\,k_{duplex} \cdot [s](t) \cdot [temp_{alone}](t) \\
\frac{d[temp_{in}]}{dt}(t) &= k_{duplex} \cdot \left([s](t) \cdot [temp_{alone}](t) + \cdot K_s \cdot [temp_{both}](t)\right) \\
&\quad - k_{duplex} \cdot [temp_{in}](t) \cdot \left([s](t) + K_s\right) - pol \cdot [temp_{in}](t) \\
\frac{d[temp_{out}]}{dt}(t) &= k_{duplex} \cdot \left([s](t) \cdot [temp_{alone}](t) + \cdot K_s \cdot [temp_{both}](t)\right) \\
&\quad - k_{duplex} \cdot [temp_{out}](t) \cdot \left([s](t) + K_s\right) \\
\frac{d[temp_{both}]}{dt}(t) &= k_{duplex} \cdot [s](t) \cdot \left([temp_{in}](t) + [temp_{out}](t)\right) \\
&\quad - 2 \cdot k_{duplex} \cdot K_s \cdot [temp_{both}](t) \\
&\quad + nick \cdot [temp_{ext}](t) - pol \cdot [temp_{both}](t) \\
\frac{d[temp_{ext}]}{dt}(t) &= pol \cdot \left([temp_{in}](t) + [temp_{out}](t)\right)
\end{aligned}
\tag{11.2}
$$

In this equation, k_{duplex} represents the (global) hybridization rate, K_s the stability of sequence s, *nick*, *pol* and *exo* the activity of the nickase, polymerase and exonuclease, respectively. A complete explanation of all parameters involved in this model are available elsewhere [35].

11.2.4 Sequence Dependence

Beyond the domain level, we can consider sequence dependence. Taking whole sequences into account might add new possible reactions such as cross-talks. A hybrid solution is to only define the couple of nucleotides at the nicking site, as they are mostly relevant to dangles [43] and stacking [44, 45]. Specifically, Santalucia *et al.* argues that the nearest-neighbor model is not significantly improved by taking into account bases further away. The full sequence can also be used to check beforehand for cross-talks (including secondary structures, that is, a compound folding onto itself) and to determine a given compound stability.

The hybrid version adds three parameters to each templates: dangle slowdowns to the denaturation of DNA strands, respectively, from the input and output position, and a stacking slowdown to the denaturation in the case both input and output are present. There is also a shared dangle slowdown for inhibitors on their target template, as the base at this position is fixed in the current design. To be more precise, it is part of the recognition site of the nicking enzyme, which means surrounding bases are also fixed. In a design where inhibitors have variable length, a template-specific value should be introduced.

11.2.5 Enzymatic Saturation

While the previous sections explored various level of details for the interactions of DNA molecules, enzymatic activity has been kept as a constant reaction rate. In this section, we explore the impact of adding enzymatic saturation in the model. Enzymatic saturation is the cause for many effects, including winner-take-all effect [35, 46, 47] and robust oscillations [48]. An example of the impact of saturation is shown in Figure 11.6. Keeping track of it adds one variable per enzyme. This comes at a slight computational cost and may affect performances as ODE can become "steep".

Saturation is done by the substracts; so the first step is to identify which substracts are valid. See Figure 11.7. A few design tricks, such as using `U` instead of `T` in nickase recognition site of the output area of a template, can remove some of those substracts or reduce their impact [13].

For all models, the exonuclease term becomes:

$$exo_s(t) = \frac{V_{m,exo}}{K^s_{m,exo}\left(1 + \sum_{s' \in seq} \frac{[s'](t)}{K^{s'}_{m,exo}}\right)} \tag{11.3}$$

where $V_{m,exo}$ is the maximum theoretical rate and $K^s_{m,exo}$ the Michaelis constant for the compounds s. Note than in our model $V_{m,exo}$ is independent of s and $K^s_{m,exo}$ can only take one of two values, depending on whether s is an inhibition compound or not.

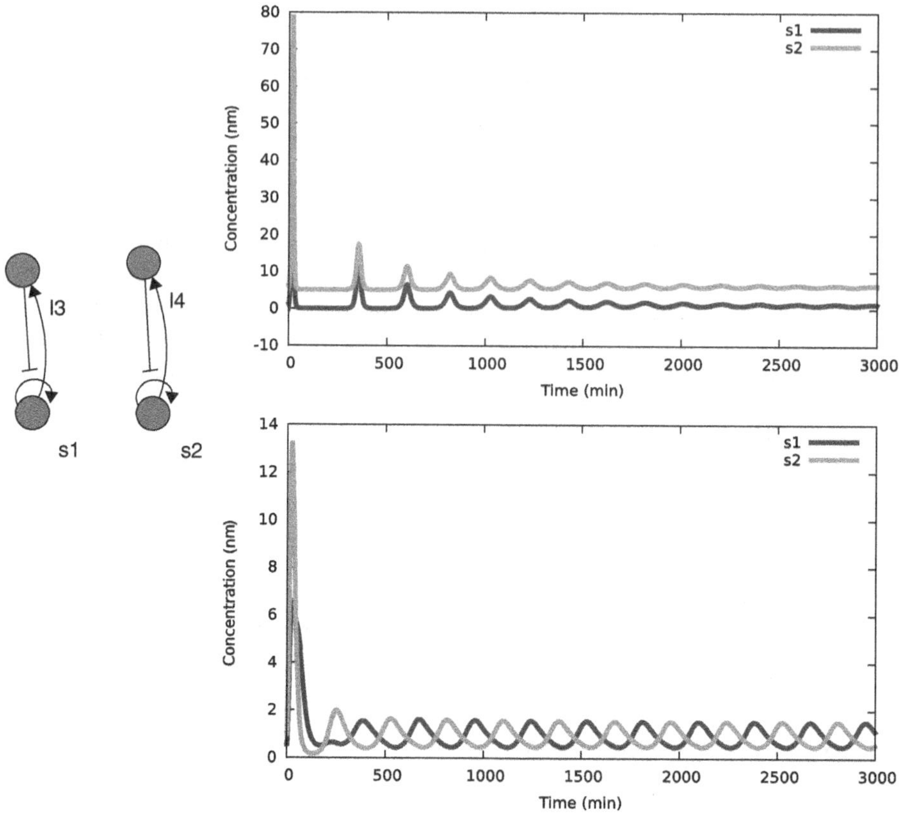

Figure 11.6 A saturation-based oscillator. Left: the topology of the system. Note that both parts are identical. The only symmetry-breaking element is the initial concentrations of the signal compound. Right, top: without saturation, the system can only perform damped oscillations. The green curve has been offset to increase the readability. Right, bottom: when enzymatic saturation is added to the model, the system can oscillate as each side alternatively gets to a high state where the polymerase gets saturated, decreasing the generation of inhibitor for the other side.

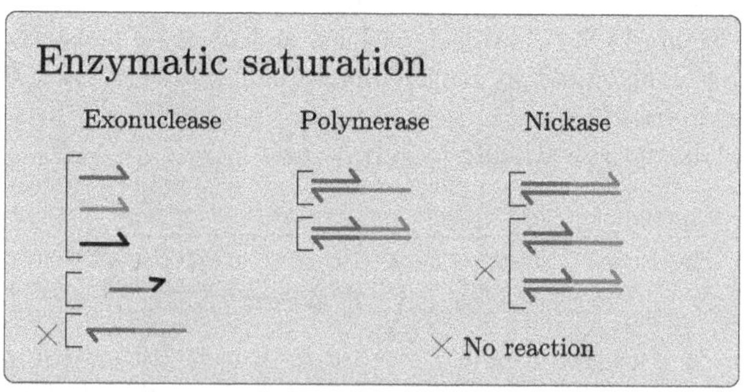

Figure 11.7 Example of substracts that can be considered to saturate the various enzymes present in the system.

For the polymerase and nickase, since all substracts are taken into account at the second level of detail, that of templates internal states, saturation can be easily expressed.

The polymerase activity is separated in two terms: *pol* for templates with input alone and pol_{displ} in the case where both input and output are present. We consider that the polymerase does not interact noticeably with other states of the template.

$$\begin{aligned} pol(t) &= \frac{V_{m,pol}}{K_{m,pol}\left(1+\sum_{temp}\frac{[temp_{in}](t)}{K_{m,pol}}+\frac{[temp_{both}](t)}{K_{m,displ}}\right)} \\ pol_{displ}(t) &= \frac{V_{m,displ}}{K_{m,displ}\left(1+\sum_{temp}\frac{[temp_{in}](t)}{K_{m,pol}}+\frac{[temp_{both}](t)}{K_{m,displ}}\right)} \end{aligned} \tag{11.4}$$

Note that $V_{m,displ}$ depends on the length of the output. It has been argued that signal strands are short enough to not be a bother (no slowdown) while inhibitor have been measured to have a slowdown of 0.2 with the Bst polymerase [4]. The nickase term is simpler. We consider that only fully double stranded templates can capture this enzyme, yielding:

$$nick(t) = \frac{V_{m,nick}}{K_{m,nick}\left(1+\sum_{temp}\frac{[temp_{ext}](t)}{K_{m,nick}}\right)}. \tag{11.5}$$

For the simplest model presented here, since we are considering that hybridization /denaturation reactions are near equilibrium, we can estimate the concentration of the various substracts of the enzymes. Additionally, *pol* and *nick* should give us a more explicit expression of α in Section 11.2.

If K_s represents the equilibrium constant of the hybridization of the DNA compound s with its complementary $\bar{s}$, then the amount of duplexed DNA $s \cdot \bar{s}$ is

$$[s \cdot \bar{s}] = K_s[s][\bar{s}]. \tag{11.6}$$

By taking the system of reactions from Figure 11.5 and assuming equilibrium for hybridization/denaturation pairs, in the simplified case where there is no inhibitor, we find:

$$\left.\frac{d[out]}{dt}\right|_{temp} = \frac{pol \cdot K_{in}^{-1} \cdot [temp]\,[in]}{1+\frac{pol+nick}{nick}\cdot K_{in}^{-1}\cdot [in]}. \tag{11.7}$$

This yields $\alpha_{temp} = pol \cdot K_{in}^{-1}$ and $\beta_{temp} = \frac{pol+nick}{nick} \cdot K_{in}^{-1}$.

In the presence of inhibitor, and under the assumption that it only reacts with free templates, we get a slightly more complicated formula:

$$\left.\frac{d[out]}{dt}\right|_{temp} = \frac{pol \cdot K_{in}^{-1} \cdot [temp]\,[in]}{1 + \frac{pol+nick}{nick} \cdot K_{in}^{-1} \cdot [in] + \frac{K_{inhib}^{-1}}{1+K_{out}^{-1}[out]} \cdot [i]}. \tag{11.8}$$

Note that the equation now depends on $[out](t)$. Also note that $\beta_i = \frac{K_{inhib}^{-1}}{1+K_{out}^{-1}[out]}$ is only approximately constant for $[out](t) \ll K_{out}$.

In the case of the saturation of the nicking enzyme, we want to find the fraction of all templates which are double-stranded.

$$sat_{nick} = \sum_{temp} \frac{[temp]\,[in]}{\frac{nick}{pol} \cdot K_{in} \cdot \left(1 + \frac{K_{inhib}^{-1}}{1+K_{out}^{-1}[out]} \cdot [i]\right) + \left(1 + \frac{nick}{pol}\right) \cdot [in]}. \tag{11.9}$$

Note that this equation contains *nick*. However, the activity in the formula should the one at the previous time step of the simulation. We might also want to add other potential substrates, such as templates with output or templates with both input and output.

In the case of the saturation of the polymerase, we want to know the fraction of all templates with input. Going through the same calculation as for the nickase, we get:

$$\mathrm{sat_{pol}} = \sum_{temp} \frac{[temp]\,[in]}{K_{in} \cdot \left(1 + \frac{K_{inhib}^{-1}}{1+K_{out}^{-1}[out]} \cdot [i]\right) + \left(1 + \frac{pol}{nick}\right)[in]}. \tag{11.10}$$

This yields the actual enzyme activities

$$\mathrm{nick} = \frac{\mathrm{nick_{base}}}{K_{\mathrm{nick}} + \mathrm{sat_{nick}}} \tag{11.11}$$

$$\mathrm{pol} = \frac{\mathrm{pol_{base}}}{K_{\mathrm{pol}} + \mathrm{sat_{pol}}} \tag{11.12}$$

were both base activities and Michaelis parameter K_{enzyme} can be determined experimentally.

11.3 RELATED WORK

In this section, we cover the previous attempts to automate the design of genetic or chemical circuits by using EC techniques.

Initially, Francois and Hakim [20] used a mutation-only genetic algorithm (GA) to evolve both the structure and reaction rates of genetic regulatory networks in the form of an idealized abstraction. The discovered circuit designs were shown to be promising, as they were functionally relevant to the organization of known biological networks. The efficiency of evolutionary computations on this initial attempt, despite the simplicity of the evolutionary scheme, encourages later research on such problems. As a result, Fujimoto *et al.* [21] used a similar algorithm to evolve gene regulatory networks that produce striped patterns of gene expression. The evolved networks were analyzed and classified into three categories which reproduce various segmentation strategies observed among arthropods. Likewise, an algorithm with only structural mutations were used by Kobayashi *et al.* [22] to discover genetic networks that display steady periodic oscillations with a given temporal period. Another simple genetic algorithm without crossover was used by Deckard and Sauro [23] to evolve chemical reaction networks with specific signal processing capabilities. As reported in their work, crossover provided worse performance. Later, Deckard and Sauro's work was extended by Paladugu *et al.* [24] to successfully evolve oscillators, bistable switches, homeostatic systems, and frequency filters. Interestingly, crossover was still not included for the possibility of being disruptive.

When it comes to difficult problems or complex models that require heavy computation, simple GAs with mutation as the only genetic operator are not efficient enough, and several workarounds have been suggested. One example is the study carried out by Marchisio and Stelling [25], in which the structure design and parameter optimization are separated from each other. For a given truth table that specifies a circuit's input–output relations, their algorithm generates and ranks several possible circuit schemes without any optimization procedure. Then, only the most feasible solution undergoes a parameter optimization aiming for reliable performance in actual wet experiments. This method relies on the existence of a rational solution, which cannot always be expected, especially in complex tasks. Moreover, the most feasible solution might require an accurate combination of both structure and parameters, advocating for their co-evolution. Cao *et al.* [26] also highlighted this issue, and efficiently addressed it with a nested evolutionary algorithm. Their algorithm first identifies the sequential structure of the target system and then optimize parameters of the selected modules. Notably, there are several works that simply freeze the structure, and only optimize parameters with the use of an efficient Evolution Strategy (ES), for example, the work of Jin and Sendhoff [27], and Hallinan *et al.* [28].

A complete GA, including mutation, elitist selection, and crossover, was introduced by Drennan and Beer [29] to successfully evolve the genetic regulatory networks that express proteins in an oscillatory manner similar to a repressilator [49]. In their algorithm, pseudo-DNA sequences are used directly as individuals' encoding and genetic operations are applied in a pseudo-biological way. Because these genetic operators are applied at the sequence level of the genomes while the network graph is encoded by a matrix, there is a high probability that connection will be broken in genetic operators, making it hard to pass on connectivity innovation to later generations. This make the approach inappropriate for problems that require precise combination of connections [31,32].

Interestingly, on the evolution of a different network structure—the artificial neural network (ANNs)—an algorithm called NEAT [31] has been proposed to efficiently solve the co-evolution of the structure and parameters, and showed better performance than the best fixed topology methods on challenging problems. The increased efficiency has been ascribed to an effective crossover strategy and structural innovation protection, combined with a parsimonious addition of new nodes. All the previous EA approaches for GRN avoided using crossover for the reason that meaningful structural crossover strategies are not obvious and worse performance is usually experienced when crossing over non-homologous networks. Because GRNs and ANNs can both be represented as graphs, we reasoned that it might be possible to apply NEAT's strategies to GRN's *in silico* evolution, with a potentially large improvement in search efficiency. The next section will describe how we adapt NEAT to come up with a new algorithm, namely ERNe, for the evolution of realistic molecular circuits targeting a given dynamic function.

11.4 FRAMEWORK FOR EVOLVING REACTION NETWORKS (ERNe)

ERNe is an efficient derivative of the NeuroEvolution of Augmenting Topologies (NEAT) algorithm [31] directed at the evolution of biochemical systems or molecular programs. The main differences between ANN and reaction networks are the addition of inhibition links and biochemical parameters. Therefore, as an extension of NEAT, ERNe encoding allows representation of inhibition, and has added parameters. In addition, ERNe's mutation and crossover operators are also modified from the original NEAT. For better understanding of ERNe algorithm, this section described in detail the encoding, the genetic operators for mutation and crossover, and the speciation process.

11.4.1 Encoding

Similar to NEAT, our genome consists of sequence genes and template genes. However, each sequence gene can represent either a signal sequence or an inhibiting sequence in the system, and consists of a sequence name, a kinetic parameter, and an initial concentration. Each template gene represents a template sequence in the system, and specifies the from-node, the to-node, the template concentration, an enable bit that indicates whether or not the template is enabled, and an innovation number to help align corresponding template genes. This innovation number is very important for the implementation of the crossover and speciation. For assigning innovation number to a template gene, we use the following rules: during the evolution, whenever a template is added to a system, we check if that specific link exists in the evolution history (using the from-node and the to-node fields), in which case it takes the original linkś innovation number. Otherwise, the next available innovation number will be assigned to the template gene. As for naming newly added nodes, the following mechanism is used: we map node names with the way they are created, for example, an entry A→A to B shows that whenever a new node is added in the middle

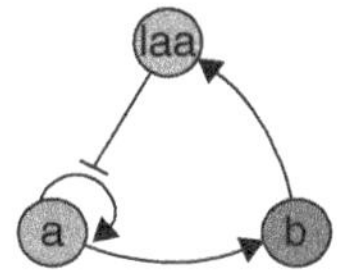

Sequence (Node) Genes

Name	a	b	I_{aa}
Kinetic parameter	3.873	3.873	0.0129
Initial concentration	1.0	0.1	0.1

Template (Connection) Genes

From	a	a	b
To	a	b	I_{aa}
Concentration	30.5	10.3	15.4
Enabled	True	True	True
Innovation	1	2	3

Figure 11.8 A graph representation and corresponding ERNe encoding of the Oligator. Nodes represent sequences while arrows represent templates; bar-headed arrows represent inhibition. a and b are signal sequences, whereas the green node Iaa is an inhibiting sequence; it inhibits the template a→a (i.e., the self-activation of a). The system has three sequences and three templates. Thus, there are three node genes and three connection genes in its ERNe encoding.

of the template A→A, it must be named B. An example of the genetic encoding is shown in Figure 11.8.

11.4.2 Mutations

In our framework, mutations can be applied to change both the parameters and the network structure. We have the following mutation rules: parameter only, disable template, switch template, add sequence, add activation, and add inhibition. Their relationship is shown in Figure 11.9. Which mutation to be applied is defined based on some probability parameters that can be set before running.

In parameter mutation, every parameter has a probability to be mutated to a new value calculated as:

$$newValue = oldValue \times (1 + f_1 \times rand_1) + f_2 \times rand_2 \qquad (11.13)$$

where *rand* are standard normal deviates. If the new value of a template concentration is above zero and below a threshold, it has a probability to be disabled. In this case, we call it disable template mutation. If that value is below zero, a switch template mutation happens. An example of switch template mutation is shown in Figure 11.10. In this example, assume template b→a's concentration is mutated to a value below zero. The following changes will be applied to the system:

- Disable template b→a.
- Search for any template that create a, if there is no such template, add the template a→a. In this example, a→a is found and selected.

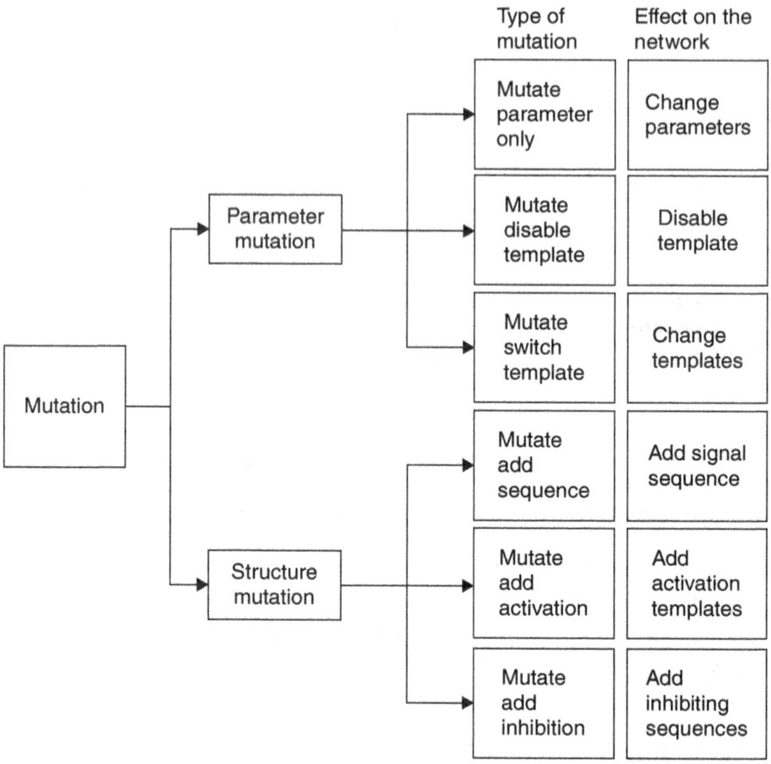

Figure 11.9 Different types of mutation used in ERNe. Note that although mutate disable template and mutate switch template belong to parameter mutation, they actually change the structure as well.

- Add the inhibiting sequence to the selected template a→a, Iaa.
- Add the template from b to Iaa, b→Iaa.

Because the parameter is mutated based on normal distribution, a template that has the concentration closer to zero has higher chance to be disabled, or be switched.

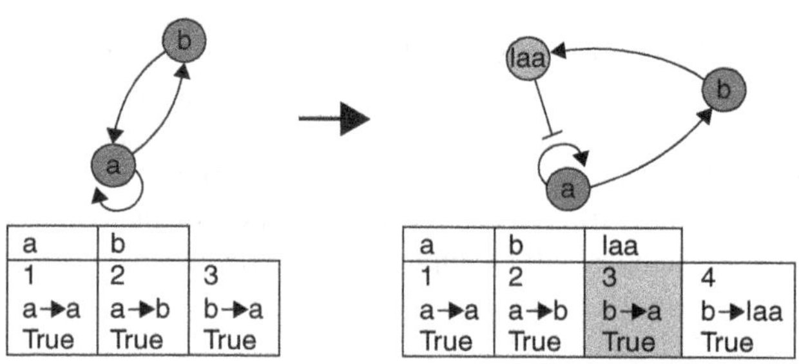

Figure 11.10 Switch template mutation. The template b→a is disabled, an inhibition node Iaa is added, then a template b→Iaa is added with a new innovation number.

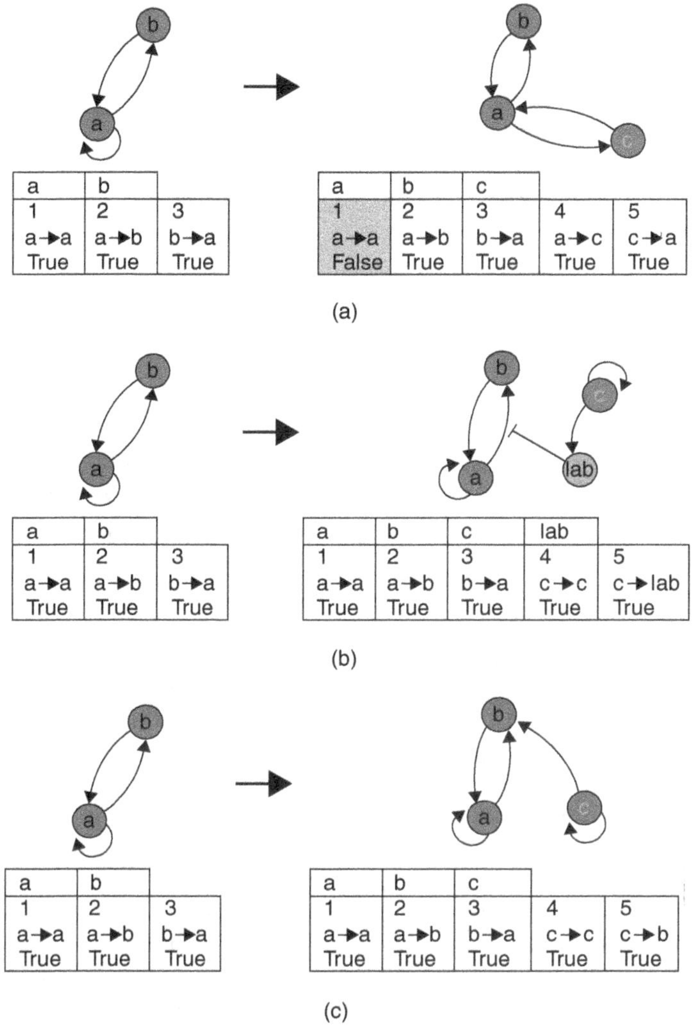

Figure 11.11 Add sequence mutation, with (a) a sequence c added in the middle of existing template a→a, (b) a sequence c added to inhibit template a→b, and (c) a sequence c is added to activate sequence b.

These mutations are based on the idea that, once a connection has no meaning (template concentration close to zero), it should be removed (disabled), or changed to a reverse polarity.

In add sequence mutation, a new signal sequence is added to the system. There are several ways to do so as described in Figure 11.11. The first way is to select an existing template, split it and place the new sequence in the middle. The second way is to add a new sequence that is connecting to itself to inhibit an existing template.

a	b	
1	2	3
a→a	a→b	b→a
True	True	True

a	b		
1	2	3	4
a→a	a→b	b→a	b→b
True	True	True	True

Figure 11.12 Add activation mutation. The template b→b is added.

In the third way, the added sequence is connected to itself and activates an existing sequence.

In add activation mutation, two sequences are chosen randomly with the condition that there is no connection between them. Then, a new template is added connecting the two selected sequences. This process is shown in Figure 11.12.

The add inhibition mutation's example is described in Figure 11.13. First, two simple sequences are chosen, and the template connecting those two simple sequences is disabled (if it exists). In the example, two sequences b and a are chosen and template b→a is disabled. Next, a random template that is generating a is selected (a→a in the example). Then, a new inhibition node is added to inhibit the selected template (Iaa). Finally, the mutation adds the template b→Iaa, completing the module in which sequence b is inhibiting the creation of sequence a.

In the mutation, whenever a template is added to the system, we have to check whether that template innovation exists in the evolution history or not. In order to keep track of that, we use a list to map the template to innovation number. For example, a→a is mapped to 1; a→b is mapped to 2. Thus, in the future, if the template a→b is somehow added to a system, its innovation number is set to 1, whereas a new template that does not exist in the map will be given the next available innovation number. This new map is then added to the list. This innovation number plays an important role in the implementation of the crossover and speciation.

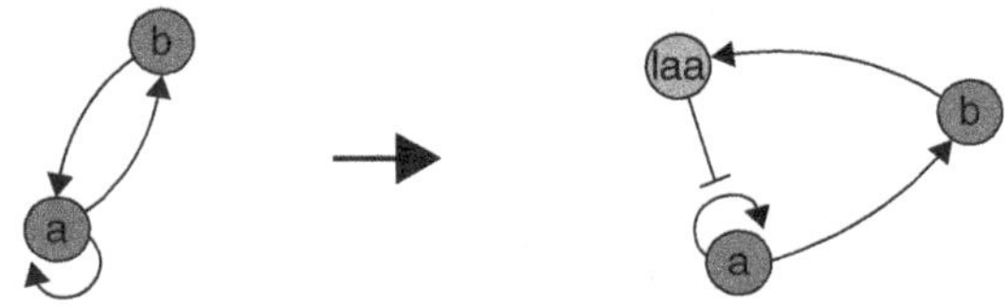

a	b	
1	2	3
a→a	a→b	b→a
True	True	True

a	b	Iaa	
1	2	3	4
a→a	a→b	b→a	b→Iaa
True	True	False	True

Figure 11.13 Add inhibition mutation. The template b→a is disabled, an inhibiting sequence Iaa is added, and a template b→Iaa is added.

	1	2	3	4	5	6	7	8	9
Parent 1	a→a True	b→b True	a→lbb False	a→b True	b→c False	c→c True	b→lcc True	c→laa True	
Parent 2	a→a False	b→b True	a→lbb True	a→b False	b→c True				c→a True
Child	a→a True	b→b True	a→lbb True	a→b False	b→c False	c→c True	b→lcc True	c→laa True	

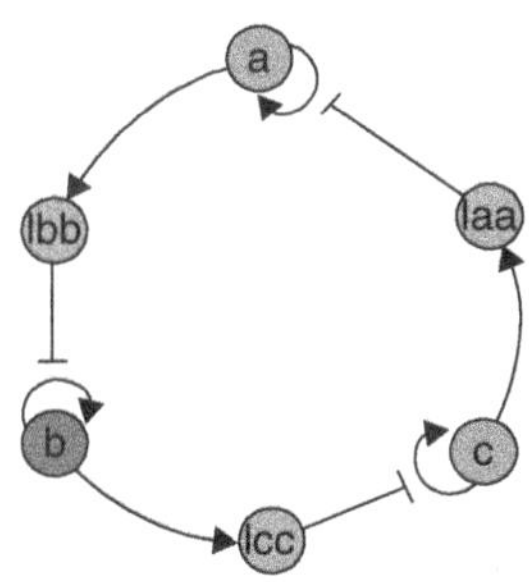

Figure 11.14 Two-point crossover of the two individuals Parent 1 and Parent 2. The template genes are lined up in the order of innovation number. The red lines show the two crossover points. The child takes the gene number 1, 2, 5, 6, 7, and 8 from Parent 1, and gene number 3 and 5 from Parent 2. As a result, we have a new topology that has some parts of both parents.

11.4.3 Crossover

The ERNe encoding and the use of innovation number make crossover straightforward. The template genes in both parents are lined up in the order of innovation number and aligned, and crossover techniques such as one-point and two-point crossover can be applied. Figure 11.14 shows an example of two different individuals' two-point crossover, in which a new topology can be efficiently constructed. We then need to decide how to create node genes for the child. Currently, we are using a simple way: we start with an empty list of node genes, and for each connection gene in the child, we add the involved nodes (from-node and to-node)—with the parameters of corresponding nodes in the parent the connection is inherited from—to the list if no such node is already there.

11.4.4 Speciation

In complex optimization problems, topological innovations are very hard to be preserved. The first reason is that smaller structures tend to optimize faster than larger structure. Moreover, adding new nodes and connections to the system usually decreases the fitness initially. Thus, newly discovered structures are hard to survive more than one generation, even though the innovations they represent might be crucial toward solving the task in the end [31]. We will divide the population into species so that individuals that have similar topologies will be put in same species. Individuals then only have to compete within their own niches instead of with the whole population. This way, topological innovations are protected and have time to optimize. Our speciation is similar to NEAT as we also use a compatibility distance δ calculated as:

$$\delta = \frac{M}{N}, \tag{11.14}$$

where M is the number of mismatched genes and N is the number of genes in the larger genome. If δ is below the speciation threshold δ_t, we say that the two individuals can be put in the same species. For every individual created in the current generation, we try to find its species by looking at all the species from all the generations, with higher priority to more recent species. If no suitable species is found, a new species is added containing the individual.

Every species at the current generation is then assigned a different number of offspring in proportion to the average fitness of its individuals. This mechanism ensures that individuals in the same species must share the fitness of their niche. Therefore, a species is considered better if it has higher average fitness from all the individuals. We also implement a capping mechanism to prevent some temporal good species from growing too fast and taking all the population space. That is, from a generation to the next one, each species can grow to at most 10% of the whole population plus its current population. In order to control the number of species at each generation, we implemented a dynamic adjustment of speciation threshold so that the number of species will move toward a specific target N_s:

$$\delta_t = \begin{cases} \delta_t + \epsilon & \text{if number of species} > N_s \\ \delta_t - \epsilon & \text{if number of species} < N_s \end{cases}$$

ϵ is called the modification step.

11.5 ERNe FOR THE DISCOVERY OF OSCILLATORY SYSTEMS

In its appearance [32], ERNe was demonstrated by evolving credible biochemical answers to challenging autonomous molecular problems: *in vitro* batch oscillatory networks that match specific oscillation shapes. Statistical results clearly show that significant improvements in performance were obtained over other approaches. In this section, we report ERNe's application to two other interesting oscillatory problems: fast-strong oscillator, and robust-fast-strong oscillator.

11.5.1 Fast-Strong Oscillator

In this experiment, the target solution is a PEN DNA toolbox system that can oscillate with high amplitude and frequency. The tracking signal sequence is the sequence a, and the fitness is given as follows:

$$fitness = \frac{amplitude \times limitcycle}{step\_size},$$

where *amplitude* is the concentration of the tracking sequence at its first peak after some skip time, *step_size* is the difference in time between the first peak and the second peak, and *limitcycle* is calculated as follows:

$$lymitcycle = \left(1 + \frac{|amplitude - lastPeakAmplitude|}{lastPeakAmplitude}\right)^{-1},$$

where *lastPeakAmplitude* is the concentration of the tracking sequence at its last peak within observation time (see Figure 11.15 for better understanding of the terms).

From the fitness function, it is clearly seen that we are looking for a system that should have the highest amplitude, the highest limitcycle (capped at 1), and the lowest step size.

For this problem, the ERNe algorithm was run 100 times using the same settings shown in Table 11.1. The best individuals from the runs were then taken for analysis. Interestingly, all the solutions are based on a long oligator [32], having the tracking signal sequence a as the autocatalysis (Figure 11.5.1). The solutions are different in the length of feedback loop (i.e., number of nodes in the feedback loop), varying from 4 to 9. We will investigate how this length affects the fitness.

Figure 11.17 shows the distribution of fitness over feedback loop's length. It can be seen that solutions with four nodes in their feedback loop has the highest median fitness. However, only 3/100 of the runs finished finding these solutions. The largest proportion of the runs (38/100), and the highest fitness solution (1.97) fall into the group with feedback loop length of 7. The number of solutions for the group with 5, 6, 7, and 9 nodes in their feedback loop are 10, 17, 24, and 8, respectively. The median fitness for these group are not significantly different.

As we have a closer look at all the solutions, their limit cycles are all equal to or very close to the capped value of 1. Thus, it is the amplitude and step size that

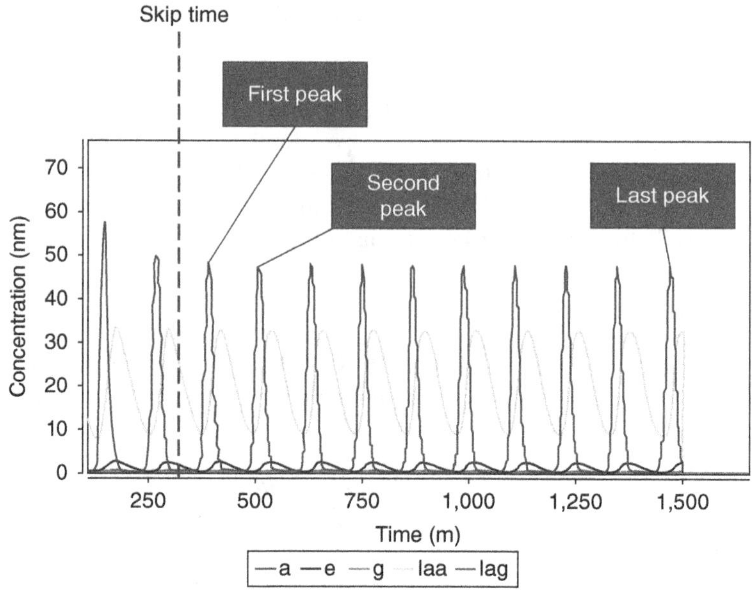

Figure 11.15 How to define first peak, second peak, and last peak.

Table 11.1 Parameters used in experiments

General parameters	
Population size	200
Number of generations	100
Selection method	Tournament (size 5)
Crossover technique	One-point crossover
Speciation parameters	
Preferred number of species	10
Starting speciation threshold	0.6
Minimum speciation threshold	0.1
Speciation modification step ϵ	0.03
Crossover and mutation parameters	
P (mutation only)	0.5
P (interspecies mating)	0.01
P (mutation after crossover)	0.75
Mutation parameters	
P (parameter only)	0.9
P (single gene mutation)	0.8
P (structure-add node)	0.2
P (structure-add activation)	0.2

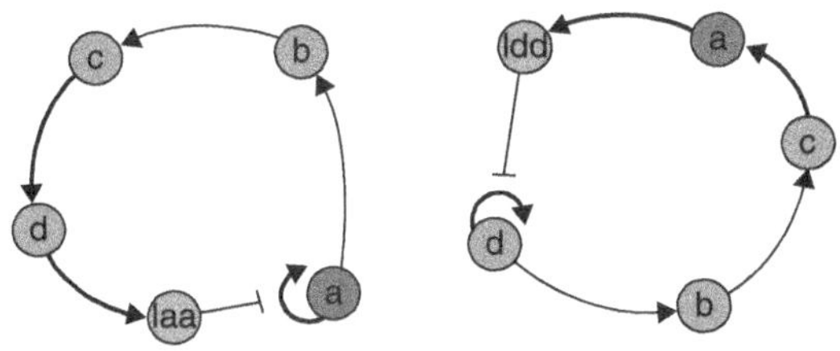

Figure 11.16 Two derivatives of the long oligator, having tracking sequence as the first (a), and the last (b) nodes of the feedback loop.

contribute to the fitness. For better understanding of the relationship between these behaviors and the feedback loop length, we have two additional plots, showing the amplitudes (Figure 11.18) and step sizes (Figure 11.19) over the feedback loop length. It can be clearly seen that, step size increases with the length of the feedback loop. As for the amplitude, we observe a similar thing, with the exception of the group of length four. This group can reach as high an amplitude as the group of length six, but cannot compete with the group of length seven or above.

11.5.2 Robust-Fast-Strong Oscillatior

Our ultimate goal is for finding direct candidates for the wet experiment. Thus, robustness is also an important factor. Similar to previous experiment, we want a PEN DNA toolbox system that can oscillate with high amplitude and frequency. In this experiment, however, we also require the system to be robust. For a solution, we generate 30 systems which are the noisy version of the solution, simply by mutating

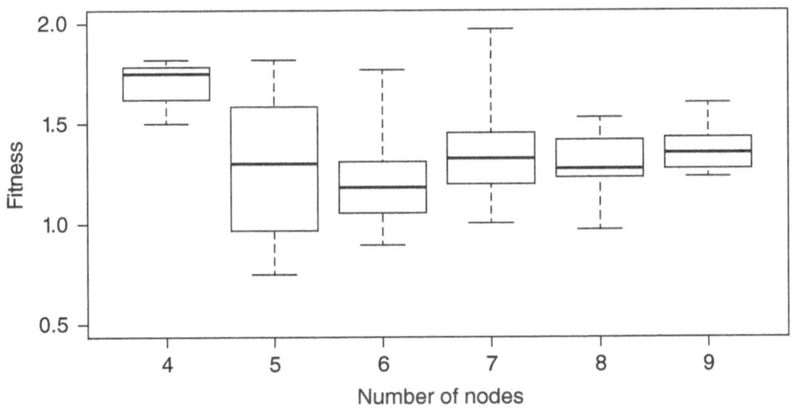

Figure 11.17 Fitness plotted against number of nodes in the feedback loop.

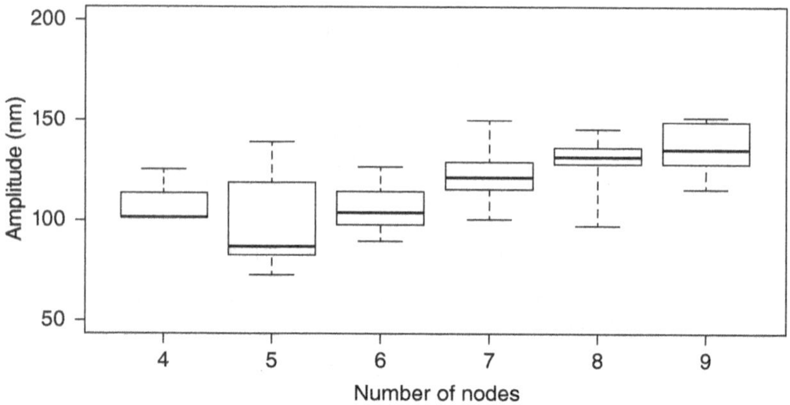

Figure 11.18 Amplitudes plotted against number of nodes in the feedback loop.

all its sequences' parameters using the parameter mutation operation. Then, the fitness of a solution is defined as:

$$fitness = \sum \frac{amplitude_i \times limitcycle_i}{step\_size_i},$$

where $amplitude_i$, $limitcycle_i$, and $step_size_i$ are the *amplitude*, *limitcycle*, and *step_size* (see previous section) of the *i*th noisy version of the solution.

This way of evaluating the robustness is efficient, but it is also costly in computational power. For this reason, we only have the ERNe algorithm run for five times using the settings shown in Table 11.1. Again, all runs converged to a topology based on the long oligator of length five, this time having the tracking signal sequence positioned in the end of the feedback loop (Figure 11.5.1).

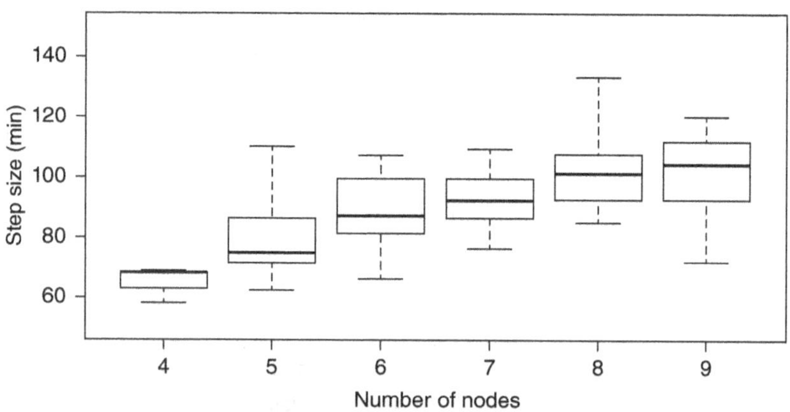

Figure 11.19 Step sizes plotted against number of nodes in the feedback loop.

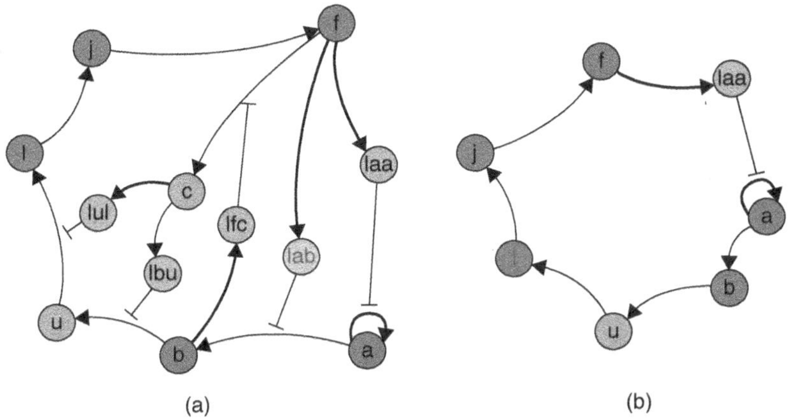

Figure 11.20 Original (a) and pruned (b) versions of the best solution for the Fast-strong oscillator.

11.6 DISCUSSION

The results reported for the fast-strong oscillator shows that high amplitude can be obtained by simply increasing the length of the feedback loop. However, longer feedback loop tends to have slower oscillation. As a result, it is non-trivial to know the design of a best fast-strong oscillator. It can be concluded from our study that a long oligator starting from the tracking signal sequence with the length of the feedback loop between 4 to 9—with their parameters optimized—can do the job fairly well. From the plot shown in Figure 11.17 also, it can be seen that fitness slightly improves with the complexity, but we never found a solution with the length of feedback loop greater than nine. It means that the bloating problem is controlled well with ERNe without any need to implement fitness penalty to the complexity.

Besides the length of the feedback loop, the solutions are also different in the way side branches are forged. At the moment, it is impossible for us to know in a systematic way how such branches contribute to the fitness. Instead, we take the best solution (Figure 11.20a) with fitness of 1.97 (amplitude is 149.70 nm and step size is 76 min) and remove its side branches, to form the topology in Figure 11.20b. Its new observed fitness dropped significantly to 0.77 (amplitude is 83.88 nm and step size is 108 min). This shows that the side branches contribute greatly to both the amplitude and frequency, therefore boosting the fitness score.

It is also interesting that we found the completely different topology as the solution for the robust-fast-strong oscillation. This means that robustness can not be expected to be easily compatible with fast and strong characteristics in oscillators. Nevertheless, there are still some difficulties in analyzing the results coming out of the ERNe algorithm. We know that they are most feasible candidates for the actual wet experiments, but we do not know by which mechanism could they obtain the desired behavior. We believe that, by actually implementing them we could get some insights.

Another thing we would like to discuss is that, even though ERNe is designed for the evolution of biochemical reaction networks, it can be used on GRNs without any difficulties. Indeed, the ERNe encoding can be used directly to encode the structure and parameters of GRNs, and ERNe's crossover and speciation can also be used directly. The mutation rules, though, should be redefined, as in our previous paper [32] we discovered that the physical nature of the information processing medium should be taken into account in the definition of relevant mutation operators. In the future, we would like to explore the mixed use of the model levels, the use of local search [50], multi-objective optimization [51, 52], and automatic parameter tuning [53, 54].

11.7 CONCLUSION

In this chapter, we introduced an up-to-date research in *in vitro* molecular programming: a GRN-inspired system called the PEN DNA Toolbox. One of its advantages is that it offers simulation in various levels of details, with the highest level accurate enough to enable direct transfer from *in silico* design to *in vitro* wet experiment. Using the most efficient technique which is the ERNe algorithm, we discovered some designs for the two interesting problems: fast-strong oscillator, and robust-fast-strong oscillator. The study clearly shows that the amplitude and the frequency of oscillations cannot be expected at the same time, and that a long oligator with feedback loop length 4 to 9 should be best for the job. In the future, we would like to explore the use of mixed model levels, the use of local search [50], multi-objective optimization [51, 52], and automatic parameter tuning [53, 54].

REFERENCES

1. F. Lienert, J. J. Lohmueller, A. Garg, and P. A. Silver, "Synthetic biology in mammalian cells: next generation research tools and therapeutics," *Nature Reviews Molecular Cell Biology*, 15(2), 95–107 (2014).
2. K. Roberta, "Five hard truths for synthetic biology," *Nature*, 463(7279), 288–290 (2014).
3. K. Montagne, R. Plasson, Y. Sakai, T. Fujii, and Y. Rondelez, "Programming an *in vitro* DNA oscillator using a molecular networking strategy," *Molecular Systems Biology*, 7(1) (2011).
4. A. Padirac, T. Fujii, and Y. Rondelez, "Bottom-up construction of *in vitro* switchable memories," *Proceedings of the National Academy of Sciences*, 109(47), E3212–E3220 (2012).
5. L. Qian, E. Winfree, and J. Bruck, "Neural network computation with DNA strand displacement cascades," *Nature*, 475, 368–372 (2011).
6. A. Padirac, T. Fujii, and Y. Rondelez, "Nucleic acids for the rational design of reaction circuits," *Current Opinion in Biotechnology*, 24(4), 575–580 (2012).
7. M. N. Stojanovic and D. Stefanovic, "A deoxyribozyme-based molecular automaton," *Nature Biotechnology*, 21, 1069–1074 (2003).

8. J. Kim, K. S. White, and E. Winfree, "Construction of an *in vitro* bistable circuit from synthetic transcriptional switches," *Molecular System Biology*, 2(68) (2006).
9. G. Seelig, D. Soloveichik, D. Y. Zhang, and E. Winfree, "Enzyme-free nucleic acid logic circuits," *Science*, 314(5805), 1585–1588 (2006).
10. T. Fujii and Y. Rondelez, "Predator-prey molecular ecosystems," *ACS Nano*, 7(1), 27–34 (2013).
11. A. Padirac, T. Fujii, A. E. Torres, and Y. Rondelez, "Spatial waves in synthetic biochemical networks," *Journal of the American Chemical Society*, 135(39), 14586–14592 (2013).
12. D. Soloveichik, G. Seelig, and E. Winfree, "DNA as a universal substrate for chemical kinetics," *Proceedings of the National Academy of Sciences*, 107(12), 5393–5398 (2010).
13. A. Baccouche, K. Montagne, A. Padirac, T. Fujii, and Y. Rondelez, "Dynamic DNA-toolbox reaction circuits: a walkthrough," *Methods*, 67(2), 234–249 (2014).
14. S. Tavazoie, J. D. Hughes, M. J. Campbell, R. J. Cho, G. M. Church, "Systematic determination of genetic network architecture," *Nature Genetics*, 22(3), 281–285. (1999).
15. T. I. Lee, N. J. Rinaldi, F. Robert, D. T. Odom, Z. Bar-Joseph, G. K. Gerber, N. M. Hannett, C. T. Harbison, C. M. Thompson, I. Simon, J. Zeitlinger, E. G. Jennings, H. L. Murray, D. B. Gordon, B. Ren, J. J. Wyrick, J-B Tagne, T. L. Volkert, E. Fraenkel, D. K. Gifford, R. A. Young, "Transcriptional Regulatory Networks in Saccharomyces cerevisiae," *Science*, 298(5594), 799–804 (2002).
16. D. Peer, A. Regev, G. Elidan, and N. Friedman, "Inferring subnetworks from perturbed expression profiles," *Bioinformatics*, 17(Suppl 1), S215–S224 (2001).
17. K. Basso, A. A. Margolin, G. Stolovitzky, U. Klein, R. Dalla-Favera, and A. Califano, "Reverse engineering of regulatory networks in human B cells," *Nature Genetics*, 37(4), 382–390 (2005).
18. N. Kasabov, "Knowledge-based neural networks for gene expression data analysis, modelling and profile discovery," *Drug Discovery Today: BIOSILICO*, 2(6), 253–261 (2004).
19. Z. S. Chan, N. Kasabov, and L. Collins, "A two-stage methodology for gene regulatory network extraction from time–course gene expression data," *Expert Systems with Applications*, 30(1), 59–63 (2006).
20. P. Francois and V. Hakim, "Design of genetic networks with specified functions by evolution in silico," *Proceedings of the National Academy of Sciences*, 101(2), 580–585 (2004).
21. K. Fujimoto, S. Ishihara, and K. Kaneko, "Network evolution of body plans," *PLOS ONE*, 3(7) (2008).
22. Y. Kobayashi, T. Shibata, Y. Kuramoto, and A. Mikhailov, "Evolutionary design of oscillatory genetic networks," *European Physical Journal B*, 76, 167–178 (2010).
23. A. Deckard and H. M. Sauro, "Preliminary studies on the in silico evolution of biochemical networks," *ChemBioChem*, 5(10), 1423–1431 (2004).
24. S. R. Paladugu, V. Chickarmane, A. Deckard, J. P. Frumkin, M. Mc-Cormack, and H. M. Sauro, "In silico evolution of functional modules in biochemical networks," *Systematic Biology (Stevenage)*, 153(4), 223–235 (2006).
25. M. A. Marchisio and J. Stelling, "Automatic design of digital synthetic gene circuits," *PLoS Computers in Biology*, 7(2), (2011).
26. H. Cao, F. J. Remoro-Campero, S. Heeb, M. Cmara, and N. Krasnogor, "Evolving cell models for systems and synthetic biology," *Systems and Synthetic Biology*, 4(1), 55–84 (2010).

27. Y. Jin and B. Sendhoff, "Evolving *in silico* bistable and oscillatory dynamics for gene regulatory network motifs," *IEEE Congress on Evolutionary Computation*, 386–391 (2008).
28. J. S. Hallinan, G. Misirli, and A. Wipat, "Evolutionary computation for the design of a stochastic switch for synthetic genetic circuits," *Conference Proceedings of the IEEE Engineering in Medicine and Biology Society*, 768–774 (2010).
29. B. Drennan and R. Beer, "Evolution of repressilators using a biologically-motivated model of gene expression," *Artificial Life X: Proceedings of the Tenth International Conference on the Simulation and Synthesis of Living Systems*, 22–27 (2006).
30. A. H. Chau, J. M. Walter, J. Gerardin, C. Tang, and W. A. Lim, "Designing synthetic regulatory networks capable of self-organizing cell polarization," *Cell*, 151(2), 320–332 (2012).
31. K. O. Stanley and R. Miikkulainen, "Evolving neural networks through augmenting topologies," *Evolutionary Computation*, 10(2), 99–127 (2002).
32. Q. H. Dinh, N. Aubert, N. Noman, T. Fujii, Y. Rondelez, and H.Iba "An effective method for evolving reaction networks in synthetic biochemical systems," *IEEE Transactions on Evolutionary Computation*, 19 (3), 374–386 (2015).
33. L. Qian and E. Winfree, "Scaling up digital circuit computation with DNA strand displacement cascades," *Science*, 332(6034), 1196–1201 (2011).
34. R. Plasson and Y. Rondelez, "Synthetic biochemical dynamic circuits," *Multiscale Analysis and Nonlinear Dynamics: From Genes to the Brain*, p. 113–145 (2013).
35. N. Aubert, C. Mosca, T. Fujii, M. Hagiya, and Y. Rondelez, "Computer-assisted design for scaling up systems based on DNA reaction networks," *Journal of the Royal Society Interface*, 11(93), (2014).
36. P. D. Dans, A. Zeida, M. R. Machado, and S. Pantano, "A coarse grained model for atomic-detailed DNA simulations with explicit electrostatics," *Journal of Chemical Theory and Computation*, 6(5), 1711–1725 (2010).
37. T. E. Ouldridge, "Springer Theses: Coarse-Grained Modelling of DNA and DNA Self-Assembly," *Springer*, (2012).
38. N. R. Markham and M. Zuker, "Dinamelt web server for nucleic acid melting prediction," *Nucleic Acids Research*, 33(Suppl 2), W577–W581 (2005).
39. P. Yakovchuk, E. Protozanova, and M. D. Frank-Kamenetskii, "Base-stacking and base-pairing contributions into thermal stability of the DNA double helix," *Nucleic Acids Research*, 34(2), 564–574 (2006).
40. Y. Bansho, N. Ichihashi, Y. Kazuta, T. Matsuura, H. Suzuki, and T. Yomo, "Importance of parasite RNA species repression for prolonged translation-coupled RNA self-replication," *Chemistry and Biology*, 19(4), 478–487 (2012).
41. N. Aubert, Q. H. Dinh, M. Hagiya, H. Iba, T. Fujii, N. Bredeche, and Y. Rondelez, "Evolution of cheating DNA-based agents playing the game of rock-paper-scissors," *Advances in Articial Life, ECAL*, 12, 1143–1150 (2013).
42. D. Y. Zhang and E. Winfree, "Control of DNA strand displacement kinetics using toehold exchange," *Journal of the American Chemical Society*, 131(47), 17393–17314 (2009).
43. S. Bommarito, N. Peyret, and J. SantaLucia Jr., "Thermodynamic parameters for DNA sequences with dangling ends," *Nucleic Acids Research*, 28(9), 1929–1934 (2000).

44. V. A. Vasiliskov, D. V. Prokopenko, and A. D. Mirzabekov, "Parallel multiplex thermodynamic analysis of coaxial base stacking in DNA duplexes by oligodeoxyribonucleotide microchips," *Nucleic Acids Research*, 29(11), 2303–2313 (2001).
45. D. V. Pyshnyi and E. M. Ivanova, "The influence of nearest neighbors on the efficiency of coaxial stacking at contiguous stacking hybridization of oligodeoxyribonucleotides," *Nucleosides, Nucleotides and Nucleic Acids*, 23(6–7), 1057–1064 (2004).
46. J. Kim, J. Hopfield, and E. Winfree, "Neural network computation by in vitro transcriptional circuits," *Advances in Neural Information Processing Systems*, p. 681–688 (2004).
47. A. J. Genot, T. Fujii, and Y. Rondelez, "Computing with competition in biochemical networks," *Physical Review Letters*, 109(20) (2012).
48. Y. Rondelez, "Competition for catalytic resources alters biological network dynamics," *Physical Review Letters*, 108(1) (2012).
49. M. B. Elowitz and S. A. Leibler, "Synthetic gene oscillatory network of transcriptional regulators," *Nature*, 403, 335–338 (2000).
50. R. Storn and K. Price, "Differential evolution - a simple and efficient heuristic for global optimization over continuous spaces," *Journal of Global Optimization,* 11(4), 341–359 (1997).
51. K. Deb, "Multi-Objective Optimization Using Evolutionary Algorithms," *NJ: Wiley*, (2001).
52. A. Warmflash, P. Francois, and E. D. Siggia, "Pareto evolution of gene networks: an algorithm to optimize multiple fitness objectives," *Physics and Biology*, 9(5) (2012).
53. A. E. Eiben, R. Hinterding, and Z. Michalewicz, "Parameter control in evolutionary algorithms," *IEEE Transactions on Evolutionary Computation,* 3(2), 124–141 (1999).
54. A. E. Eiben and S. K. Smit, "Parameter tuning for configuring and analyzing evolutionary algorithms," *Swarm and Evolutionary Computation*, 1(1), 19–31 (2011).

IV

APPLICATION OF GRN WITH EAs

12

ARTIFICIAL GENE REGULATORY NETWORKS FOR AGENT CONTROL

Sylvain Cussat-Blanc, Jean Disset, Stéphane Sanchez and Yves Duthen
University of Toulouse – IRIT – CNRS UMR5505, Toulouse, France

12.1 INTRODUCTION

Gene regulatory networks (GRNs) are biological structures that control the internal behavior of living cells. They regulate gene expression by enhancing and inhibiting the transcription of certain parts of the DNA. However, they can be used as agent controllers: for example, instead of regulating gene expressions, they can be used to regulate either agent actuators or high-level complex behaviors. This chapter summarizes three applications of a computational model of gene regulatory network to control virtual agents. The aim is to provide an overview of problems that gene regulatory networks can address in reference to some recent works done in this area.

When used to simulate gene expression regulation, a GRN is usually encoded within a bit string, as DNA is encoded within a nucleotide string. As in real DNA, a gene sequence starts with a particular sequence, called the *promoter* in biology [19]. In the real DNA, this sequence is represented with a set of four proteins: *TATA* where T represents the thymine and A the Adenine. Torsten Reil is one of the first to propose a biologically plausible model of gene regulatory networks [21]. The model is based on a sequence of bits in which the promoter is composed of four bits *0101*. The gene is coded directly after this promoter whereas the regulatory elements are coded before the promoter. To visualize properties of these networks, he uses

Evolutionary Computation in Gene Regulatory Network Research, First Edition.
Edited by Hitoshi Iba and Nasimul Noman.

graph visualization to observe concentration variations of the different proteins of the system. He points out three different kinds of behavior from randomly generated gene regulatory networks: stable, chaotic, and cyclic. He also observes that these networks are capable of recovering from random alterations of the genome, producing the same pattern when they are randomly mutated. In 2003, Wolfgang Banzhaf formulated a new gene regulatory network, also inspired from biology [2]. He uses genomes composed of multiple 32-bit integers encoded as a bit string. Each gene starts with a promoter coded by any integer ending with the sequence "XYZ01010101" (XYZ can be any kind of sequence that complement the 32-bit integer). This sequence occurs with a 2^{-8} probability (0.39%). The gene following this promoter is then coded in five 32-bits integers (160 bit) and regulatory elements are coded upstream to the promoter by two integers, one for enhancing and one for inhibiting kinetics. Banzhaf's model confirms the hypothesis pointed out by Reil's one; the same properties emerges from his model. From these seminal models, many computational models have been initially used to control cells of artificial developmental models [11, 12, 17]. They simulate the very first stage of living organism embryogenesis and more particularly cell differentiation mechanisms. One of the initial problem of this field of research is the French Flag problem [25] in which a virtual organism has to produce a rectangle that contains three strips of different colors (blue, white, and red). This simulates the capacity of cells to differentiate in a spatial environment. Many models addressed this benchmark with cells controlled by a gene regulatory network [4, 17, 18]. More recently, gene regulatory networks have proven their capacity to regulate complex behaviors in various situations: they have been used to control virtual agents [7,16,20], real swarm [15] or modular robots [5].

However, these models are not easy to evolve due to the nature of their genetic encoding. This chapter shows how an artificial gene regulatory network, designed for computational purposes, can be used to control various kinds of virtual agents. Therefore, through three different experiments, we will see two ways to use GRNs: directly connected to sensors and actuators of the agent or as a system to regulate high-level behaviors. For this purpose, this chapter is organized as follows. First, we will detail the computational model of the GRN and how we evolve it with a genetic algorithm. Then, we will visualize the behavioral abilities of the GRN with images, which show the variation of output expressions according to input values, and with videos, which add the temporal aspect of gene regulation. This chapter then presents the use of this GRN to control three kinds of agents. The first experiment shows its application to control virtual cells in an artificial embryogenesis process. Then, the same GRN is directly connected to a virtual race car and is taught to drive on a track. Finally, the last experiment uses the GRN to regulate a set of high-level scripted behaviors for a team of agents to defend a target against incoming threats.

12.2 COMPUTATION MODEL

The gene regulatory network used in this work is a simplified model based on Banzhaf's model. Whereas Banzhaf's model uses a bitstream with promoters that separate genes, here we use abstract proteins assimilable to objects. Proteins are designed to interact but the interaction network is not directly encoded in the genome: protein properties are used to determine these interactions. Therefore, in this model, a set of these proteins simply represents a gene regulatory network. The number of proteins as well as their properties are evolved with a genetic algorithm. The remaining of this section describes these proteins and how they build a gene regulatory network. It also presents the dynamical aspect of regulation and how the protein set is embedded in a genome to be evolved by a genetic algorithm.

12.2.1 Representation of the Proteins

This model is composed of a set of abstract proteins. A protein a is composed of three tags:

- the *protein tag* id_a that identifies the protein,
- the *enhancer tag* enh_a that defines the enhancing matching factor between two proteins, and
- the *inhibitor tag* inh_a that defines the inhibiting matching factor between two proteins.

Each tag is coded with an integer in $[0, p]$ where the upper bound p can be tuned to control the network precision. In addition to these tags, a protein is also defined by its concentration that will vary over time with the dynamics described later. A protein can be of three different types:

- *input* protein, whose concentration is provided by the environment, which regulates other proteins but is not regulated,
- *output* protein with a concentration used as network output, which is regulated but does not regulate other proteins, and
- *regulatory* protein that regulates and is regulated by others proteins.

12.2.2 Dynamics

With this structure, the GRN's dynamics are computed by using protein tags. They determine pairwise interaction between two proteins and, based on this interaction, the production rate of each protein. Thus, the affinity of a protein a for another protein b is given by the enhancing factor u^+_{ab} and the inhibiting factor u^-_{ab}. They are calculated as follows:

$$u^+_{ab} = p - |enh_a - id_b|; u^-_{ab} = p - |inh_a - id_b|. \tag{12.1}$$

The proteins are then compared pairwise according to their enhancing and inhibiting factors. For a protein a, the total enhancement g_a and inhibition h_a are given by:

$$g_a = \frac{1}{card(N)} \sum_{b}^{N} c_b e^{\beta(u^+_{ab} - u^+_{max})}; \; h_a = \frac{1}{card(N)} \sum_{b}^{N} c_b e^{\beta(u^-_{ab} - u^-_{max})}, \qquad (12.2)$$

where N is the set of the GRN's proteins, c_b is the concentration of protein b, u^+_{max} is the maximum observed enhancing factor, u^-_{max} is the maximum observed inhibiting factor and β is a control parameter which will be detailed hereafter. At each timestep, the concentration of a protein a changes with the following differential equation:

$$\frac{dc_a}{dt} = \frac{\delta(g_a - h_a)}{\Phi},$$

where Φ is a normalization factor to ensure that the total sum of output and regulatory protein concentrations remains equal to 1. β and δ are two constants that influence reaction rates of the network. β affects the matching factors' importance and δ is used to modify the proteins' production level in the differential equation. In summary, the lower both values are, the smoother the regulation is; the higher the values are, the more sudden the regulation is.

Figure 12.1 summarizes how the model functions. Edges represent enhancing (in green) and inhibiting (in red) matching factors between two proteins. Their thickness represents the distance value: the thicker the line, the closer the proteins.

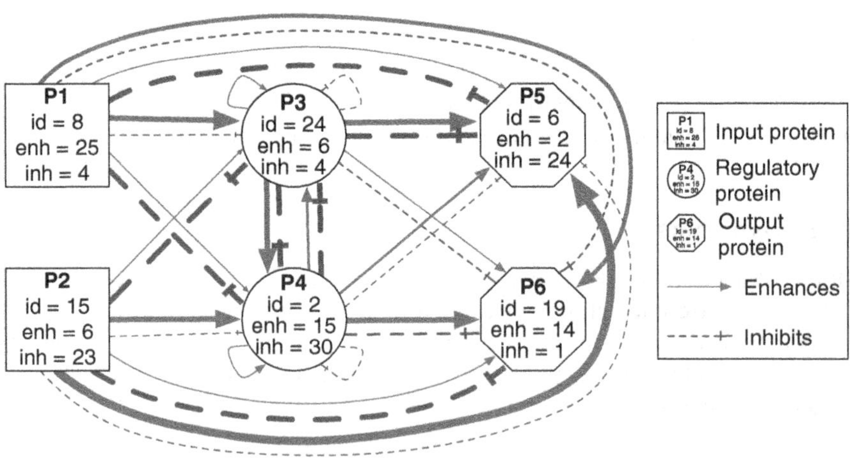

Figure 12.1 Graphical representation of a GRN. Nodes are proteins and edges represent enhancing and inhibiting affinities between two proteins. The bigger the edges, the closer the proteins.

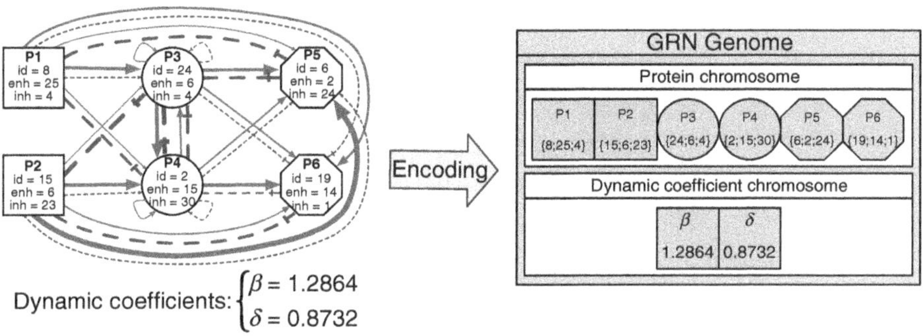

Figure 12.2 Encoding of a GRN in a genome: the first chromosome is a set of indivisible proteins and the second chromosome contains the dynamics coefficients.

12.2.3 Encoding and Genetic Evolution

In all problems this gene regulatory network is used, protein tags and dynamics coefficients need to be optimized. For this purpose, we use a standard genetic algorithm. To encode the GRN, genomes contain two independent chromosomes. The first one is a variable length chromosome of indivisible proteins. Each protein is encoded with three integers between 0 and p that correspond to the three protein tags. Usually, p is set at 32, which is a sufficient precision for most of the problems this GRN has been used on, and proteins are organized in the chromosome with input proteins first, followed by output proteins and then regulatory proteins. Figure 12.2 depicts this encoding. Other genome organizations could be imagined: for example, one could evolve the protein types and use the protein tags in order to dynamically link an input or output protein to a specific sensor or actuator. This organization can allow more dynamics by multiplying the regulation flows for inputs and outputs.

This chromosome requires particular crossover and mutation operators (represented in Figure 12.3):

- a *crossover* can only occur between two proteins and never between two tags of the same protein. This ensures integrity of each subnetwork composed of each half of both parent genomes. When reassembling the genomes parts to produce the offsprings, local connections in the subnetworks are kept with this operator and only new connections between the two networks are created.
- three *mutations* can be equiprobably used: add a new random regulatory protein, remove one protein randomly selected in the set of regulatory proteins, or mutate a tag within a randomly selected protein.

A second chromosome is used to evolve the dynamics variables β and δ. This chromosome consists of two double-precision floating-point values and uses standard mutation and one-point crossover methods. These variables are evolved in the interval $[0.5, 2]$. Values under 0.5 produce unreactive networks whereas values over 2 produce

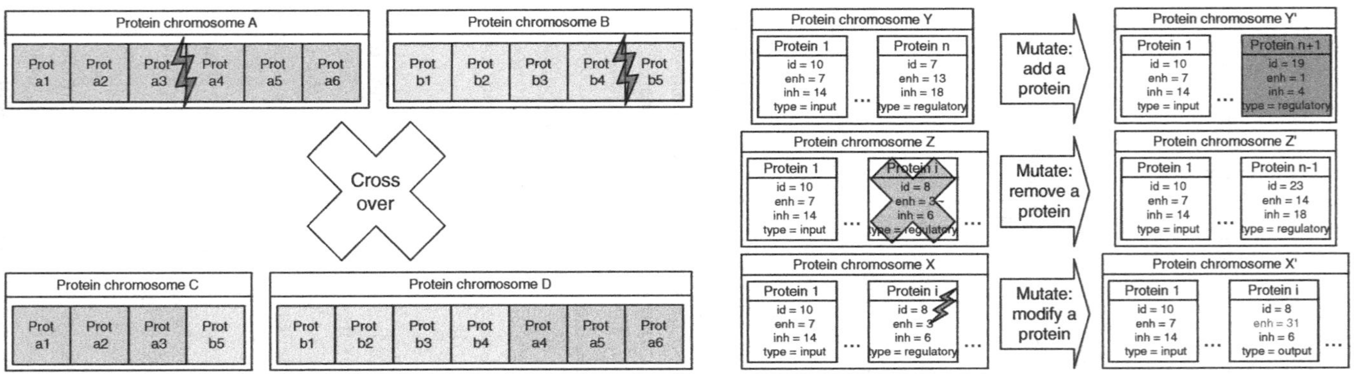

Figure 12.3 Crossover and mutation operators applied to the protein chromosome. A crossover (on the left-hand side) can only occur between two proteins and a mutation (on the right-hand side) consists in adding, removing, or changing a protein. Reprinted with permission from Ref. [22], copyright 2014 Springer Science and Business Media.

very unstable networks. These values are chosen empirically through a series of test cases.

12.3 VISUALIZING THE GRN ABILITIES

In order to evaluate the complexity of behaviors generated by GRNs, we propose a first application where the GRN is used to generate pictures and videos. Technically, the GRN, duplicated in every pixel of the picture, calculates the RGB components of each pixel of the picture. As depicted in Figure 12.4, each GRN uses the pixel coordinates (input proteins) to compute each color component (output proteins). The coordinate (x, y) of a pixel are transformed into proteins concentrations so that they do not overflow the network:

$$in_x = \frac{0.1x}{width}; in_y = \frac{0.1y}{height}, \tag{12.3}$$

where in_x (resp. in_y) is the concentration of the protein associated with the abscissa x (resp. the ordinate y) of the current pixel, *width* and *height* define the picture size.

The resulting RGB component values are given by the following equations:

$$out_r = \frac{255 * c_r}{max_r}; out_g = \frac{255 * c_g}{max_g}; out_b = \frac{255 * c_b}{max_b}, \tag{12.4}$$

where out_r (resp. out_g and out_b) is the value of the red (resp. green and blue) component for the current pixel, c_r (resp. c_g and c_b) is the concentration of the output protein associated with the red (resp. green and blue) component in the GRN (this concentration is always between 0 and 1) and max_r (resp. max_g and max_b) is the maximum concentration observed in the picture for the red (resp. green and blue) component. Since a GRN includes a temporal component (protein concentrations vary over time), one image can be generated at each timestep. Therefore, a GRN can

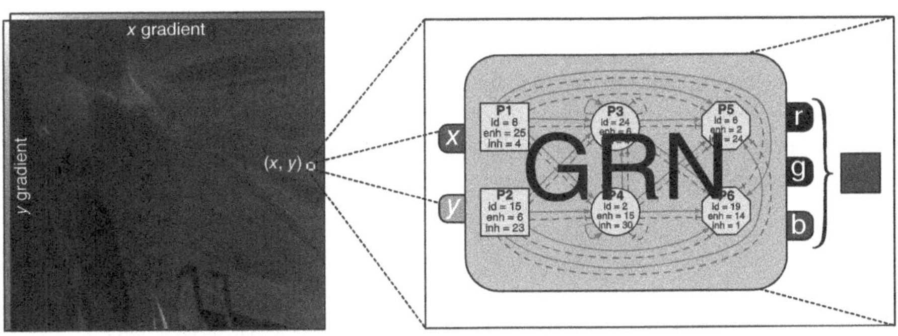

Figure 12.4 GRNs can be used to generate pictures and videos. To do so, pixel coordinates are connected to input proteins and color components are connected to output proteins.

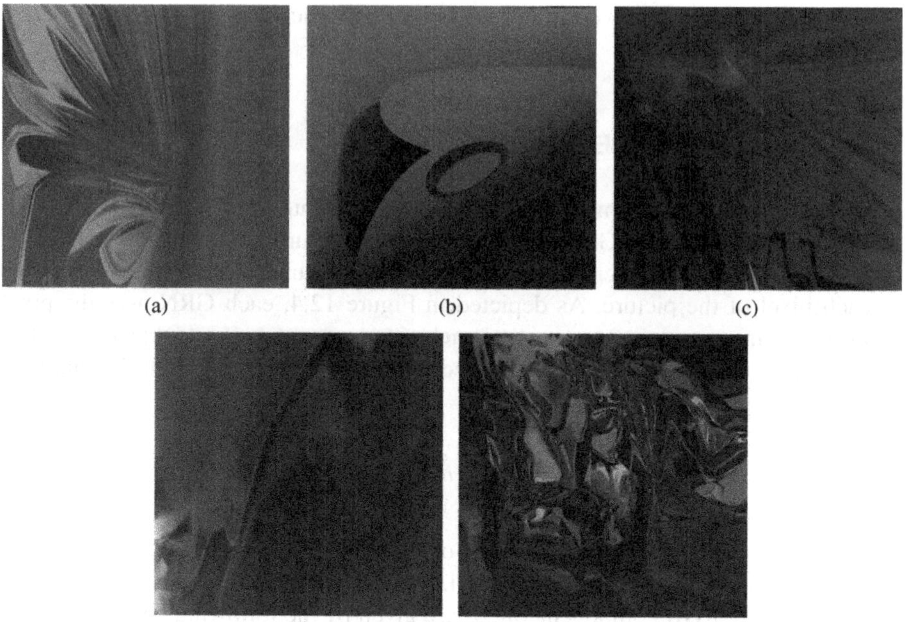

(a) (b) (c)

(d) (e)

Figure 12.5 Examples of pictures generated with a gene regulatory network. Reprinted with permission from Ref. [10], copyright 2014 The MIT Press.

generate either a video (a sequential set of pictures) or one particular image which corresponds to a snapshot of network's concentrations at a particular timestep.

With the encoding described in previous section and the use of an interactive genetic algorithm,[1] various pictures and videos have been generated. Some pictures are presented in Figure 12.5, and Figure 12.6 presents example of videos.[2] They allow us to evaluate the kind of possible behavioral structures generated by gene regulatory networks. For example, in Figure 12.5(b), repetitive patterns can be observed with some modifications. That shows the capacity of gene regulatory networks to produce modular patterns. Figures 12.5(b)–12.5(d) also show the ability of the GRN to produce either smooth or abrupt transition with color transitions. Finally, Figure 12.5(e) depicts the GRN capacity to produce extremely complex behaviors, with very different outputs with close input values. More details about properties highlighted by the pictures can be found in Ref. [6]. Videos add the temporal aspect of gene regulation. As presented in Figure 12.6, oscillatory behaviors can be easily visualized. Other

[1] Users can select one or multiple pictures or videos they "like". The selected GRN are crossed over and mutated in order to generate a new set of pictures or videos, which are displayed for evaluation. Users are therefore providing the fitness function and the selection algorithm of the genetic algorithm.

[2] Additional and colored pictures, videos and the software to generate the pictures and videos can be found online: http://www.irit.fr/~Sylvain.Cussat-Blanc/ColorfulRegulation/index_en.php

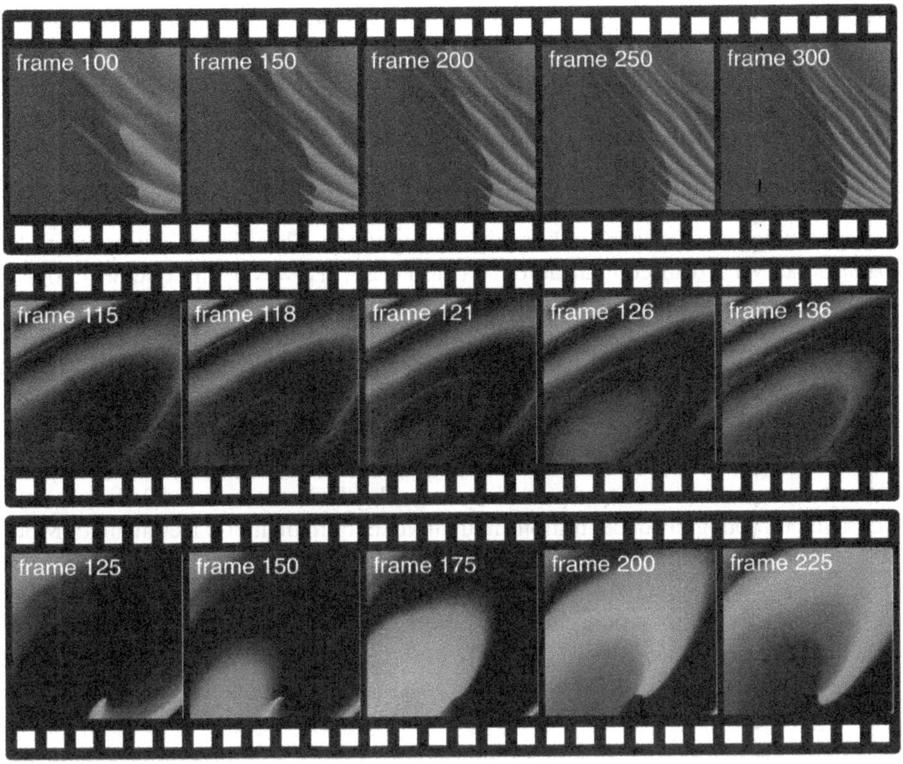

Figure 12.6 Examples of videos generated with a gene regulatory network.

videos show chaotic and steady behaviors, which are the two other main behaviors of gene regulatory networks in addition to oscillations.

With a computational model of gene regulatory network that can be efficiently evolved by a evolutionary algorithm, these networks can generate a broad set of complex behaviors, as we saw in this section. In the remaining of this chapter, this gene regulatory network is used to control different kinds of virtual agents such as cells, a car, or collaborative agents. Next section presents how the GRN is connected to virtual cells that have to collaborate to survive in a complex environment.

12.4 GROWING MULTICELLULAR ORGANISMS

One of the most obvious use of a gene regulatory network is the control of virtual cells. As in nature, a gene regulatory network can manage cell actions in a simulated physical and chemical environment. Another objective of growing multicellular organisms is to study the specialization mechanisms of cells. Specialization allows emergence of organs, that is, large sets of cells with one or multiple common functions. Before showing how a gene regulatory network can be used to control cells and can be optimized to grow multicellular organisms with high-level objectives, the simulated environment must be described.

Virtual cells can either act in a discrete or a continuous environment. In case of a discrete environment, either 2D or 3D matrix is generally used. Each voxel of the matrix represents the cell state. This one can either be binary (dead or alive) or more complex, expressing the cell's specialization state, for example. When a cell is reproducing, it "switches on" a neighboring voxel. This kind of environment has been broadly used in cellular automata [8, 14, 24]. However, in cellular automata, switching cells on or off is mainly based on neighborhood rules and cells are not controlled by a cellular intelligence such as a GRN. A gene regulatory network can be used to generate specialization behaviors (usually graphically represented with voxel colorations) for cells [3, 9, 13]. Some models also use morphogen gradients to position the cells in the environment. Morphogens are diffused by the cells in the environment and allow a form of indirect communication between cells. In a discrete environment, morphogens diffuse gradually in the matrix. Cells can either sense the morphogen concentrations in their voxel or in their von Neumann or Moore neighborhood.

Using a continuous environment improves the physical simulation quality and thus cells' abilities. Within a continuous environment, cells can interact physically with mechanical pressure and move precisely in the environment. Mechanical structures of the cells are usually simulated with a mass-spring-damper system [11, 17]: this mechanics is precise enough for most of digital organism simulations and computationally efficient. Other systems based on microfluidic droplets can also be used because they react closer to reality but they are also extremely computationally expensive. If the purpose of the simulation is to evolve cells' behaviors to produce complex multicellular organisms, it is important to carefully evaluate the environment simulation precision so that cells can have access to a complex enough environment keeping the simulation time reasonable. In a continuous space, morphogens can also be simulated by using a diffusion grid (to reduce the diffusion computational cost) or by computing the distance to the morphogen source.

In a continuous environment, because cells have more degrees of freedom, functional specialization can be added: some cells can react differently to physical stress (compression, shocks, etc.) or trigger specific actions such as release proteins in the environment, attach or detach from neighboring cells, etc. Specialization states can be managed with a specialization tree either handwritten (to reduce the search space complexity) or evolved through an evolutionary process.

In such a complex environment, cells need a decision system capable of selecting appropriate actions. A gene regulatory network is perfectly suited for this kind of problem: the temporal aspect of the GRN allows it to naturally integrate the variation of morphogen concentrations sensed by cells over the time. In such a use of the GRN, each cell of the multicellular organism has a copy of genetically identical GRN (only genetically identical since protein concentrations can be different according to the cell state and neighborhood and therefore produce different behaviors). The GRN inputs are the set of morphogens sensed by the cells as well as the parameters from

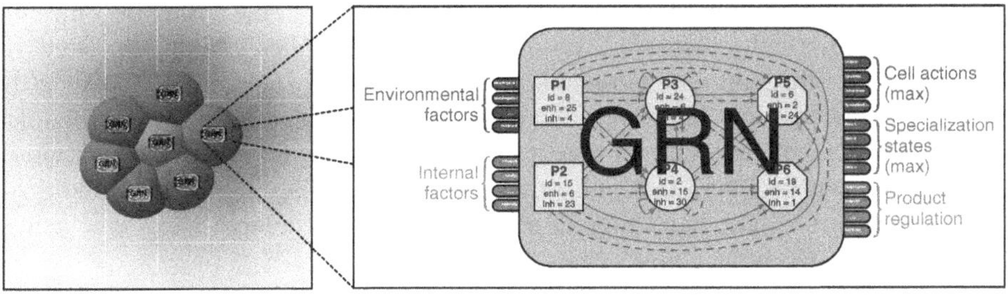

Figure 12.7 Each cell of the organism has a copy of the GRN encoded in the genome. This GRN takes environmental factors (concentration of morphogens, of nutrients, etc.) and internal factors (mechanical pressure, level of energy, current specialization state, etc.) to regulate their behavior, both the cell lifecycle actions (division, quiescence, apoptosis, specialization state) and chemical product regulation (production of energy, diffusion of morphogens and ambrosia, etc.).

current cell state such as the energy level of the cell, its current differentiation state, etc (see Figure 12.7). Theses inputs can vary according to the specialization state of the cell. The GRN outputs set contains at least the four main actions from the real cell cycle (division, quiescence, specialization and apoptosis). When time comes to decide between one of these four outputs, the action associated with the output protein with the higher concentration is selected. Other output proteins can be added to control morphogen production, specific output from the current specialization state, etc.

To produce virtual organisms with emergent properties, the gene regulatory network must be evolved as presented in the previous section. In this problem, the GRN genome usually starts with inputs first, followed by outputs and ends with regulatory proteins. Each input is associated with a specific sensor (either morphogen or cell state) and each output is associated with a specific cell actuator (cell action, morphogen production, etc.). The genetic algorithm evolves the connectivity between these input and output proteins with regulatory proteins as well as the internal network structure of regulatory proteins (number of proteins and affinities between them). The most complex part of the evolution of this kind of system is the fitness function. It has to be carefully designed to reach the general purpose of the virtual organism (proliferate, use a particular resource in the environment, etc.) without reducing the creativity of the developmental process. In order to do so, an idea is to complexify the environment instead of creating a complex fitness function. By doing so, the creativity of the growth process is left to evolution instead of being controlled by the fitness function. In our opinion, evaluating this process can reduce the capacity of the system to produce unexpected solution and lead evolution to propose a solution "engineered" by the fitness function.

With this aim in mind, we have run some experiments with virtual cells, acting in a 2D continuous environment and controlled by the GRN model presented in this chapter. A mass-spring-damper system simulates cellular mechanics: cells are the mass of the mechanical model and they are all linked together by a spring and its damper (that reduces oscillations). This method, broadly used in the literature, allows a realistic clustering of cells with a limited computational effort. To improve this system, cells can unlink themselves when the pressure on a spring becomes too high: that allows cell clusters to split. However, this implies adding adhesions (to aggregate two clusters) and a collision system. In this model, in order to keep the computational cost low, adhesions and collisions are simulated with two additional springs added when two cells get closer: adhesion springs tend to bring the cells closer and collision springs tend to repulse them away. By setting up adhesion and these springs' properties and their range of actions, this creates a large set of mechanical interactions between the cells.

The environment includes a simplistic chemical simulator that provides the cells an energy chain based on nutrients that diffuses in the environment. Cells cannot control these nutrients: they can only absorb them from their local neighborhood. However, cells can transform nutrients into a controllable protein, named *ambrosia*. Ambrosia can either be stored for later use, diffused to neighboring cells (direct transmission, not through the environment) or transformed into energy which can be used to trigger actions (divide, specialize, etc.). This very simple chemical system requires cells to learn how to manage such an energy chain and provides different specialization strategies to the cells by managing the quantity of cells in the organism specialized to produce ambrosia from nutrient, to store this ambrosia more or less accurately, etc. More details about this model can be found in Ref. [10].

12.4.1 Resisting to Extern Aggressions

To validate this model, we have run a simple experiment where cells are attacked by harmful particles present in the environment. The aim of cells in this unwelcoming environment is to organize themselves into a structured multicellular organism so that they can survive as long as possible. Therefore, they have to specialize to one of the two following different types:

- *nutritive cells* can absorb nutrients and transform them very efficiently into ambrosia but they are inefficient at resisting the harmful particles,
- *defensive cells* absorb particles efficiently (they destroy up to 10 particles before dying) but cannot produce ambrosia.

The idea behind this experiment is the necessity for cells to collaborate and find a aggregation strategy in order to survive in the environment. To do so, each cell of the organism is controlled by its own gene regulatory network: cells use their GRN to decide which action to trigger (divide, specialize, wait, or die) as well as their level of production of energy, ambrosia and morphogens, their level of diffusion of ambrosia, their level of absorption of nutrients. To decide, they need information (translated to input concentrations for the GRN): they sense information from the environment such as morphogens produced by other cells from the organism and information from their internal state (level of energy, level of ambrosia, etc.). Figure 12.7 depicts how cells use their GRN to regulate their behavior. Another key idea of the model is to avoid the design of a complex fitness function. Therefore, the gene regulatory

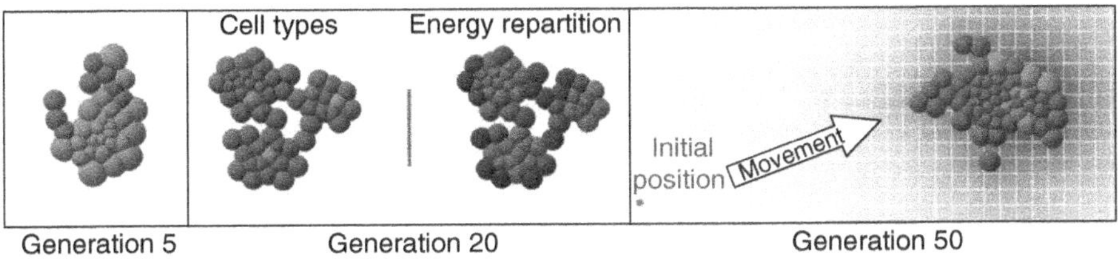

Figure 12.8 Three developmental stages emerge from evolution. During the early stage of evolutions (generation 5), the organisms are not organized: they produce many nutritive cells with no protections. Then (generation 20), they organize into clusters in which nutritive cells are surounded by defesive ones. Later in the evolutionary process (generation 20), the start moving out of the center of the environment, where particles are more concentrated. The movement emerges with an asymetrical proliferation of the cells, directed by a morphogen gradient created by dying cells.

network is evolved with the aim to keep the organism alive as long as possible: the fitness function (to be maximized) is the number of steps the organism survives in the environment from the moment the first cell is added to the moment the last cell dies.

Results show a global increase in the creatures' lifetime over generations. Three key developmental stages (depicted on Figure 12.8) emerge from evolution:

1. *Constant renewing*. At the beginning of the evolution, cells tend to constantly divide and specialize into defensive cells. This constant renewing makes up for defensive cells dying because of the particle aggressions and also because they receive less ambrosia when they are far from nutritive cells.
2. *Clusters*. After a few generations, cells start to organize themselves in fuzzy clusters where a few nutritive cells are surrounded by protective cells. This strategy seems to be an efficient way to use the specialization states: energy providing cells, which are vulnerable, stay protected from particles by feeding a shell of protective cells.
3. *Shifting from the center*. In this particular environment, we observe that, over the dozen different evolution runs we made, a special behavior tends to emerge: *movement*. Cell clusters tend to avoid the center of the environment which is more dangerous than the outskirts because of the concentric pattern formed by the particles. They do so by using an unbalanced division ratio: a gradient of morphogen is produced by the cell, which modifies their behaviors. It causes the cells with lower concentration of morphogen to divide more. This moves the whole organism out of the center of the environment. This behavior is depicted by Figure 12.9.

12.4.2 Resisting to Aggression and Starvation

The environment can be complexified in various ways. For example, cells can be subject to nutrients starvation. Nutrient can only be available in very low quantity from time to time in a specific area of the environment. In this case, a new cell state

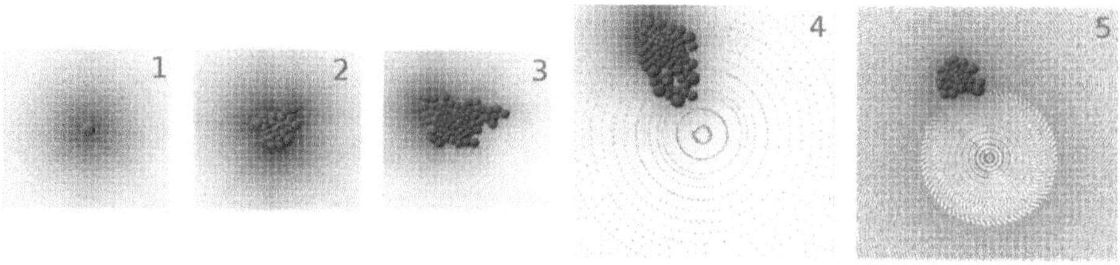

Figure 12.9 Successive steps of the development of one of the best individuals. Nutritive cells (in orange) surround themselves with a shield of protective cells (in blue) which they feed. Movement toward the outside of the environment can also be noticed (in the direction of the magenta morphogen gradient). Harmful particles become intentionally denser and denser until no organism can possibly survive. This allows for a constant constraint increase throughout the simulation. Reprinted with permission from Ref. [10], copyright 2014 The MIT Press.

can be imagined: in addition to nutritive and defensive states, *storage cells* store ambrosia very efficiently in comparison to the two other states but are bad to produce ambrosia and resist to particles.

With the same previous fitness (time of the organism survival), the best GRN we obtained was able to complexify the cell behaviors in order to maintain a global organism that is able to both survive starvation and resist particles. Figure 12.10 presents the best organism at different stages of its development. First, lots of nutritive cells are created from the initial cell. Secondly, a shield of protective cells starts to appear. At this stage of the simulation, nutrients are still abundant and the main concern for the organism is to protect these nutritive cells from particles. Then, as nutrient availability starts to become a problem, most organisms start specializing

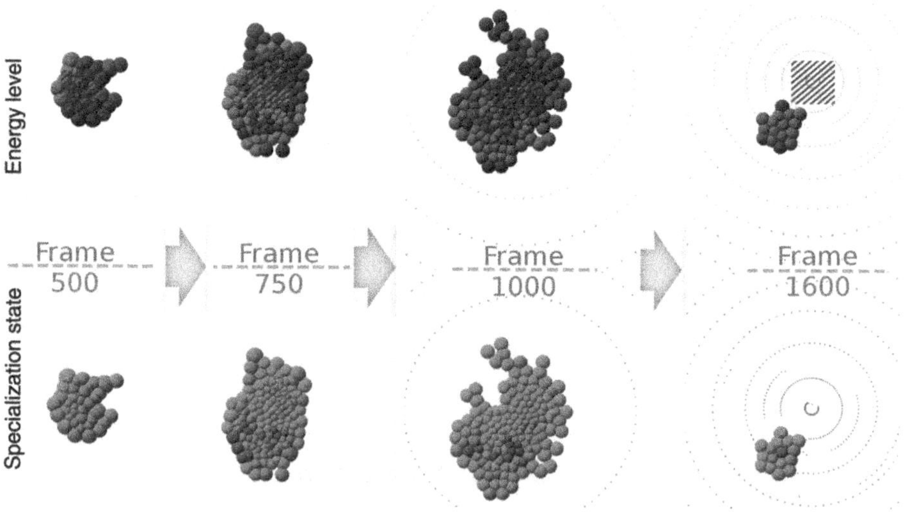

Figure 12.10 Development of the best organism experiment, showing a four-step developmental strategy. Nutritive cells are orange, defensives are green and storage ones are blue. Striped area indicates where nutrients appear. The greener the cells, the higher their energy.

some of their nutritive cells into storage ones. Storage cells are positioned away from the center of the circle of particles, but the organism maintains a network of nutritive cells surrounded by defensives near the center, where nutrients still appear.

We can see this behavior between frame 750 and 1000 of Figure 12.10: nutritive cells that are waiting on the extremity of the organism specialize into storage cells in anticipation of starvation. Finally, with the simulation moving forward, regeneration of nutrients become extremely rare, and the organisms shrink to a tiny cluster composed of a few storage cells diffusing the ambrosia they collected to a shield of defensive cells.

12.5 DRIVING A VIRTUAL CAR

In the previous problem, we used a GRN to control virtual cells, which is a very direct natural way to use it. In this second application, the GRN is used as a virtual car controller, directly plugged to the car sensors and actuators. Indeed, we show that the GRN can be seen as any kind of computational controller: it computes inputs provided by an agent and returns values for the agent's effectors. In this section, the aim of the GRN is to drive a simulated car on complex tracks, with various kinds of turns and with different surfaces (asphalt, sand, rock, etc.). To do so, we used a fully simulated car racing simulator, TORCS [1]. It provides full car physics and an easy way to connect any kind of artificial intelligence to the car sensors and effectors. By using a GRN to drive a car in this simulator, we will see that the GRN can generate a generalized adaptive behavior with the very interesting property of being naturally resistant to noise.

In this particular application, another aim is to keep the connection between the GRN and the car sensors and actuators as simple as possible: the GRN should be able to handle the reactivity necessary to drive a car, the possible sensor noise and unexpected situations. The car simulator provides 18 track sensors as well as many other sensors such as car fuel, race position, motor speed, distance to opponents, etc. However, all of the sensors are not required to drive the car. They are more useful for race strategies, which is not the aim of this work. Reducing the number of inputs directly reduces the complexity of the GRN optimization. Therefore, only a subset of sensors provided by the TORCS simulator are used:

- nine track sensors that provide the distance to the track border in nine different directions,
- longitudinal speed and transversal speed of the car.

Figure 12.11 represents the sensors used by the GRN to drive the car. Before being computed by the GRN, each sensor value is normalized to [0, 1] with the following formula:

$$norm(v(s)) = \frac{v(s) - min_s}{max_s - min_s} \tag{12.5}$$

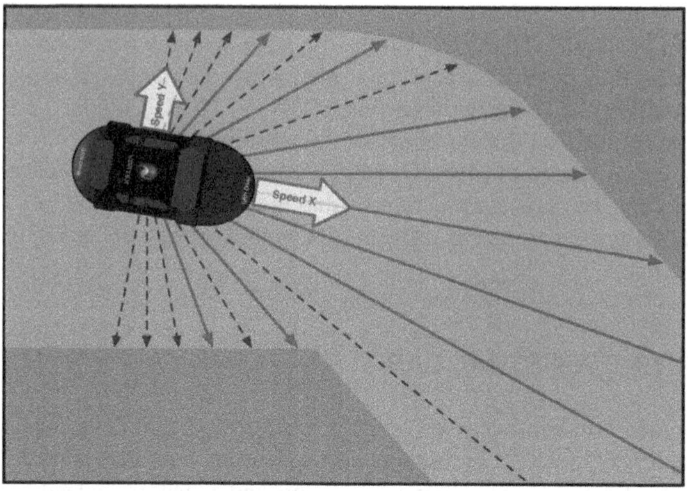

Figure 12.11 Sensors of the car connected to the GRN. Red plain arrows are used to track sensors whereas gray dashed ones are track sensors also available in the simulator but not used by the GRN. The plain arrows *Speed X* and *Speed Y* are respectively the longitudinal and the transversal car speeds.

where $v(s)$ is the value of sensor s to normalize, min_s is the minimum value of the sensor and max_s is the maximum value of the sensor.

Once the concentrations of the input proteins updated, the GRN's dynamics are run one timestep in order to propagate concentration modifications to the whole network. The output protein concentrations are then used to regulate the car actuators. Four output proteins are necessary: two proteins o_l and o_r for steering (left and right), one protein o_a for the accelerator and one o_b for the brake. The final values provided to the car simulator are computed as follows:

$$steer = \frac{c(o_l) - c(o_r)}{c(o_l) + c(o_r)} \tag{12.6}$$

$$accel = \begin{cases} 0 & \text{if } ab <= 0 \\ ab & \text{otherwise} \end{cases} \tag{12.7}$$

$$brake = \begin{cases} -ab & \text{if } ab <= 0 \\ 0 & \text{otherwise} \end{cases} \tag{12.8}$$

$$\text{with } ab = \frac{c(o_a) - c(o_b)}{c(o_a) + c(o_b)},$$

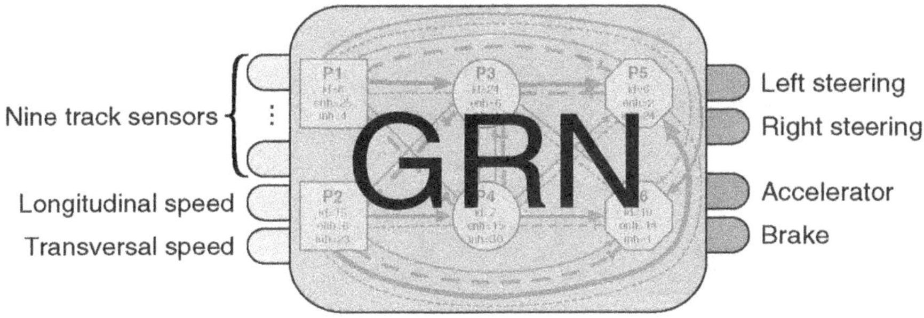

Figure 12.12 The GRN uses nine track sensors and longitudinal and transversal speeds to compute the car steering, acceleration, and brake.

where *steer* is the final steering value of the car in $[-1, 1]$, *accel* is the final acceleration value in $[0, 1]$, *brake* is the final brake value in $[0, 1]$, $c(o_*)$ is the concentration of the output protein o_*. Figure 12.12 shows the connection of the GRN to the virtual car.

To simplify the problem, the gear value is handwritten as it is a very simple script to develop; when the motor is over a given threshold that depends of the current gear, the driver shifts up. Under another threshold, the driver shifts down.

Figure 12.13 presents how the protein chromosome of the GRN is ordered: the first 11 proteins are input proteins, the next four are output proteins and the remaining ones are regulatory proteins. Whereas the input and the output proteins can be crossed over or mutated, they cannot be deleted.

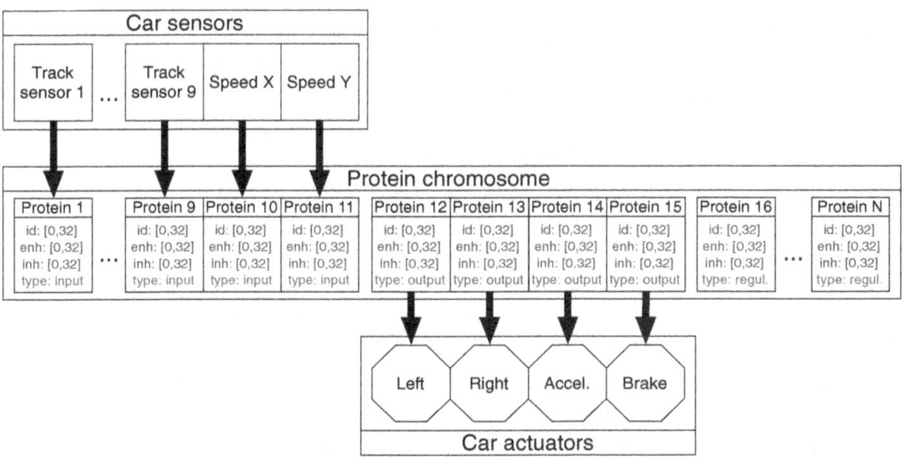

Figure 12.13 Organization of the protein chromosome and link to the car sensors and actuators: protein tags (in red) are evolved by the genetic algorithm whereas protein types (in green) are fixed and always plugged to the same car sensors (for input proteins) and the same car actuators (for output proteins).

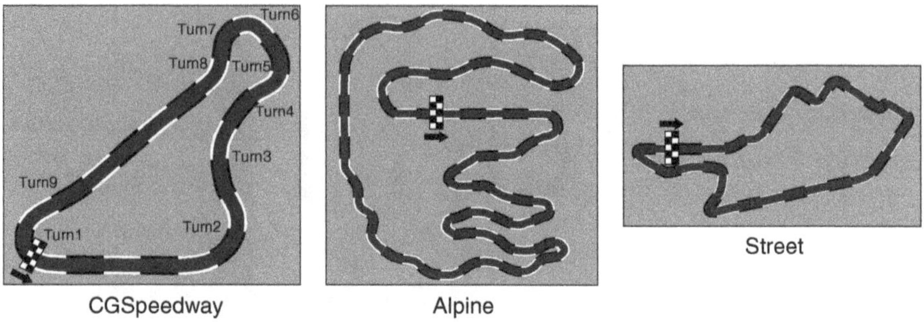

Figure 12.14 Tracks used to train the GRN. All of them are provided by TORCS.

In order to optimize the GRN to drive a car, we use an incremental evolution in three stages.[3] During these stages, the same parameters have been used to tune the genetic algorithm. Only the fitness function is modified. The genetic algorithm parameters are

- population size: 500,
- mutation rate: 15%,
- crossover rate: 75%,
- GRN Size: [4, 20] regulatory proteins plus inputs and outputs.

The first stage consists of training the GRN to drive as far as possible, with a minimum speed, on one track. We use CGSpeedway, the left-hand side track provided by the car simulator (see Figure 12.14), which is simple with long turns and straight lines. This track is interesting for learning to drive because it is a relatively simple track with long fast turns and with fast straight lines to learn how to steer and to accelerate, and with more difficult short turns to learn how to slow down and to brake. Each GRN is tested on this track for 31 kilometers (about 10 laps) maximum. The simulation is stopped as soon as the car leaves the track or gets damaged (by hitting a rail for example). To ensure the car is driving fast enough, we use a ticket system in which the GRN must cover $500 * nLap$ meters per 1000 simulation steps, where *nLap* is the current lap number. This pushes the GRNs that go far to accelerate. If a GRN cannot reach this goal, the simulation is stopped. When the simulation ends, the fitness function is given by the distance covered by the GRN along the central line of the track. If a GRN has traveled all 31 kilometers, a bonus is added. The bonus is inversely proportional to the number of simulation steps needed to complete the race (the faster, the better).

[3] Videos of this evolution are available online: http://www.irit.fr/~Sylvain.Cussat-Blanc/GRNDriver/index_en.php.

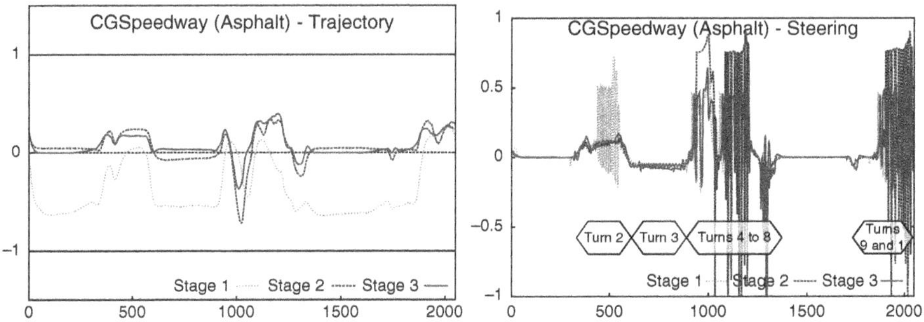

Figure 12.15 Evolution of the car behavior during the three different stages of evolution. The left-hand side plot represents the car position on the track along the distance from the start line: 0 means the car is on the track centerline, −1 means the car is on the right edge of the track and 1 means the car is on the left one. The right-hand side plot is the steering output value along the distance from start. −1 means the steer is fully rotated to the right and 1 means fully rotated to the left.

After evolving GRN with a genetic algorithm, we obtain a GRN that is able to drive endlessly on this track. It is a safe driver that regulates its speed so that it can go through all the turns of this track. However, we observed that this GRN is not able to drive on other tracks. In fact, as presented on Figure 12.15, the car is slightly shifted to the right side of the track. That might explain why the GRN cannot generalize its driving: most of the turns on the training track are left-handed. Thus, staying on the right side is better. However, on tracks with hard right turns, this position can be dangerous, the angle for right turns being closed. Moreover, some significant oscillations on the steering can be noticed. Even if they do not imply oscillations on the car track position, this behavior is unwanted and can be harmful in a car race. The aim of the next evolution stages is to correct these defects.

From these observations, we evolved the GRN to safely cover all possible tracks, with all possible kinds of turns. With this aim in mind, we evolved the previous GRN a second time with the same evolutionary process but on three different tracks. The tracks used are CGSpeedway (in order not to lose the driving capacity of the previous GRN), Alpine and Street, whose layouts are presented in Figure 12.14. The fitness function consists of summing the fitness value defined during the first evolution stage successively applied to the three tracks.

At the end of this evolution, the best GRN is able to drive on every possible track. It drives very safely, going at a suitable speed to go through every kind of turn and braking when it detects a turn. However, the best GRN has an oscillatory behavior and is slightly shifted to the right-hand side of the track. Whereas oscillatory behaviors are common in gene regulatory networks, both these issues could be harmful during a car race. This oscillation can be observed in Figure 12.15 where the trajectories and the steering of the car are plotted during the second lap on one of learning tracks (CGSpeedway). This oscillatory behavior can still be noticed on the steering plot in which the blue curve, which represents the second stage of evolution, strongly oscillates. The result is some parasitic behavior of the car on the trajectory, especially at the end of the turns. The final cleaning stage aims to reduce these parasitic behaviors.

To minimize oscillatory behaviors, we re-optimize the best GRN one last time. This time we add to the fitness function another test case that penalizes continuous oscillations of the car on straight lines and long turns or fast multiple steering changes from full right to full left. As with the ticket system or the damage control used in the previous fitness functions, we simply stop the evaluation if we detect oscillatory behaviors.

The green line on Figure 12.15 shows the steering values of the best GRN on CGSpeedway track at the end of this evolution stage. The steering spikiness of the previous evolution (blue curve) that is visible in the first two fast curves is smoothed and the steering does not oscillate anymore from full right to full left in the track section from turn 4 to turn 8 and from turn 9 to turn 1 (see Figure 12.14).

It can be noted that this last evolution stage reinforces the generalization stage by improving the central position of the car. The GRN is also faster than before because the oscillations slow the car down. These multiple evolution stages were then strongly efficient to produce a GRN able to drive the car efficiently on most of the tracks. Another property emerges from the evolved GRN: it is also able to adapt to various kinds of track surfaces such as rock and sand, with no reoptimization. The GRN is also resistant to sensor noise. This is an uncommon result since most other approaches used to control a car are based on noise reduction systems and, even with these systems, are still impacted by the noise (more details can be found in Ref. [22]). Both adaptativity and noise resistance are crucial properties for agent control since they allows the agent to handle uncertainty in the environment. Moreover, these are necessary to bridge the reality gap when such a system is used in real robots.

12.6 REGULATING BEHAVIORS

In this last example of application of the gene regulatory network, instead of connecting the GRN directly to the agent sensors and actuators, we propose to use the GRN to regulate high-level behaviors in each agent of a team. For this purpose, we have designed a defense/interception mission in which a team of agents has to defend a target and intercept incoming threats. Multiple defending agents have to coordinate their behavior with local information only (e.g., distance to the target to defend, to the next threat, to teammates, etc.) in order to intercept as much as possible threats coming sequentially or simultaneously from random position. The objective of the threats is to destroy the target. Because this study focuses on the defending agent behaviors, the threat behavior has been kept very simple: they are driving to the base at various speeds, following random checkpoints.

The behavioral architecture is based on a three-layer structure: the top layer, named *behavior* layer, is a set of high-level behaviors that provides specialized functionalities to the agent; the intermediate layer, named *control* layer, is an aggregation layer, where all decisions taken by the high-level modules are regulated to generate the global behavior of the agent; the bottom layer, named *actuation* layer, is used to actuate the agent in the environment and to apply the actions generated by the upper layer.

Applied to our defense problem, defending agents have access to four driving behaviors (top layer) which are: defense (drives to the target to defend), interception (drives to the threat), scattering (drive away from defending group gravity center), and grouping (drive to the gravity center). This architecture uses the GRN described in Section 12.2 as behavior aggregator. In this approach, instead of controlling agents at a very low level (e.g., control the wheel rotation speed), the GRN weights each driving behavior by the mean of one output protein per driving behavior.

To generate the driving direction, the GRN regulates the defending, intercepting, grouping, and scattering behaviors by weighting the vectors generated by these behaviors. The final movement $\vec{F}_a$ of agent a is given by $\vec{F}_a = w_1\vec{D}_a + w_2\vec{I}_a + w_3\vec{G}_a + w_4\vec{S}_a$, where $\vec{D}_a$ (resp. $\vec{I}_a$, $\vec{G}_a$ and $\vec{S}_a$) is the vector provided by the defense (resp. interception, grouping and scattering) behavior. Each weigth w_i is calculated with the concentration of four output proteins: $w_i = \frac{c_i}{\sum_{j=1}^{4} c_j}$ where c_i is the concentration of the protein associated with the weight w_i. Figure 12.16 depicts the use of a GRN to regulate defending agents' driving behaviors.

To evaluate this approach, we imagine a mission in which the agents have to coordinate their strategies in order to defend a target against a sequence of incoming threats. The threats are incoming sequentially in the environment from random positions. The complexity of the task also increases with the arrival of new threats: the speed of each new threat is increased so that it gets harder to intercept. Therefore, the main aim of this experiment is to coordinate the defender team in order to intercept as many threats as possible and produce global strategies in order to intercept fast threats.

To tackle this problem, the GRN uses the following inputs: a defender–enemy distance morphogen, the defender/threat speed ratio, an enemy–target distance morphogen, an agent–allies distance morphogen, the team interception level (average level of interception of all the defending agents) and the position of the agent in the

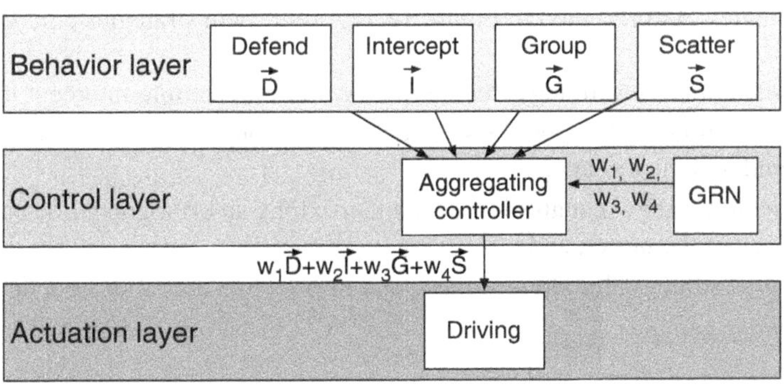

Figure 12.16 Integration of the high-level behaviors and of the GRN to the three-layer behavioral architecture.

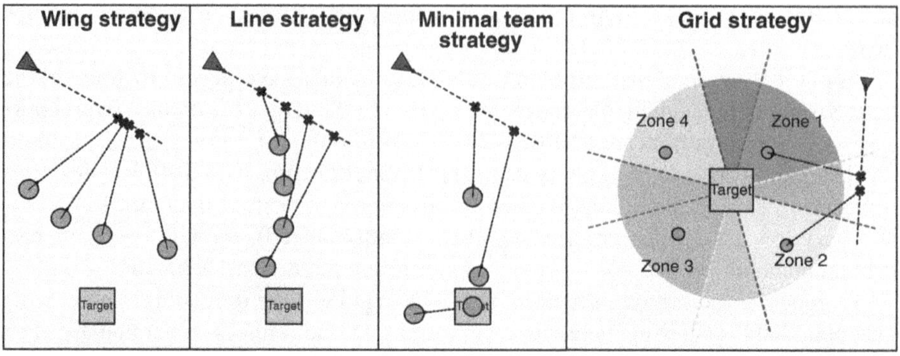

Figure 12.17 Strategies that emerge from independent training of the GRN. From left to right, the first one consists in creating a "wing" between the target to defend and the threat. In the second strategy, defenders are distributed on the line between the target and the threat. The third strategy consists of sending one defender at the time to intercept the threat and, if it fails, other defenders are available to be sent in back up. Finally, the last strategy consists of dividing the space into sectors and have one defender in each sector. These sectors are dynamical: they change according to the threat position, dangerousness and direction. This improves the capacity of the defenders to intercept the threats.

team according to their distances to the threat. These inputs allow each defender to evaluate their position in the team, their position relatively to the target to defend and to the threats and the dangerousness of the current threat. They provide the concentration of the GRN's input proteins.

A standard genetic algorithm evolves the GRN. Each individual is evaluated on three random scenarios with a sequence of 100 threats. Threats' speeds linearly increase from 70% to 120% of defender speeds. The fitness of a GRN is the sum of the fitness obtained on three random scenarios. Each scenario is evaluated by a fitness function that takes into account the lifetime of the threat before interception. This function is designed to promote controllers that intercept threats as far as possible from the target to defend.

The genetic algorithm is run 48 independent times in order to produce diverse team strategies. As presented in Figure 12.17,[4] four main strategies emerge:

- the *minimal team* strategy that builds the minimal team to intercept the threats;
- the *line* strategy that positions the defenders along a line between the target to defend and the threat;
- the *wing* strategy that aligns the defenders along an orthogonal line between the threat and the target; and
- the *grid* strategy that divides the space into sectors and allocate a defender per sector.

[4]Videos are available online: www.irit.fr/~Sylvain.Cussat-Blanc/Behavior_regulation/index_en.php

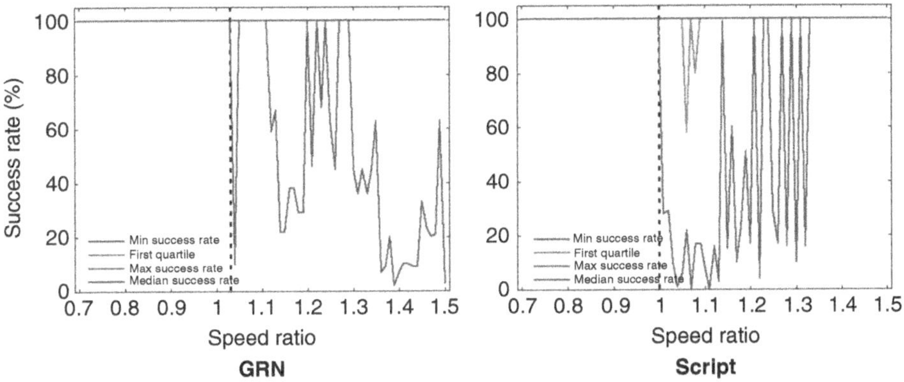

Figure 12.18 Comparison of the best GRN (left-hand side) approach and the scripted approach (right-hand side) on 30 random scenarios composed of 100 threats to intercept with various fixed speed ratios (abscissa). The minimum success rates (worst-case scenario, in red) drops faster with the scripted approach than with the GRN (vertical dashed line): the GRN adapts better to higher speed ratios than the scripted approach.

We have compared the GRN-based approach to a script provided by defense experts. The script allocates the closest defender to the threat between the target to defend and the threat as an *interceptor*, driving it to the interception point as fast as possible. The second and third closest, named the *wingers*, are positioned on both sides of the interceptor at a distance equal to the distance separating the threat and the interception point of the interceptor. These defenders produce a wing-like formation. The fourth defender, called the *protector*, is positioned between the threat and the target to defend. If the team is composed of more than four defenders, other teammates are kept on the target to defend. If the team is smaller, the wingers are deleted first before the protector. Figure 12.18 compares the results of the GRN-based approach against the scripted one. The comparison consists of 30 random scenarios. A scenario is initialized with 100 threats to intercept with a four-defender team. The threats are randomly positioned in the environment and have a random checkpoint list to follow before colliding on their target. Each scenario is played with various speed ratios (the higher the ratio, the faster the threats in comparison to the defenders) to evaluate the capacity of the approaches to handle simple and complex situations. For both approaches, we provide the maximum (in blue), the minimum (in red), the median (in purple), and the first quartile (in green) of the success rates on the pool of scenarios. Because the problem is to defend a target, the worst-case scenario is the most important measure. With this measure, we can observe that the GRN is doing better than the script: the minimum success rate is kept to 100% with higher speed ratios with the GRN than with the script. Moreover, the first quartile is dropping down with the scripted approach with a speed ratio between 1.05 and 1.10 whereas the GRN approach is able to keep it to 100% for all speed ratios. This means the GRN is better at adapting to faster threats than the scripted approach due to the capacity of the GRN to adapt to unknown situations.

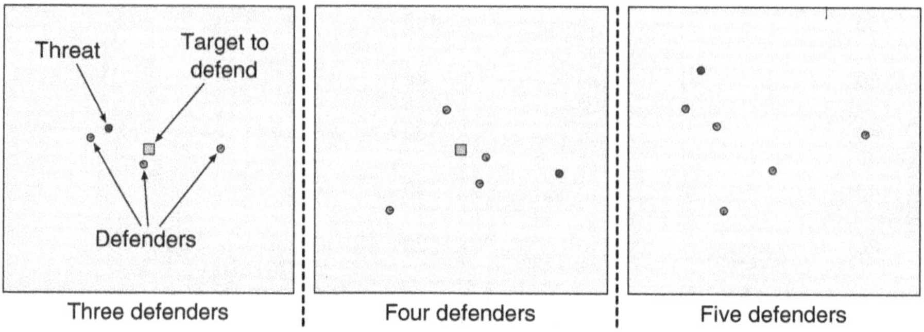

Figure 12.19 Example of system adaptation to the team size: the same GRN is used to handle teams composed of three, four, or five defenders. The team covers the environment according to the team size.

Another property of the GRN approach is its ability to adapt the strategy to the size of the team with no further learning. Figure 12.19 shows this property on the grid strategies where defenders modify their positions in the environment according to allies positions: the switch from a square to a triangle or a pentagon. Moreover, the structure is twisted according to the threat position. Comparable observations can be made on the wing, the line, and the minimal team strategies. This generalization on the number of defenders is inherent to the system because all the data used by a defender to take its decision (via the GRN) are local and relative. Therefore, defenders can be added or removed from the system without impacting the decision process. This team size adaptation can even be done during the simulation without perturbing the system.

12.7 CONCLUSION

In this chapter, we have presented three applications of a computational model of a gene regulatory network. We showed how to directly connect the network's inputs and outputs to the agent sensors and actuators in the artificial embryogenesis and the car driving problems. Then, we showed that the GRN can also be used at a higher level, to regulate agent behaviors in a cooperation mission. In all these experiments, crucial properties for virtual agents emerge from the system: the GRN can generate behaviors both resistant to noise and adaptive to unknown environment.

The domain of applications of gene regulatory networks is extremely wide. It can be easily applied to many agent-based problems. Indeed, we showed in this chapter that the GRN is able to generate very complex behaviors. However, the complexity can be double-edged: evolving gene regulatory networks with standard evolutionary algorithms can be difficult. Developing an evolutionary algorithm dedicated to evolve gene regulatory networks might help to handle this complexity, as NEAT was meant to evolve neural networks [23]. Currently, most evolutionary algorithms used to evolve gene regulatory networks are mainly based on mutation and few of them use crossover. Actually, crossover usually induces too many modifications in the network structure to be efficient. Therefore, finding a crossover that conserves the global structure of the network would improve the convergence of the evolutionary algorithm. As used in NEAT, historical markers that identify key genes, present in the network for a long time, could be a first line of research since similar mechanisms exist in nature. That could help the GRN to address more complex problems making it capable of exploiting more inputs to regulate more outputs.

REFERENCES

1. Torcs: The open source racing car simulator. http://torcs.sourceforge.net
2. W. Banzhaf. Artificial regulatory networks and genetic programming. In R. L. Riolo and B. Worzel, editors, *Genetic Programming Theory and Practice.* Springer, Chapter 4, pages 43–62, 2003.
3. A. Chavoya and Y. Duthen. A cell pattern generation model based on an extended artificial regulatory network. *BioSystems*, 1–2:95–101, 2008.
4. S. Cussat-Blanc, N. Bredeche, H. Luga, Y. Duthen, and M. Schoenauer. Artificial gene regulatory networks and spatial computation: a case study. In *Proceedings of the European Conference on Artificial Life (ECAL'11).* MIT Press, Cambridge, MA, 2011.
5. S. Cussat-Blanc and J. Pollack. A cell-based developmental model to generate robot morphologies. In *Proceedings of the 14th Annual Conference on Genetic and Evolutionary Computation.* ACM New York, NY, 2012.
6. S. Cussat-Blanc and J. Pollack. Using pictures to visualize the complexity of gene regulatory networks. In *Artificial Life*, volume 13. MIT Press, pages 491–498, 2012.
7. S. Cussat-Blanc, S. Sanchez, and Y. Duthen. Simultaneous cooperative and conflicting behaviors handled by a gene regulatory network. In *2012 IEEE Congress on Evolutionary Computation (CEC)*, pages 1–8. IEEE, 2012.
8. H. de Garis. Artificial embryology and cellular differentiation. In Peter J. Bentley, editor, *Evolutionary Design by Computers.* Morgan Kaufmann, pages 281–295, 1999.
9. F. Dellaert and R. Beer. Toward an evolvable model of development for autonomous agent synthesis. In *Artificial Life IV*, Cambridge, MA, 1994. MIT press.
10. J. Disset, S. Cussat-Blanc, and Y. Duthen. Self-organization of symbiotic multicellular structures. In *ALIFE 14: The Fourteenth Conference on the Synthesis and Simulation of Living Systems*, volume 14. MIT Press, pages 3–5.
11. R. Doursat. Organically grown architectures: creating decentralized, autonomous systems by embryomorphic engineering. *Organic Computing.* Springer, pages 167–200, 2008.
12. P. Eggenberger Hotz. Combining developmental processes and their physics in an artificial evolutionary system to evolve shapes. *On Growth, Form and Computers.* Academic Press, page 302, 2003.
13. N. Flann, J. Hu, M. Bansal, V. Patel, and G. Podgorski. Biological development of cell patterns: characterizing the space of cell chemistry genetic regulatory networks. *Lecture Notes in Computer Science*, 3630:57, 2005.
14. M. Gardner. The fantastic combinations of John Conway's new solitaire game life. *Scientific American*, 223:120–123, 1970.

15. Hongliang Guo, Yan Meng, and Yaochu Jin. A cellular mechanism for multi-robot construction via evolutionary multi-objective optimization of a gene regulatory network. *BioSystems*, 98(3):193–203, 2009.

16. M. Joachimczak and B. Wróbel. Evolving gene regulatory networks for real time control of foraging behaviours. In *Proceedings of the 12th International Conference on Artificial Life*. MIT Press, 2010.

17. M. Joachimczak and B. Wróbel. Evolution of the morphology and patterning of artificial embryos: scaling the tricolour problem to the third dimension. In *Advances in Artificial Life. Darwin Meets von Neumann*, pages 35–43. Springer, 2011.

18. J. Knabe, M. Schilstra, and C.L. Nehaniv. Evolution and morphogenesis of differentiated multicellular organisms: autonomously generated diffusion gradients for positional information. *Artificial Life XI*, 11:321, 2008.

19. R.P. Lifton, M.L. Goldberg, R.W. Karp, and D.S. Hogness. The organization of the histone genes in drosophila melanogaster: functional and evolutionary implications. In *Cold Spring Harbor Symposia on Quantitative Biology*, volume 42, pages 1047–1051. Cold Spring Harbor Laboratory Press, 1978.

20. M. Nicolau, M. Schoenauer, and W. Banzhaf. Evolving genes to balance a pole. In A. I. Esparcia-Alcazar, A. Ekart, S. Silva, S. Dignum, and A. Sima Uyar, editors, *Proceedings of the 13th European Conference on Genetic Programming, EuroGP 2010*, volume 6021 of *LNCS*. LNCS, pages 196–207, 2010.

21. T. Reil. Dynamics of gene expression in an artificial genome-implications for biological and artificial ontogeny. *Lecture Notes in Computer Science*, 457–466, 1999.

22. S. Sanchez and S. Cussat-Blanc. Gene regulated car driving: using a gene regulatory network to drive a virtual car. *Genetic Programming and Evolvable Machines*, 15(4):477–511, 2014.

23. K. O. Stanley and R. Miikkulainen. Evolving Neural Networks through Augmenting Topologies. *Evolutionary Computation*, 10(2):99–127, 2002.

24. J. Von Neumann and A.W. Burks. *Theory of Self-reproducing Automata*. University of Illinois Press, 1966.

25. L. Wolpert. Positional information and the spatial pattern of cellular differentiation. *Journal of Theoretical Biology*, 25(1):1, 1969.

13

EVOLVING H-GRNs FOR MORPHOGENETIC ADAPTIVE PATTERN FORMATION OF SWARM ROBOTS

Hyondong Oh

Loughborough University, Loughborough, UK

Yaochu Jin

Department of Computing, University of Surrey, Guildford, UK

13.1 INTRODUCTION

A swarm robotic system is a multi-robot system consisting of large numbers of autonomous but small-sized simple mobile robots, each having often limited communication and computation capabilities. In recent years, due to the rapid advancements of robotics technology and its inherent attractive features such as cost efficiency, robustness to a single robot failure, and high adaptability, considerable attention has been paid to develop and to efficiently operate these swarm robotic systems. Shape construction and pattern formation using swarm robots are of special interest since they provide a basic functionality to accomplish various collective missions such as search and rescue, environment mapping and monitoring, efficient deployment of sensor networks, and object transportation. Pattern formation in swarm robotic

Evolutionary Computation in Gene Regulatory Network Research, First Edition.
Edited by Hitoshi Iba and Nasimul Noman.

systems is defined as the coordination of a group of robots to maintain a formation with a certain shape of either a predefined pattern or adaptively formed pattern in a self-organized way with the local interaction between robots and with the environment. Swarm pattern formation algorithms have been widely studied, and can be largely divided into four categories: behavior-based control [1, 2], leader-follower and virtual structures [3,4], potential field [5,6], and bio-inspired approaches [7–10], as suggested in Ref. [11].

One significant limitation of most existing algorithms is that the external target shape needs to be predefined and they form only a small number of simple geometric shapes, which may be inadequate for dealing with unknown complex environments. To alleviate this issue, Belta and Kumar [12] proposed a formation control method based on the abstraction of the team to a small dimensional manifold. In this work, although target shapes can be adaptable to environmental changes by changing the shape and size, only simple shapes are considered, and the final formation results in the loose distribution of robots inside the shape. Cheah et al. have introduced a region-based shape control for robot swarms based on potential field, where each robot stays within a moving region while maintaining a minimum distance from each other [13, 14]. They considered various shapes using a combination of inequality functions; however, the complexity of the target shapes is still limited, and the target shape must be predefined. Flocking algorithms might also be relevant to pattern formation allowing robots to move as a single entity, often resulting in a certain form of complex patterns [15–17]. However, these algorithms are typically more concerned on a control architecture than pattern formation itself in a decentralized way using local sensing and interaction. Arbitrary patterns arise as an emergent phenomenon rather than being able to generate task-driven or adaptable complex shapes.

A biologically inspired approach provides us with more flexible ways to generate an adaptive and robust pattern to cope with unknown environmental changes and disturbances. Among others, a morphogenetic approach to pattern formation has been of use due to its direct relevance to biological morphogenesis, which is known to be highly adaptive and robust against real-world disturbances and noises [18, 19]. Morphogenesis refers to the biological process in which cells divide, grow and differentiate, and finally result in the mature morphology under gene regulatory networks (GRNs) and morphogens [11]. Here, GRNs are models of genes and the interaction of gene products that describe the gene expression dynamics [20]. Morphogen gradients, concentration gradients of substances present in the environment, play an important role to trigger the specific gene expression in morphogenesis. Applying this morphogenetic approach to a pattern formation problem, Mamei et al. [21] proposed the use of simple morphogen gradient diffusion to achieve pattern formation, where patterns are polygonal shapes and robots are assumed to be able to pass through each other. Shen et al. [7] and Ikemoto et al. [22] used a Turing's reaction–diffusion model [23], which utilizes the interaction between two morphogens (activator and inhibitor) as well as diffusion into the environment to generate self-organizing patterns. Gene regulatory networks are also adopted for swarm pattern formation by exploiting a reaction–diffusion process among a number of genes in a cell. Taylor et al. [24] proposed a hybrid pattern formation control using GRNs and cellular adhesion molecules for a swarm of spherical robots. A hierarchical GRN (H-GRN) concept is introduced for target entrapment [11] based on the fact that hierarchy in GRNs plays a central role in the evolution of developmental GRNs, resulting in better adaptability and evolvability [25]. Although the H-GRN algorithm generates adaptive patterns according to the position of targets, this pattern is not adaptable to environments as it uses a fixed GRN structure for pattern generation. To address unknown environmental inputs, a network motifs-based GRN (NM-GRN) model is proposed, which evolves predefined network motifs—frequently recurring network patterns—as basic building blocks [26].

Extending previous work on GRNs [11,26] for more adaptable and robust patterns, this chapter first introduces a morphogenetic approach to a target entrapping problem by integrating a two-layer H-GRN with the region-based shape control for swarm robots. The upper layer of the proposed H-GRN is responsible for entrapping pattern generation that provides a morphogen gradient to the lower layer. The lower layer implements a region-based shape control strategy that guides all robots into the desired region with the morphogen gradient from the upper layer, and maintains a minimum distance among robots. Obstacle avoidance is done by a local repulsive control in the lower layer without changing the entrapping pattern. Then, instead of relying on a fixed H-GRN structure, we introduce an evolving network motifs-based H-GRN (EH-GRN) concept for flexibility and adaptability on designing target entrapping patterns. A generic GRN model is proposed with primitive network motifs as basic building blocks for the upper layer, and then an evolutionary algorithm is applied to evolve parameters and the structures. In particular, by introducing obstacles as one of the environmental inputs along with targets, we address the weakness of the pattern of the H-GRN not being adaptable to obstacles; in this work, a target entrapping pattern itself changes as an obstacle approaches toward the swarm in the upper layer. Thus, the proposed EH-GRN enables all the robots in the swarm to perform both of their missions at the same time: entrapping targets and avoiding obstacles. Note that although the lower layer of the EH-GRN is similar to that of the H-GRN, boundary coverage is considered rather than region coverage to show this entrapping pattern adaptation more clearly.

The rest of this chapter is structured as follows. Section 13.2 presents a problem statement including the connection between multi-cellular organisms and multi-robot systems. Section 13.3 introduces an H-GRN model with a region-based shape control strategy, consisting of an upper layer for region generation and a lower layer for robot guidance into the region. Section 13.4 presents an EH-GRN model with an evolving GRN strategy. Conclusions and future work are given in Section 13.5.

13.2 PROBLEM STATEMENT

This chapter considers a dynamic target entrapment problem with obstacle avoidance using a two-layer H-GRN concept for a swarm of robots. Based on the location of targets and obstacles, an adaptive pattern or region is generated by the upper layer of the H-GRN, and the robots are deployed on or into the generated pattern/region by the lower layer GRN dynamics. Similar to the work in Ref. [11], there are organizing robots that can detect at least one target and are responsible for the pattern generation, and non-organizing robots that have not yet detected any target. Non-organizing robots follow their neighboring organizing robots while obtaining information about a target until they detect targets themselves. Patterns used in this study refer to one-dimensional curves embedded in a two-dimensional (2D) workspace, and the region is the area inside the generated closed-form pattern. Targets are the static or dynamic objects in the environment to be entrapped by the robots. Target patterns refer to the patterns generated by the organizing robots that can entrap/surround targets. This target pattern is generated by the H-GRN upper layer based on the positions of the detected target and/or incoming obstacles.

Several assumptions are made to employ the H-GRN for adaptive pattern formation in this study: (i) the robots can localize themselves in a local coordinate system with their onboard sensors and local communication among the robots; (ii) area coverage or search algorithm to find targets is not considered in this study, and all targets in the region of interest can be detected by at least one robot; (iii) organizing robots have the computation capability to generate spline curves; (iv) the movement of robots is much faster than the targets and obstacles so that robots can follow or avoid others; (v) all robots have a limited sensing and communication range; and (vi) communication bandwidth is wide enough to transmit required data (detected target information and robot position) between robots.

As we exploit the GRNs found in multi-cellular organisms to control swarm robots in the lower layer of the H-GRN, a metaphor between cells and robots needs to be established. To this end, each cell is considered as a single robot where protein concentrations of genes in the cell correspond to the position and internal states (representing the analog of the velocity) of robots. Each protein has the following three roles in controlling the robot's behavior: (i) auto-regulation, which regulates the gene expression level that produces the protein, (ii) reaction to a certain morphogen gradient from the environment (in this chapter, from targets to be entrapped), and (iii) diffusion into other cells to avoid collision of the robots. In Section 13.3, we will discuss in detail how to use these GRN functions for a target entrapping problem.

13.3 H-GRN MODEL WITH REGION-BASED SHAPE CONTROL

In this study, a two-layer H-GRN model is combined with region-based shape formation for a swarm of robots to cover the target entrapping region as illustrated in Figure 13.1 [27]. The upper layer of this H-GRN is responsible for an entrapping region generation using local sensory inputs of detected targets. The protein concentration p is governed by the environment input, that is, detected target positions, and it serves as the input of the upper GRN activating g_1, g_2, and g_3. In particular, the concentration of g_3 takes the role of a morphogen gradient to form the desired entrapping region, which will be transmitted to the lower layer. It is assumed that the dynamics of the GRN in the upper layer is activated only in the organizing robots

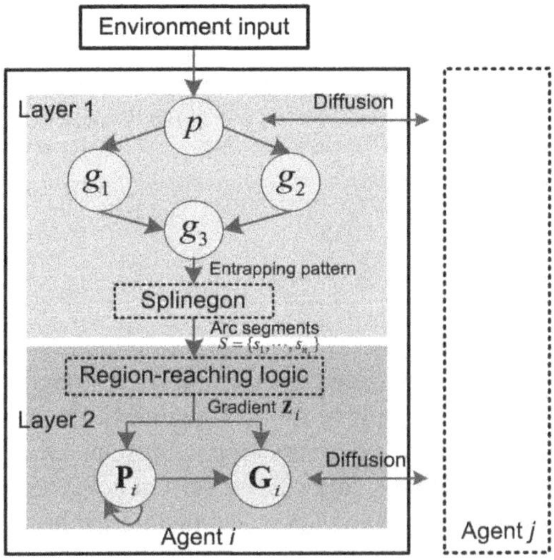

Figure 13.1 Illustration of a two-layer hierarchical gene regulatory network structure for target entrapping. Reprinted with permission from Ref. [27], copyright 2014 Springer Science and Business Media.

that can detect targets, while the non-organizing robots simply follow neighboring organizing robots.

Once the entrapping pattern is generated by the upper layer, it will function as the input to the lower layer to trigger its dynamics. The lower layer is for region-based shape control that guides all robots into the desired region (the area inside the pattern) while maintaining a specified minimum distance among the robots. These functions are implemented by a single GRN where proteins **G** and **P** are used to represent the current positions and internal states of the robots. Thanks to diffusion of proteins **P** and **G**, the neighboring robots can share the target information and maintain a desired distance among robots to avoid robot collision as well as to cover the region as evenly as possible.

13.3.1 Upper Layer: Region Generation

13.3.1.1 Target Entrapping Pattern For generation of the target entrapping pattern, each organizing robot will utilize the following gene regulatory dynamics to generate the concentrations:

$$\frac{dp_j}{dt} = -p_j + \nabla^2 p_j + \gamma_j, \tag{13.1}$$

$$p = \sum_{j=1}^{n_t} p_j, \tag{13.2}$$

$$\frac{dg_1}{dt} = -g_1 + \text{sig}(p, \theta_1, k), \tag{13.3}$$

$$\frac{dg_2}{dt} = -g_2 + [1 - \text{sig}(p, \theta_2, k)], \tag{13.4}$$

$$\frac{dg_3}{dt} = -g_3 + \text{sig}(g_1 + g_2, \theta_3, k), \tag{13.5}$$

$$\text{sig}(x, z, k) = \frac{1}{1 + e^{-k(x-z)}}, \tag{13.6}$$

where γ_j represents the obtained target information (through the sensor or local communications with other organizing robots) that holds a positive constant value in which the target is present in the environment. p_j represents the protein concentration produced by the jth target input γ_j, and p is the sum of concentrations of all detected n_t targets. The integrated protein p will activate the internal protein concentrations. k is a positive constant, which determines the slope of sigmoid function, and g_1, g_2, and g_3 are protein concentrations, where g_3 defines target entrapping pattern. Note that protein g_3 is regulated by both g_1 and g_2, which are regulated by p only within a certain range of its concentration. This leads to a band of circle shapes depending on the thresholds of a sigmoid function θ_1, θ_2, and θ_3, as shown in Figure 13.2. This figure shows the examples of a target entrapping pattern from the protein concentration level of g_3 according to different thresholds of θ_3 with fixed θ_1 and θ_2 values where a target is located at $(x, y) = (10, 10)$ producing the protein concentration p.

13.3.1.2 Splinegon Representation For the generated entrapping pattern to be used in the lower layer of the H-GRN, the contour of the target pattern needs to be extracted. To this end, Splinegons [28] are used to create a set of vertices that are connected by line segments of constant curvature, representing the curved nature of the target pattern as a set of arc segments. This allows us to form shape inequality constraints as arc equations, which will be used for a region-based shape control. A few representative points whose g_3 concentration is higher than a threshold on the pattern are selected as extraction points to describe the target pattern. These points are sorted in an ascending order of the line-of-sight angle (from $-\pi$ to π) with respect to the center position of the targets, in order to generate a smooth and closed polygon around targets. Then, neighboring two points are connected by two segments of constant curvature by introducing an intermediate point with C^1 contact, which represents that the first derivative is continuous, as shown in Figure 13.3. The mathematical details for constructing the Splinegon can be found in Ref. [28].

Figure 13.4 shows the procedure for the Splinegon representation from a given target pattern using selected representative points and constant curvature generation. The generated region from the upper layer of the H-GRN can then be modeled as a set of circular arc segments $s_l \in S = \{s_1, \dots, s_{n_s}\}$. For each arc segment, only the center position and its curvature will be used for region-based shape control in Section 13.3.2.

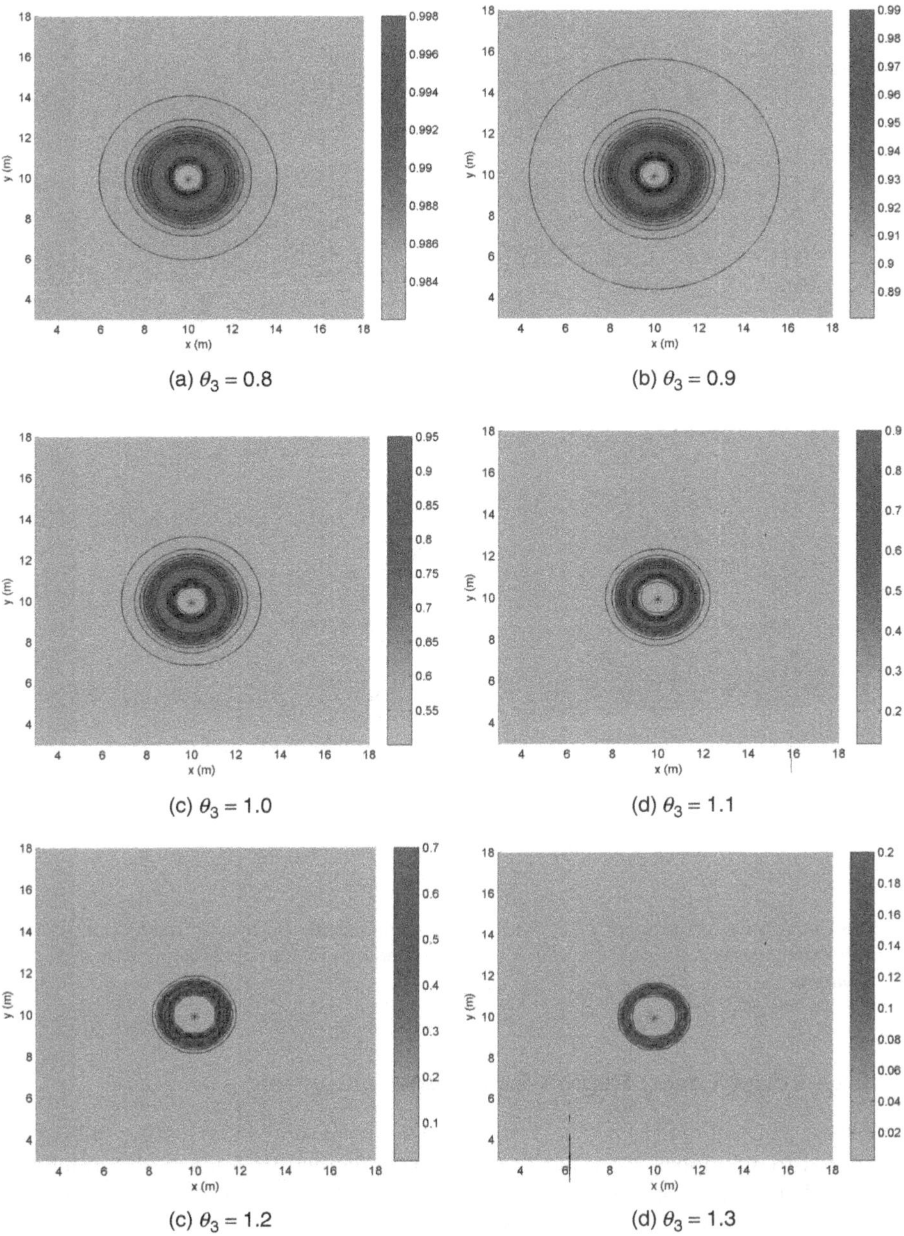

Figure 13.2 Examples of a target entrapping pattern from protein concentration g_3 generated by the upper layer of the H-GRN according to different thresholds θ_3 with fixed $\theta_1 = 0.25$ and $\theta_2 = 0.3$ where a target is located at $(x, y) = (10, 10)$.

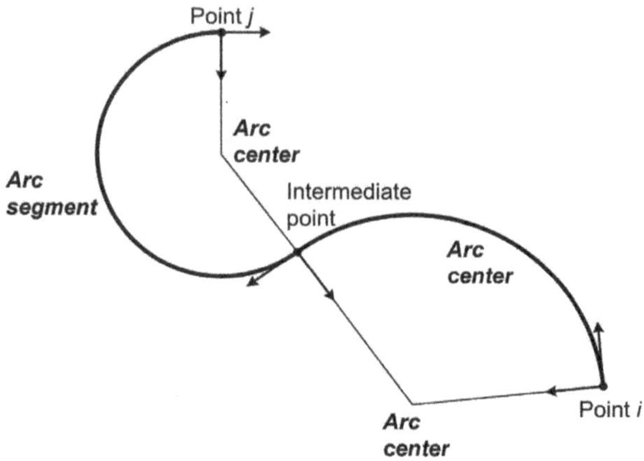

(a) Orientation 1

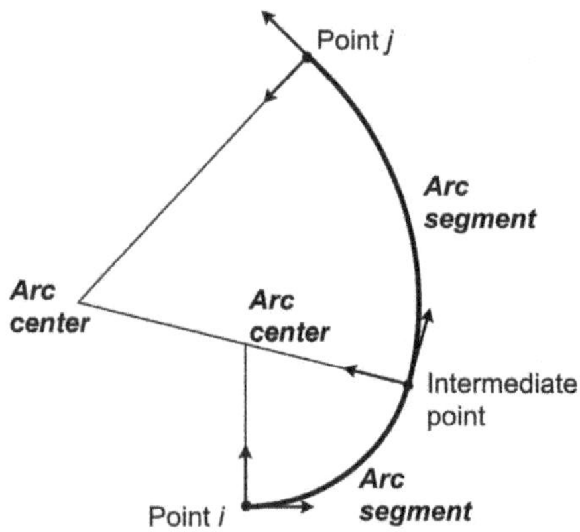

(b) Orientation 2

Figure 13.3 Arc segments with C^1 contact and intermediate point to connect neighboring points. Two arc configurations are possible with the same (orientation 2) or opposite (orientation 1) sign of curvature.

13.3.2 Lower Layer: Region-Based Shape Control

The lower layer of the H-GRN is designed to guide all robots into the desired region generated by the upper layer and to maintain a specified (but adaptable) minimum distance between robots. In the following, we first describe the GRN dynamics for driving the robots to the entrapping region, and then present the region-based shape control logic. Moreover, at the end of this section, several implementation issues to use the region-based shape control within the GRN framework will be discussed.

13.3.2.1 GRN Dynamics The general structure of GRN in the lower layer is adopted from our previous work [11], which was a modified GRN model

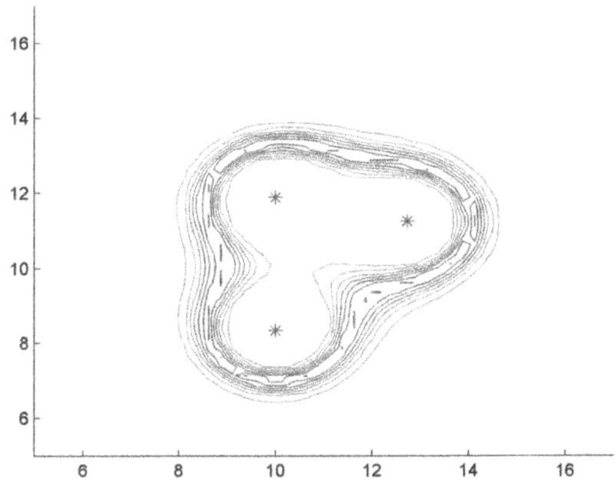

(a) Gradient value (protein concentration) of *g3*

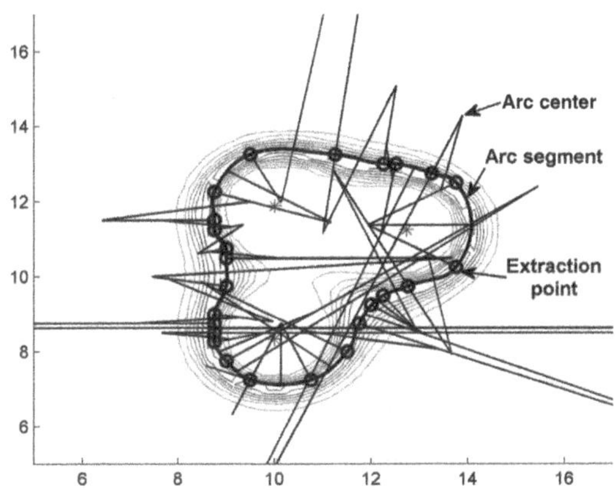

(b) Arc curvature segment generation using extraction points

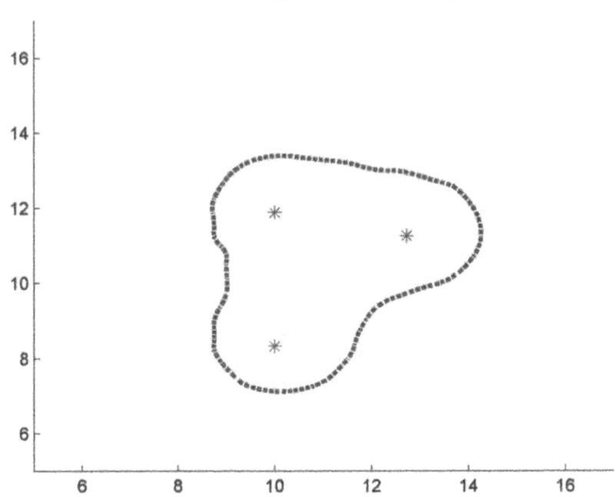

(c) Splinegon

Figure 13.4 Splinegon representation procedure from a target entrapping pattern generated by the upper layer where targets are represented as a blue star.

used in Ref. [29]:

$$\frac{d\mathbf{G}_i}{dt} = -a\mathbf{z}_i + m\mathbf{P}_i, \tag{13.7}$$

$$\frac{d\mathbf{P}_i}{dt} = -c\mathbf{P}_i + rf(\mathbf{z}_i) + b\mathbf{D}_i, \tag{13.8}$$

where $i \in \{1, \ldots, N_o\}$ is the index of an organizing robot, and protein types $\mathbf{G}_i$ and $\mathbf{P}_i$ correspond to a 2D position and internal state vector of robot i, respectively. a, m, c, r, and b are constants to be optimized depending on the objectives of the task. $\mathbf{D}_i$ represents the concentration of protein $\mathbf{G}$ diffused out of the cell, indicating the density of robots and obstacles in the neighborhood:

$$\mathbf{D}_i = \sum_{j=1}^{n_i} \mathbf{D}_i^j, \tag{13.9}$$

where n_i denotes the number of robots in the neighborhood of robot i and $\mathbf{D}_i^j$ represents the diffused protein concentration vector from robot j:

$$\mathbf{D}_i^j = \frac{\mathbf{G}_i - \mathbf{G}_j}{\|\mathbf{G}_i - \mathbf{G}_j\|}. \tag{13.10}$$

The diffusion process is activated only when the distance to the neighbor is less than a threshold r_n. In order to embed the desired shape into the regulatory dynamics, $f(\mathbf{z}_i)$ is defined as the following sigmoid function:

$$f(\mathbf{z}_i) = \frac{1 - e^{\alpha \mathbf{z}_i}}{1 + e^{\alpha \mathbf{z}_i}}, \tag{13.11}$$

where $\mathbf{z}_i$ represents a gradient value at the robot's current position, and $\alpha > 0$ determines the slope of the sigmoid function. This $\mathbf{z}_i$ regulates the concentration of both proteins $\mathbf{G}$ and $\mathbf{P}$ as a feed forward input in the GRN dynamics as in Equation (13.8) so that robots could form a desired shape, which is defined as follows:

$$\mathbf{z}_i = \frac{dh_i}{d\mathbf{G}_i}, \tag{13.12}$$

where h_i is the desired shape on which the robots need to be deployed, defined by the upper layer. For instance, if the robots need to be deployed onto a circle of a radius R at a point $\mathbf{c} = [c_x, c_y]^T$, the shape function can be defined as

$$h_i = [(G_{i,x} - c_x)^2 + (G_{i,y} - c_y)^2 - R^2]^2. \tag{13.13}$$

Note that if a circle radius R approaches to zero, then the robot is to be deployed onto a point $\mathbf{c}$ rather than a circle.

13.3.2.2 Region-Based Shape Control Logic The region-based shape control logic is developed in which the desired region is specified by an inequality function, as provided in the following:

$$F(X) \leq 0, \tag{13.14}$$

where $F(X)$ is a continuous scalar function with continuous first partial derivatives. In this chapter, the target pattern is represented as a set of arc segments:

$$S_l(\mathbf{G}_i) = (G_{i,x} - c_{l,x})^2 + (G_{i,y} - c_{l,y})^2 - \left(\frac{1}{\kappa_l}\right)^2 = 0, \tag{13.15}$$

where $[c_{l,x}, c_{l,y}]^T$ and κ_l are the center position and the curvature of the lth arc segment, respectively. Thus, inequality functions for region shape control can be defined by $S_l(\mathbf{G}_i) \leq 0$ for keeping robots inside an arc segment or $-S_l(\mathbf{G}_i) \leq 0$ for pushing robots toward the outside of an arc, given by

$$\pm S_l(\mathbf{G}_i) = (G_{i,x} - c_{l,x})^2 + (G_{i,y} - c_{l,y})^2 - \left(\frac{1}{\kappa_l}\right)^2 \leq 0. \tag{13.16}$$

Inequality functions for the entire complex shape using the Splinegon representation can then be expressed as follows:

$$F(\mathbf{G}_i) = [\pm S_1(\mathbf{G}_i), \pm S_2(\mathbf{G}_i), \dots, \pm S_{n_s}(\mathbf{G}_i)]^T \leq 0. \tag{13.17}$$

In this study, for an efficient region shape control, each robot finds its nearest arc segment $s_{l_i^*}$ at its current location rather than considering all arc segments consisting of the Splinegon:

$$s_{l_i^*} = \arg\min_{s_l \in S} \left(\|\mathbf{G}_i - s_l^m\|\right), \tag{13.18}$$

where s_l^m is the middle point of lth arc segment. Depending on the sign of the arc curvature and the position of targets, the shape control logic is divided into three cases with a gradient $\mathbf{z}_i$ and the inequality function of the nearest arc segment as follows:

Case 1: $\kappa_{l_i^*} \leq 0$,

$$\mathbf{z}_i = \begin{cases} 0, & \text{if} S_{l_i^*}(\mathbf{G}_i) \leq 0, \\ \frac{dh_{1,i}}{d\mathbf{G}_i}, & \text{otherwise}, \end{cases} \tag{13.19}$$

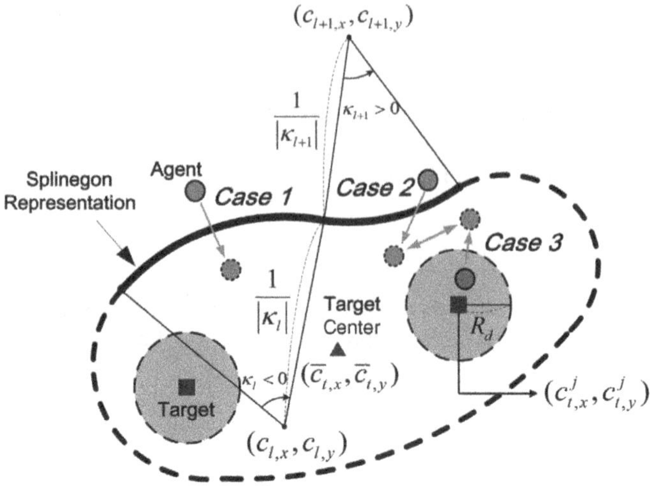

Figure 13.5 Geometric relations for the region-based shape control logic. Reprinted with permission from Ref. [27], copyright 2014 Springer Science and Business Media.

where $h_{1,i} = [(G_{i,x} - c_{l_i^*,x})^2 + (G_{i,y} - c_{l_i^*,y})^2]^2$. This gradient drives a robot toward the center of its nearest arc segment, and once the robot enters a circular sector of a desired Splinegon, the gradient becomes zero, as illustrated in Figure 13.5.

Case 2: $\kappa_{l_i^*} > 0$,

$$\mathbf{z}_i = \begin{cases} 0, & \text{if} - S_{l_i^*}(\mathbf{G}_i) \leq 0 \text{ and} \\ & \|\mathbf{G}_i - \bar{\mathbf{c}}_t\| \leq \|\mathbf{G}_i - \mathbf{c}_{l_i^*}\|, \\ \frac{dh_{2,i}}{d\mathbf{G}_i}, & \text{otherwise}, \end{cases} \tag{13.20}$$

where $h_{2,i} = [(G_{i,x} - \bar{c}_{t,x})^2 + (G_{i,y} - \bar{c}_{t,y})^2]^2$. Here, $\bar{\mathbf{c}}_t = [\bar{c}_{t,x}, \bar{c}_{t,y}]^T$ is the center position of a target group. This gradient brings a robot toward the center of a target group instead of an arc segment, which is outside the Splinegon. Only if a robot is outside a circular sector ($-S_{l_i^*}(\mathbf{G}_i) \leq 0$) as well as closer to the center of the target group than that of an arc (i.e., $\|\mathbf{G}_i - \bar{\mathbf{c}}_t\| \leq \|\mathbf{G}_i - \mathbf{c}_{l_i^*}\|$), the gradient is zero.

Case 3: $\|\mathbf{G}_i - \mathbf{G}_t^j\| < R_d$ (for all $j \in \{1, \ldots, n_t\}$, regardless of the sign of $\kappa_{l_i^*}$),

$$\mathbf{z}_i = \frac{dh_{3,i}}{d\mathbf{G}_i}, \tag{13.21}$$

where $h_{3,i} = [(1 + \epsilon)R_d^2 - (G_{i,x} - c^j_{t,x})^2 - (G_{i,y} - c^j_{t,y})^2]^2$. Here ϵ is a positive constant. If a robot is inside a certain circular boundary of R_d around the target, the robot is pushed out of that area (which is larger than R_d, $(1 + \epsilon)R_d$), in order to avoid any collision with the targets.

In the following, we will show that all robots will converge to the static desired region under the control of the GRN dynamics.

Lemma 13.3.1 *The system states $\mathbf{G}_i$ will converge to the desired region defined by the Splinegon, and the states $\mathbf{P}_i$ of all robots will converge to zero for all $i \in \{1, \dots, N_o\}$, provided that $m \cdot r \leq a \cdot c$ and a neighbor distance threshold r_n are small enough for all robots to be in the region.*

Proof: First, in all three cases of the region-based shape control logic, the robots are guided to the specified points by the gradient $\mathbf{z}_i$ with $h_{1,i}, h_{2,i,}$ and $h_{3,i}$ until they enter the desired region. This convergence to a time-invariant constant is shown in Appendix A.13.1 using Lyapunov theory, given that $m \cdot r \leq a \cdot c$ is satisfied. Once a robot is inside the desired region, the convergence of $\mathbf{P}_i$ to zero needs to be proven so that the robot is stable anywhere in the region. Let us define a positive definite Lyapunov function for the ith robot in the following:

$$V_i = \frac{1}{2}\mathbf{P}_i \cdot \mathbf{P}_i. \tag{13.22}$$

Since $\mathbf{z}_i$ is always zero in the region, differentiating V_i with respect to time with Equation (13.8) gives

$$\frac{dV_i}{dt} = -c\|\mathbf{P}_i\|^2 + b\mathbf{P}_i\mathbf{D}_i. \tag{13.23}$$

If $|c\mathbf{P}_i| \geq |b\mathbf{D}_i|$, which means that autoregulation of the internal state is stronger than the effect of diffusion, then $\frac{dV_i}{dt} \leq 0$; however, this will prevent the efficient distribution of robots in the region. A more reasonable option for convergence is that D_i becomes zero so that $\frac{dV_i}{dt}$ becomes equal to or less than zero, which completes the convergence proof. If the neighborhood size threshold r_n representing the activation area of the diffusion is small enough for all the robots to be in the desired region, the robots will push away from each other until the diffusion is no longer effective (i.e., $\mathbf{D}_i = 0$ for all $i \in \{1, \dots, N_o\}$), while maintaining the minimum distance of r_n among them. Otherwise, there will be chattering between pushing robots in (by a gradient $\mathbf{z}_i$) and out (by $\mathbf{D}_i$ to maintain the minimum distance) of the region, which does not guarantee the convergence. Note that the distant r_n cannot be determined analytically as the target region is unknown and the desired target can be very complicated. This brings a need to introduce a neighborhood size (distance) adaptation mechanism described in Section 13.3.3. ■

13.3.3 Implementation Issues

In order to implement the proposed algorithm in real robots with a limited capability in a changing environment, there are several issues to be addressed. First, as mentioned in Section 13.2, the robots should be able to localize themselves with their own onboard sensors such as encoders or an inertial navigation system. Note that this localization can be performed in a local coordinate system via robot–robot communications by choosing a reference point as the origin. The position of the robot, which first detects a target, would be a simple example of the reference point, and various other methods can be employed depending on the task objectives. Second, as the total number of robots and targets is unknown to each robot, an appropriate neighborhood size for the diffusion process in Equation (13.9) cannot be predefined in practice to ensure the convergence of the system or distribute the robots in the region as evenly as possible. This study adopts a similar method used in Ref. [30], which adjusts the initial guess of the distance with two physical constraints of the robot: the bumper range d_{min} and the maximum sensor range d_{max}. If the current neighbor distance d is found to be small, then it is updated with a half-sum of d and d_{max} resulting in an increase in distance, whereas if the distance is large, it is updated with d_{min} resulting in a decrease in distance. By checking the average number of neighbors within d for several iterations, the optimal neighbor distant d^* is determined for robots to stabilize and evenly distributed inside the region. Last, the effect of communication/sensing noise and delay should be carefully considered as this might lead to performance degradation or even instability of the shape control.

13.3.4 Numerical Simulations

Numerical simulations have been performed using scenarios containing either static or moving targets to validate the feasibility and benefit of the proposed algorithm. The number of robots used in the simulation is 100. Parameters for the upper layer of the H-GRN are set up as $\theta_1 = 0.25, \theta_2 = 0.3, \theta_3 = 1.2$, and $k = 20$, and for the lower layer as $a = 6.50, m = 4.24, c = 9.90, r = 4.29$, and $b = 3.54$. These values are obtained by an evolutionary optimization run as explained in Ref. [11] and fine-tuned for stabilization of robots inside the desired region within a reasonable time.

13.3.4.1 Forming a Predefined Target Region A predefined simple ring shape is first employed for the target entrapment where the target is assumed to be at the center of a circle. Radii 0.2 and 0.5 m are used for inner and outer circles, respectively, to define a ring-shaped region. A desired distance between the robots is 0.065 m, and the one between robots and obstacles is 0.3 m. Figure 13.6 shows snapshots of a situation where robots are randomly distributed in the space, form the target region driven by the H-GRN, and then avoid two moving obstacles. By using a diffusion term **D** between robots and obstacles as in Equations (13.8)–(13.10), the GRN dynamics inherently adapts itself to the environmental changes, that is, the moving obstacles.

In the following, we consider entrapment and tracking of a single moving target while maintaining a desired shape. If the trajectory and the velocity of the target are known or can be estimated by the estimation technique, it can be embedded in the GRN dynamics by modifying Equation (13.7) as

$$\frac{d\mathbf{G}_i}{dt} = -a\mathbf{z}_i + m\mathbf{P}_i + \mathbf{v}_t, \tag{13.24}$$

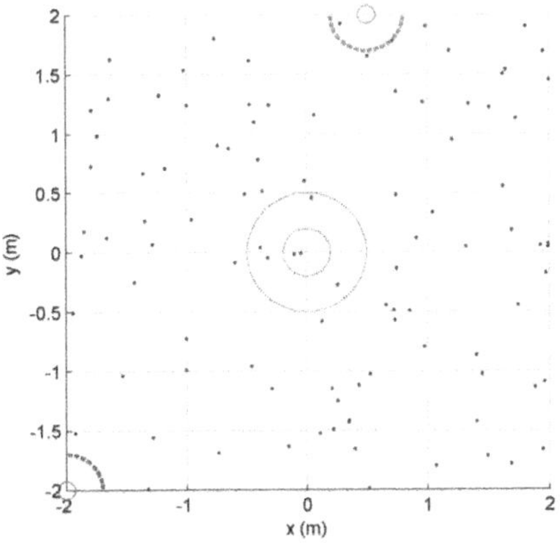

(a) T = 0s

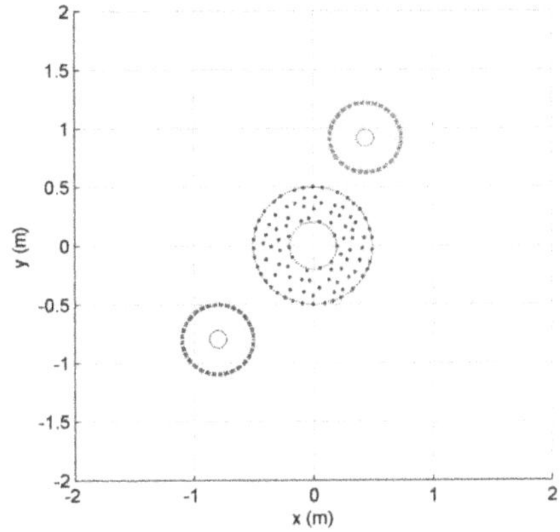

(b) T = 6s

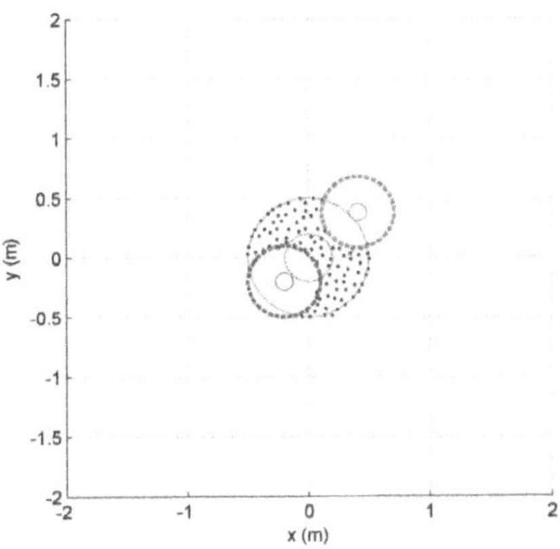

(c) T = 8s

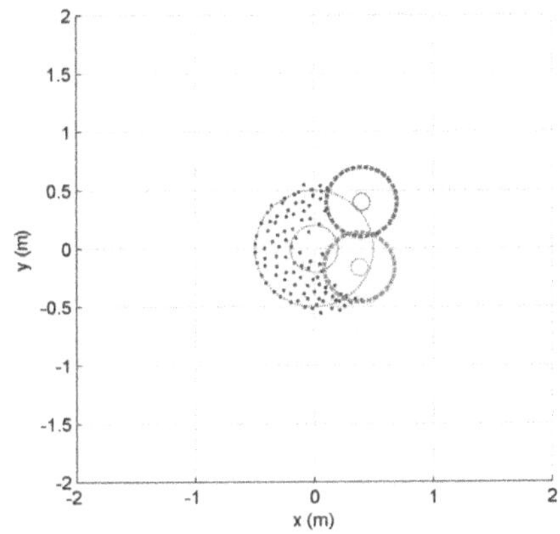

(d) T = 12s

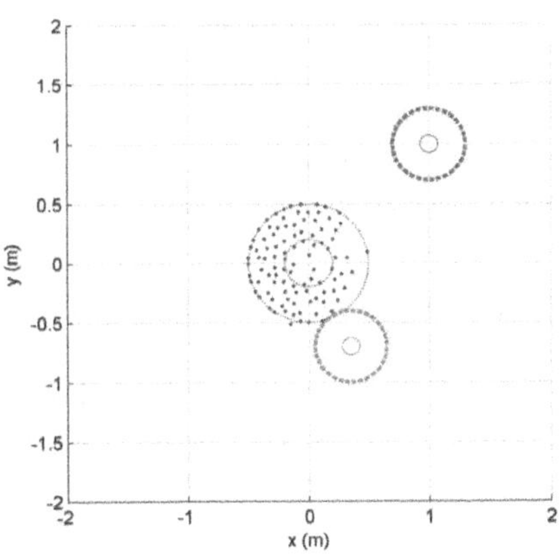

(e) T = 16s

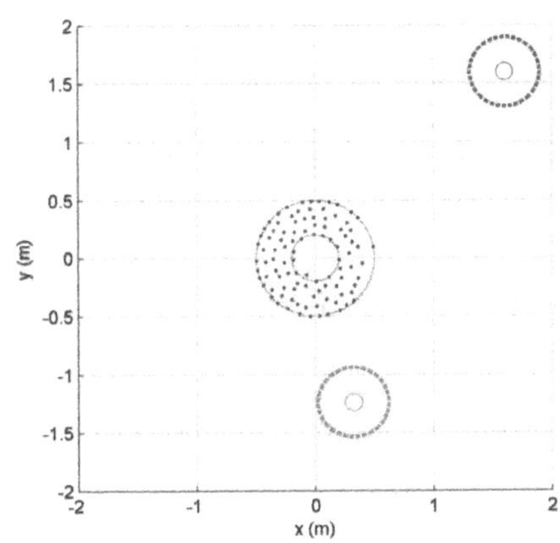

(f) T = 24s

Figure 13.6 One hundred robots (blue points) forming a target shape denoted as a ring and avoiding moving obstacles, which are denoted by a red and a green small circle.

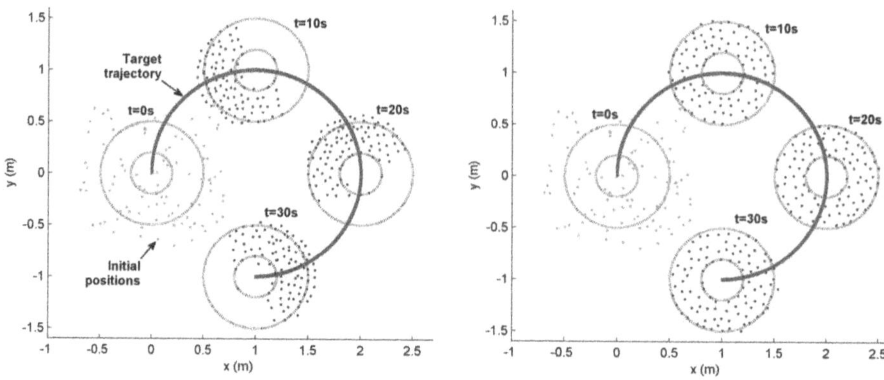

(a) Known target position only (b) Estimated position and velocity

Figure 13.7 Forming a ring-shaped region while tracking a moving target.

where $\mathbf{v}_t$ is the velocity of the target to be followed. Figure 13.7 shows the numerical simulation results on tracking a moving target using a ring shape with or without information about the target movement. The trajectory of the moving target is a circular orbit, denoted by a red line. It is assumed that each robot is equipped with a sensor that gives the range and direction to the target at 10 Hz. Sensory noise is set to zero-mean Gaussian noise with a standard deviation of $\sigma_r = 0.2$ m for the range and $\sigma_\phi = 2°$ for the direction. To obtain accurate estimates of the target position and velocity, the decentralized extended information filter (DEIF) [31] is applied with a constant velocity target model as explained in Appendix A.13.2.

If the target velocity is unknown, it appears to be difficult for the swarm of robots to closely follow the moving target while forming the desired ring shape, even with the perfectly known target position, as shown in Figure 13.7(a). By contrast, if the target position and velocity can be estimated by the DEIF filter, the robots successfully entrap the moving target, as shown in Figure 13.7(b). Inaccuracy in position and velocity estimation due to sensory noise and uncertainty in the model (as shown in Figure 13.8) might affect the stability of pattern formation inside the region; this aspect needs to be carefully investigated for better tracking performance.

13.3.4.2 Adaptive Formation of Complex Regions To entrap multiple targets, the desired region generated by the upper layer of the H-GRN may become very complex and needs to be represented using Splinegon. Figure 13.9 shows the entrapment of three static targets using the Splinegon representation with 100 robots. An initial guess of a neighborhood size of 0.5 m fails to stabilize the motion inside the region and maintain the given distance between robots as shown in Figure 13.9(a). By adjusting the neighborhood size with $d_{\max} = 0.8$ m and $d_{\min} = 0.2$ m through iterations of adaptation as shown in Figure 13.9(c), the final neighborhood size d^* of 0.4133 m is obtained resulting in the even distribution of the robots, as shown in Figure 13.9(b).

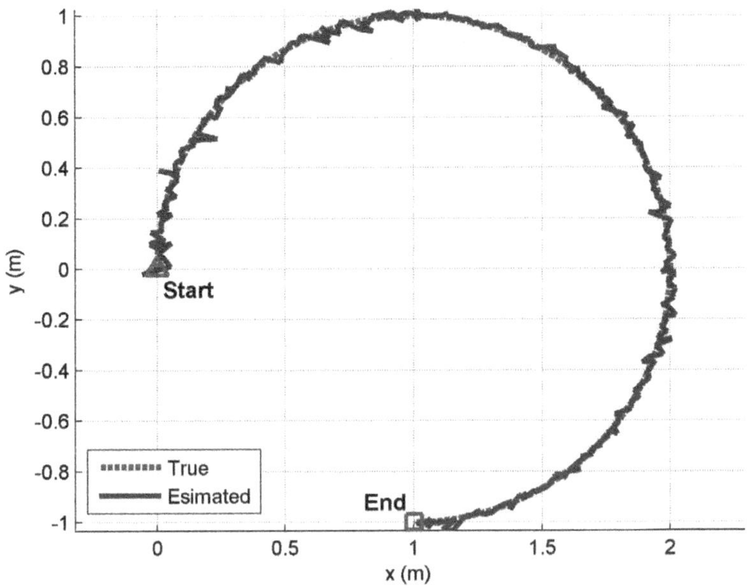

(a) Position

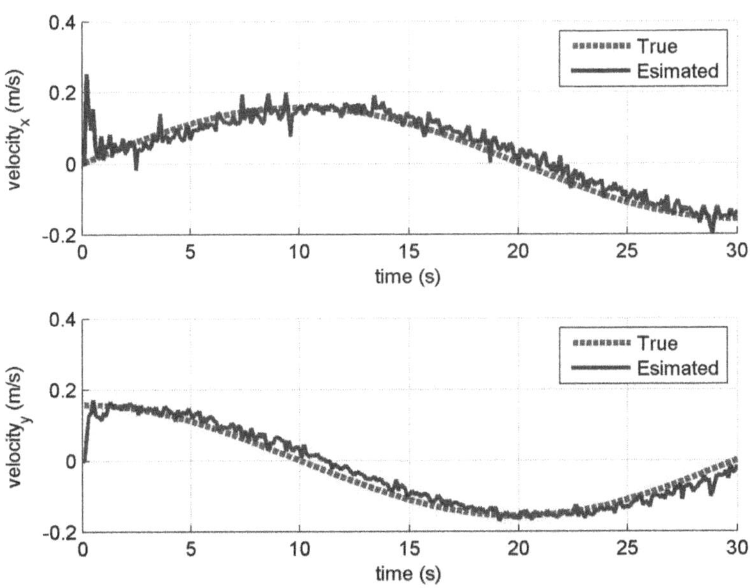

(b) Velocity

Figure 13.8 Position and velocity estimation results using the decentralized extended information filter.

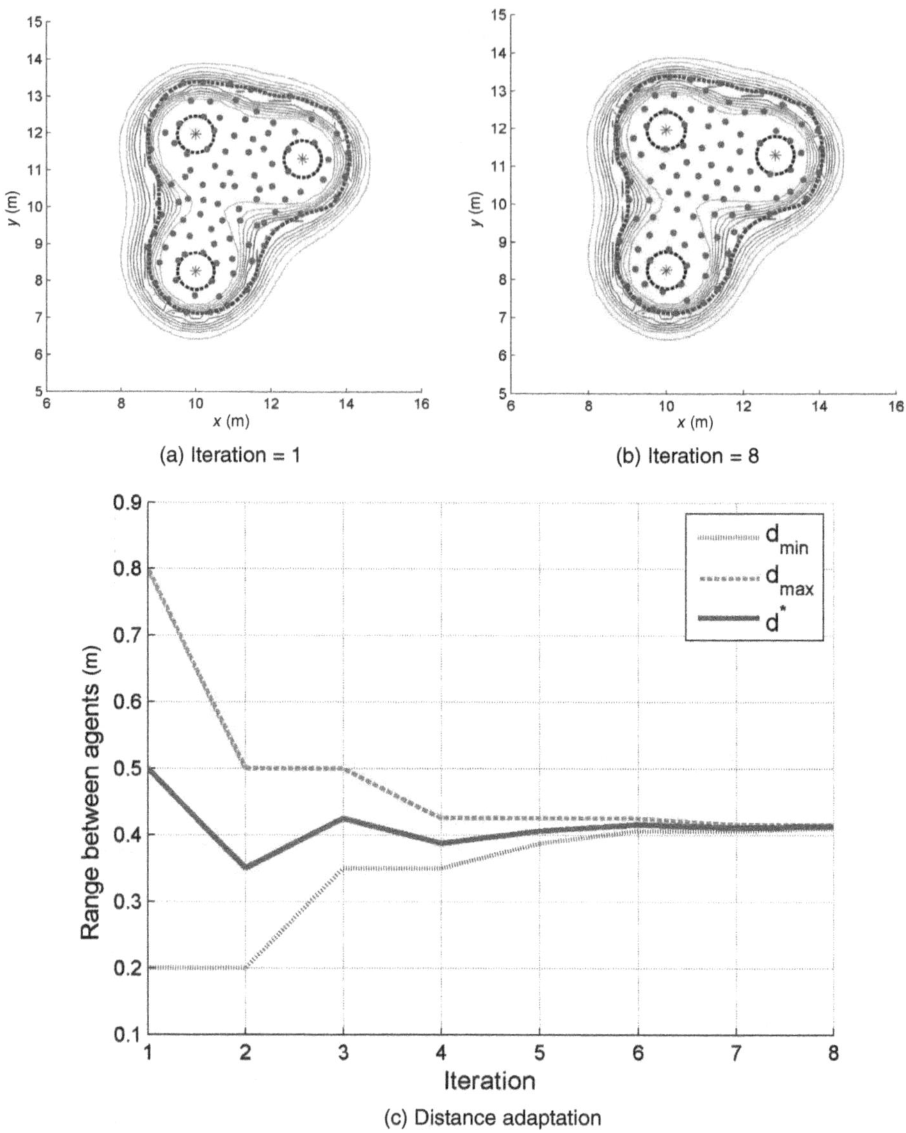

Figure 13.9 Entrapping of multiple static targets with neighborhood size adaptation. By adjusting the neighborhood size through iterations using the bumper range $d_{\min}$ and the maximum sensor range $d_{\max}$, the final size of 0.4133 m is obtained at eighth iteration.

Finally, Figure 13.10 shows the entrapment of multiple moving targets. As the targets move further away from each other, the desired shape is dynamically changed, and robots react accordingly to enter and remain in the generated region. After 20 s, once the targets stop moving, the robots are able to organize themselves evenly into a region that surrounds the targets, as shown in Figure 13.10(f).

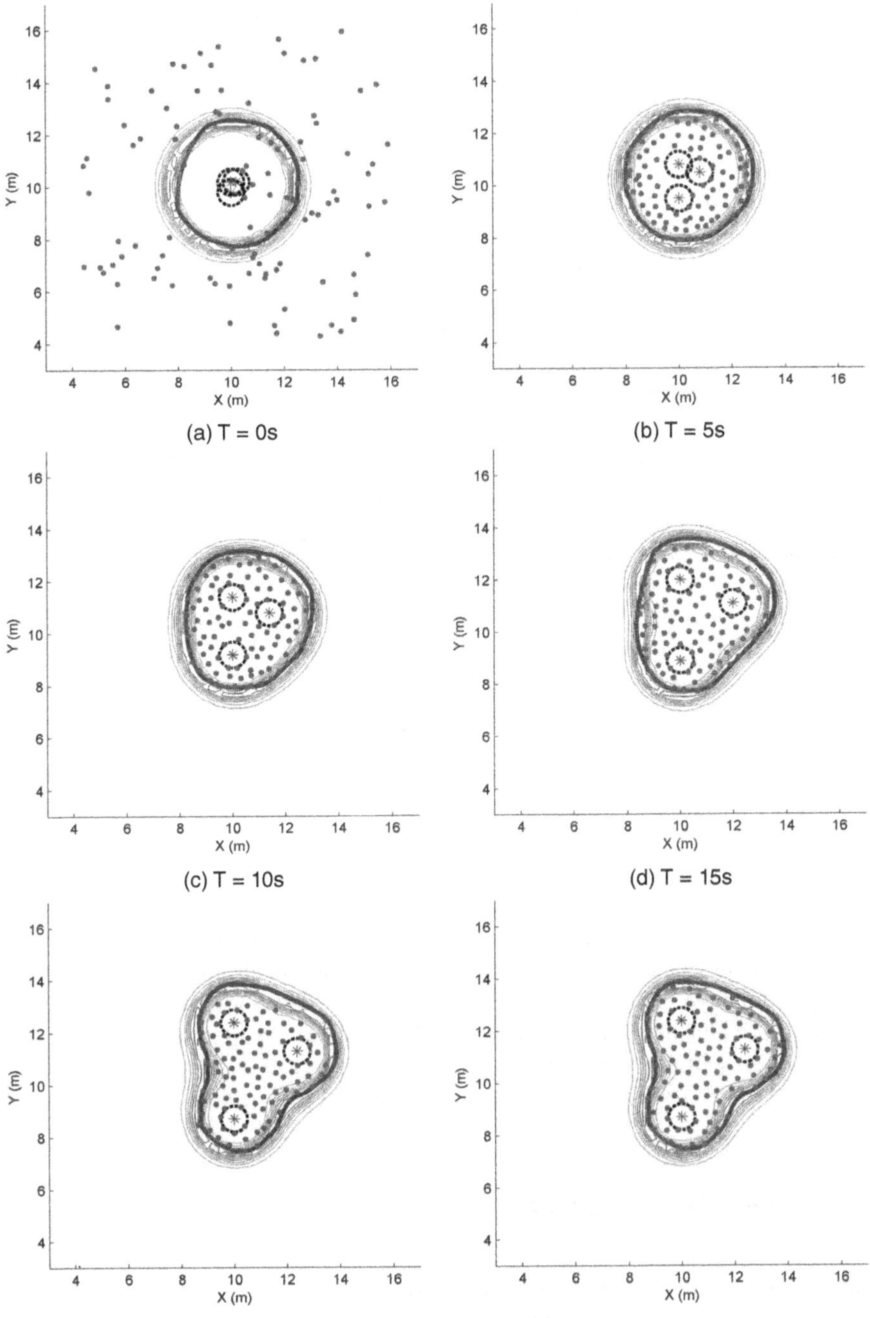

Figure 13.10 Entrapping of multiple moving targets with a changing complex region. As the targets move away from each other, an entrapping shape is changed accordingly, and consequently the robots are organizing themselves inside a shape.

13.4 EVOLVING H-GRN USING NETWORK MOTIFS

Inspired by the recent finding that GRNs often show the use of frequently recurring patterns termed network motifs, this section proposes a morphogenetic approach to target entrapping based on an EH-GRN and network motifs, extending our previous work [32]. The EH-GRN consists of two layers: the upper layer is for adaptive pattern generation where the GRN model is evolved by basic network motifs with genes and environmental inputs, and the lower GRN layer is for driving robots onto the target pattern generated by the upper layer. Separability aspect (splitting or merging) of multiple patterns resulting from target movement is addressed by the inherent feature of the upper layer and the *k*-means clustering algorithm [33].

This GRN framework enables us to generate patterns adaptive to target movement as well as environmental changes. In order to avoid incoming obstacles, the previous H-GRN relies on a repulsive avoidance scheme using a diffusion term of robots and obstacles in a GRN model. This leads to the situation while most robots remain at their positions; only robots closer to obstacles move away from their desired entrapping positions to avoid obstacles. Meanwhile, in this study, by introducing obstacles as one of environmental inputs along with targets, a target entrapping pattern from the upper layer is regenerated so that all the robots in a swarm can reconfigure their positions accordingly as an obstacle gets closer. This allows a swarm of robots to continue both of their missions: entrapping targets and avoiding obstacles simultaneously.

General structure of the proposed EH-GRN using network motifs to be evolved is illustrated in Figure 13.11. The upper layer of the EH-GRN is for adaptive pattern generation where the GRN model is evolved by the combination of basic network motifs with genes and environmental inputs. In Figure 13.11, p_1 and p_2 represent the protein concentrations produced by the environmental inputs (i.e., target and obstacle, if any) and diffused spatially. These will serve as the inputs to the upper GRN layer and activate gene g_1, g_2, and g_3. In particular, the concentration of M takes the role of morphogen to form the desired shapes around targets, which will be transmitted to the lower layer. As a preliminary study, this work only evolves the upper layer with a fixed number of genes. However, the entire structure including the lower layer and varying number of genes can be used with the same methodology exploiting basic network motifs introduced in this section. Once the target pattern is generated by the upper layer, it will function as the input of the lower layer to trigger its dynamics to control the robots in a similar way as done in the H-GRN of the previous section.

13.4.1 Basic Network Motifs

Recent biological research findings suggest that network motifs, patterns of interconnections, occur in real complex networks at numbers significantly higher than those in randomized networks [34]. These frequently recurring network motifs can also be found in GRNs of a multi-cellular organism [35]. Inspired by these findings, this study utilizes five basic regulations: positive, negative, AND, OR, and XOR regulations as the basic network motifs to construct GRN networks.

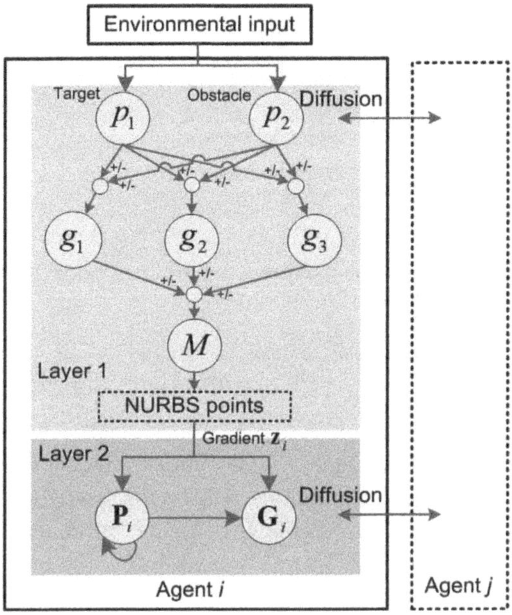

Figure 13.11 Illustration of an evolving two-layer H-GRN structure for target entrapping.

First, a positive (or negative) regulation is defined as gene X activates (or inhibits) or poses a positive (or negative) feedback to gene Y. The mathematical description of the positive regulation from X to Y can be represented as follows:

$$\frac{dy}{dt} = -y + \text{sig}(x, \theta), \tag{13.25}$$

$$\text{sig}(x, \theta) = \frac{1}{1 + e^{-k(x-\theta)}}, \tag{13.26}$$

where x and y represent the expression levels of genes X and Y, respectively, and θ is a regulatory parameter for the gene expression. Similarly, the negative regulation is represented as

$$\frac{dy}{dt} = -y + (1 - \text{sig}(x, \theta)). \tag{13.27}$$

In the AND regulation, gene Y expresses only if both genes X_1 and X_2 express, as defined by

$$\frac{dy}{dt} = -y + \text{sig}(g_1 \cdot g_2, \theta), \tag{13.28}$$

where g_1 and g_2 are the expression levels of genes X_1 and X_2, respectively, and the expression level of each gene is governed by a positive or negative regulation with a regulatory parameter θ_1 or θ_2. In the OR regulation, gene Y expresses if either gene X_1 or X_2 expresses, as defined by

$$\frac{dy}{dt} = -y + \text{sig}(g_1 + g_2, \theta). \tag{13.29}$$

Last, in the XOR regulation, gene Y expresses only if genes X_1 and X_2 have disparate expression levels, as defined by

$$\frac{dy}{dt} = -y + \text{sig}(g_1 \cdot (1 - g_2), \theta) + \text{sig}((1 - g_1) \cdot g_2, \theta). \tag{13.30}$$

Thus, if gene X_1 expresses and gene X_2 does not express (or the other way round), then gene Y expresses. These regulations will be the basic building blocks of our GRN structure, and a regulatory parameter θ will be optimized by an evolutionary algorithm.

13.4.2 Upper Layer of the EH-GRN

The generation process of an adaptive target entrapping pattern consists of two steps: (i) forming a morphogen gradient space according to the evolving GRN model and extracting candidate points whose gradient value is higher than a threshold and (ii) representing candidate points with a non-uniform rational B-splines (NURBS) [36] to generate a target pattern for robots to follow, as illustrated in Figure 13.12.

A morphogen gradient space, the concentration level of M in the environment, is first generated by the regulation parameters and environment inputs using the upper layer as shown in Figure 13.12(a). Then, the points whose gradient value is higher than a certain percentage of the maximum value are selected as candidate points, as shown in Figure 13.12(b). Here, requirement of the desired entrapping pattern is to have as many candidate points as possible by (i) satisfying not too close (d_{min}) to or far away (d_{max}) from targets and (ii) maintaining a minimum distance ($d_{\text{min}}^{\text{obs}}$) to the obstacle. The fitness function for the evolutionary algorithm can then be set up as

$$\begin{aligned} f = &\sum_{i=1}^{N_p} \sum_{j=1}^{N_t} \frac{\text{sig}(d_{pt}^{ij}, d_{\text{max}}, k_1) + \text{sig}(d_{\text{min}}, d_{pt}^{ij}, k_2) - A_{ij}}{N_p N_t} \\ &+ \sum_{i=1}^{N_p} \sum_{k=1}^{N_o} \frac{\text{sig}(d_{\text{min}}^{\text{obs}}, d_{po}^{ik}, k_3)}{N_p N_o}, \end{aligned} \tag{13.31}$$

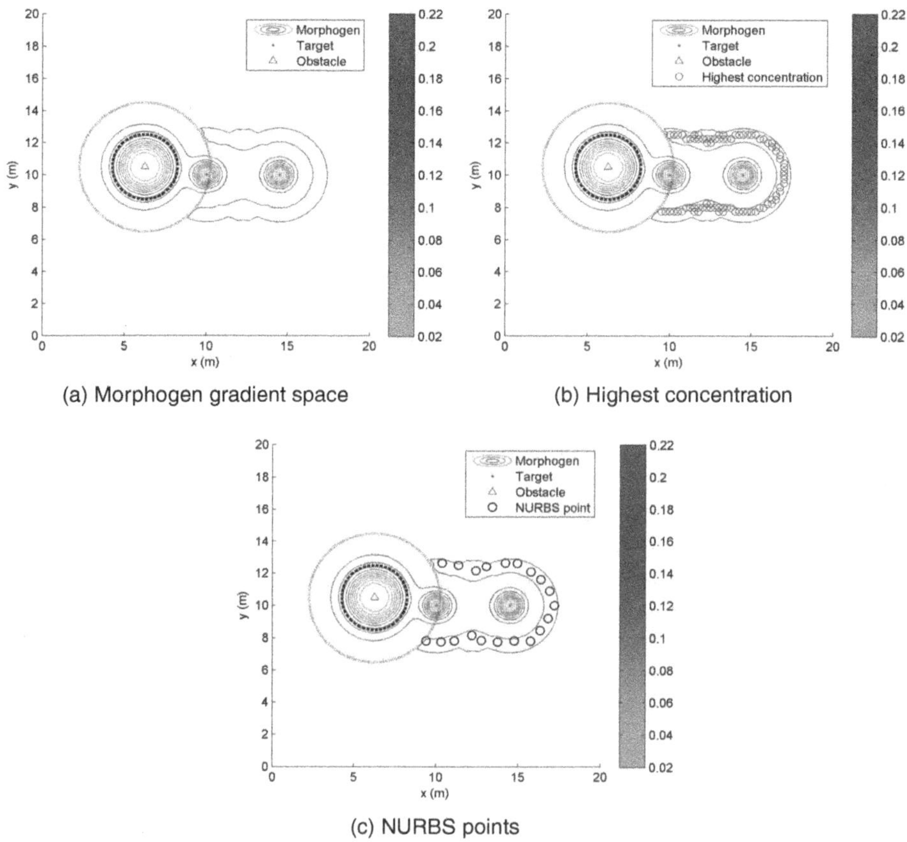

Figure 13.12 Non-uniform rational B-splines (NURBS) representation of a target entrapping pattern.

where

$$A_{ij} = \begin{cases} 1, & \text{if } d_{\min} \leq d_{pt}^{ij} \leq d_{\max}, \\ 0, & \text{otherwise,} \end{cases} \tag{13.32}$$

and N_p, N_t, and N_o represent the number of candidate points, targets, and obstacles, respectively. d_{pt}^{ij} denotes a distance between ith candidate point and jth target, d_{po}^{ik} denotes a distance between ith candidate point and kth obstacle.

For the upper layer, there are 26 parameters to be optimized: θ_i $(i = 1, \ldots, 13)$ for a regulatory parameter between environmental inputs p_i $(i = 1, 2)$ and genes g_i $(i = 1, 2, 3)$, and genes and morphogen gradient M as shown in Figure 13.11; t_i $(i = 1, \ldots, 9)$ for a positive or negative regulation; and c_i $(i = 1, \ldots, 4)$ for AND or OR regulation. XOR regulation is not used in this work for simplicity. t_i can be 0 (positive regulation) or 1 (negative regulation), and c_i can also be 0 (AND regulation) and 1 (OR regulation). For instance, if $t_1 = 1$, $t_2 = 0$, and $c_1 = 1$, then the partial dynamics between target p_1, obstacle (p_2), and gene 1 (g_1) can be described as

$$\frac{dg_{11}}{dt} = -g_{11} + (1 - \text{sig}(p_1, \theta_1)), \tag{13.33}$$

$$\frac{dg_{12}}{dt} = -g_{12} + \text{sig}(p_2, \theta_2), \tag{13.34}$$

$$\frac{dg_1}{dt} = -g_1 + \text{sig}(g_{11} + g_{12}, \theta_7), \tag{13.35}$$

where g_{11} and g_{12} are internal genes between environmental inputs and g_1. Entire upper layer dynamics of the EH-GRN can be constructed with a similar manner using defined 26 parameters.

The evolution strategies (ESs) with (μ, λ) selection [37,38] are used to optimize the regulatory parameters. The population size is set to 30 for parents and 150 for offspring, and the ES was run for 200 generations. In this simulation, as there may exist many local optima due to the unconventional fitness formulation (such as extracting candidate points), relatively big population and generation are used. The data type of chromosomes are double for regulation parameters θ_i, Boolean for t_i, and integer for c_i. The resulting evolved structure of the upper layer is shown in Figure 13.13 with the following parameter values: $\theta_1 = 0.1873$, $\theta_2 = 0.6411$, $\theta_3 = 0.0386$, $\theta_4 = 0.5795$, $\theta_5 = 0.9495$, $\theta_6 = 0.3202$, $\theta_7 = 1.3249$, $\theta_8 = 3.4224$, $\theta_9 = 0.3799$, $\theta_{10} = 0.4833$, $\theta_{11} = 1.6287$, $\theta_{12} = 0.2084$, and $\theta_{13} = 1.6868$.

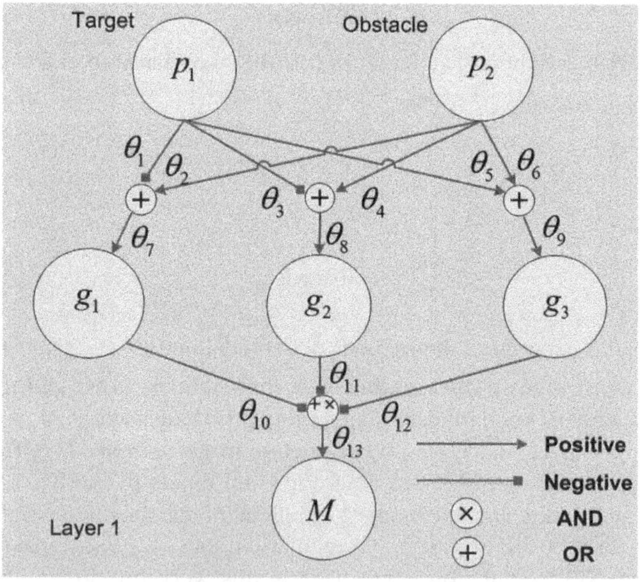

Figure 13.13 Illustration of an evolved two-layer H-GRN structure for target entrapping.

Figure 13.13 shows the evolved H-GRN structure with corresponding regulations. The morphogen M that defines a gradient space for the entrapping pattern is inhibited by (i) g_1 and g_2 activated by p_2 and inhibited by p_1, and (ii) g_3 activated by both p_1 and p_2. Thus, gene g_3 lowers the morphogen gradient value where there is a target or obstacle expression around that results in a pattern not too close to both targets and obstacles. The role of g_1 and g_2 is more subtle since it is a combination of activation by the target and inhibition by the obstacle, which results in the pattern as a band of ellipses as shown in Figure 13.12(a).

As the targets move away from, or get closer, to each other, entrapping patterns from gradient space might be merged or split. This pattern separation aspect is addressed by the k-means clustering algorithm to group candidate points based on attributes into predefined k number of groups [33]. The k-means algorithm is one of the unsupervised learning methods, and the grouping is done by minimizing the sum of squares of distances J between points and the corresponding cluster centroids. As there is no initial knowledge on the number of patterns, the number of cluster k starts from one and increases by one while analyzing the cost J in order to find the best k.

The extracted candidate points (whose gradient value is higher than the certain percentages of the maximum value from a morphogen gradient space M as shown in Figure 13.11) are then used to construct a target pattern using NURBS [36] to be used by the EH-GRN lower layer. The NURBS is a mathematical model commonly used in computer graphics and structural design for generating and representing curves and surfaces. NURBS is considered here (rather than the Splinegon as used in the previous section) since it can offer the following features for pattern generation. First, it provides a common mathematical form for both analytical and free-form (open and closed) shapes. Second, it is a parameterized representation independent of an absolute coordinate system; once the parameter in the NURBS curve is fixed, a corresponding point on the NURBS curve can be determined without a global coordinate system. Last, NURBS can be easily segmented with their parametric input variables so that they can provide desired points (of the same number of robots involved) on the target pattern to be followed by robots, as shown in Figure 13.12(c). A NURBS curve can be defined as a combination of a set of piecewise rational basis functions with $n + 1$ control points $\mathbf{p}_i$ and the associated weights w_i as in Ref. [36]:

$$\mathbf{C}(u) = \frac{\sum_{i=1}^{n} \mathbf{p}_i w_i B_{i,k}(u)}{\sum_{j=1}^{n} w_j B_{i,k}(u)}, \tag{13.36}$$

where n is the number of control points, u is a parametric variable, and $B_{i,k}(u)$ are B-spline basis functions where i corresponds to the ith control point and k corresponds to the degree of the basis function. More details on the NURBS can be found in Ref. [36].

13.4.3 Lower Layer of the EH-GRN

The lower layer of the EH-GRN is for guiding and controlling robots to the corresponding patterns generated by the upper layer. The GRN in the lower layer adopts the same equations (Equations (13.7)–(13.11)) used in the previous section. Only morphogen gradient $\mathbf{z}_i$ is different as this $\mathbf{z}_i$ regulates the concentration of both proteins $\mathbf{G}$ and $\mathbf{P}$ in the GRN dynamics so that robots could form a desired shape, defined as

$$\mathbf{z}_i = \mathbf{G}_i - \mathbf{C}(u_i), \tag{13.37}$$

where h_i is the desired target shape on which the robots need to be deployed as defined by the upper layer, and u_i should satisfy the following condition:

$$u_i = \arg\min_{0 \le u \le 1} \parallel \mathbf{G}_i - \mathbf{C}(u_i) \parallel . \tag{13.38}$$

Since it is difficult to have an analytic function to represent of the NURBS $\mathbf{C}(u_i)$, u is segmented in the range of [0, 1] by the number of robots in the environment as $u_d = \{0, 1/(n-1), 2/(n-1), \ldots, 1\}$, and to give

$$u_i = \arg\min_{u_i \in u_d} \parallel \mathbf{G}_i - \mathbf{C}(u_i) \parallel . \tag{13.39}$$

Thus, only segmented points $\mathbf{C}(u_i)$ on the continuous NURBS need to be communicated from the organizing robots. Note that, depending on the number and length of NURBS for the case of multiple target entrapping patterns, the segmentation size of u needs to be varied accordingly. For instance, if there are two NURBS curves for two separate entrapping patterns, the segmentation size is a half compared to the case for one curve so that robots can be equally distributed to each pattern.

13.4.4 Numerical Simulations

Numerical simulations have been performed using scenarios containing static/moving targets as well as a moving obstacle to validate the feasibility and benefits of the proposed EH-GRN algorithm. The number of robots used in the simulation is 20. The distances to be maintained relative to the target and the obstacle set to $d_{\text{min}} = 2$, $d_{\text{max}} = 4$ and $d_{\text{min}}^{\text{obs}} = 4$. Slope values for the sigmoid in the fitness function in Equation (13.31) are $k_1 = 1$, $k_2 = 2$, and $k_3 = 2$.

The entrapping of two stationary targets with an incoming obstacle is first tested as shown in Figure 13.14. From their random initial locations (Figure 13.14a), the robots are guided onto the generated pattern by the upper layer of the EH-GRN to entrap targets, while keeping the constraints of being not too close or far away from them. As the obstacle gets closer, the entrapping pattern changes adaptively, and the robots are guided into the new positions accordingly. Once the obstacle gets further away from the robots, robots are redeployed to the previous target entrapping pattern.

Figure 13.15 illustrates the difference between the previous H-GRN and the proposed EH-GRN for the same obstacle avoidance scenario. In Figure 13.15(a), obstacle avoidance is done by local reactive control from the lower layer only, which uses a diffusion term to avoid the collision as in Equation (13.8). As an obstacle gets closer,

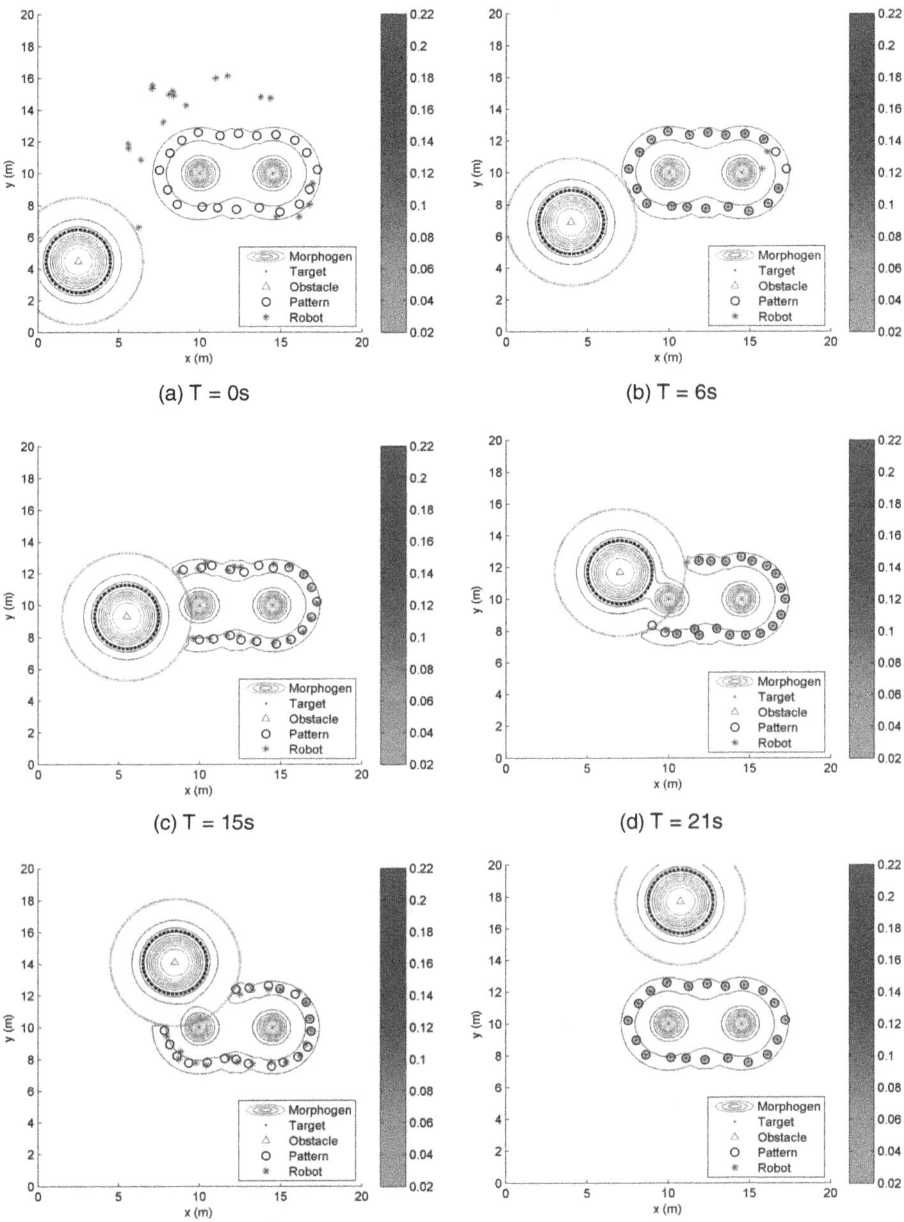

(a) T = 0s (b) T = 6s

(c) T = 15s (d) T = 21s

(e) T = 27s (f) T = 36s

Figure 13.14 Entrapping of two targets with a moving obstacle.

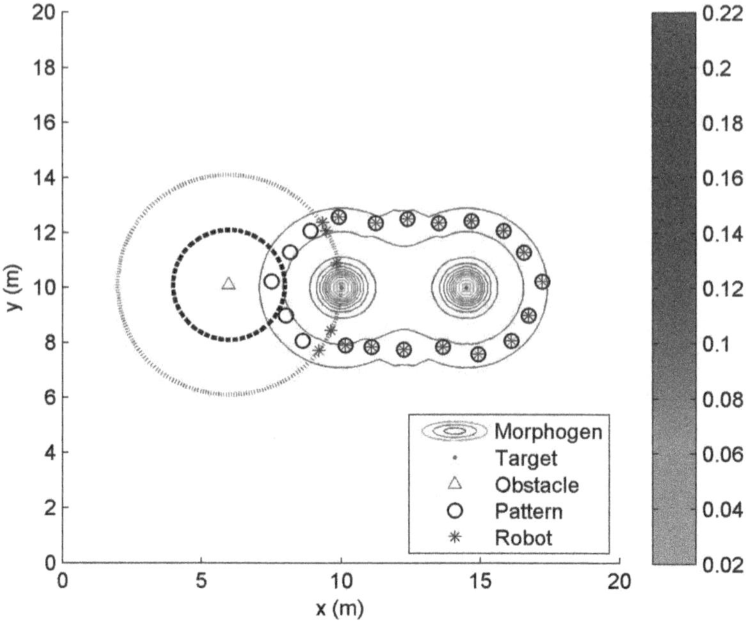

(a) Local reactive control by the lower layer

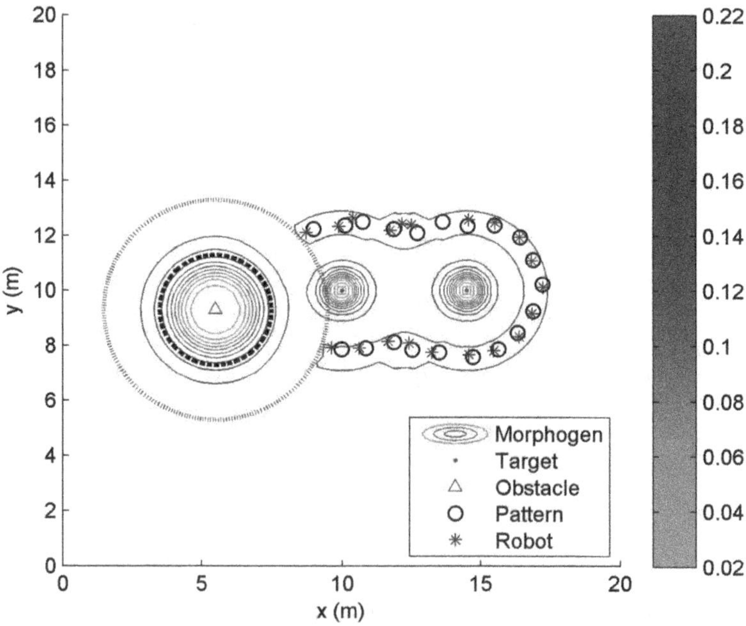

(b) Adaptive pattern generation by the upper layer

Figure 13.15 Comparison between different obstacle avoidance methods.

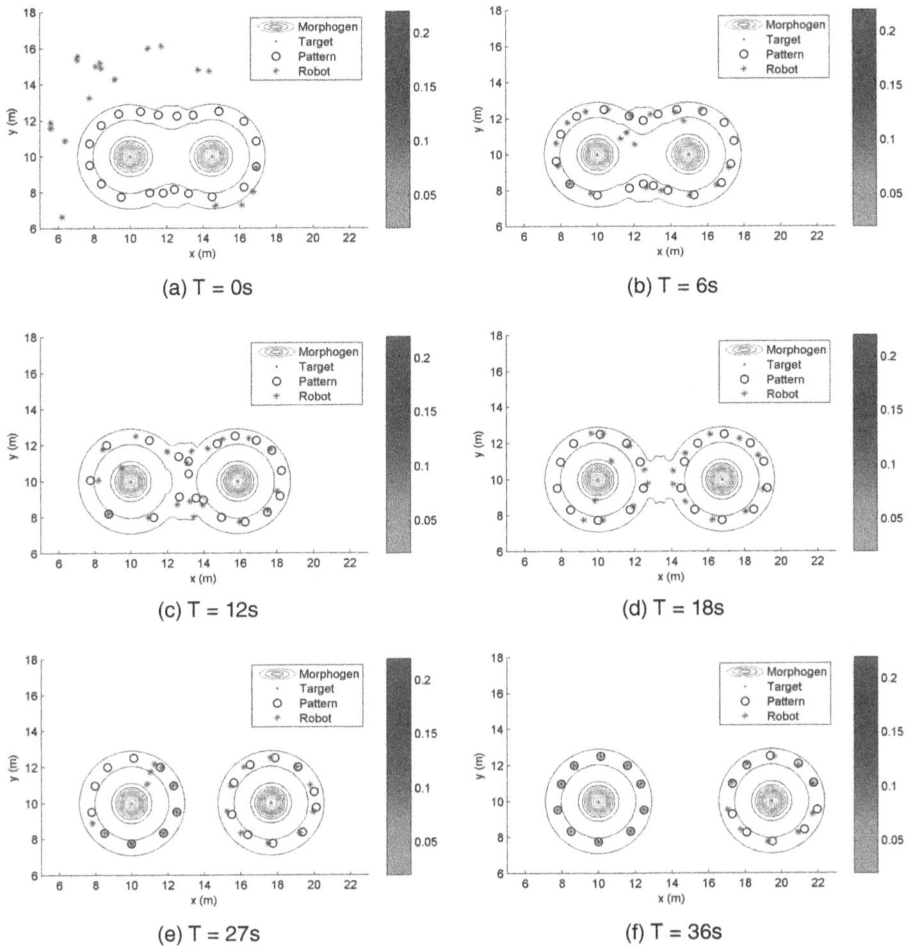

Figure 13.16 Entrapping pattern separation as targets move away from each other.

some of the robots, which lie within the effective range of the obstacle, (illustrated by the red dash) are pushed away from their original desired positions, while the remaining robots (outside the obstacle range) hold the position to entrap the targets. Thus, this scheme leads some of the robots to avoid the obstacle only but not carrying out the original target entrapping task for a while. Meanwhile, in Figure 13.15(b), as the effect of the obstacle is considered in the upper layer as well as in the lower layer, it results in generating an adaptive pattern and desired positions for the whole robot swarm change continuously reflecting environmental changes.

Last, Figure 13.16 shows the entrapping pattern separation as targets move away from each other. By the k-means clustering algorithm and cluster analysis as explained in the previous section, the number of entrapping patterns is determined, and the robots are distributed to the patterns accordingly. All these simulations verify the robustness and adaptability of the proposed algorithm to the environmental changes such as the target and obstacle movement.

13.5 CONCLUSIONS AND FUTURE WORK

This paper has presented a morphogenetic approach to pattern formation for entrapping targets using swarm robots. The proposed H-GRN with region-based shape control algorithm does not require a centralized control for the position of the individual robots inside the region, providing a practical flexibility for self-organization of a large number of robots. Besides, the proposed evolving GRN framework improves the flexibility of the pattern generation to be applied for various tasks as it has an open structure of gene regulation. By introducing obstacles as one of environmental input sources along with one of the targets, we address the weakness of the previous H-GRN pattern not being adaptable to obstacles. Numerical simulations considering static/moving targets and obstacles have demonstrated that the proposed approaches are able to automatically generate complex patterns highly adaptable and robust to dynamic and unknown environments. This capability of entrapping stationary or moving targets by forming a region can be applied to a variety of tasks such as contaminant/hazardous material boundary monitoring or isolation and transporting/herding targets into a predefined pattern or goal position.

As future work, a proof-of-concept experiment will be performed to evaluate the proposed pattern formation algorithm using *e*-puck education robots [39] in a real-world environment. A fully open structure for evolving H-GRN (i.e., the number of genes is not fixed) including the lower layer will be considered to be a more flexible and adaptable framework for various applications using swarm robots. Moreover, relaxation of the assumption in the present model that relies on organizing robots will be studied for a more realistic distributed self-organizing system.

ACKNOWLEDGMENT

This work was funded by the European Commission 7th Framework Program, Project No. 601062, SWARM-ORGAN.

APPENDIX

A.13.1 Convergence Proof

In this appendix we will theoretically prove the convergence of the system states of the GRN dynamics described in Equations (A13.7) and (A13.8) to the desired points (and, consequently, inside the desired region) following Ref. [30]. Similarly, the diffusion term is omitted in the proof, and subscript i is removed since all robots share the same GRN dynamics under this assumption. According to the Lyapunov theory [40], the system will be convergent if a Lyapunov function V exists, satisfying the following conditions: (1) $V(\mathbf{G}, \mathbf{P})$ is positive definite; (2) $\dot{V}(\mathbf{G}, \mathbf{P})$ is negative semidefinite. Let us then consider the following candidate Lyapunov function, which is positive definite:

$$V(\mathbf{G}, \mathbf{P}, s) = h(\mathbf{G}) + \frac{1}{2}s\mathbf{P} \cdot \mathbf{P}, \tag{A13.1}$$

where $s = \frac{a \cdot c}{r^2} > 0$. Differentiating V with respect to time and using the fact that $||f(\mathbf{z})|| \leq ||\mathbf{z}||$ gives

$$\begin{aligned}
\frac{dV}{dt} &= \frac{dh}{d\mathbf{G}} \cdot \frac{d\mathbf{G}}{dt} + s\mathbf{P} \cdot \frac{d\mathbf{P}}{dt} \\
&= \mathbf{z} \cdot (-a\mathbf{z} + m\mathbf{P}) + s\mathbf{P} \cdot (-c\mathbf{P} + rf(\mathbf{z})) \\
&= -a||\mathbf{z}||^2 - cs||\mathbf{P}||^2 + m\mathbf{z} \cdot \mathbf{P} + rs\mathbf{P} \cdot f(\mathbf{z}) \\
&\leq -a||\mathbf{z}||^2 - cs||\mathbf{P}||^2 + m||\mathbf{z}|| \cdot ||\mathbf{P}|| + rs||\mathbf{P}|| \cdot ||\mathbf{z}|| = \mathbf{x}^T A \mathbf{x},
\end{aligned} \tag{A13.2}$$

where

$$A = \begin{bmatrix} -a & \frac{1}{2}(rs+m) \\ \frac{1}{2}(rs+m) & cs \end{bmatrix}, \text{ and } \mathbf{x} = \begin{bmatrix} ||\mathbf{z}|| \\ ||\mathbf{P}|| \end{bmatrix}. \tag{A13.3}$$

One can easily show that given the condition $m \cdot r \leq a \cdot c$, eigenvalues of matrix A are less than or equal to zero; thus, $\frac{dV}{dt}$ is negative semidefinite. Then, given this condition, $\mathbf{z}$ converges to the invariant set satisfying $\frac{dV}{dt} = 0$ at which $\mathbf{z}$ becomes zero. The fact that $\mathbf{z}$ is zero if and only if $\mathbf{G}$ is on the desired point or inside the desired region. This completes the convergence proof.

A.13.2 Position and Velocity Estimation

To estimate the position and velocity of the target, first the target dynamics is defined as the discrete linear state equation at time step k:

$$\mathbf{x}_{k+1}^t = F_k \mathbf{x}_k^t + \eta_k, \tag{A13.4}$$

where $\mathbf{x}_k^t = (x_k^t, \dot{x}_k^t, y_k^t, \dot{y}_k^t)^T$, η_k is zero-mean white Gaussian noise with covariance Q_k, and the state transition matrix F_k can be represented as

$$F_k = \begin{bmatrix} 1 & T & 0 & 0 \\ 0 & 1 & 0 & 0 \\ 0 & 0 & 1 & T \\ 0 & 0 & 0 & 1 \end{bmatrix}, \tag{A13.5}$$

where T is the time step. The sensor measurement $\mathbf{z}_k$ can be defined as the following nonlinear function of the target position $(x_k^t, y_k^t)^T$ and the robot position $(x_k, y_k)^T$ as follows:

$$\mathbf{z}_k = \begin{bmatrix} r_k \\ \phi_k \end{bmatrix} = h(\mathbf{x}_k^t) + v_k = \begin{bmatrix} \sqrt{(x_k^t - x_k)^2 + (y_k^t - y_k)^2} \\ \tan^{-1} \frac{y_k^t - y_k}{x_k^t - x_k} \end{bmatrix} + v_k, \quad \text{(A13.6)}$$

where r_k and ϕ_k represent the range and direction of the target with respect to the robot, respectively, v_k is a measurement noise vector, and its noise covariance matrix is defined as

$$V[v_k] = R_k = \begin{bmatrix} \sigma_r^2 & 0 \\ 0 & \sigma_\phi^2 \end{bmatrix}. \quad \text{(A13.7)}$$

In case multiple robots detect the same target from different locations, sensor fusion techniques can be applied to optimally fuse the individual estimates in order to improve estimation accuracy. Considering the fact that $h(\mathbf{x}_k^t)$ is nonlinear as shown in Equation (A13.6) and the advantage of using information form in multi-sensor system with its less communication load, the estimation of a target is designed by the DEIF [31]. Each robot computes local predictions based on previous and locally determined information estimates using its process model, in the following form:

Prediction

$$\mathbf{y}_{i,k|k-1}^t = Y_{i,k|k-1} F_k Y_{i,k-1|k-1}^{-1} \mathbf{y}_{i,k-1|k-1}^t, \quad \text{(A13.8)}$$

$$Y_{i,k|k-1} = (F_k Y_{i,k-1|k-1}^{-1} F_k^T + Q_k)^{-1}, \quad \text{(A13.9)}$$

where $Y_{i,k} = (P_{i,k})^{-1}$ and $\mathbf{y}_{i,k}^t = Y_{i,k}\mathbf{x}_{i,k}^t$ represent the information matrix and information state vector of the ith robot. The predictions are combined with information from local observations to compute local estimates. After exchanging these information data, information estimates for the ith robot are updated with information obtained from the other robots represented by $i_{j,k}$ and $I_{j,k}$ to give the following.

Update

$$\mathbf{y}_{i,k|k}^t = \mathbf{y}_{i,k|k-1}^t + \sum_{j=1, j\neq i}^{N} g_k^{ij} i_{j,k}, \quad \text{(A13.10)}$$

$$Y_{i,k|k} = Y_{i,k|k-1} + \sum_{j=1, j\neq i}^{N} g_k^{ij} I_{j,k}, \quad \text{(A13.11)}$$

where

$$i_{j,k} = H_{j,k}^T (R_{j,k})^{-1} [\mathbf{z}_{j,k} - h_j(\mathbf{x}_{j,k|k-1}^t) + H_{j,k}\mathbf{x}_{j,k|k-1}^t], \quad \text{(A13.12)}$$

$$I_{j,k} = H_{j,k}^T (R_{j,k})^{-1} H_{j,k}. \quad \text{(A13.13)}$$

Here subscript j is associated with the jth robot and the matrix $H_{j,k}$ is a Jacobian of h_j with respect to the state $\mathbf{x}_{j,k|k-1}^t$. g_k^{ij} represents the communication link between robots i and j. If there is an unidirectional information exchange link between the robots, then g_k^{ij} is 1, otherwise it is 0.

To reduce computational cost, this study fuses estimates from only two randomly chosen neighboring robots for the same target rather than using all available neighbors. Note that the performance of sensor fusion depends on the geometric relationship of robots to the target (e.g., orthogonal separation between two robots might be optimal for direction sensor) [41, 42]. Therefore, how to optimally select neighbors for the sensor fusion will be tackled in the future work.

REFERENCES

1. J.R. Lawton, R.W. Beard, and B.J. Young. A decentralized approach to formation maneuvers. *IEEE Transactions on Robotics and Automation*, 19(6):933–941, 2003.
2. T. Balch and R.C. Arkin. Behavior-based formation control for multirobot teams. *IEEE Transactions on Robotics and Automation*, 14(6):926–939, 1998.
3. L. Consolini, F. Morbidi, F. Prattichizzo, and M. Tosques. Leader-follower formation control of nonholonomic mobile robots with input constraints. *Automatica*, 44(5):1343–1349, 2008.
4. H. Oh, S. Kim, A. Tsourdos, and B. White. Decentralised standoff tracking of moving targets using adaptive sliding mode control for UAVs. *Journal of Intelligent and Robotic Systems*, 76(1):169–183, 2014.
5. M. Hsieh, V. Kumar, and L. Chaimowicz. Decentralized controllers for shape generation with robotic swarms. *Robotica*, 26(5):691–701, 2008.
6. A.R. Pereira and L. Hsu. Adaptive formation control using artificial potentials for eulerlagrange agents. In: *17th IFAC world congress*, COEX, Korea, 2008.
7. W.-M. Shen, P. Will, and A. Galstyan. Hormone-inspired self-organization and distributed control of robotic swarms. *Autonomous Robots*, 17:93–105, 2004.
8. K. Yeom and J.-H. Park. Artificial morphogenesis for arbitrary shape generation of swarms of multi agents. In: *IEEE Fifth International Conference on Bio-Inspired Computing: Theories and Applications (BIC-TA)*, Liverpool, UK, pp. 509–513, 2010.
9. A. Kondacs. Biologically-inspired self-assembly of two-dimensional shapes using global-to-local compilation. In: *International Joint Conference on Artificial Intelligence*, San Francisco, CA, pp. 633–638, 2003.
10. H. Xu, H. Guan, A. Liang, and X. Yan. A multi-robot pattern formation algorithm based on distributed swarm intelligence. In: *International Conference on Computer Engineering and Applications*, IEEE, Chengdu, China, 2010.

11. Y. Jin, H. Guo, and Y. Meng. A hierarchical gene regulatory network for adaptive multirobot pattern formation. *IEEE Transactions on Systems, Man, and Cybernetics-Part B: Cybernetics*, 42(3):805–816, 2012.
12. C. Belta and V. Kumar. Abstraction and control for groups of robots. *IEEE Transactions on Robotics*, 20(5):865–875, 2004.
13. C.C. Cheah, S.P. Hou, and J.J.E. Slotine. Region-based shape control for a swarm of robots. *Automatica*, 45:2406–2411, 2009.
14. S.P. Hou and C.C. Cheah. Multiplicative potential energy function for swarm control. In *IEEE/RSJ International Conference on Intelligent Robots and Systems*, pp 4363–4368, St. Louis, USA, 2009.
15. R. Olfati-Saber. Flocking for multi-agent dynamic systems: algorithms and theory. *IEEE Transactions on Automatic Control*, 51(3):401–420, 2006.
16. H.G. Tanner, A. Jadbabaie, and G.J. Pappas. Stable flocking of mobile agents. Part II: dynamic topology. In: *Conference on Decision and Contol*, Hawaii, USA, IEEE, Piscataway, NJ, 2003.
17. H. Sayama. Robust morphogenesis of robotic swarms. *IEEE Computational Intelligence Magazine*, 5(3):43–49, 2010.
18. R. Doursat, H. Sayama, and O. Michel. A review of morphogenetic engineering. *Natural Computing*, 12:517–535, 2013.
19. Y. Jin and Y. Meng. Morphogenetic robotics: an emerging new field in developmental robotics. *IEEE Transactions on Systems, Man, and Cybernetics, Part C: Applications and Reviews*, 41(2):145–160, 2011.
20. H. De Jong. Modeling and simulation of genetic regulatory systems: a literature review. *Journal of Computational Biology*, 9(1):67–103, 2002.
21. M. Mamei, M. Vasirani, and F. Zambonelli. Experiments of morphogenesis in swarms of simple mobile robots. *Applied Artificial Intelligence*, 18:903–919, 2004.
22. Y. Ikemoto, et al. Gradual spatial pattern formation of homogeneous robot group. *Information Sciences*, 171:431–445, 2005.
23. A. Turing. The chemical basis of morphogenesis. *Philosophical Transactions of the Royal Society of London. Series B: Biological Sciences*, 237:37–72, 1952.
24. T. Taylor, P. Ottery, and J. Hallam. Pattern formation for multi-robot applications: robust, self-repairing systems inspired by genetic regulatory networks and cellular self-organisation. *Informatics Research Report*, School of Informatics, University of Edinburgh, Edinburgh, UK, 2006.
25. D. Erwin and E. Davidson. The evolution of hierarchical gene regulatory networks. *Nature Reviews Genetics*, 10(2):141–148, 2009.
26. Y. Meng and H. Guo. A gene regulatory network based framework for self-organization in mobile sensor networks. In: *IEEE World Congress on Computational Intelligence*, Brisbane, Australia, June 2012.
27. H. Oh and Y. Jin. Adaptive swarm robot region coverage using gene regulatory networks. In: *15th Towards Autonomous Robotic Systems (TAROS)*, Birmingham, UK, September 2014.
28. B.A. White, A. Tsourdos, I. Ashokaraj, S. Subchan, and R. Zbkowski. Contaminant cloud boundary monitoring using network of UAV sensors. *IEEE Sensors Journal*, 8(10):1681–1692, 2008.

29. I. Salazar-Ciudad, J. Jernvall, and S. Newman. Mechanisms of pattern formation in development and evolution. *Development*, 130(10):2027–2037, 2003.
30. H. Guo, Y. Jin, and Y. Meng. A morphogenetic framework for self-organized multirobot pattern formation and boundary coverage. *ACM Transactions on Autonomous and Adaptive Systems*, 7(1):15.1–15.23, 2012.
31. A.G.O. Mutambara. *Decentralized Estimation and Control for Multisensor Systems*. CRC Press, Boca Raton, FL, 1998.
32. H. Oh and Y. Jin. Evolving hierarchical gene regulatory networks for morphogenetic pattern formation of swarm robots. In: *IEEE Congress on Evolutionary Computation*, Beijing, China, July 2014.
33. C.M. Bishop. *Neural Networks for Pattern Recognition*. Oxford University Press, Oxford, 1995.
34. R. Milo, S. Shen-Orr, S. Itzkovitz, N. Kashtan, D. Chklovskii, and U. Alon. Network motifs: simple building block of complex networks. *Science*, 298:824–827, 2002.
35. N. Kashtan and U. Alon. Spontaneous evolution of modularity and network motifs. *Proceedings of National Academy of Sciences of the United States of America*, 102:13773–13778, 2005.
36. L. Piegl and W. Tiller. *The NURBS Book*. Springer-Verlag, New York, 1997.
37. T. Back. *Evolutionary Algorithms in Theory and Practice*. Oxford University Press, Oxford, 1996.
38. M. Kreutz, B. Sendhoff, and C. Igel. Ealib: a c++ class library for evolutionary algorithms in shark library. Available at: http://shark-project.sourceforge.net/2.1.0/doc/EALibManual/EALib.pdf, July 2008 (last access date: September 23, 2015).
39. F. Mondada, École Polytechnique Fédérale de Lausanne (EPFL), e-puck education robot, Available at: http://www.e-puck.org/ (last access date: September 23, 2015).
40. H.K. Khalil. *Nonlinear Systems*, 3rd ed., Prentice Hall, Upper Saddle River, NJ, 2002.
41. S. Martinez and F. Bullo. Optimal sensor placement and motion coordination for target tracking. *Automatica*, 42(4):661–668, 2006.
42. G. Gu, P. Chandler, C. Schumacher, A. Sparks, and M. Pachter. Optimal cooperative sensing using a teams of UAVs. *IEEE Transactions on Aerospace and Electronic Systems*, 42(4):1446–1458, 2006.

14

REGULATORY REPRESENTATIONS IN ARCHITECTURAL DESIGN

Daniel Richards and Martyn Amos

School of Computing, Mathematics and Digital Technology, Manchester Metropolitan University, Manchester, United Kingdom

14.1 INTRODUCTION

In this chapter, we consider the application of evolutionary algorithms and regulatory networks to problems in *architectural design*. The "cybernetic theory of architecture" dates back to 1969, when Pask [49] predicted that "Various computer-assisted (or even computer-directed) design procedures will be developed into useful instruments" for the design of buildings and cities. Pask's ideas about control and communication were developed by architects such as John Frazer, who were interested in how concepts of adaptation might be applied to the design, construction and evolution of building form and performance. In *An Evolutionary Architecture*, Frazer sets out how natural processes might be harnessed as "the generating force for architectural form" [19]. Importantly, he seeks to go beyond the rigid and specific "blueprint" metaphor, and to develop a "genetic language of architecture", in which form-generating processes give rise to structure and behavior. Since then, evolutionary design in architecture has become well-established [6,7,16]. However, many such syntheses have still been modelled on what Oxman [48] calls the cyclical "generate and test" paradigm, in which conceptual design generation is followed by performance evaluation. Oxman

Evolutionary Computation in Gene Regulatory Network Research, First Edition.
Edited by Hitoshi Iba and Nasimul Noman.

argues that this approach naturally follows from the traditional way in which architectural designs have been conceived, designed and built, where form takes priority over performance. The "form-first" approach, in which "real world" performance (such as structural integrity or temperature regulation) is considered relatively late, can lead to designs that are wasteful in terms of material, are much less integrated, and are inherently less sustainable. An alternative *performative* [48, 66] paradigm seeks to invert this approach, so that "features such as material properties, fabrication constraints and environmental performance are allowed to influence the early development and synthesis of form" [55]. In order for such an approach to be successful, we require new forms of representation for computational design synthesis, and in this chapter, we discuss recent developments in this area. We describe previous foundational work, and motivate and support the need for such representations, before describing our developmental mapping process. We then investigate specific properties of our "evo-devo" model, which, we argue, allows us to exploit neutral mutations and thus produce evolutionary innovation. We present extensive experimental results to support our claim before concluding with a discussion and suggestions for future work.

14.2 BACKGROUND

Computer technologies have transformed architectural design. The proliferation of *computer-aided design* (CAD) software has significantly enhanced the control and analysis of three-dimensional designs [5], while *computer-aided manufacturing* (CAM) has created entirely new opportunities for large-scale construction [23]. CAD/CAM technologies provide designers with new abilities to design and construct complicated three-dimensional geometries. However, an important open question for design research is: can these technologies do more? Principally, can advanced computational design techniques facilitate new forms of architecture that are highly efficient and increasingly performance-oriented?

Within architectural design, a key motivation is "sustainability." That is, to address issues such as climate change and rapid urbanization, architects and engineers require new methods of designing built forms that minimize or mitigate environmental impact [65]. There exists much current debate over what constitutes a "sustainable" solution [41], but the basic argument is that designers should use CAD/CAM technologies to do *more with less*. This means that architects should seek to design *performance-driven* structures that are more efficient and more multifunctional, while also being less energy intensive and less wasteful in their use of material resources. The problem, however, is that designing complex multifunctional structures is significantly challenging. To address this, recent work has advocated the application of bio-inspired approaches to design that aim to simulate natural processes of formation in order to generate efficient material-based structures [37, 47].

Biological structures have extraordinary physical properties, and have inspired designers and engineers for generations [18]. However, biological structures are not *designed*; rather, they *emerge* through a synthesis of complex natural processes.

That is, by combining Darwinian evolution with complex and stochastic physical or developmental processes, nature assembles extremely efficient material structures that are economical, multifunctional and often (to human eyes) beautiful. These structures may be much more complex than anything we can currently design (we are currently nowhere near synthesizing even the simplest microbe), but the ability to harness the *underlying processes* for computation would provide significant advances in the field of architectural design.

The idea of using bio-inspired algorithms to generate useful architectural structures within computers (or "*in silico*") represents a growing area of interest [26, 60, 76]. So far, however, existing work remains either entirely theoretical, or describes bio-inspired analogies that relate to semi-automated design processes. The critical question for this emerging area of research is: *what type of computation is required to synthesize complex and multifunctional material structures?* In the context of this chapter, the term "computation" refers to a process of transforming a series of inputs (such as material properties, structural performance or fabrication constraints) into three-dimensional outputs. Hence the question could be restated as: what types of algorithmic processes (digital or analogue) are required to transform a set of known constraints and physical properties into complex and multifunctional architectural forms? Once we address this, we then naturally seek to understand *how* these processes facilitate the evolution of form. In what follows, we present investigations into both questions.

14.3 THE NEED FOR REGULATORY REPRESENTATIONS

Biological organisms are complex adaptive systems that demonstrate two remarkable, yet seemingly opposing characteristics. First, biological organisms demonstrate the extraordinary capacity to *innovate* and evolve novel phenotypic traits to exploit specific environmental niches. Second, biological organisms are *robust* to a great deal of genetic and environmental perturbations, showing a profound ability to persist when conditions change or parts fail. A key challenge for evolutionary computation (and associated engineering domains) is to consider *which* principles of biological systems can be extracted and used to build similarly efficient artificial systems [33, 63]. Indeed, the question of how to exploit self-organizing processes in order to build efficient artificial systems is arguably *the* "meta-problem of engineering" [36]. Critically, for many engineering domains, an ideal evolutionary system would be *robust* enough to allow fine-grained optimization of well-constrained phenotypic attributes; but it would also be *flexible* enough to explore a diverse set of solutions when prior information about the design problem is unavailable [2]. That is, an ideal system would (perhaps paradoxically) be robust and well constrained, yet flexible and capable of radical innovation. However, constructing *encodings* (genotypes) and *representations* (genotype–phenotype mappings) that facilitate both types of opposing properties is a challenging problem [12, 63]. Typically, representations tend to be either: (1) highly constrained (i.e., rigid), enabling efficient parameter optimization [30]; or (2) highly flexible, yet brittle, enabling "creative" exploration of diverse solutions that may, or *may not*, be buildable in the real world [8,45,46]. As Roudavski explains [60], a design space is defined by (and thus restricted to) the set of variables that are accessible to a designer, and "unconventional, lateral, associative moves are often necessary to expand this space and to find in it innovative outcomes."

We have previously argued [53–55] that *evolutionary design systems*, which simultaneously *invent* and *calibrate* well-constrained variables, will lead to *next-generation* CAD software and facilitate game-changing opportunities in numerous engineering domains. Consequently, our ongoing research has sought to bridge this gap between "highly constrained" and "highly flexible" representations in order to facilitate the evolution of diverse three-dimensional morphologies for real-world problem domains. We have previously presented a novel representation inspired by gene regulatory networks and used it to evolve three-dimensional structures for architectural design [55]. In this work, our evolved structures address multiple real-world performance objectives (such as structural efficiency and capacity to provide solar shading), while the material properties and size of elements were constrained to the machining limitations of specific digital fabrication equipment (Figure 14.1). Our findings demonstrated that our gene regulatory network-inspired representation could generate functional three-dimensional structures in response to a simple multi-objective design problem, and—critically—that the evolved structures had performance qualities comparable to those of a similar human designed solution. That is, the evolved solutions obtained similar levels of structural efficiency (i.e., ability to resist deflection under loading) and solar performance (ability provide a specific uniform level of shading over a period of time), yet often used much less material, and were therefore more cost-effective. However, the significance of this proof-of-principle study was not the direct performance comparison between the human-designed solutions and evolved designs, but that our gene network representation could discover and then optimize functional three-dimensional characteristics *without explicit parameterization*. For example, the arch-like characteristics of evolved designs (as shown in Figure 14.1) were not explicitly parameterized but emerged from evolved interactions within our gene networks.

We believe that this work represents an important first step toward next-generation CAD tools. Specifically, we think the model holds important clues for creating sophisticated *indirect* representations that can solve truly complex design problems by eliminating reliance on highly- constrained *direct* representations that are known to be limited as problems increase in scale and complexity [63,67].

By examining the evolutionary dynamics of our model, this chapter demonstrates how our regulatory representation can facilitate the desirable, yet seemingly opposed properties of being highly constrained *and* able to innovate. First, we describe our encoding and developmental representation. Second, we describe our testable hypothesis that our model exploits redundancy and neutrality within robust canalized genomes to facilitate enhanced evolutionary innovations. Third, we present our analysis and results, and finally, we conclude with a discussion of ongoing challenges and exciting opportunities for further work.

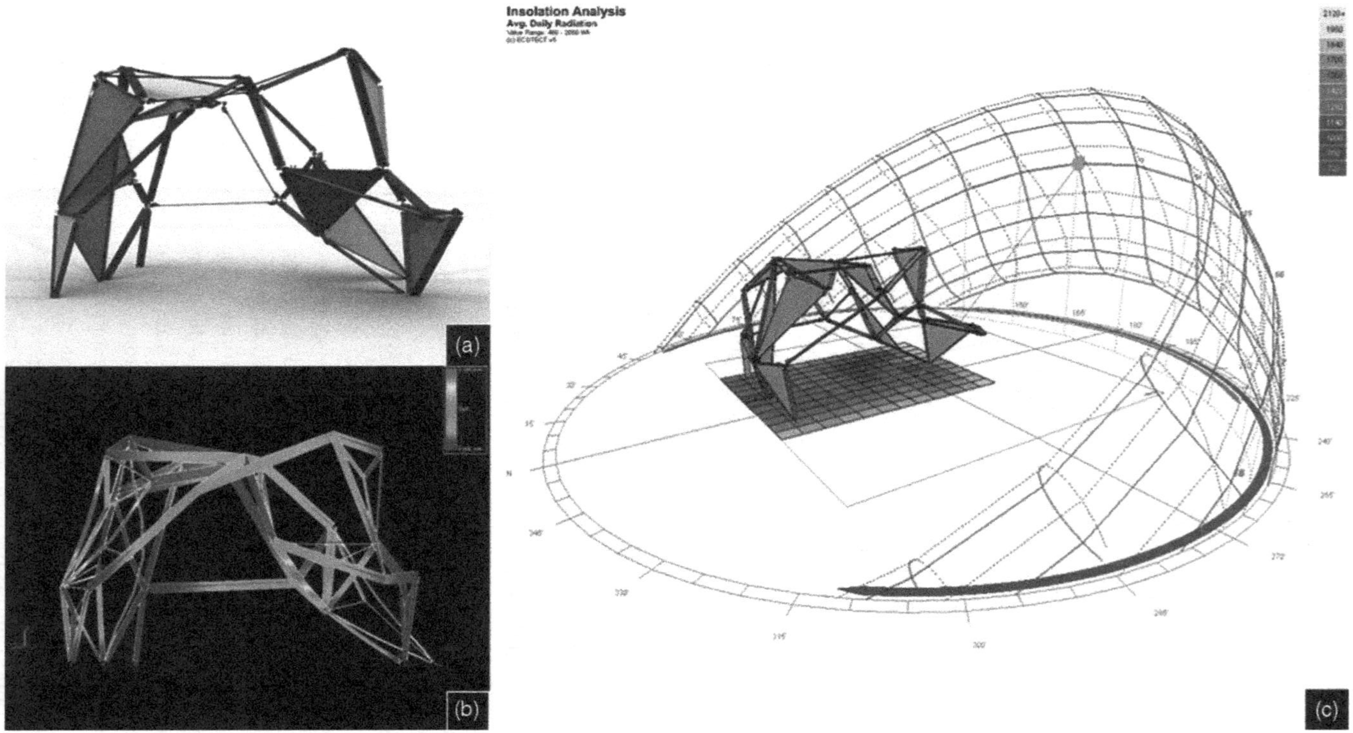

Figure 14.1 (a) Evolved canopy, (b) structural analysis, and (c) solar analysis. Reprinted with permission from Ref. [55], copyright 2012 Association for Computing Machinery, Inc.

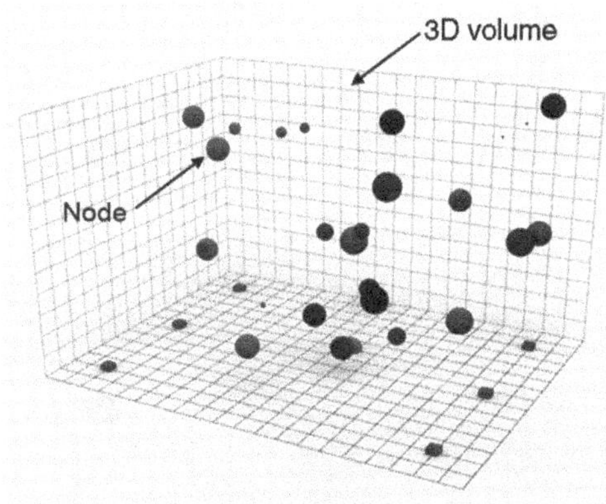

Figure 14.2 Model setup. A three-dimensional volume contains a fixed number of node objects. These nodes are encoded with simple growth instructions, which are used during the developmental representation to grow three-dimensional designs. Reprinted with permission from Ref. [55], copyright 2012 Association for Computing Machinery, Inc.

14.4 DEVELOPMENTAL MAPPING

Our model controls how a fixed number of nodes interact within a volume to create buildable three-dimensional structures (Figure 14.2). The encoding is a simple fixed string genome and the representation, which takes inspiration from gene regulatory networks, works by sequentially activating each node's associated growth instructions to construct three-dimensional designs that can then be subjected to various performance evaluations (such as structural analysis).

14.4.1 Encoding

Each node is described by a 10-digit gene, which can be broken down into four gene attributes (Figure 14.3a). These four gene attributes define four aspects of how nodes grow local connections and ultimately assemble larger network structures. First, each node has a "range of influence," *ROI*, which defines a radial dimension (or neighborhood) within which it can communicate with other nodes (Figure 14.3b). Second, each node has an explicit Cartesian position, *XYZ*, within the three-dimensional volume (Figure 14.3c). Third, nodes are able to make different geometric connections, *G*, between their neighboring nodes (i.e., within their ROI). In our initial experiment, we limited these different geometries to simple struts between neighboring nodes (Figure 14.3d), closed rings (Figure 14.3e) and "petal"-shaped components (Figure 14.3f). Finally, nodes can create connections that have various material properties, *M*, and/or different size cross-sectional areas. As shown in Figure 14.2, our model also has explicitly defined *anchor* points (denoted by the cubes), these points are necessary when performing structural analysis calculations, and simply provide

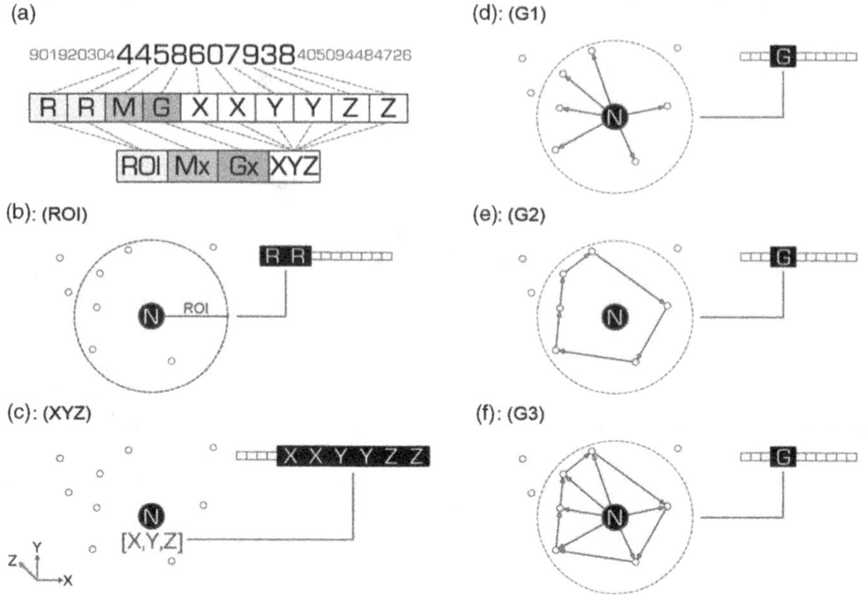

Figure 14.3 Genotype structure: (a) Each node is described by four genes, which define four aspects of its development: ROI, M, G, and $[X, Y, Z]$. (b) The *range of influence* (*ROI*), a radial dimension, described by a two-digit gene (in the range 0–99), within which nodes can perceive, communicate, and connect to other nodes. (c) The *position* of the node within 3D space as described by a six-digit gene (each digit in the range 0–9) that defines the Cartesian coordinates $[XX, YY, ZZ]$. (d–g) The library of *geometry types*, *G*, which the node uses to construct structural connections within its *ROI*. This is specified using a one-digit gene (in the range 0–9) to select either: $G1, G2$, or $G3$. Reprinted with permission from Ref. [55], copyright 2012 Association for Computing Machinery, Inc.

fixed supports that nodes can connect with in order to distribute loads to the ground plane (see Ref. [55] for full details).

14.4.2 Representation

The developmental mapping activates the growth instructions of all individual nodes to assemble a larger three-dimensional structure. During this process, each node operates as a spatially embedded *genetic switch* that, when activated, constructs various types of geometric, *G*, and material, *M*, connections between nodes within its *ROI*. However, grown network structures represent physical three-dimensional designs and, consequently, the connections have a structural depth as well as two key fabrication and assembly constraints that need to be respected. First, there can only ever be *one* connection between any two nodes (to prevent intersecting geometry) and second, depending on the fabrication method, connections must also not intersect while crossing in three-dimensional space.

To enforce these physical constraints, we *sequentially* activate nodes, using a fixed order of activation defined by node index. This means that the order in which nodes construct connections is significant. Critically, nodes that are activated early in the growth process indirectly modify the growth of subsequent connections via the process of *positive* and *negative* regulation. Positive and negative regulation is observed in biological processes of gene expression [75]. Put simply, genes are switched "on" or "off" by the existence, or non-existence, of specific proteins (in reality, genes are rarely binary). If a gene is switched "on," or *expressed*, a protein is produced, which, in turn, will control the expression or repression of further genes through *positive* regulation. Conversely, if the gene is switched "off," or *repressed*, no protein is transcribed, which influences the expression or repression of further genes by, what is termed, *negative* regulation. While existing works use biologically analogous models of gene regulation to evolve complex morphologies [4, 14], our model makes no such attempt to accurately model gene regulatory networks. However, we do utilize the basic mechanism of positive and negative regulation to produce three-dimensional structures that are described by complex regulatory dependencies.

To ensure that there is only ever one connection between any two nodes, each node is given a small "memory space" which stores information about all other nodes to which it is currently connected. Before a node creates a connection with another node (i.e., within its ROI defined neighborhood), it first checks that it does not already have an *imprint* of that node (analogous to a protein) in its memory space (analogous to a cis-regulatory region of DNA). If the node finds an imprint, then no connection is made, but if no imprint is found then a new connection is made and each node exchanges a copy of their unique imprint to prevent further connections being made. To illustrate this process, Figure 14.4 describes the developmental growth and imprint change system in more detail. For clarity, the nodes are set on a two-dimensional plane and each node is defined using only three gene attributes: *ROI*, *M*, *G*. Each node's memory space is shown as square brackets that hold the node index value (imprint) of nodes that they are connected to. Notably, Figure 14.5 illustrates that small changes to gene attributes can cause significant alterations to the phenotypes through positive and negative regulation.

To address the second physical constraint and ensure buildable phenotypes, we perform a maximum of two further processes to prune unbuildable connections. First, if physical connections are fabricated using a discrete assembly of parts, such as steel struts, we prune connections that intersect within three-dimensional space (Figure 14.6a). Note that this process may not be necessary when using certain fabrication methods such as additive manufacturing (i.e., 3D printing). Second, disconnected parts of the structure that are not supported within space (i.e., floating parts) are removed (Figure 14.6b). Critically, following the sequential activation of growth instructions and potential pruning steps, phenotypes are *always* buildable and suitably constrained for physical assembly and digital simulation processes.

14.4.3 Experimental Results

To test the evo-devo representation, we set it the task of designing an economical, free-standing structure that can provide controlled (passive) solar shading of a space during the summer months, and may be fully fabricated using 1 mm CNC cut aluminium

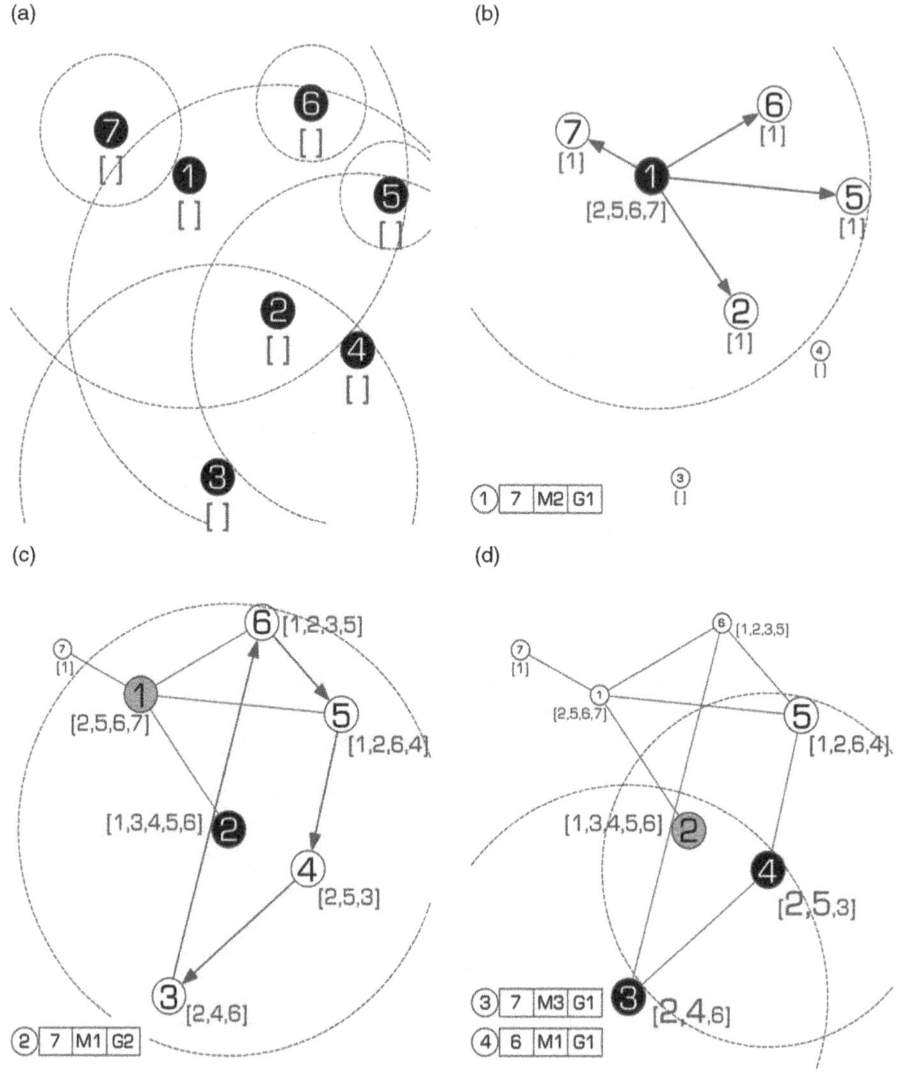

Figure 14.4 Developmental growth. (a) Shows a random arrangement of nodes with empty memory spaces. (b) Node 1 grows *G*1 connections between itself and nodes within its *ROI*. (c) Node 2 grows *G*2 connections between surrounding nodes omitting Node 1 as it already has a connection. (d) Nodes 3 and 4 have been repressed by Node 2's earlier growth. Reprinted with permission from Ref. [55], copyright 2012 Association for Computing Machinery, Inc.

sheets. For brevity, full details of the experiments are supplied in Refs. [54, 55], and the results are shown in Figure 14.1.

Briefly, the fitness of each phenotype is calculated using the weighted sum of four performance measures: first, the average nodal deflection of structures under loading was considered. Second, a measure of the average height of structural nodes

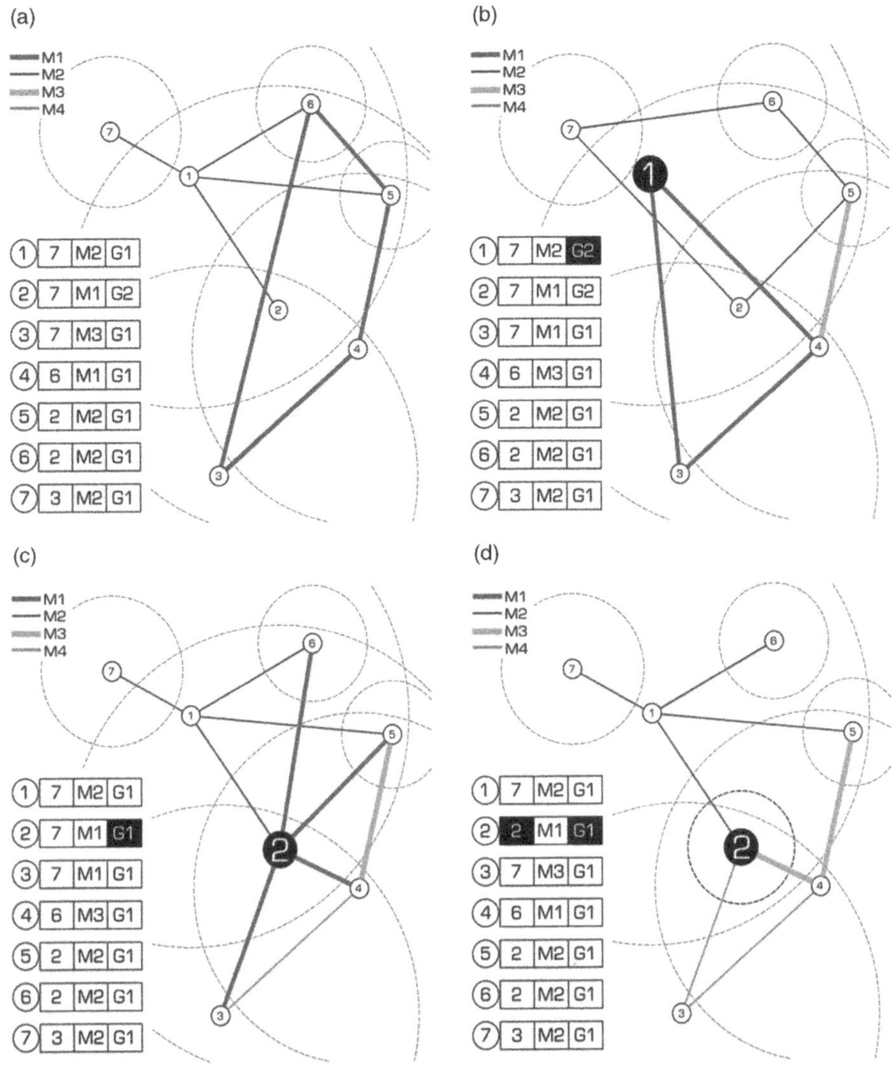

Figure 14.5 Phenotype variability. Small changes to node *ROI* or geometry type can dramatically influence phenotype formation through positive and negative regulation. (a) Phenotype produced from Figure 14.4. (b) Mutation of Node 1 geometry type|*G*1 to *G*2. (c) Mutation of Node 2 geometry type|*G*2 to *G*1. (d) Mutation of Node 2 *ROI* - *ROI* = 7 to *ROI* = 2. Reprinted with permission from Ref. [55], copyright 2012 Association for Computing Machinery, Inc.

was used. Third, total fabrication cost was considered, based on material usage, and finally, a measure of daily solar performance was used. Structural fitness (*StF*) is calculated using physical simulation. The average combined deflection in $[x, y, z]$ position of structural nodes (*AvD*) is compared with an acceptable nodal deflection value (*AcD*) and an unacceptable deflection value (*UaD*) that defines the breaking

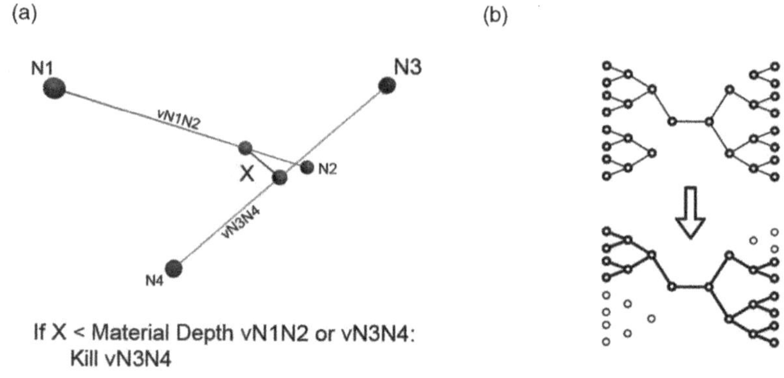

Figure 14.6 Pruning process. (a) Following the developmental growth process, if any components intersect the last component to be formed is deleted. (b) Following the intersection test, a connectivity test is used to identify and remove any components that are disconnected from the main structure. Reprinted with permission from Ref. [55], copyright 2012 Association for Computing Machinery, Inc.

point of the component when subjected to a combination of dead weight of structure and an imposed 0.2 kN wind loading:

$$\tau = 1 - \frac{AvD - AcD}{UaD - AcD}$$

$$StF = \begin{cases} 1 & \text{if } \tau > 1 \\ 0 & \text{if } \tau < 0 \\ \tau & \text{otherwise} \end{cases}$$

Height fitness (HeF) is defined by the average height of structural nodes (AvH) in relation to the maximum build volume height (BvH). In this experiment, tall models are favored in an attempt to consider how accommodation requirements of structures might be evolved in larger models:

$$HeF = \frac{AvH}{bvH}.$$

The total fabrication cost (CoF) is defined by the required amount of aluminium sheeting (ReA) to fabricate the structure. This is compared against a defined fabrication budget (FaB) and a maximum overspend allowance (MoS). For illustrative purposes the $FaB = £300$ and $MoS = £500$.

$$\gamma = 1 - \frac{ReA - FaB}{MoS - FaB}.$$

$$CoF = \begin{cases} 1 & \text{if } \gamma > 1 \\ 0 & \text{if } \gamma < 0 \\ \gamma & \text{otherwise} \end{cases}$$

Solar fitness (*SoF*) is defined by the average difference (*AvDiff*) between each solar analysis grid cell and a desired measure of solar performance (*DeS*). Full exposure of any grid cell during analysis returns 2400 kWh. For illustrative purposes, $DeS = 1500$ kWh.

$$\sigma = 1 - \frac{AvDiff - Des}{2400 - Des}$$

$$SoF = \begin{cases} 0 & \text{if } \sigma < 0 \\ \sigma & \text{otherwise} \end{cases}$$

Finally, the overall fitness of each phenotype (F) is calculated by the weighted sum of StF, HeF, CoF, and SoF. The relative weightings correspond to the importance of each performance-based attribute in the architectural solution

$$F = \frac{StF \cdot w1 + HeF \cdot w2 + CoF \cdot w3 + SoF \cdot w1}{45},$$

where $w1 = 1.5, w2 = 1, w3 = 0.5$.

The genetic algorithm (GA) parameter settings used in this section are shown below:

Population size	100
Generations	60
Independent runs	10
Selection type	Tournament
Tournament size	5
Crossover type	One-point
Crossover rate	100%
Mutation rate (per gene)	0.005%

14.5 ROBUSTNESS AND EVOLUTIONARY ADAPTATION IN BIOLOGICAL SYSTEMS

The link between robustness and evolutionary innovations within biological systems is not well understood and represents an ongoing area of investigation with many open questions [1, 22, 34, 42, 74]. Various works demonstrate that biological systems are often extremely robust to genotypic and environmental perturbation [61]. Indeed, phenotypes often seem *protected*, or buffered, against a large amount of internal (genetic) and external (environmental) variations. This mutational buffering is termed "canalization" [70]. As noted by Stanley and Miikkulainen [63][p. 113], Waddington's use of the term *canalization* draws an analogy between "the way water running down a hill eventually carves out regular streams in the surface, and the way development slowly settles on a set of conventions that become ingrained in the genome." Critically, this means that a canalized genome is less sensitive to certain perturbations and thus more likely to produce consistent phenotypic traits. The origins of mutational robustness and canalization remain the subject of investigation [1, 17, 62, 68]. Yet, existing work suggests they can originate from *adaptive evolution* [71], or be an *emergent property of gene regulatory networks* [29]. Canalization is clearly an important property of biological structures because it preserves important phenotypic traits in response to (genetic and environmental) variation and perturbation. Consequently, Stanley and Miikkulainen [63] suggest that biologically analogous mechanisms of canalizing "brittle" (yet flexible) artificial encodings may benefit evolutionary algorithms by providing a safer search through genotypic space. Critically, this ability to optimize highly flexible representations and address real-world problems is a key goal for next- generation CAD tools; therefore, models that can exploit bio-inspired processes and enhance mutational robustness may significantly benefit future design software.

Interestingly, existing work shows that mutational robustness may provide another desirable property and may actively facilitate evolutionary innovations [32,73]. In the late 1960's, Kimura [31] proposed that during evolution, many mutations that occur at the molecular level are "silent" or neutral (i.e., do not affect phenotype fitness) and consequently, these mutations are selectively propagated throughout populations via the random process of genetic drift. More recently, research has suggested that these neutral mutations could be the source of innovation and beneficial evolutionary adaptations in biological organisms [13, 15, 74]. Put simply, this means that over generational time canalization allows organisms to accumulate hidden genetic variation [35], which may eventually become non-neutral and facilitate useful evolutionary adaptation [25, 73]. This leads to an interesting question of whether certain types of genome are more likely to acquire novel functionality through genetic change. Put simply, does the ability to evolve itself evolve? From a CAD perspective, this idea is extremely exciting, because representations that could manipulate their sensitivity to mutations might have the capacity to perform both: flexible, explorative search and fine grained optimization of robust three-dimensional parts. The term *evolvability* is often used to discuss how well an organism can adapt to changing conditions and produce evolutionary innovations; however, we note that the term does not have a generally agreed definition or common methods for measurement [50, 52]. Indeed, computational studies have shown that artificial gene regulatory networks may possess useful properties relating to "evolvability" [1, 10], while further work has also shown that gene regulatory networks may increase mutational robustness simply as a by-product of "developmental stability" [62]. Interestingly, existing work has also explored possible connections between the properties of evolvability, redundancy, and neutrality in evolution. However, this topic remains a contentious topic, with experimental results finding both positive [77] and negative [9] effects of encouraging neutrality and redundancy during evolution search [21, 28].

14.5.1 Hypothesis

Following on from our previous work [55], we developed the working hypothesis that our gene networks become more robust following evolution, and that this increased robustness enables genomes to exploit phenotypically neutral mutations and produce better evolutionary innovations. We now briefly outline two observations from our previous investigations that led us to this hypothesis. First, we observed that mutations to random gene networks seemed to cause more damage (as assessed visually) to the resulting phenotypes than evolved networks. We speculated that this *phenotype variability* could be associated with disconnected elements in the gene networks (Figure 14.7a) and chaotic cascades of gene expression. That is, because network connections are grown using relative information, rather than relying on explicitly encoded index values in the genomes, the developmental process used to build the gene networks may be fragile and susceptible to disruption. Subsequently, we hypothesized that an evolutionary trait that actively reduces this susceptibility for disruption

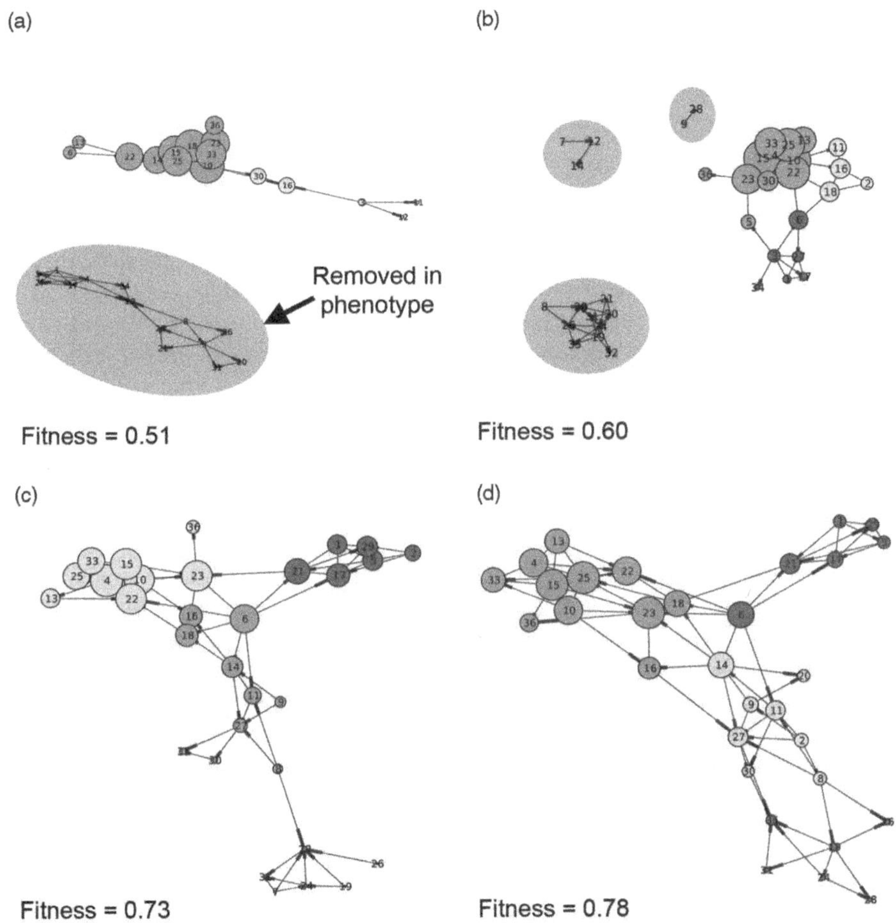

Figure 14.7 Growth and differentiation of gene regulatory networks in two-dimensions. Networks begin with small disconnected connections (a–b) and through evolution become larger functional structures (c–d). Nodes are positioned using Blondel *et al.*'s [3] method in Python using Thomas Aynaud's "community" module for NetworkX. Nodes are then sized according to their eigenvector centrality.

(thus increasing mutational robustness) may emerge "for free" in our system. This means that our gene networks may develop methods of "buffering" against deleterious mutations (which destabilize, or disrupt the developmental growth of connections) by compensating for certain perturbations using evolved regulatory logics. This idea is consistent with the work of Siegal and Bergman [62], who found that evolutionary selection of developmental stability is enough to evolve mutational robustness in computational gene regulatory networks. Additionally, our regulatory network model shares several similarities with Julian Miller's Cartesian Genetic Programming (CGP) model [40], which, interestingly, has been used to evolve robust cellular models that can self-repair after perturbation [39]. Indeed, we suggest that Miller's self-repairing behavior may be the result of a similar need for developmental stability during a growth phase, which would further support our hypothesis that the model may indirectly evolve this form of mutational robustness.

Second, we speculated that our evolved genomes might contain a high degree of *neutrality* and *redundancy*. That is, due to mutual inhibition of genes during the growth of the gene networks, evolved genomes may have a significant number of redundant genes, which are completely *silent* and have no phenotypic effect. To test this idea, we replaced our standard genetic algorithm with a simpler (1+9) evolutionary strategy (as favored by Miller's CGP approach [40]), and found that this considerably improved our search performance. Critically, we hypothesized that this change may produce better evolutionary adaptations by allowing our gene networks to accumulate hidden genetic variation, through the random process of genetic drift, and exploit this neutral "tinkering" via changes to highly connected hub genes. Notably, this idea is consistent with Vassilev and Miller's findings, which show that neutral mutations improve the evolution of digital circuits when using a CGP approach [69]. Additionally, the idea that biological systems exploit hidden genetic variation to facilitate evolutionary innovations has been widely advocated in recent research [13, 15, 25, 73, 74].

The following section interrogates two key features of our model to explore the role of mutational robustness and evolvability. For clarity, the two features that we expect to play important roles in the evolutionary dynamics of our model are the *sequential activation of nodes* and *pruning of unbuildable connections*. First, recall that our encoding does not define specific connections between genes (i.e., no indexed targeting of nodes), but rather specifies a relative *connective potential* based on four attributes of the gene and a sequential growth process. The significant implication of this *sequential growth* process is that earlier activation of genes (to grow network connections) can influence the later formation of connections through a primitive form of *regulation*. Second, the three-dimensional structures described by gene networks are not guaranteed to be build-able. This means that gene networks, as shown in Figure 14.8(b), simply represent an "unzipped" version of the string genome. In order to translate the gene network into a viable three-dimensional phenotype with specific material dimensions and assembly tolerances, we use a secondary pruning process, whereby unbuildable elements are removed, in a similar manner to Nolfi and Parisi's [44] rationalization of "grown neural networks" (Figure 14.8).

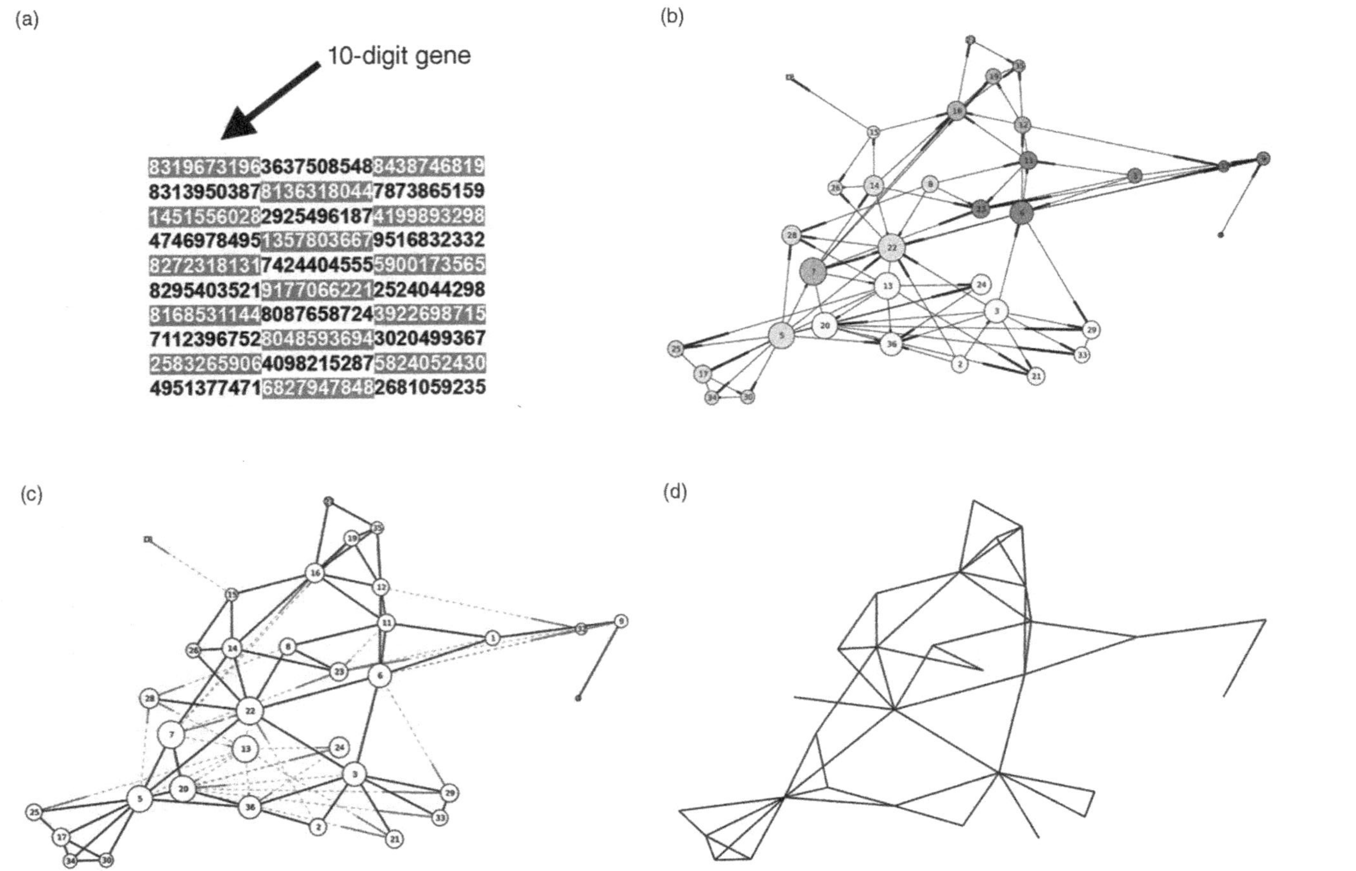

Figure 14.8 Genotype–phenotype mapping process. (a) The genotype is a fixed string genome where all characters are integers between 0 and 9. (b) Directed three-dimensional network of regulated connections between genes–here shown in two-dimensions. (c) Network connections that are unbuildable are removed–shown as red dotted lines. (d) The phenotype represents all buildable three-dimensional connections of the regulated gene network–here shown in two-dimensions for clarity.

14.5.2 Experimental Results

To test our hypothesis, we investigate three properties of the model. First, we show that evolved solutions are more robust to mutations than random gene networks. Second, we show that neutral mutations improve evolutionary search and allow us to generate better three-dimensional designs. Third, we demonstrate *how* hidden genetic variation facilitates better evolutionary innovations within our gene networks.

The experimental set-up is the same as that described in Section 14.4.3, with the following parameters:

Population size	10
Generations	600
Independent runs	20
Selection type	Elitism
Crossover type	None
Mutation rate (per gene)	0.02%

14.5.3 Canalization of Gene Networks

We first consider *mutational robustness* and *developmental stability* by measuring how well solutions are able to retain phenotypic traits following genetic perturbation. To do this, changes in "locality" are considered. *Locality* is a measure of how neighboring genotypes correspond to neighboring phenotypes [59]. When small changes to genotypes correspond to small changes in phenotypes, the genotype-phenotype mapping is said to have high locality. Conversely, if small changes to genotypes represent large jumps across phenotype space, the mapping can be considered more volatile and is described as having low locality. High locality has shown to improve evolutionary search [58], but it is also an indicator that solutions are more robust to mutations because they are better able to retain phenotypic traits following genetic perturbation. To establish the similarity between individuals, a measure of phenotypic distance, *pDist*, is used, which captures the semantic difference between two solutions:

$$pDist(A,B) = \frac{Size_A}{Comm_{AB} - Diff_{AB}},$$

where *SizeA* is the total connections in phenotype A, *CommAB* is the number of common connections between phenotype A and B, and *DiffAB* is the number of different connections between phenotype A and B. Additionally, if $(CommAB - DiffAB) \leq 0$ then $pDist(A,B) = 1$. Note that we use fixed string genomes for all individuals, therefore, $Size_A$ is a constant, and the distance between all solutions is symmetric.

To understand how individuals respond to genetic perturbation, we use Raidl and Gottlieb's [51] measure of mutation innovation, *MI*, to calculate how much "innovation" (or difference) is introduced into the solution by perturbing genotype

$$MI = pDist(X, X^m),$$

where X^m is the result of perturbing gene, m, in solution X. Thus, MI represents the semantic difference between solution X and the perturbed solution X^m. Notably, MI is directly related to locality [51].

To understand the relative influence of gene m in phenotype X, it is necessary to consider each fixed string genotype (Figure 14.8a) as its associated gene network (Figure 14.8b). That is, to appreciate which changes to genes represent small or large *genotypic* adaptions, it is important to consider the architecture of the gene networks. To do this, the *importance* of genes can be measured using their eigenvector centrality. Note the influence of any node within a simple network can be measured by how many incoming and outgoing connections it has, where this number is called the *degree centrality*. However, a measure of eigenvector centrality takes into account the fact that *all nodes are not equal*. Consequently, nodes are not measured simply by their number of connections, but also by the influence or "quality" of the nodes with which they share connections [43]. We calculate the eigenvector centrality of nodes using the NetworkX package for Python.

To test robustness of solutions in response to genetic perturbation, we perform gene knockout tests on random and evolved solutions and measure the correlation between MI and eigenvector centrality of the removed gene. Figure 14.9(a) illustrates the MI obtained by removing each gene from 20 *random* solutions. Figure 14.9(b) then demonstrates the MI obtained by removing each gene, m, in 20 *evolved* solutions. These results show that random solutions are (1) extremely brittle and (2) contain many redundant genes. This is evidenced in Figure 14.9(a), whereby removal of genes that have fewer important connections (i.e., low eigenvector centrality) can lead to completely different phenotypes (i.e., $MI = 1$). Conversely, genes that appear to have many influential connections (i.e., high eigenvector centrality) may lead to unbuildable parts of the phenotype (that will be pruned during the developmental mapping); therefore, the elimination of these genes can have zero phenotypic effect (i.e., $MI = 0$). This behavior indicates that random solutions have unstable developmental processes, which can be easily disrupted by genetic perturbation. However, as shown in Figure 14.9(b), evolved solutions appear to increase *locality* and thereby increase mutational robustness in response to genetic perturbation. Indeed, here we see an emerging correlation between eigenvector centrality and MI. This means that perturbation of influential genes tends to produce larger phenotypic changes, and conversely, perturbation of less influential genes tend to produce smaller phenotypic effects. Figure 14.9(c) illustrates that evolved solutions are better able to retain phenotypic traits following genetic perturbation and have increased developmental stability and mutational robustness.

To understand how the genomes of evolved solutions create phenotypes that are more robust to genetic perturbation we plot how individual genes modify their eigenvector centrality over a typical evolutionary run (Figure 14.10). Here we see that following a short period of fluctuation (around 50–80 generations), the gene network appears to become canalized. Significantly, we see the emergence of "hub genes" that have high eigenvector centrality and seem to preserve phenotypic traits by buffering against large and potentially deleterious mutations. We think that this type of mutation buffering or "heuristic bias" [51] emerges simply as a by-product of requiring

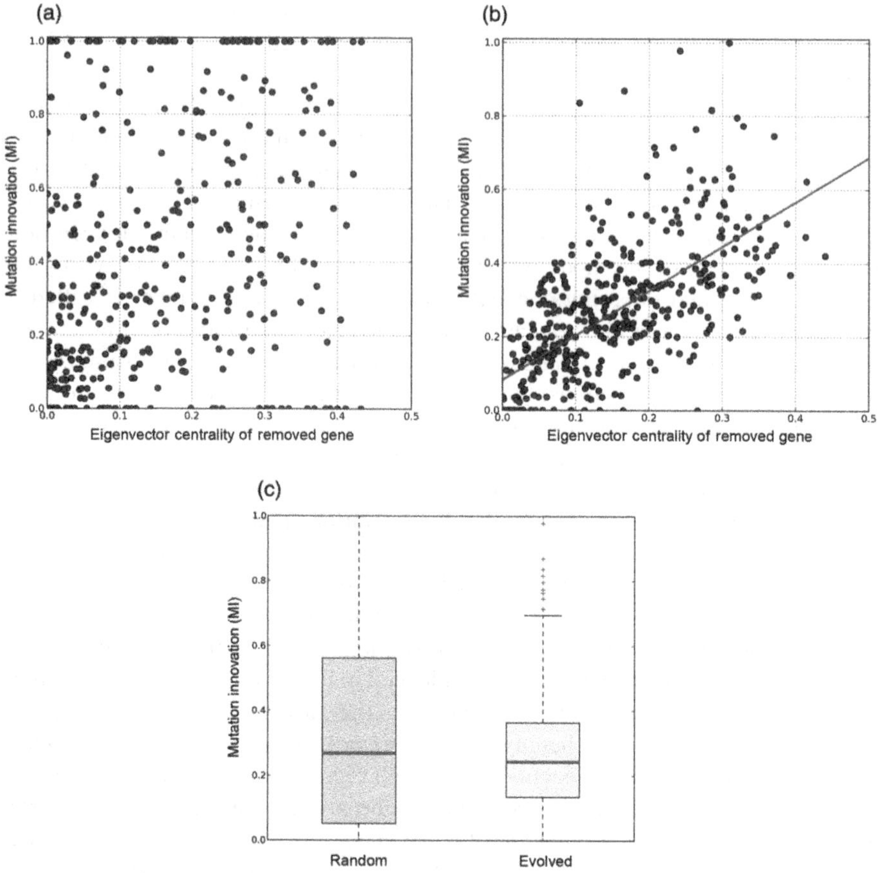

Figure 14.9 Changes in locality and mutational robustness. (a) Low locality in random solutions. No correlation between mutation innovation and eigenvector centrality of genes in random solutions. (b) Increased locality in evolved solutions. Correlation between mutation innovation and eigenvector centrality in evolved solutions. (c) Increased mutational robustness in evolved solutions. Cumulative mutation innovation in both random and evolved solutions.

developmental stability to optimize gene networks, and notably, supports the earlier findings of Siegal and Bergman [62].

The significance of this finding is that gene networks can indirectly control how mutations affect phenotypes and this may make them easier to optimize. Critically, by increasingly the locality of the genotype-phenotype mapping, evolved solutions establish a higher correlation between small changes in gene networks and small changes in phenotypes. Higher locality in traditional bit-string encodings is known to improve evolutionary search [58,59]. However, it is less well known how locality affects "tree" or network-based encodings in methods such as genetic programming (GP). In GP, previous strategies of improving search have focused on varying mutation rates in order to mediate between explorative search (high mutation rate when fitness

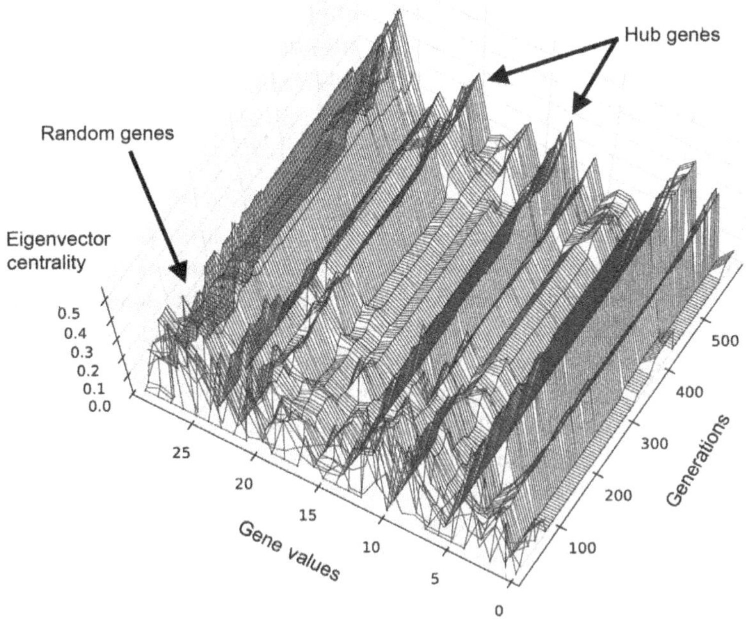

Figure 14.10 Typical canalization of a genome. Canalization typically occurs within 100–150 generations and produces noticeable hubs in the gene network. Canalized genomes are better able to retain phenotypic traits following gene knockouts.

is low) and optimization (low mutation rate when fitness is high) [20]. However, through canalization of the gene networks, our approach seems capable of implicitly controlling the effect of mutations on phenotypes, without explicit variations to mutation rates or operators.

14.5.4 Neutral Shaping of Canalized Gene Networks

This section demonstrates that our gene networks perform better evolutionary search when they are able to accumulate hidden genetic variation via neutral mutations and random genetic drift. To show this we compare two sets of results by running the model on the same problem (as detailed in Ref. [55]), with and without neutral mutations, for 20 independent runs. For the purposes of this analysis, it is sufficient to know that our objective function favored tall and structurally stable designs.

The first treatment, which does allow neutral mutations to occur (and thus enable the accumulation of hidden genetic variation via genetic drift), follows Miller's [39, 40] CGP approach, using a simple (1+9) evolutionary strategy that has one significant feature. That is, during selection, if two or more phenotypes obtain an equally good fitness score, the phenotype selected to seed the next generation is not the *current* best. Put simply, the phenotype selected will be *"equally fit, but genetically different"* from the phenotype selected in the previous generation [38]. In contrast, the second treatment, which does not allow neutral mutations, uses the same evolutionary strategy but only selects phenotypes that are not the *current best* if they increase fitness. Figure 14.11(a) shows that over 20 test runs, the gene networks that allow neutral mutations perform better. Significantly, genetic drift produces better solutions in around half the time. Figure 14.11(b) illustrates that the model that enables neutral mutations does in fact accumulate hidden genetic variation over time. This is shown by plotting the Hamming distance between consecutive genotypes that increase phenotype fitness throughout evolutionary runs. This plot illustrates, as do Vassilev and Miller's findings [69], that Hamming distance increases during periods of phenotypic stasis, which is a strong indicator of genetic drift caused by neutral mutations. Notably, during one of the evolutionary runs, we observed that following a period of stasis, the Hamming distance between consecutive genotypes was 59. Note that since each fixed string genome has only 120 gene attributes (comprising *ROI*, *M*, *G*, and *XYZ*); this value of 59 represents around 50% genetic difference caused by neutral "tinkering"! To illustrate that neutral mutations are in fact making *structural changes* to the gene network architecture, Figure 14.11(c) shows the cumulative changes to network topology over an average evolutionary run. Interestingly, this figure demonstrates that neutral mutations constantly sculpt the topology of canalized gene networks over generational time. We suggest that the ability to exploit genetic drift allows our solutions to better avoid local optima. However, our results show that genetic drift enhances the early exploration of solutions during the first 100 generations (Figure 14.11a), which is also when phenotypes are known to be more susceptible to mutations (Figure 14.10). Figure 14.12 illustrates how evolutionary adaptations appear during typical evolutionary runs, with and without neutral mutations. Figures 14.12a and 14.12b show how the eigenvector centrality of each gene changes over time (i.e., a two-dimensional projection of the graph shown in Figure 14.10). These figures both highlight short periods of phenotypic stasis, which are punctuated by points of evolutionary adaptation (increases in fitness). Figure 14.12(a) shows a typical run where neutral mutations have been actively prevented. Note that here periods of phenotypic stasis are also periods of genotypic stasis. Figure 14.12(b) shows a typical run with neutral mutations. Here neutral mutations continually adapt the topology of gene networks and eventually accumulate to facilitate useful evolutionary adaptions. Figure 14.12(c) illustrates how the neutral mutations shown in Figure 14.12(b) (generations 253–316) appear in relation to specific gene values. Here we see that neutral mutations slowly change gene network topology, whereas evolutionary adaptations represent larger changes to the eigenvector centrality of genes.

14.5.5 Neutral Mutations Contribute to Evolutionary Innovations

We have shown that (A) phenotypically neutral mutations improve evolutionary search (Figure 14.11a), (B) using neutral mutations allows genomes to accumulate hidden genetic variation (Figure 14.11b), and (C) that hidden variation allows gene networks to continually sculpt their topology over generational time (Figure 14.12).

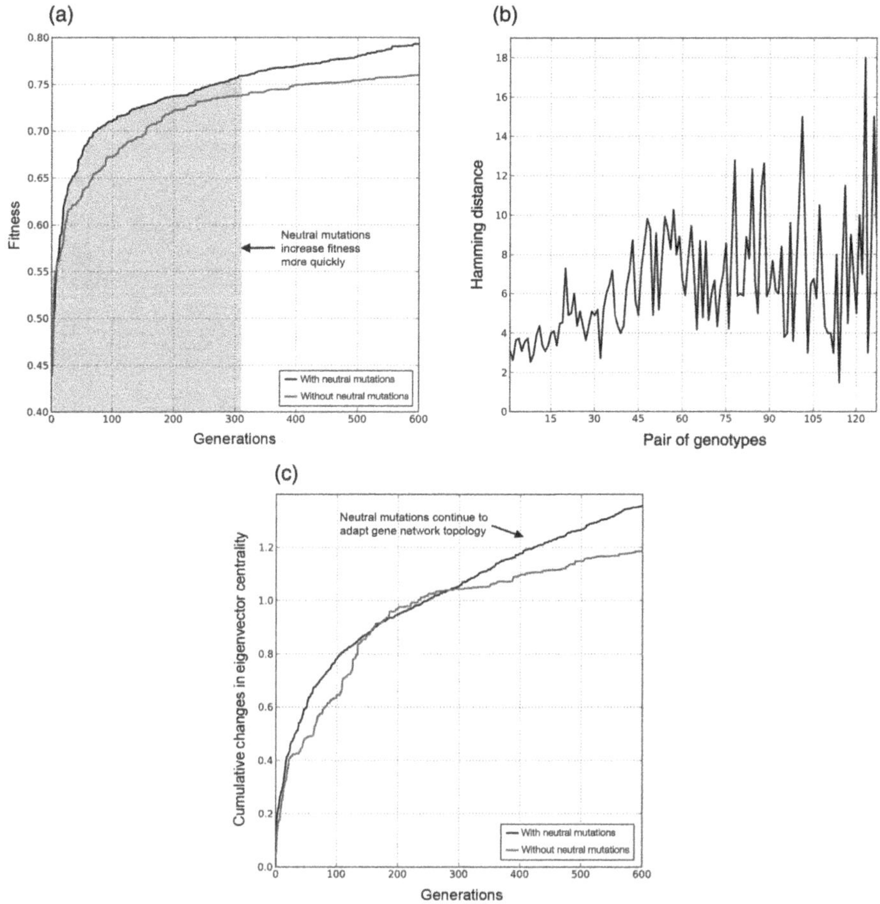

Figure 14.11 (a) Neutral mutations improve evolutionary search. Comparative fitness of model with and without neutral mutations. (b) Hamming distance of model with neutral mutations. This graph shows the average Hamming distance of consecutive genotypes which increase fitness throughout evolutionary runs. (c) Phenotypically neutral mutations continue to adapt gene network topology. This graph shows the average cumulative change in eigenvector centrality of the gene networks over generational time.

In this section, we show that this neutral sculpting of canalized gene networks actively contributes to evolutionary innovations. To describe how neutral mutations influence the emergence of evolutionary innovations, we use two different time scales. First, we focus on Figure 14.13, which investigates mutations that occur over entire evolutionary runs. Then later, Figures 14.16 and 14.17 focus on a small period of phenotypic stasis from a typical evolutionary run to give a more detailed view of *how* the model successfully combines canalization and genetic drift to produce better evolutionary adaptations. Figure 14.13(a) shows which gene attributes (*ROI*, *M*, *G*, and *XYZ*) tend to facilitate evolutionary adaptions shown by plotting the cumulative number of

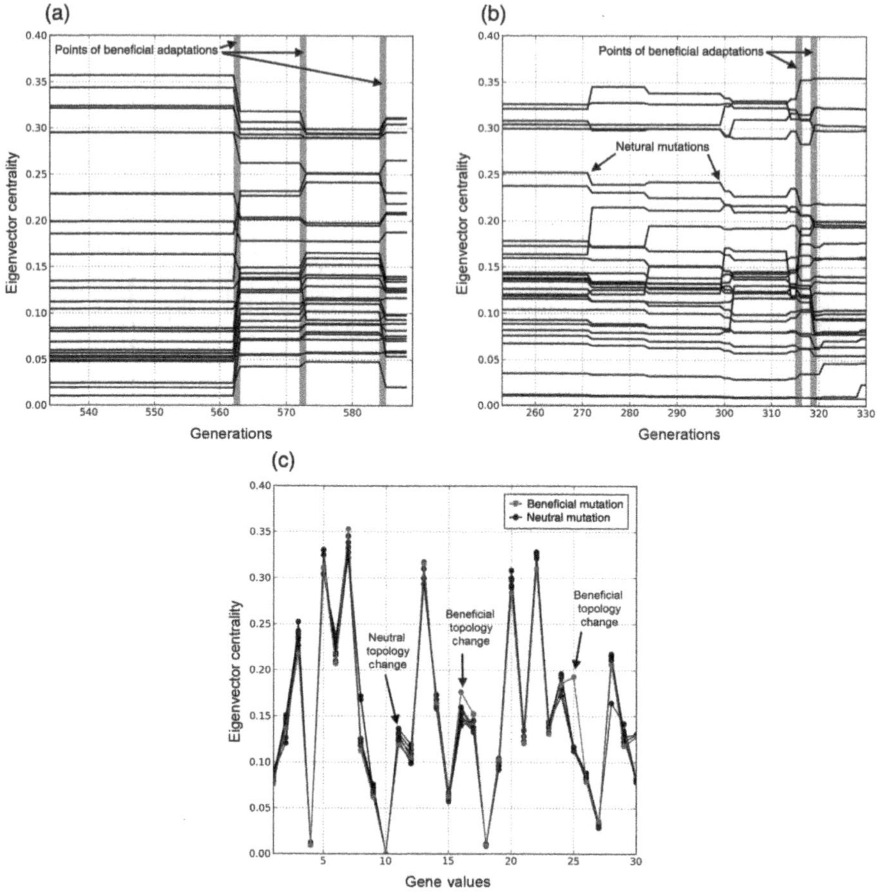

Figure 14.12 Changes in eigenvector centrality of gene networks during phenotypic stasis. (a) Model without neutral mutations. Each line represents a gene within the genotype. (b) Model with neutral mutations. This graph shows neutral "tinkering" of the gene networks during phenotypic stasis. (c) Model with neutral mutations showing how eigenvector centrality deviates through phenotypic stasis in relation to individual genes.

mutations that cause beneficial mutations over the evolutionary runs. Figure 14.13(b) shows what type of mutations cause evolutionary innovations following periods of phenotypic stasis. Figure 14.13(c) illustrates the cumulative number of neutral mutations during evolutionary runs. Finally, Figure 14.13d shows what type of neutral mutations occur during periods of phenotypic stasis.

These results show three interesting properties of our model. First, Figures 14.13c and 14.13d show that neutral mutations occur consistently during generations. Second, Figures 14.13a and 14.13b show that mutations to certain type of gene attributes are more likely to be selected by evolution than others. For example, the *XYZ* attribute of each gene (which specifies the Cartesian position of each node in space) is least

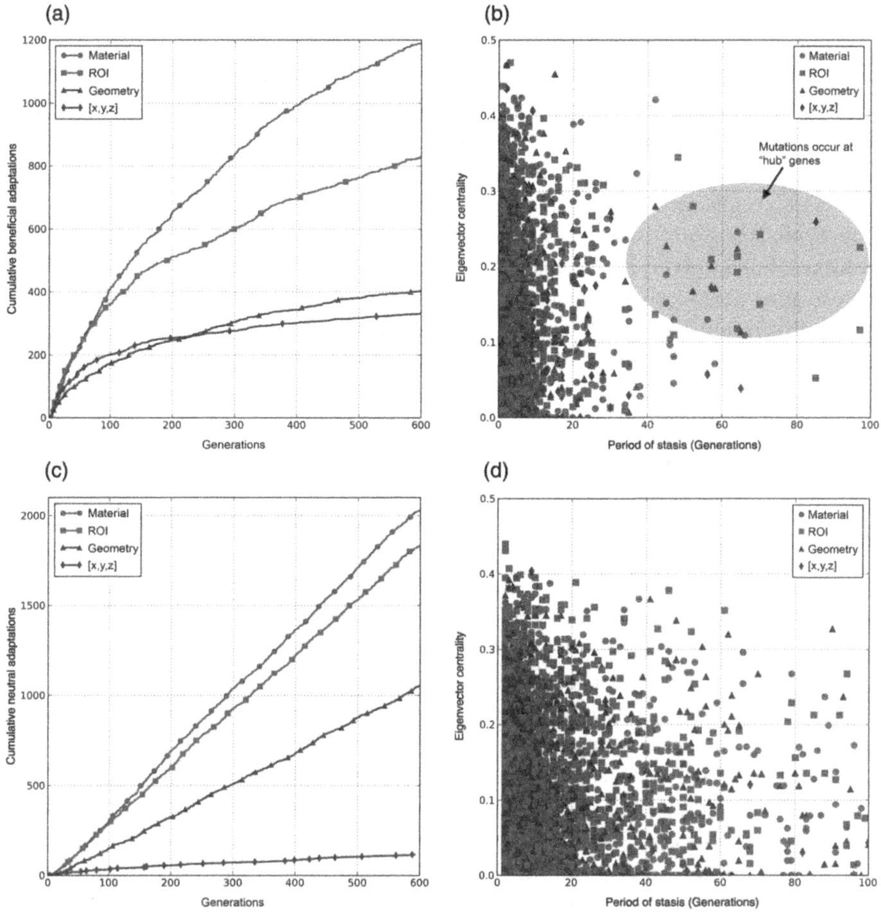

Figure 14.13 (a) Cumulative beneficial adaptations of specific gene attributes. (b) This plot shows all mutations which end periods of phenotypic stasis during the evolutionary runs. (c) Cumulative neutral mutations of specific gene attributes. (d) This plot shows all neutral mutations which occur in periods of stasis, during evolutionary runs.

likely to provide beneficial adaptations and is also the least likely to cause neutral mutations. This is perhaps not surprising, because changes to the position of nodes are likely to cause major disruptions during development because of the relative positional information used to create connections. That is, changes to the *XYZ* attribute of any gene will likely cause the largest *developmental instability*. Similarly unsurprising is that changes to the material properties attribute, *M*, of any gene (analogous to the weighting of a neural network connection) is shown to be the more likely to produce beneficial adaptions and neutral mutations. Indeed, this gene attribute does not directly change network topology and is therefore the least disruptive using development. However, these figures also show that the *ROI* gene attribute, which defines the size of each node's neighborhood, is also often a source of evolutionary adaptions and neutral mutations. Yet, this gene attribute does have the potential to cause significant developmental instability by altering the topology of gene networks. However, it appears that deleterious mutations to this gene attribute are also more likely to be buffered by earlier activation of regulatory *hub genes*.

Third, Figure 14.13(b) indicates the existence of these types of "hub genes" that can act as simple genetic switches. Here we see that mutations that end periods of phenotypic stasis often occur at genes with higher than average eigenvector centrality (average of an evolved node is around 1.4). As shown in Figure 14.9, perturbations to these genes are likely to introduce more "innovation" (difference) into the solution and yet this innovation is likely to be deleterious as solutions approach local optima. We acknowledge that this behavior is observed in both treatments, that is, with and without neutral mutations. However, we suggest that the ability to accumulate "potentially useful junk" [24] that is eventually activated by influential hub genes provides a significantly more efficient evolutionary search (Figure 14.11a). Figures 14.14 and 14.15 illustrate the solutions obtained over a typical evolutionary run (with neutral mutations) and demonstrate the evolution of a structurally efficient dome-like structure to fulfil various performance criteria.

Neutral mutations shape gene network topology during periods of phenotypic stasis. However, an important question is how do neutral mutations contribute toward evolutionary innovations in our model? Critically, are the individual mutations that increase phenotype fitness fundamentally reliant upon neutral tinkering or would the same mutations produce beneficial adaptions without them? For example, consider a solution, X, that begins a period of phenotypic stasis. This period of stasis is ended by a new solution, X^{nm}, which contains some neutral mutations, X^n, and at least one beneficial mutation, X^m. Critically, here the fitness, F, of X^{nm} must be greater than X to end stasis. However, if the phenotypically neutral mutations, X^n, are also *genotypically* neutral, then the fitness of X^m should also be greater than X. Figure 14.16 illustrates that this is not the case, and that neutral mutations are *integral* to evolutionary adaptions.

Figure 14.16(a) shows that solution X and X^m produce the same gene network and exactly the same phenotype (i.e., $MI = 0$), therefore the fitness of solution X^m is equal to the fitness of solution X. Figure 14.16(b) illustrates that solution X^{nm} (with neutral mutations) does produce a different phenotype and increases fitness. Here we see that one new phenotypic connection is created (between gene 25 and gene 7, i.e., 25 > 7), but also four new gene network connections are created (2 > 24; 13 > 30; 28 > 7; 16 > 25). Interestingly, these four additional gene network connections are removed during the pruning process (as shown by the dotted line); yet ultimately facilitate this phenotypic innovation. To understand how these changes are represented in the fixed string genome, Figure 14.17 visualizes how gene properties are altered during this period of phenotypic stasis. Each genome is represented as a horizontal line of 30 spheres. Each sphere represents one 10-digit gene where the size and shade is defined by the associated gene attribute. Stacking multiple genomes on top of one another reveals the hidden genetic variation that is accumulated over the period of stasis and the beneficial mutation to gene 25 which ultimately increases

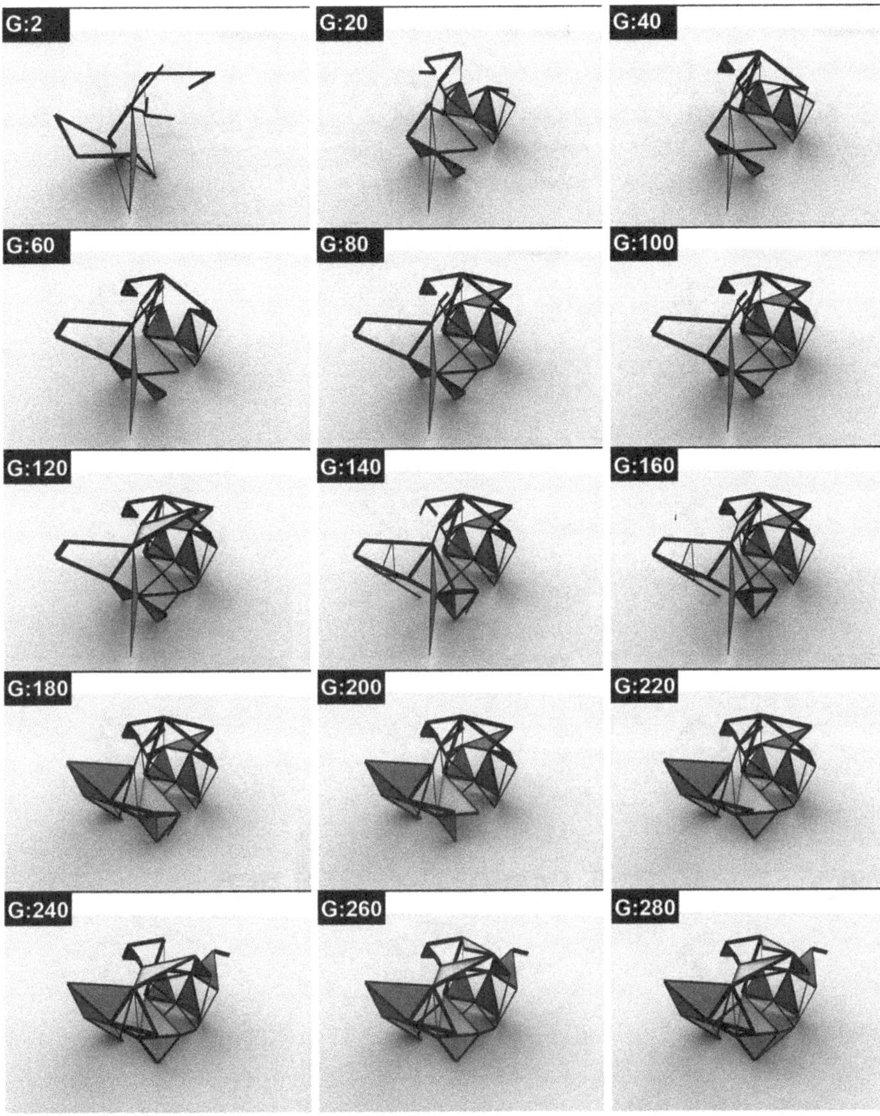

Figure 14.14 Typical evolutionary run for generations, $G = 2$ to $G = 280$. This figure shows that in the early stages of evolution changes to genes tend to introduce more "innovation" (or difference) into the phenotype. Significantly, this allows the evolutionary algorithm to explore novel solutions during the early stages and discover useful "fundamental" properties which can be protected (by canalization) and subsequently optimized.

phenotype fitness. Figure 14.17(a) shows the entire genetic history of best solutions. Notably, the early canalization of the genome can be seen here whereby fluctuating gene values are eventually replaced by noticeable columns that represent regulatory hub genes. Figure 14.17(b) supplements Figure 14.16(b) and provides a method of

Figure 14.15 Typical evolutionary run for generations, $G = 300$ to $G = 580$. This figure illustrates the latter stages of an evolutionary run (following canalization). Here the architectural structure is adapted by exploiting hidden genetic variation to make smaller changes than shown in the early stages of the evolutionary process. Significantly, this allows the solutions to become slowly optimized over time to produce well-performing architectural structures (see $G = 580$).

visualizing how neutral changes to the genome eventually allow gene 25 to contribute a beneficial mutation. From observation, it appears that neutral changes to genes 13 and 17 alter the regulatory sequence of inhibition in the gene network without changing phenotypic connections and these changes make it possible for a beneficial

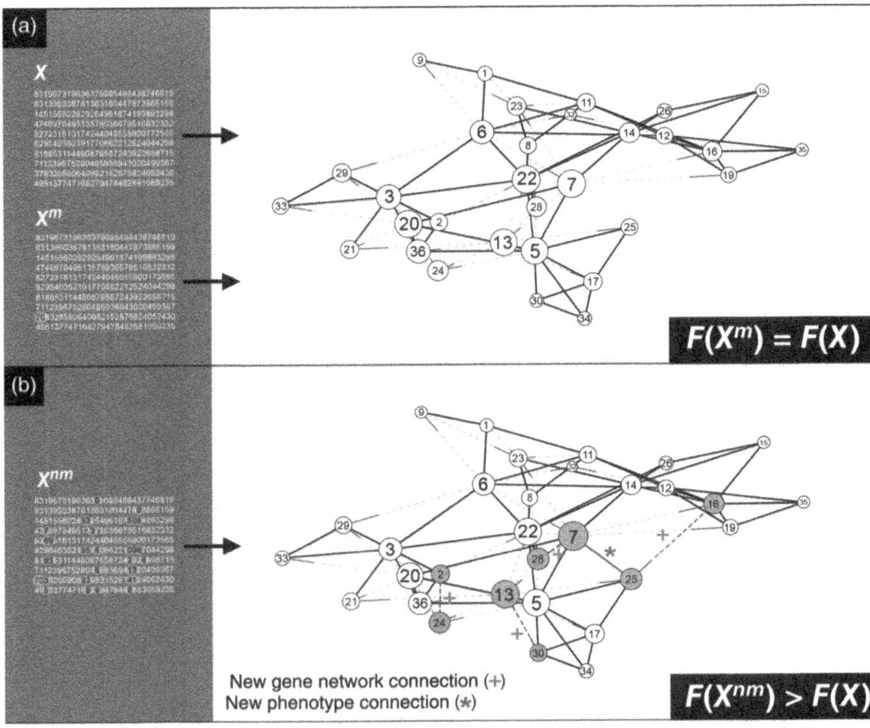

Figure 14.16 This diagram shows that neutral mutations can be integral to evolutionary innovations. (a) shows that when neutral "tinkering" is removed from the genome, beneficial mutations are canalized. (b) shows that the inclusion of evolved neutral "tinkering" enables an evolutionary adaptation to occur which improves the fitness of the solution: $F(X^{nm})$.

mutation to occur at gene 25. Further investigations are required to fully understand how neutral mutations improve evolutionary search in this model. However, we have shown that hidden genetic variation is an integral component of the model that aids performance.

14.6 CONCLUSIONS AND DISCUSSION

Advanced CAD software tools that enable designers to explore large search spaces, exploit emerging fabrication opportunities, and ultimately discover new types of high-performance *tailored* materials, will have significant benefits to many engineering fields. Yet, as we argue in this chapter, a major challenge for developing these types of tools is how to create alternative *representation* schemes that are both robust and evolvable. Critically, we propose that indirect representations, which can exploit primitive regulatory interdependencies, will be useful in this area.

In analyzing the dynamics of our model, we demonstrate that randomly generated gene networks are easily disrupted by changes to the genome. This means that small

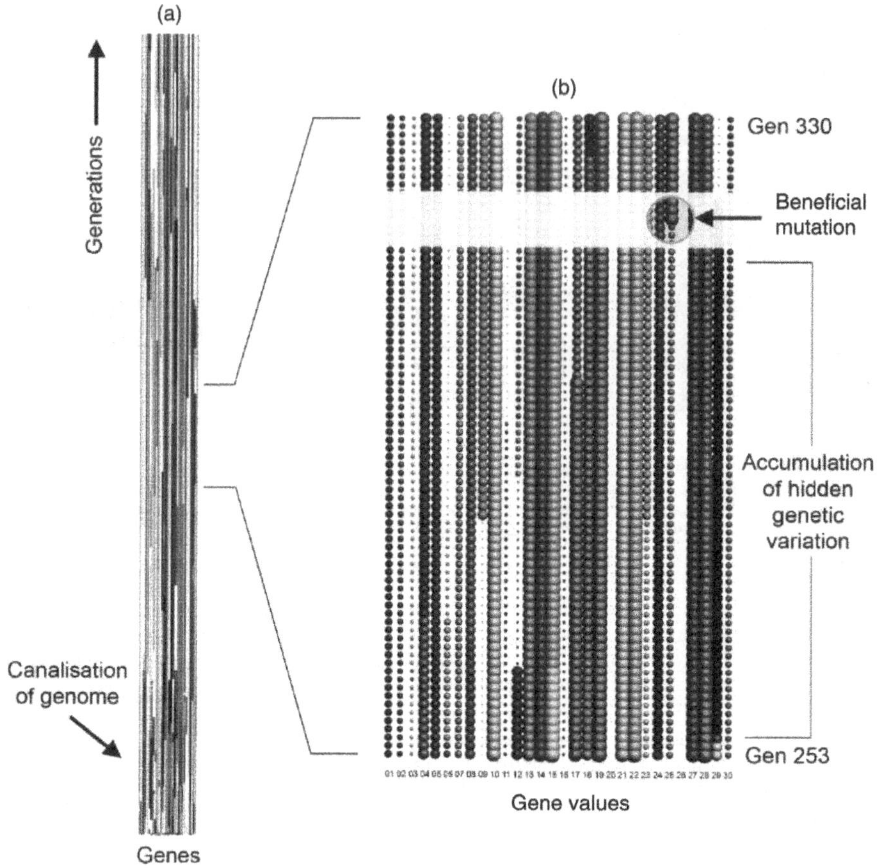

Figure 14.17 Changes made to the genome over generations. Each genome is represented as a horizontal line of 30 spheres (representing each gene). Size and shade of each sphere is defined by an attribute of the gene it represents. Size represents the ROI attribute of each gene. Shades represent different geometry types and material properties of the connections (analogous to weights of connections). Note that some genes do not change very much following canalization—these usually represent "hubs" of regulation in the gene network. (a) Entire selected genomes from a typical evolutionary run. (b) Zoomed-in view of a period in the "fossil record" which showed phenotypic stasis.

changes to the genome can cause big phenotypic changes in random solutions. The implication being that, while this behavior is potentially useful for speculatively exploring *diverse* three-dimensional designs, the genome is brittle and thus not well suited for optimization. Simply put, if random solutions have poor performance and small changes to the genome effectively re-randomize the solution, the model will find it hard to improve fitness using evolutionary search. Indeed, in this situation it is desirable for genomes to become more *robust* to mutations (increase locality) and in doing so, make the developmental process more stable. Our findings demonstrate that this *developmental stability* emerges "for free" in the gene network model and occurs due to canalization. The process of canalization allows gene networks to create more redundancy in the genome and establish highly connected "hub genes" that increase locality of solutions and ultimately protect phenotypic innovations via compensatory growth rules. Critically, this means that our initially volatile gene networks, capable of generating highly diverse solutions, quickly become suitable for more fine grained optimization of phenotypic traits. This ability to control how gene mutations affect phenotypes relates to *evolvability* [52] and is potentially significant for many design and engineering domains because existing CAD tools cannot do this. Interestingly, our analysis then shows that the ability of genomes to accumulate hidden genetic variation (via neutral mutations) significantly improves evolutionary search. Here increased mutational robustness allows our genomes to act as *evolutionary capacitors*, capable of storing and releasing genetic variation. The benefit of hidden genetic variation in this situation is that when populations enter periods of "stasis" (i.e., numerous generations when fitness does not increase) neutral mutations allow populations to avoid getting stuck at local optima and thereby more easily generate beneficial innovations.

The key insight of this work is that alternative representations that exploit simple regulatory dynamics and evolutionary capacitance may provide exciting new capabilities for future CAD tools, enabling designers to simultaneously *explore* and *optimize* diverse physical designs. Indeed, enhanced CAD tools are a key component of exploiting advances in manufacturing technologies and creating next-generation materials and structures for various engineering domains, including but not limited to: aerospace applications, military armors, improved prosthetic limbs, medical implants, advanced structural and civil engineering, high-performance machine parts, and soft robotics. Further interdisciplinary collaborative work is required to progress this important and exciting area of engineering research. Therefore, we conclude by highlighting two key limitations of our model and further trajectories for investigation.

There are two key limitations of our gene network model: scalability and the ability to generate geometrically regular features. First, while our model does not require additional information as new connections are added to the solution, the genome does grow in size as new nodes are added to the solution. This is a significant limitation because truly complex designs may well require large numbers of nodes, and increased genome sizes will lead to worse search performance. Second, while the gene networks are able to discover functional morphologies, the solutions they produce do not exhibit regular geometric features such as symmetries or repeating structural motifs (see Figure 14.18). This is potentially important as biological structures do exhibit these types of regularities, as too do many existing complicated engineering products. To illustrate this, consider the structure shown in Figure 14.18. To create this structure, we set our gene network model within a simple physics environment, whereby nodes are free to move under the force of gravity (Figure 14.19). By simply selecting for structures that are tall (using the average height of any connections made), we quickly evolve solutions such as Figure 14.18. Notably, here we use a time-based physics engine instead of a structural analysis method to ensure buildable

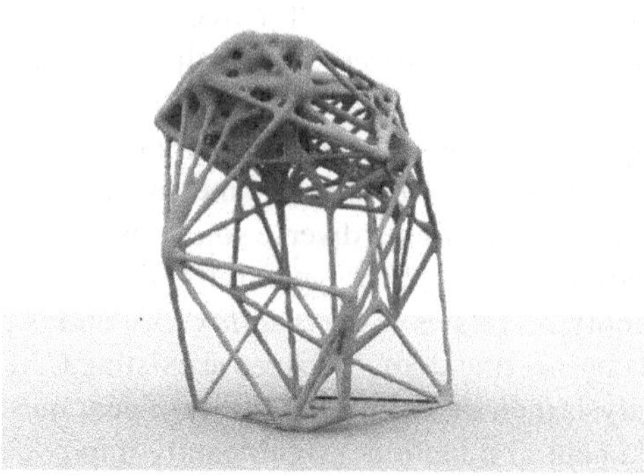

Figure 14.18 Evolved solution that does not contain geometric regularities.

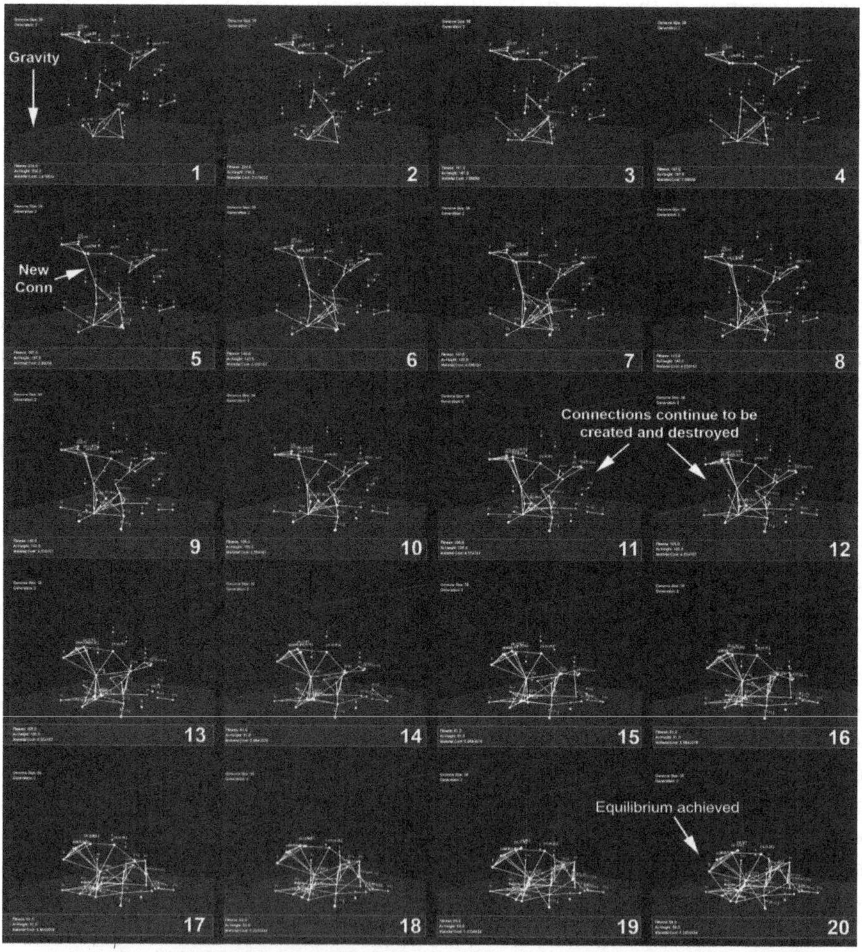

Figure 14.19 Solutions evolved within a simple physics simulation. Nodes are positioned within Cartesian space and contain growth instructions. During growth, nodes are subjected to gravity and "settle" to form a network solution.

designs, which means that nodes continue to create and destroy connections during the lifetime of the structure (i.e., as nodes deform and move due to physics). The evolved solutions are able to obtain physical properties that allow them to be "tall" (which is "good" in relation to our objective function). However, this problem would likely be easier to solve if our model could generate geometric regularities [27].

To address these key issues, our recent work [56] extends our gene network using an alternative encoding method, replacing our cumbersome bit-string genomes with CPPNs [64], and this allows us to evolve scalable and regular designs for truss optimization problems. Additionally, further work [57] is now underway to explore the evolution of high-value structures for various engineering applications with additive manufacturing technologies (i.e., 3D printing).

ACKNOWLEDGMENTS

Daniel Richards was supported by funding from the Dalton Research Institute, Manchester Metropolitan University.

REFERENCES

1. Aldana, M., Balleza, E., Kauffman, S., and Resendiz, O. 2007. Robustness and evolvability in genetic regulatory networks. *Journal of Theoretical Biology*, 245, 433–448.
2. Bentley, P. J. and O'Reilly, U. M. 2001. Ten steps to make a perfect creative evolutionary design system. In *Proceedings of GECCO '01, Workshop on Non-Routine Design with Evolutionary Systems*, San Francisco, July 7–11, 2001.
3. Blondel, V. D., Guillaume, J. L., Lambiotte, R., and Lefebvre, E. 2008. Fast unfolding of communities in large networks, *Journal of Statistical Mechanics*, 10, doi:10.1088/1742-5468/2008/10/P10008.
4. Bongard, J. and Pfeifer, R. 2003. Evolving complete agents using artificial ontogeny. In Hara, F. and Pfeifer, R (Eds), *Morpho-Functional Machines: The New Species (Designing Embodied Intelligence)*, Berlin: Springer Verlag, pp. 237–258.
5. Burry, M. 2011. *Scripting Cultures: Architectural Design and Programming*. Sussex: John Wiley & Sons.
6. Caldas, L. G. 2001. *An Evolution-Based Generative Design System: Using Adaptation to Shape Architectural Form*. Ph.D. thesis, Massachussetts Institute of Technology.
7. Caldas, L. 2008. Generation of energy-efficient architecture solutions applying GENE_ARCH: an evolution-based generative design system. *Advanced Engineering Informatics*, 22, 59–70.
8. Clune, J. and Lipson, H. 2011. Evolving three-dimensional objects with a generative encoding inspired by developmental biology. In *Proceedings of ECAL 2011*, Paris, pp. 141–148.
9. Collins, M. 2005. Finding needles in haystacks is harder with neutrality. In *Proceedings of GECCO '05*, Washington, June 25–29, 1613–1618.

10. Crombach, A. and Hogeweg, P. 2008. Evolution of evolvability in gene regulatory networks. *PLoS Comput Biol*, 4(7): e1000112. doi:10.1371/journal.pcbi.1000112

11. Davis, D., Burry, M., and Burry, J. 2011. Untangling parametric schemata: enhancing collaboration through modular programming, In *Proceedings of the 14th International Conference on Computer Aided Architectural Design Futures*, Liege (Belgium), 4–8 July 2011, 55–68.

12. Doncieux, S., Mouret, J. B., Bredeche, N., and Padois, V. 2011. *Evolutionary Robotics: Exploring New Horizions.* Studies in Computational Intelligence. Vol 341, Springer, pp. 3–25.

13. Draghi, J. A., Parsons, T. L., Wagner, G. P., and Plotkin, J. B. 2010. Mutational robustness can facilitate adaptation. *Nature Letters*, 263, 353–355.

14. Eggenberger, P. 1997. Evolving morphologies of simulated 3D organisms based on differential gene expression, In Husbands, P. and Harvey, I. (Eds), *Proceedings of the Fourth European Conference on Artificial Life*, Brighton, UK, July 28–31, 1997, MIT Press, pp. 205–213.

15. Espinoso-Soto, C., Martin, O. C., and Wagner, A. 2011. Phenotype plasticity can facilitate adaptive evolution in gene regulatory circuits. *BMC Evolutionary Biology*, 11(5): doi:10.1186/1471-2148-11-5.

16. Evins, R. 2013. A review of computational optimisation methods applied to sustainable building design. *Renewable and Sustainable Energy Reviews*, 22, 230–245.

17. Flatt, T. 2005. The evolutionary genetics of canalization. *The Quarterly Review of Biology*, 80(3), 287–316.

18. Forbes, P. 2006. *The Gecko's Foot: How Scientists are Taking a Leaf from Nature's Book.* London: Harper Perennial.

19. Frazer, J. 1995. *An Evolutionary Architecture.* London, UK: Architectural Association.

20. Galvan-Lopez, E., McDermott, J., O'Neil, M., and Brabazon, A. 2010. Towards an understanding of locality in genetic programming. In *Proceedings of GECCO '10*, July 7–11, 2010, Portland, USA. ACM Press, 901–908.

21. Galvan-Lopez, E., Poli, R., Kattan, A., O'Neil, M., and Brabazon, A. 2011. Neutrality in evolutionary algorithms what do we know? *Evolving Systems* 2(3), 145–163.

22. Gibson, G. and Wagner, G. 2000. Canalization in evolutionary genetics: a stabilizing theory? *BioEssays* 22(4), 372–380.

23. Glynn, R. and Sheil, B. (Eds). 2011. *Fabricate: Making Digital Architecture.* Ontario: Riverside Architectural Press.

24. Harvey, I. and Thompson, A. 1997. Through the labyrinth evolution finds a way: a silicon ridge, In Higuchi, T. Iwata, M., and Liu, W. (Eds), *Proceedings of 1st International Conference on Evolvable Systems: From Biology to Hardware*, Berlin, Germany: Springer-Verlag, Vol. 1259, 406–422.

25. Hayden, E. J., Ferrada, E., and Wagner, A. 2011. Cryptic genetic variation promotes rapid evolutionary adaptation in an RNA enzyme. *Nature*, 474(6), 92–95.

26. Hensel, M., Menges, A., and Weinstock, M. 2010. *Emergent Technologies and Design: Towards a Biological Paradigm for Architecture*, Oxon: Routledge.

27. Hornby, G. 2004. Functional scalability through generative representations: the evolution of table designs. *Environment and Planning B*, 3, 569–587.

28. Hu, T., Payne, J. L., Banzhaf, W., and Moore, J. H. 2011. Robustness, evolvability and accessibility in linear genetic programming, In Silva, S., Foster, J., Nicolau, M., Machado,

P., and Giacobini, M. (Eds), *Proceedings of EuroGP 2011*, Evostar, Springer's Lecture Notes in Computer Science, vol. 6621, 13–24.
29. Kauffman, S. A. 1993. *The Origins of Order: Self?Organization and Selection in Evolution*. Oxford: Oxford University Press.
30. Kicinger, R., Arcissewski, T., and Jong, K. 2005. Evolutionary computation and structural design: a survey of the state of the art. *Computers and Structures*, 83, 1943–1978.
31. Kimura, M. 1968. Evolutionary rate at the molecular level. *Nature*, 217, 624–626.
32. Kitano, H. 2004. Biological robustness. *Nature Reviews Genetics*, 5(11), 826–837.
33. Kumar, S. and Bentley, P. J. (Eds). 2003. *On Growth, Form and Computers*. London: Elsevier.
34. Lenski, R. E., Barrick, J. E., and Ofria, C. 2006. Balancing robustness and evolvability, *PLoS Biology*, 4(12): e428. doi:10.1371/journal.pbio.0040428
35. Le Rouzic, A. and Carlborg, O. 2007. Evolutionary potential of hidden genetic variation. *TRENDS in Ecology and Evolution*, 23(1), 33–37.
36. Lipson, H. 2005. Evolutionary design and open ended design automation. *Biomimetics*, 129–155.
37. Menges, A. 2012. Higher integration in morphogenetic design. *Architectural Design*, 82(2).
38. Miller, J. F. 2003. Evolving developmental programs for adaptation, morphogenesis and self-repair, In *Proceedings of ECAL '03*, Dortmund, 2003, 256–265.
39. Miller, J. 2004. Evolving a self-repairing, self-regulating, french flag organism. In *Proceedings of GECCO '04*, Washington, June 26–30, 129–139.
40. Miller, J. F. 2011. Cartesian genetic programming, In Miller, J. F. (Ed), *Cartesian Genetic Programming*, Berlin: Springer Verlag, 17–34.
41. Moore, S. A. 2007. *Alternative Routes to the Sustainable City*. Plymouth: Lexington Books.
42. Muller, G. B. 2007. Evo-devo: extending the evolutionary synthesis. *Nature Reviews Genetics*, 8(12), 943–949.
43. Newman, M. E. J. 2008. The mathematics of networks. In Blume and S. N. Durlauf (Eds.), *The New Palgrave Encycolpedia of Economics*, Basingstoke: Palgrave Macmillan.
44. Nolfi, S. and Parisi, D. 1991. Growing neural networks (Tech. Rep. PCIA-91-15). Rome: Institute of Psychology, C.N.R.
45. O'Neil, M., McDermott, J., Swafford, J. M., Byrne, J., Hemberg, E., Brabason, A., Shotton, E., McNally, C. and Hemberg, M. 2010. Evolutionary design using grammatical evolution and shape grammars: designing a shelter. *International Journal of Design Engineering*, 3(1), 4–24.
46. O'Reilly, U. M. and Hemberg, M. 2007. Integrating generative growth and evolutionary computation for form exploration. *Genetic Programming and Evolvable Machines*, 8(2), 163–186.
47. Oxman, N. 2010. Structuring materiality: design fabrication of heterogeneous materials. *Architectural Design*, 80(4), 78–85.
48. Oxman, R. 2009. Performative design: a performance-based model of digital architectural design. *Environment and Planning B: Planning and Design*, 36, 1026–1037.
49. Pask, G. 1969. The architectural relevance of cybernetics. *Architectural Design*, 7(6), 494–496.

50. Pigliucci, M. 2008. Is evolvability evolvable? *Nature Reviews Genetics*, 9(1), 75–82.
51. Raidl, G. R. and Gottlieb, J. 2005. Empirical analysis of locality heritability and heuristic bias in evolutionary algorithms: A case study for the multidimensional knapsack problem. *Evolutionary Computation*, 13(4), 441–414.
52. Reisinger, J., Stanley, K. O., and Miikkulainen, R. 2005. Towards an empirical measure of evolvability, In *Proceedings of GECCO '05*, June 25–29, 2005, Washington, DC, 257–264.
53. Richards, D. 2011. Towards morphogenetic assemblies. In Herr, C.M., Gu, N., Roudavski, S., Schabel, M. A. (Eds), *Circuit Bending, Breaking and Mending: Proceedings of the 16th International Conference on Computer-Aided Architectural Design Research in Asia* (CAADRIA), 515–524.
54. Richards, D. 2013. *Automatic Synthesis of Architectural Structures using an Evo-Devo Approach to Design*. Ph.D. thesis, Manchester Metropolitan University, UK.
55. Richards, D., Dunn, N., and Amos, M. 2012. An evo-devo approach to architectural design. In *Proceedings of GECCO 12*, July 7–11, 2012, Philadelphia, USA. ACM Press, 569–576.
56. Richards, D. and Amos, M. (2014a). Evolving morphologies with CPPN-NEAT and a dynamic substrate. In Hiroki Sayama, John Rieffel, Sebastian Risi, Ren Doursat and Hod Lipson (Eds), *Proceedings of ALIFE 14, the Fourteenth International Conference on the Synthesis and Simulation of Living Systems*, July 30–August 2, 2014, New York, USA., 255–262, MIT Press.
57. Richards, D. and Amos, M. (2014b). Designing with gradients: bio-inspired computation for digital fabrication. Accepted for presentation at *ACADIA 2014: Design Agency*, October 23–25, 2014, University of Southern California, Los Angeles, USA.
58. Rothlauf, F. and Goldberg, D. E. 2000. Pruefer numbers and genetic algorithms: a lesson on how the low locality of an encoding can harm the performance of GA's. In *Proceedings of the 6th International Conference on Parallel Problem Solving from Nature* (PPSN), London: Springer Verlag, 395–404.
59. Rothlauf, F. 2011. *Design of Modern Heuristics Principles and Application*. Heidelberg: Springer.
60. Roudavski, S. 2009. Towards morphogenesis in architecture. *International Journal of Architectural Computing*, 7(3), 345–374.
61. Rutherford, S. L. 2000. From genotype to phenotype: buffering mechanisms and the storage of genetic information. *BioEssays*, 22(12), 1095–1105.
62. Siegal, M. L. and Bergman, A. 2002. Waddington's canalization revisited: Developmental stability and evolution. *Proceedings of the National Academy of Sciences of the United States of America*, 99(16), 10528–10532.
63. Stanley, K. O. and Miikkulainen, R. 2003. A taxonomy of artificial embryogeny, *Artificial Life*, 9(2), 93–130.
64. Stanley, K. O. 2007. Compositional pattern producing networks: a novel abstraction of development. *Genetic Programming and Evolvable Machines*, 8(2), 131–162.
65. Tanzer, K. and Longoria, R. (Eds). 2007. *The Green Braid: Towards an Architecture of Ecology, Economy and Equity*. Oxon: Routledge.
66. Turrin, M., von Buelow, P., and Stouffs, R. 2011. Design explorations of performance driven geometry in architectural design using parametric modelling and genetic algorithms. *Advanced Engineering Informatics*, 25(4), 656–675.
67. Ulieru, M. and Doursat, R. 2011. Emergent engineering: a radical paradigm shift. *International Journal of Autonomous and Adaptive Communications Systems*, 4(1), 39–60.

68. van Dijk, A. D. J., van Mourik, S., and van Ham, R. C. H. J. 2012. Mutational robustness of gene regulatory networks. *PLoS ONE*, 7(1): e30591. doi:10.1371/journal.pone.0030591.
69. Vassilev, V. K. and Miller, J. F. 2000. The advantages of landscape neutrality in digital circuit evolution, In *Proceedings of ICES '00*, April 17–19, 2000, Edinburgh, 252–263.
70. Waddington, C. H. 1942. Canalization of development and the inheritance of acquired characters. *Nature*, 150, 563.
71. Waddington, C. H. 1957. *The Strategy of the Genes: A Discussion of Some Aspects of Theoretical Biology*. London: Allen and Unwin; New York: MacMillan.
72. Wagner, A. 2005. Robustness, evolvability, and neutrality. *FEBS Letters* 579(8), 1772–1778.
73. Wagner, A. 2011. *The Origins of Evolutionary Innovation: A Theory of Transformative Change in Living Systems*, Oxford: Oxford University Press.
74. Wagner, A. 2012. The role of robustness in phenotypic adaptation and innovation, *Proceedings of the Royal Society B: Biological Sciences*, 279, 1249–125.
75. Watson, J. D., Baker, T. A., Bell, S. P., Gann, A., Levine, M., and Losick, R. 2007. *Molecular Biology of the Gene*. New York: Benjamin Cummings.
76. Weinstock, M. 2006. Self-organisation and the structural dynamics of plants, *Architectural Design*, 76(2), 26–33.
77. Yu. T. and Miller. J F. 2002. Finding needles in haystacks is not hard with neutrality. In Foster, J. A., Lutton, E., Miller, J. F., Ryan, C., and Tezzamanzi, A. G. B. (Eds), *Proceedings of EuroGP 2002*. LNCS, Vol 2278, 13–25.

15

COMPUTING WITH ARTIFICIAL GENE REGULATORY NETWORKS

Michael A. Lones
School of Mathematical and Computer Sciences, Heriot-Watt University, Edinburgh, Scotland, UK

15.1 INTRODUCTION

Gene regulatory networks (GRNs) are the fundamental mechanisms through which biological organisms control their growth, their dynamical behavior, their interaction with their environment, and which underlie much of the complexity we see in the biosphere. Biological complexity has long been an inspiration to computer scientists, and many seek to model the biological processes that generate this complexity, using these models to generate complex behavior that can then be used to solve problems in computer science and engineering. In this chapter, the term "artificial gene regulatory network" (AGRN) is used to refer to computational models of GRNs that are used in this manner.

Unlike other computational models of GRNs, AGRNs are predominantly used to solve computational and engineering problems, not biological problems. Hence, AGRNs often differ from these other models. This chapter reviews current understanding of AGRNs, discussing what is known about their computational properties, detailing how they have previously been applied to computational problems, and speculating about how they may be used in the future. Reflecting the theme explored in this book, we focus on approaches that have used evolutionary algorithms (EAs)

Evolutionary Computation in Gene Regulatory Network Research, First Edition.
Edited by Hitoshi Iba and Nasimul Noman.

to design AGRNs, a task for which they are presumably well suited given the close relationship between biological evolution and biological GRNs.

The chapter is organized as follows. Section 15.2 discusses what we know about biological GRNs, and the implications this has for the design of AGRNs. Section 15.3 presents the different motivations behind the development of AGRN models. Section 15.4 discusses the modeling decisions that have to be made when developing AGRN models. Sections 15.5 and 15.6 review previous work on using AGRNs for computation, focusing on theoretical and applied perspectives, respectively. Section 15.7 considers future directions for AGRN research, and Section 15.8 concludes.

15.2 BIOLOGICAL GRNs

Before discussing computational models of GRNs, it is instructive to first consider the system, which is being modeled—biological GRNs. The key feature of biological GRNs, and one that is captured in all computational models, is that a group of genes regulate one another's expression. However, something that is not often captured is the considerable heterogeneity and complexity of these regulatory processes. In a eukaryotic cell, all the following factors play important roles in regulating a gene's expression:

Gene accessibility Deoxyribonucleic acid (DNA) is supercoiled within a protein complex known as chromatin. Genes may only be expressed if they are accessible to the cellular machinery that carries out transcription, and this is only possible in regions where the chromatin-bound DNA has been unwound. Winding and unwinding, in turn, are subject to regulation by proteins expressed by other genes. These act in a number of different ways, for instance, through epigenetic processes such as DNA methylation [35].

Transcription preinitiation The transcription of all genes involves a core collection of proteins, which includes ribonucleic acid (RNA) polymerase and a group of stabilizing general transcription factors (TFs) that assemble at a gene's transcription start site (TSS). All these proteins are required to achieve a baseline level of transcription. Since they are finite in number, the transcription rate of a gene is also affected by the number and transcription rates of genes elsewhere within the cell's nucleus [24, 67].

Transcription initiation The preinitiation complex is insufficient in most cases, and is supplemented by special-purpose TFs that bind to regulatory regions upstream of the TSS, and act to up- or down-regulate the gene by either increasing or decreasing the stability of the transcription complex. TFs are proteins and, hence, the products of other genes. The number of TFs involved, and the complexity of their interactions, varies considerably between genes [75].

Transcription elongation Initiation is just the first stage of transcription. Regulation also takes place during transcription, typically through RNA-binding proteins that interfere with the progress or stability of the transcription complex. Again, these are the products of other genes [47].

Post-transcriptional modification Once a gene has been transcribed, the resulting strand of messenger RNA is acted upon by other proteins in order to form a mature transcript that is ready for translation. Regulation of this process is possible at all stages. Most prominent is the process of alternative splicing, where proteins differentially cleave away parts of the transcript, leading to the expression of different variants of the encoded protein [28].

Gene silencing During the last decade, it has become evident that RNA plays an important role in gene regulation. This process, known as gene silencing or RNA interference, involves microRNAs that are encoded by other genes, and which interfere with translation by binding to messenger RNA transcripts. This binding process remains poorly understood, but it is known that the specificity of individual microRNAs varies considerably, resulting in effects ranging from the silencing of individual genes to individual pathways to large-scale changes in gene expression [11].

Transcription and post-transcription Regulatory interactions also occur during protein synthesis, transportation, and degradation. An example of this is the regulation of the availability of chaperones, helper proteins that are required to fold certain proteins into their mature, functional, form [33,41].

Metabolic availability The cellular machinery, including those sub-systems responsible for translation and transcription, requires various metabolites. For example, transcription requires a ready supply of transfer RNA and its associated amino acids. The availability of these metabolites places a global constraint on the cell's activities, and may also be regulated at a more local level [16].

These regulatory processes occur on different temporal and spatial levels. Hence, it is unlikely that a model operating on a single temporal or spatial level could accurately capture the overall behavior of a biological GRN. Furthermore, it is important to note that biological GRNs are only partially understood. It is quite likely that our knowledge of them is biased toward those components which are easiest to recognize and model, and quite possible that we do not yet understand all the fundamental mechanisms underlying their behavior. For example, the involvement of microRNA in genetic regulation is a relatively recent discovery, and understanding of its biochemical interactions remains incomplete. Nevertheless, although this incomplete understanding is an issue for research into AGRNs, it is also an opportunity, since the behavior of AGRNs could provide valuable insights into the working of their biological analogs.

15.3 COMPUTATIONAL MODELS

Historically, the development of computational models of GRNs has focused on their role in understanding biological systems. Typical uses include inferring computational models from measurement data in order to determine the structure of a biological GRN, simulating computational models in order to understand the dynamics of a biological GRN, and using computational models to understand systems-level properties of biological pathways. However, for the area of research considered in this chapter, the justification for using computational models of GRNs is quite different. Rather than trying to understand how actual biological circuits work, the aim is to apply the principles of biological circuits to the design of computational and engineered systems. Typical motivations for this include the following:

Structures Biological systems are structurally complex, and this complexity is known to be a direct consequence of the behavior of GRNs. This has led to an interest in how computational models of GRNs can be used to generate intricate structures [9, 12, 13, 15, 21, 29, 36, 37, 43, 68, 72, 77, 81], something that is of interest to both engineers and people working in creative disciplines.

Dynamics Biological systems are also dynamically complex. They respond robustly and intelligently to their environment, and this is known to be a direct consequence of the signal processing and control behaviors exhibited by biological circuits. Hence, there is an interest in how computational models of GRNs can be used to perform these functions within engineered systems [5, 8, 14, 25, 26, 38, 45, 46, 52, 54, 56, 59, 61, 64, 73, 78].

Computability Biological systems appear to process information and carry out behaviors that are analogous to computation. However, the manner in which they achieve this is quite different to conventional computers. This has promoted interest in studying the computational characteristics of GRN models [7, 10, 31, 40, 51, 58]; for instance, their apparent ability to represent complex computations in a compact and robust form.

Evolvability Evolutionary algorithms have become a successful approach for carrying out design and optimization. They are modeled upon biological evolution. Since GRNs can be considered the "solution representation" used by biological evolution, and they are known to exhibit characteristics that make them evolvable, this has led to interest in using computational models of GRNs to represent artifacts (e.g., computer programs) evolved by EAs [4, 10, 53].

The diverse motivations for computational models of GRNs have led to an equally diverse range of computational models. Some of these, such as S-systems [66], are used exclusively for biological modeling. Some, such as our work on artificial biochemical networks (ABNs) [53, 55], are intended exclusively for computational applications. Others, such as Boolean networks, have been used at both ends of this spectrum [10, 15, 39, 64]. While it is beyond the scope of this chapter to review all the modeling approaches used in the literature (various examples can be seen elsewhere in this book), it is important to be aware that there are diverse models, that different models are appropriate for solving different problems, and, moreover, that we do not necessarily know which is the best model for a particular task. For applications in computer science and engineering, the latter point reflects both the infancy of the field and the incompleteness of biological knowledge.

15.4 MODELING DECISIONS

We know that biological GRNs are capable of generating complex behavior, but it is less clear which elements of biological GRNs are important or even necessary for achieving this. This uncertainty is reflected in the array of AGRN models that have been used in computational applications [4,8,15,55,81]. These vary considerably in the level of abstraction at which they model regulatory networks, and also vary in the accuracy with which they model individual biological processes and mechanisms. Before going on to talk in detail about these models and their applications, we briefly summarize some of the decisions faced when doing this kind of modeling, and the choices available.

Expression levels In biological systems, gene expression can take place at many different rates depending on the stability of the transcription complex and the presence of downstream regulatory processes. Hence, it may seem biologically accurate to capture gene expression as a continuous-valued state. Nevertheless, for many regulatory circuits it is known that the qualitative dynamics can be approximated by modeling gene expression as an on/off process [1], meaning that it is also feasible to capture gene expression as a binary state. While the consequences of choosing binary or continuous-valued expression levels are not fully understood, it is likely to represent a trade-off between model expressiveness and model simplicity, and the implementation consequences of the latter (for instance, ease of implementation in hardware [82]). This is a rather fundamental modeling choice, since the range of states determines the choice of regulatory functions. In practice, both decisions have been taken.

Regulatory functions Regulatory functions can be captured in a number of different ways. Most common is to use an additive or integrative function to combine the expression levels of the gene's regulatory inputs [5,63]. This function may be weighted, so that some inputs are more influential than others. Inputs may also be negative weighted in order to represent negative regulatory interactions between genes. The regulatory function may be linear (e.g., simple addition) or nonlinear (e.g., sigmoidal). Some AGRN models (notably Boolean networks) use combinatorial functions, which are a more accurate model of many biological regulatory functions [80]. In particular, these allow logical and canalizing interactions between genes to be captured. However, in the general case, gene interactions are neither additive nor combinatorial. Rather, they are a result of complex physical interactions between many different biochemicals that occur over many different timescales. It is rarely feasible to model this process in detail, though some AGRN models attempt to capture notions of its complexity by using various kinds of complex nonlinear functions [8,53].

Gene products The process by which one gene regulates another can be captured at differing levels of abstraction. From a high-level perspective, genes can be viewed as directly regulating one another, and indeed this is a simplification often used by biologists when modeling gene regulation [40]. A number of

AGRN models capture this direct interaction, with the expression level of a regulating gene becoming an input of the regulated gene's regulatory function [40, 52]. Other approaches explicitly model the intermediate gene products involved in gene regulation [4, 8]. Typically, this is done using analogs of TFs, using an additive or integrative function to combine the expression levels of all genes which are tagged as expressing a particular TF. The resulting value, which is an analog of chemical concentration, is then used as an input to the regulatory functions of genes which are tagged as being regulated by this TF. In one sense, these two representations are equivalent, since indirect regulatory interactions can be rewritten as direct interactions, following a suitable change to the regulatory function. Nevertheless, some authors have argued that an indirect representation brings advantages in terms of evolvability [5]. However, the explicit modeling of gene products has a computational overhead, and also reduces the apparent simplicity of the model.

Genome organization From an abstract perspective, a genome is an array of genes. In reflection of this, it is commonplace to encode an AGRN in an array structure when using an EA to optimize it. However, viewed through the lens of natural evolution, biological systems are not static entities. Rather, they are dynamically changing over the course of evolutionary history, and this is reflected in their genome, which is organized in a way that promotes meaningful change [42]. Since we are generating AGRNs using an evolutionary approach, it also makes sense to consider how elements of this biological evolvability may be captured within these models. Reflecting this, a number of modelers have considered the role of non-coding elements, and the manner in which biological complexity grows through a process of duplication and divergence [20, 81].

Time and space Another important modeling choice is how to capture temporal and spatial aspects of gene regulation. Time, in biological systems, is continuous. Genes do not express proteins synchronously, or even necessarily over the same timescales. This means that biological systems are able to make use of processes happening at different timescales, and this may be important for generating certain dynamical patterns. This kind of temporal variation can be captured through continuous models, such as systems of differential equations [5]. However, the simulation of differential equations is computationally demanding, especially when compared to discrete models, such as difference equations. Hence, in practice, it may be preferable to approximate continuous dynamics using discrete models, and this design decision is seen in many AGRNs [8,52,81]. Spatial organization is also an important factor in the behavior of biological GRNs, including its role in transcription complex formation, the role of DNA conformation and intergenic distances, and spatial processes such as diffusion and crosstalk. However, simulation of spatial processes is often computationally demanding. For this reason, it is often not considered in AGRNs, and when it is, it often appears in the form of a behavioral approximation [26].

Wider biological interactions GRNs do not exist within a vacuum. They interact with wider cellular and biological systems, and in many cases it is difficult, and arguably not useful, to delineate GRNs from their interactions with other systems. For example, the cell's GRN is closely integrated with its signaling and metabolic networks. It is also closely integrated with higher-level processes such as cell division. This has led to a number of models in which the AGRN is either generalized to model a broader system [53, 55], or hybridized with other forms of computational model, such as cellular automata (CA). The latter is especially commonplace in artificial development models, where the aim is often to capture a biological GRN's role in forming spatial structures [15, 21].

15.5 COMPUTATIONAL PROPERTIES OF AGRNs

Given the complexity of biological systems, and the key role that GRNs play in achieving this complexity, we expect AGRNs to show interesting computational properties. Nevertheless, we are still in the early stages of understanding the computational abilities of these models, and conversely their computational limits. This is for several reasons. First, historically there has not been much focus on this question, since most work on computational models of GRNs has been associated with biological understanding rather than computational understanding. Second, as the previous section attests, AGRN models are very diverse, making it unclear how computational understanding of one model applies to others. Furthermore, many AGRN models are descriptively complex, having many parameters, and therefore exhibiting many degrees of freedom. This makes it infeasible to exhaustively explore, or even sample with reasonable coverage, the space of possible networks, and therefore makes experimental investigations of their behavioral space intractable. Hence, our insight into their computational potential mostly comes from testing whether they can be optimized to carry out specific computational tasks. While sometimes revealing, this depends upon the suitability of the model for being optimized, making it difficult to distinguish between those behaviors that are theoretically possible and those that happen to be reachable using a particular optimization algorithm.

Nevertheless, some understanding of their computational potential can be uncovered by looking at the properties of relatively simple AGRN models, and their relationships to other, better understood, computational models. This is typically done using Boolean networks [39] (see Figure 15.1). These are a relatively simple and relatively abstract model of biological GRNs. Despite this, they have found uses in biological simulation [1], studies of systems-level understanding [2], and as AGRNs applied to computational applications [10, 64]. The model reduces gene expression to two values: true (expressed) and false (not expressed). Regulatory functions are then modeled as Boolean functions, such as AND, OR, and NOT, and gene expression updates are normally performed synchronously and deterministically.

The relationship between Boolean networks and CAs is important for understanding their computational potential. In particular, they are a generalization of elementary CAs (i.e., one-dimensional binary CAs), since the latter can be readily implemented

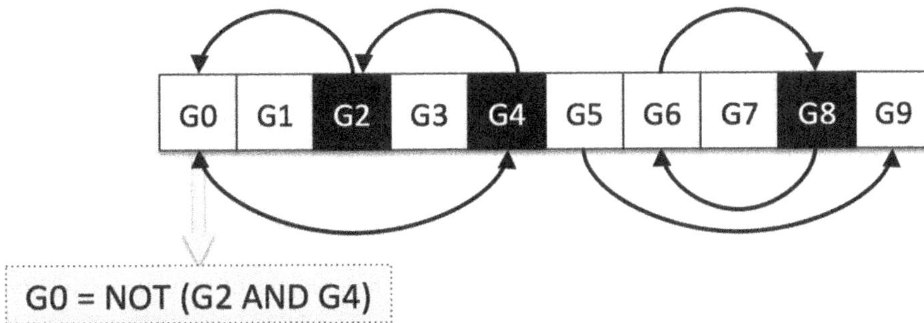

Figure 15.1 A Boolean network. Each gene has a binary state (shown in black or white), inputs from other genes, and a Boolean regulatory function (shown for G0) that determines its next state based upon the states of its inputs.

as a Boolean network by limiting connections to immediately neighboring genes and by using a single Boolean function for every gene [27]. It is well known that the use of certain update rules (such as Wolfram's rule 110 [76]) allow elementary CAs to simulate a Turing machine, and it follows from this that Boolean networks must likewise be capable of universal computation. Nevertheless, it should be taken into consideration that while assuring us that Boolean networks are computationally expressive, Turing completeness only tells us that there exists a Boolean network that can implement a particular computation. It does not tell us how small this network is, or how easy it is to construct, both of which are important in a practical sense.

In Ref. [58], the authors showed that Boolean networks perform better than CAs on tasks that measure the ability of distributed architectures to achieve global behaviors. Other work has given insights into the computational properties of particular kinds of Boolean network. For example, in Ref. [51], the authors noted that networks with a small-world topology exhibit a beneficial trade-off between information storage and propagation. In Ref. [40], it was hypothesized that computational abilities were maximized in the critical region between order and chaos, showing this to be the case when Boolean networks have an average connectivity of two inputs per gene. Various studies have observed a similar relationship between criticality and connectivity within evolved Boolean networks [31]. The relationship between criticality and evolvability in Boolean networks has also been highlighted [70]. Robustness is also an important issue, and it has been shown that certain classes of Boolean network can be strongly resistant to perturbations [3]. However, it has also been observed that changes to the Boolean network model can have a large impact upon its behavior. For instance, the introduction of stochastic or probabilistic elements considerably changes the appearance and stability of attractors [27]. The choice of functions and topologies also has significant consequences [32]. This suggests that care must be taken when trying to generalize these computational results to a wider range of AGRN models.

Various studies have considered the number and size of attractors found in Boolean network models [18]. These may give an indication of the computational capacity of a particular network, and are of significant interest to researchers who use Boolean networks to study biological systems [34]. However, the relationship between attractors and computation remains unclear. In part, this may depend upon whether attractors are explicitly used during computation. For instance, if the inputs to a problem are encoded in the initial expression state of the network, the attractor to which the network converges can be used to signal a particular computational outcome. For cyclic attractors, the repeating series of states, which the network visits, may also be interpreted as a computational outcome; for example, in a robotics application, the series of expressions at particular nodes may be read as a pattern of actuator activations. Nevertheless, in these situations, the computation is arguably being performed by the transients rather than the attractors, since it is these that *compute* the attractors from the initial state (or from other attractors if inputs are introduced during execution).

15.6 AGRN MODELS AND APPLICATIONS

Knowledge of theoretical computational properties is important, since it can guide us in choosing an appropriate model for a particular task. However, from an engineering perspective, we might not care about whether or not a model is computationally universal, so long as it can be used to effectively and efficiently solve a particular task. In this respect, previous experimental results are an important indicator of future potential. In this section, we review some of the work done on developing AGRN models and applying them to real-world computational problems, and consider some of the perceived strengths and weaknesses of the different approaches.

The most popular application area for AGRNs is arguably in the development of control systems, mirroring one of the prominent roles of biological GRNs. Table 15.1 summarizes a number of previous contributions in this area. The majority of these involve controlling robots or other autonomous agents, performing tasks such as obstacle avoidance, target following, foraging, multirobot coordination, and legged robot locomotion. A number of researchers have also tackled classic control problems, notably several variants of the inverted pendulum control task. However, despite this apparent focus on control, AGRNs have also been applied to other problems. Table 15.2 lists some of these. Many of these applications are motivated by the observed ability of GRNs to generate complex patterns, both static and dynamically varying, and how these patterns can be mapped to physical or computational structures. For example, a popular application of this approach is in the design of electronic circuits [29, 81].

In this section, we discuss some of the AGRN models that have been used in applied work. We begin with Boolean networks. We then discuss artificial genome models, a group of AGRN models that developed from research in evolutionary computing and artificial life. Following this, we review work in the area of artificial development, where AGRNs are predominantly used for pattern generation. Then we discuss the fractal GRN, an AGRN model that attempts to capture the complexity of biochemical interactions. Finally, we discuss our own work on extending AGRN models to include wider biochemical interactions.

Table 15.1 Selected control applications of AGRNs

First author, Year [Ref.]	Problem	GRN model
Dellaert, 1996 [15]	Obstacle avoidance and path following tasks	Boolean network, developmental
Quick, 2003 [61]	Temperature regulation and robotic control	Artificial genome, discrete time, continuous valued
Bentley, 2004 [8]	High-level control of a simulated hexapod robot	Fractal GRN [8]
Kumar, 2004 [46]	Obstacle avoidance for simulated and real robots	Artificial genome, discrete time, continuous valued
Taylor, 2004 [68]	Simple patterning of an underwater robotic swarm	Artificial genome based on Ref. [63]
Lee, 2009 [48]	Inference of robotic controllers from data	RNN-based model
Joachimczak, 2010 [38]	Control of simulated foraging behavior	Artificial genome, operons
Krohn, 2010 [45]	Single and joint inverted pendulum problems	Fractal GRN [8]
Lones, 2010 [52]	Chaos control in numerical dynamical systems	Artificial biochemical networks
Nicolau, 2010 [59]	Control of an inverted pendulum	Artificial genome, based on Ref. [5] with minor changes
Trefzer, 2010 [71]	Obstacle avoidance in a simulated and real robot	Artificial genome, discrete
Zahadat, 2010 [78]	Control of a modular robot	Fractal GRN [8]
Lones, 2011 [56]	Locomotion of a simulated quadruped robot	Coupled artificial biochemical networks
Roli, 2011 [64]	Light following and avoidance in robots	Boolean network
Cussat-Blanc, 2012 [14]	Controller for intelligent agents in a video game	See Ref. [59]
Jin, 2012 [36]	Control of robotic swarm patterning and formation	Developmental, hierarchical model with diffusion
Joachimczak, 2012 [37]	Control of soft-bodied multicellular animats	Developmental, based on Ref. [38]
Fuente, 2013 [25]	Locomotion of quadruped robot on rough terrain	Artificial signaling network
Turner, 2013 [73]	Control of coupled inverted pendulums	Artificial epigenetic network
Yao, 2014 [77]	Control of robot swarm in dynamic environment	Developmental, condition-dependent model

GRN, gene regulatory network; AGRN, artificial gene regulatory network; RNN, recurrent neural network.

15.6.1 Boolean Networks

There have been a number of studies in which Boolean networks have been used to solve computational problems [10, 19, 31, 60, 64, 79]. An example of this is the work of Roli et al., who have explored the use of Boolean networks in robotic control. In

Table 15.2 Other applications of AGRNs

First author, Year [Ref.]	Problem	GRN model
Gordon, 2003 [29]	Digital circuit design	Artificial genome, developmental, discrete
Koopman, 2004 [43]	Hardware fault tolerance	Artificial genome, developmental, discrete
Bentley, 2005 [7]	Fault-tolerant code	Fractal GRN [8]
Bowers, 2005 [9]	Pattern generation	Artificial genome, developmental
Gordon, 2005 [30]	Scalable digital circuits	Artificial genome, developmental, discrete
Mattiussi, 2007 [57]	Analog circuit design	Artificial genome
Bull, 2009 [10]	Multiplexer design	Boolean networks
Zhan, 2009 [81]	Digital circuit design	Artificial genome, discrete
Liu, 2010 [50]	System on chip reconfiguration	See Ref. [71]
Trefzer, 2010 [72]	Image compression	See Ref. [71]
Cussat-Blanc, 2012 [13]	Evolutionary art	See Ref. [59]
Lones, 2013 [54]	Diagnosis of Parkinson's disease	Artificial biochemical networks

GRN, gene regulatory network; AGRN, artificial gene regulatory network.

Ref. [64], for instance, they describe a problem in which a robot must move toward or away from a light source, switching between these two behaviors based upon the perception of a sound. By optimizing a Boolean network, they were able to solve this problem in simulation and then use this solution on a physical robot. The ability to transfer a solution directly from a simulated environment to a real-world environment is particularly notable, and differs from the experience of many working in the field of evolutionary robotics, where it is commonplace for evolved solutions to overfit the simulation environment [44]. This may be a reflection of the way in which GRNs work, that is, by following robust attractors, rather than using arguably less robust mechanisms such as decision rules. This is also reflected in the authors' analysis of the controllers, which they found to be robust to external interference [64].

Another notable use of Boolean networks within an evolutionary context is the work of Bull et al. [10], who used this approach to optimize digital circuits, such as multiplexers. However, compared to the other models discussed in this section, there has been relatively little applied work using Boolean networks. In part, this may be due to the inherent restrictions of working with binary values. Although Boolean logic is efficient and can be readily implemented in hardware, the requirement for inputs and outputs to be binary encoded can be problematic. This is especially the case when there are many inputs and outputs to encode, or where each input or output requires a large number of bits (e.g., a floating-point number), since this requires either a correspondingly large number of input and output nodes, or for inputs to be delivered over a number of time steps.

There is also some concern regarding the accuracy with which particular behaviors can be evolved. For instance, while Boolean networks are able to model the qualitative dynamics of many biological regulatory circuits, it has been recognized that they are not always able to capture the quantitative dynamics. This has prompted some authors to suggest the use of multivalued logics [22], which have been shown better able to capture these aspects of biological circuits. Nevertheless, it should be noted that the architectural simplicity of Boolean networks is a substantial advantage, since it promotes models that are relatively easy to comprehend, implement, and analyze.

15.6.2 Artificial Genome Models

Within the EA and artificial life communities, there has been significant interest in using AGRN models that capture a more detailed view of biological GRNs. An early example of this approach is the work by Riel [63], whose "artificial genome" (illustrated in Figure 15.2) introduced explicit analogs of genome organization and TFs. In this model, a genome is represented as a string of characters, analogous to DNA bases. Genes can be identified in this string by a sequence of characters that represent a transcription initiation site, with the sub-string immediately to the right of this sequence representing a TF and sub-strings to the left representing TF binding sites. It is possible for multiple genes to encode the same TF, and this leads to higher concentrations. Through a process of string matching, TFs are matched to regulatory elements, and the degree of this match, in tandem with the concentration of the TF, determines the gene's expression level. This model captures a number of elements that are missing from Boolean networks, such as the manner in which genetic sequences evolve, the indirect nature of gene interactions, and the presence of

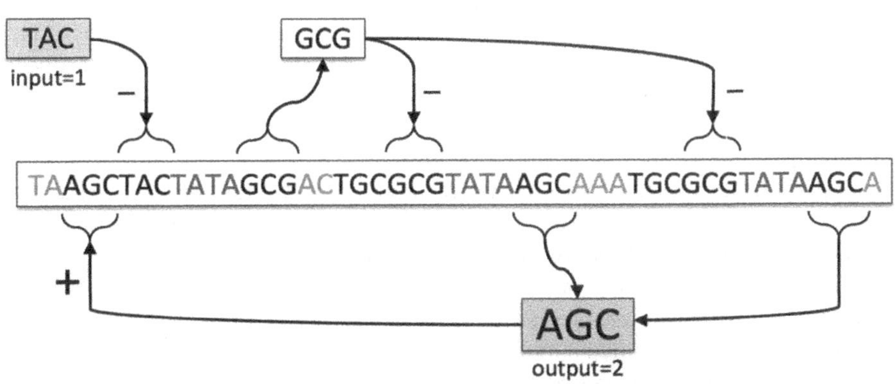

Figure 15.2 Example of an artificial genome model, loosely based on the model described in Ref. [63]. Genes are identified by the pattern TATA. The three letters to the right of this are interpreted as a gene product, and the six to the left are the regulatory region, divided into an activator and a repressor. Gray letters are comparable to non-coding DNA. Note that two genes produce the transcription factor AGC. Assuming all genes are initially expressed at the same level, this will lead to a relatively high concentration of AGC. Note also how external inputs can be encoded in the concentrations of transcription factors (TAC in this example) and how external outputs can be read from the concentrations of designated gene products (AGC).

differential expression levels. However, it does not model the combinatorial nature of gene regulation that is captured by Boolean networks, since regulation is implemented additively.

Reil originally used this model to study the dynamics of regulatory networks. The artificial genome approach was then adopted by Banzhaf [4, 5], who used a continuous-time variant and applied it to a number of computational problems. Particularly notable is the work described in Ref. [59], where the authors successfully evolved AGRNs that could control an inverted pendulum. This is a common benchmark in the control systems literature, and hence an important proof of concept. However, the authors did note that simulation time was an issue, a factor that is likely exacerbated by the use of a continuous-time model.

In more recent work, Cussat-Blanc et al. [14] applied a similar AGRN model to the problem of designing intelligent agents for use in video games and, in Ref. [65], analyzed the properties of controllers evolved to drive a virtual racing car. In common with Roli et al. [64], they found the controllers to be robust to added sensor noise, despite being evolved in the absence of noise. Again, this suggests that AGRN models are naturally robust, and that they capture at least some of the important properties of biological GRNs.

15.6.3 Artificial Development

Many of the earliest AGRN models emerged from the field of artificial development [15,21]. Central to this approach is the use of AGRNs to generate structures. Typically, this is achieved using gene products (which may or may not also be TFs) that encode various kinds of developmental signals. These signals become expressed in a certain pattern over the course of the AGRN's execution, and this pattern is then interpreted by a hybridized cellular growth model in such a way that a structure is produced. In addition, these models often capture spatial aspects of gene expression, such as diffusion, that allow communication between different parts of the developing structure; see Figure 15.3 for a simple example.

An early application of Boolean networks involved using this approach to generate the structure of a neurocontroller, which was then used to control the trajectory of an autonomous agent [15]. More recently, AGRNs have been used to generate ongoing dynamical behavior in addition to generating a structure. A good example of this is the work by Joachimczak et al. [37], who used AGRNs to both generate the morphology of a soft-bodied simulated creature and then control its locomotion by manipulating springs embedded within the resulting structure. Ideas such as this have also been extended to the case of controlling interactions between multiple autonomous elements. A prominent example is the work of Jin et al., who used AGRNs to develop patterns, which were then used to configure robot swarms [36] and modular robotic systems [77].

Another popular application area for developmental AGRN models is within the field of evolvable hardware, where EAs are used to design electronic systems [30, 43, 50, 81]. Often this involves evaluating evolved solutions directly within

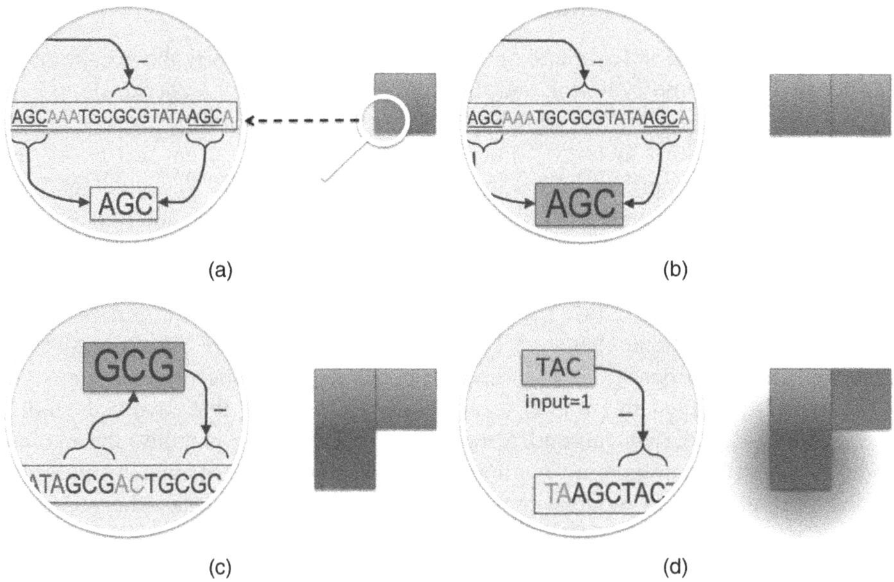

Figure 15.3 Example of how the AGRN from Figure 15.2 might be used to carry out artificial development: (a) maternal cell containing an AGRN; (b) when a designated gene product reaches a certain concentration, the cell divides; (c) different gene products can cause growth in different directions and using different cell types; and (d) AGRNs in different cells can affect each other using diffusive gene products.

reconfigurable hardware devices, and this requires AGRN models that are suitable to be efficiently implemented in this medium. Typically, this involves constraints on the complexity of the regulatory function, and the range and precision of numerical values. Elements of the patterns produced by these AGRNs can then be interpreted as electronic components, allowing them to be used for circuit design. Something that has been widely explored is the ability of developmental AGRNs to produce scalable generalized solutions, such as n-bit adders [30]. Another important aspect is the ability of the AGRN to perform ongoing dynamic behavior, allowing circuits to regenerate in the presence of faults [43], or to allow intelligent reconfiguration in system-on-chip devices [50]. Again, this makes use of the robustness present within AGRN models.

An interesting application outside of autonomous systems control and evolvable hardware is the work of Trefzer et al. [72], who used a developmental AGRN model to perform image compression. This worked by evolving the model so that it generates a time series of 8×8 pixel images, which are then used as a codebook for performing JPEG-like compression. This allows an image to be encoded as an AGRN alongside a sequence of time indices corresponding to elements of the codebook. The authors found that this produced a higher level of compression than JPEG while maintaining a similar level of quality.

15.6.4 Fractal Gene Regulatory Networks

The artificial genome and artificial development models described above capture the complexity of genome evolution and the role of intermediate biochemicals in gene regulation. However, they are arguably less successful at capturing the complexity of regulatory functions. Bentley [8] describes an interesting variant that goes some way toward addressing this. In this model, termed a fractal GRN, regulatory functions are implemented by interactions between regions of the Mandelbrot set. The intention behind this is to capture the degree of complexity of physico-chemical interactions between biochemicals without the inefficiency of implementing actual physico-chemical models. In a certain respect, this brings together AGRN models with the field of artificial chemistry [17], which aims to capture the expressiveness of chemical interactions within a purely computational or mathematical system.

Bentley [8] applied this model to a robotic control task that involved guiding a physical hexapod robot around a maze-like environment. Notably, the evolved controllers exhibited repeating nonlinear patterns of movement, which the author attributed to an implicit modularity present within the system. In another work [7], the author evolved fractal GRNs that could approximate the square root function. Significantly, it was noted that only AGRNs with a fractal chemistry could implement the requisite behavior: those without a complex regulatory function appeared unable to generate sufficient nonlinearity in their responses. Solutions to this problem were also found to be robust, degrading gracefully when intentionally damaged. Other authors have also applied fractal GRNs to control problems, including inverted pendulums in Ref. [45] and robotic modules in Ref. [78].

15.6.5 Artificial Biochemical Networks

In biological systems, genetic networks are closely coupled to other biochemical networks, and arguably it is the interaction between these different types of network that underlies the complexity seen within higher organisms. Our own work [25, 52–56, 73] focuses on modeling these wider interactions, and for this reason we refer to our models using the more generic term artificial biochemical network (or ABN). Our work is particularly motivated by the ability of biological systems to both express dynamics that are complex and diverse, and to switch robustly and sensitively between these dynamics in response to changes in their environment. This is similar to the situation faced in many real-world control tasks, where there is a need for controllers that can intelligently be transitioned between different robust control strategies in response to changes in the controlled system.

There are three main classes of biochemical network within a biological cell: genetic, metabolic, and signaling. The metabolic network, which emerges from interactions between enzyme-mediated reactions, determines the functional and biochemical properties of a cell, that is, the cell type. Since enzymes are gene products, there is a constructional relationship between the genetic and the metabolic network, with the former determining the make-up of the latter. Because of this, it is believed that transitions between the attractors of a GRN can lead to transitions between cell types [40].

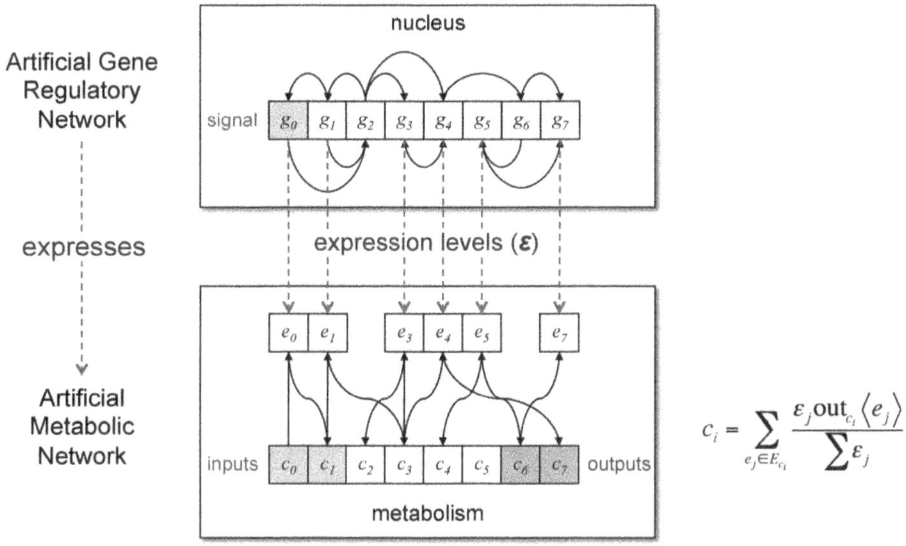

Figure 15.4 Example of an artificial biochemical network model in which an AGRN is coupled to an artificial metabolic network (AMN). The AMN models how a group of enzyme-mediated reactions modulate the concentrations of a group of chemical species. The AGRN, in turn, modulates the expression of the functional components of the AMN (i.e., its enzyme analogs), in effect switching different metabolic pathways on and off; see Ref. [53] for more details.

In Refs. [53, 56], we considered an ABN model in which an AGRN was used to express an artificial metabolic network (AMN, a simple artificial chemistry described in Ref. [52])—see Figure 15.4. This architecture was intended to capture the constructive relationship between genetic and metabolic networks, with the AGRN switching between different AMNs in response to an external signal, and the AMN then implementing the system's input–output mapping, that is, its current computational behavior. We applied this coupled ABN model to various control tasks, all of which required the ability to express multiple control behaviors and transition between these behaviors rapidly and in a context-sensitive manner. For example, in Ref. [53], we applied it to problems in chaos control and robotic locomotion, finding that coupled ABN-based controllers were generally easier to evolve and performed better than standalone AGRNs or AMNs. We also noted the value of using complex regulatory functions, in the form of discrete maps. In particular, coupled ABNs with discrete maps were observed to generate a diverse range of complex behaviors. To illustrate this, in Figures 15.5 and 15.6 we show examples of the kind of behaviors that can be generated by using coupled ABNs with discrete map regulatory functions.

In biological systems, the cell's signaling network plays an important role in instigating transitions between GRN expression states. It achieves this via a complex spatio-temporal arrangement of signaling proteins that integrate and transduce the diverse signals received from the cell's environment, eventually relaying its response to the GRN through the manipulation of one or more TFs. It is not hard to imagine

(a) Sideways roll. See http://youtu.be/hzMPB8EZGb8.

(b) Backward walk. See http://youtu.be/mT9qKZS7pds.

(c) Forward somersault. See http://youtu.be/urSNnVC1VqY.

(d) Backward somersault. See http://youtu.be/gSrQlsvbD98.

Figure 15.5 Diverse behaviors exhibited by coupled artificial biochemical networks evolved to control bipedal robots. All four controllers were found within the final population of a single multi-objective evolutionary algorithm run in which the objectives were to maximize distance covered and minimize energy used; see Refs. [53, 55] for more information about how ABNs have been used to control legged robots.

that this kind of flexible decision-making process could also be useful within real-world controllers that must respond to diverse incoming signals. Motivated by this, we considered how a particular facet of signaling network organization, crosstalk, could benefit AGRN models. Crosstalk has been identified as a means through which signaling networks achieve complex nonlinear responses. Our work with crosstalking ABNs suggests that it can carry out a similar role within a control systems context [26]. In Ref. [25], for example, we demonstrated how it can be used as the basis for

Figure 15.6 Expressive behaviors of a discrete map coupled artificial biochemical network controlling a four-legged robot with three degrees of freedom per leg. The objective was for the robot to move around as much as possible without moving away from its starting position, promoting movements that resemble dance; see: http://youtu.be/WzCmTUwtC3s.

a distributed adaptive locomotion controller for a legged robot, integrating signals from various internal and external sensors and generating a robust joint response.

Epigenetic processes also play a major role in cell fate determination, through their ability to guide the cell between different specialized states [49, 69]. Although these processes act through various biochemical mechanisms, for example, DNA methylation and histone modification, their effect is to change the accessibility of genes to the transcriptional machinery. This, in turn, causes modifications to the topology of the GRN, leading to changes in the expression state and hence the behavior of the cell. Since epigenetic modifications are the result of protein activity, this can be considered a form of self-modification, with the GRN being able to activate and deactivate different components of itself over the course of time. In Refs.[73, 74], we investigated this idea within an AGRN context by introducing epigenetic switches to an AGRN model, each being able to sense the state of the network and turn other nodes on or off as a consequence. The resulting dynamical patterns of topological self-modification appeared to work well in control problems where there was a need to switch rapidly and robustly between different behaviors. For instance, in Ref. [73], we noted how the use of epigenetic switches allowed AGRN controllers to solve larger instances of the coupled inverted pendulums problem.

15.7 FUTURE RESEARCH DIRECTIONS

15.7.1 Wider Control Applications

Previous work has demonstrated that AGRNs can be used to solve a broad range of control problems. Hence, it makes sense to speculate about the wider potential for applications in this area. Certainly, one of the advantages of using EAs to optimize AGRNs is that we do not require a theoretical understanding of the system that we are attempting to control. This is in contrast to the majority of conventional control methods in common use, which generally require a good understanding of a system before they can be applied. In this sense, the range of potential applications is very broad, and there is scope for applying these methods in areas where the conventional control community has made few inroads.

One such area is the control of executable models. Executable modeling approaches, and especially agent-based modeling [62], have become popular in fields as diverse as economics, the social sciences, biology, and medicine. The approach involves constructing a dependable model of some aspect of a naturally occurring system, usually as a collaboration between computational modelers and domain specialists. The model can then be executed, and the resulting dynamics studied in order to infer understanding about the system being modeled. In addition to observing natural dynamics, for many of these systems there is an interest in how their state can be modified. For instance, for models of disease pathways, this interest relates to the problem of designing drug interventions, moving the system from a diseased to a disease-free state. For economic models, there is an interest in how we can transition away from bad economic states, such as recession. It is plausible that there are many control strategies that can achieve these outcomes, and the use of evolvable, expressive controllers in concert with executable models could be an effective way of exploring these. It is likely too that there are different trade-offs in the space of control strategies (for instance, ease of carrying out an intervention versus effectiveness of the intervention). In this situation, multiobjective EAs could provide a means of obtaining a more complete picture of how to control a system.

15.7.2 Better Computational Understanding

Previous work also suggests that AGRNs can be applied to broader range of applications. However, there has been little work in ascertaining what kind of applications

AGRNs are most suited for. By analogy with biological systems, we can expect them to be good at control, signal processing, and pattern generation, and existing computational applications in these areas reflect this. However, the nature of biological computation is often unclear, and it is limiting to assume that apparently computational mechanisms in biological systems cannot perform a wider range of behaviors simply because we have not observed them doing these. Studies of Boolean networks indicate that these models do have a range of interesting computational properties, including computational universality. However, focused studies are required in order to develop a better understanding of their practical usefulness for different classes of problem. We also need a better understanding of how modeling choices affect computability. In almost all cases, previous applied studies have focused on a single AGRN model out of all the many possible models. Consequently, it is unclear how these results generalize to AGRNs as a whole. Improved understanding of biological GRNs may one day help with this, but at present, it is difficult to say how the many different components of biological regulation contribute to the overall behavior of GRNs.

Another difficulty in studying the computational properties of AGRNs is the key role played by the optimization algorithm. It may well be possible to express any computational behavior with a particular AGRN model, but this is academic unless we can locate the AGRN that expresses the behavior we want expressed. More expressive models are often larger (i.e., have more parameters). This means that the optimization search space is also larger, which, in turn, may mean that it is harder to find a particular model instance. For EAs, evolvability must also be taken into account. If an AGRN model is not encoded in a way that is sufficiently evolvable, then it may be infeasible to find a particular instance regardless of the expressiveness of the model. Hence, future research in this area must also take into consideration the role of evolvability, and how to achieve it, if the computational potential of AGRNs is to be fulfilled.

15.7.3 Biological Computing

AGRNs also have considerable potential for understanding biological systems. As noted above, we do not fully understand how different regulatory components contribute to the overall behavior of a GRN. However, we can study how analogs of these regulatory components contribute to the overall behavior of an AGRN, and consider what this tells us about biological GRNs. If we argue that computational problems represent a relatively wide range of dynamical environments, we might even say that this tells us more about the general characteristics of GRNs than studying a particular biological problem. We might also consider this cybernetics-like approach to be a kind of generalized systems biology, telling us things about biological mechanisms that are not apparent from their direct biological setting. This argument is perhaps even more relevant to signaling networks, whose complex temporal–spatial organization makes them very difficult to study. For example, it is currently not possible to infer the components of signaling pathways by sampling the biomolecules present within a cell, due to the vast number of possible interactions between these biomolecules. One way to guide this process would be to better understand the evolutionary pressures that constrain interactions between signaling molecules, and this could potentially be done by exploring the information processing capabilities of artificial signaling networks, identifying the topologies that are useful in a generalized, non-biological, context.

Another potentially strong application area for AGRNs is in synthetic biology [23]. Currently, this field is dominated by approaches that involve designing biological circuits based around computer engineering principles, for example, biochemical implementations of digital circuits [6]. Needless to say, biological systems are not based upon computer engineering principles. This suggests an alternative approach, which involves designing synthetic biology circuits based around the kind of computational models discussed in this chapter. There are, of course, challenges involved in this kind of work, but in many respects it would be more natural to use a controller derived from a GRN to control a biological process, rather than constructing a digital circuit to do this. It would also have the potential to be more expressive, opening up new application areas for synthetic biology, and leveraging the computational understanding of GRN models developed in AGRN studies.

15.8 CONCLUSIONS

Over the last decade, computer scientists and engineers have become increasingly interested in the role that GRNs play within the development and function of biological organisms. In particular, there has been a realization that computational models of GRNs can be used for computation, especially when the nature of this computation reflects the primary activities of biological GRNs. This has led to researchers exploring how these computational models, referred to here as AGRNs, can be used for applications that require intelligent control, pattern generation, and signal processing. The results of these investigations show considerable promise, demonstrating that AGRNs retain many of the desirable properties of their biological analogs, such as complex nonlinear behavior, robustness, compactness, and evolvability.

Nevertheless, the study of AGRNs is still in its early stages, and there is still much work to do in understanding their theoretical properties and computational potential. We have speculated about their future uses, but it is also important to gain a better understanding of what these computational models can do. By doing this, there is the possibility of informing understanding of biological systems, in addition to computer science and engineering. Biological science has made great progress in mapping and understanding GRNs, yet these systems remain only partially understood. AGRNs give us the opportunity to explore analogous behaviors within a more general setting, which, in turn, might lead to a better understanding of the general properties of GRNs. However, the reverse is also true, and it is likely that future progress in biology will continue to inform the design of AGRNs. It is also conceivable that one day the two areas of study might be unified within the context of synthetic biology.

REFERENCES

1. R. Albert. Boolean modeling of genetic regulatory networks. In: E. Ben-Naim, H. Frauenfelder, and Z. Toroczkai, editors, *Complex Networks*, volume 650 of *Lecture Notes in Physics*, pages 459–481. Springer, Berlin, 2004.
2. M. Aldana. Boolean dynamics of networks with scale-free topology. *Physica D: Nonlinear Phenomena*, 185(1):45–66, Oct. 2003.
3. J. Aracena, E. Goles, A. Moreira, and L. Salinas. On the robustness of update schedules in Boolean networks. *BioSystems*, 97(1):1–8, July 2009.
4. W. Banzhaf. Artificial regulatory networks and genetic programming. In: R. Riolo and B. Worzel, editors, *Genetic Programming Theory and Practice*, volume 6 of *Genetic Programming Series*, pages 43–61. Springer, New York, NY, 2003.
5. W. Banzhaf. On the dynamics of an artificial regulatory network. In: W. Banzhaf, J. Ziegler, T. Christaller, P. Dittrich, and J. Kim, editors, *Advances in Artificial Life*, volume 2801 of *Lecture Notes in Computer Science*, chapter 24, pages 217–227. Springer, Berlin, 2003.
6. Y. Benenson. Biomolecular computing systems: principles, progress and potential. *Nature Reviews Genetics*, 13(7):455–468, June 2012.
7. P. Bentley. Evolving fractal gene regulatory networks for graceful degradation of software. In: O. Babaoglu, M. Jelasity, A. Montresor, C. Fetzer, S. Leonardi, A. van Moorsel, and M. van Steen, editors, *Self-star Properties in Complex Information Systems*, volume 3460 of *Lecture Notes in Computer Science*, pages 21–35. Springer, Berlin, 2005.
8. P. J. Bentley. Controlling robots with fractal gene regulatory networks. In: L. Nunes de Castro and F. J. Von Zuben, editors, *Recent Developments in Biologically Inspired Computing*, pages 320–339. IGI Global, Hershey, PA, 2004.
9. C. P. Bowers. Formation of modules in a computational model of embryogeny. In: D. Corne, et al., editors, *Evolutionary Computation, 2005. The 2005 IEEE Congress on*, volume 1, pages 537–542. IEEE Press, Piscataway, NJ, 2005.
10. L. Bull and R. Preen. On dynamical genetic programming: random Boolean networks in learning classifier systems. In: L. Vanneschi, S. Gustafson, A. Moraglio, I. De Falco, and M. Ebner, editors, *Genetic Programming*, volume 5481 of *Lecture Notes in Computer Science*, pages 37–48. Springer, Berlin, 2009.
11. S. E. Castel and R. A. Martienssen. RNA interference in the nucleus: roles for small RNAs in transcription, epigenetics and beyond. *Nature Reviews Genetics*, 14(2):100–112, Jan. 2013.
12. A. Chavoya, I. R. Andalon-Garcia, C. Lopez-Martin, and M. E. Meda-Campaña. Use of evolved artificial regulatory networks to simulate 3D cell differentiation. *BioSystems*, 102(1):41–48, Oct. 2010.
13. S. Cussat-Blanc and J. Pollack. Using pictures to visualize the complexity of gene regulatory networks. In: C. Adami, D. M. Bryson, C. Ofria, and R. T. Pennock, editors, *Artificial Life 13*, pages 491–498. MIT Press, Cambridge MA, 2012.
14. S. Cussat-Blanc, S. Sanchez, and Y. Duthen. Controlling cooperative and conflicting continuous actions with a gene regulatory network. In: A. J. Fernández Leiva, et al., editors, *Computational Intelligence and Games (CIG), The 2012 IEEE Conference on*, pages 187–194. IEEE Press, Piscataway, NJ, 2012.
15. F. Dellaert and R. D. Beer. A developmental model for the evolution of complete autonomous agents. In: P. Maes, M. J. Mataric, J.-A. Meyer, J. Pollack, and S. W. Wilson,

editors, *From Animals to Animats 4: Proceedings of the Fourth International Conference on Simulation of Adaptive Behavior (SAB 1996)*, pages 393–401, MIT Press, Cambridge, MA, 1996.

16. K. A. Dittmar, J. M. Goodenbour, and T. Pan. Tissue-specific differences in human transfer RNA expression. *PLoS Genetics*, 2(12):e221, Dec. 2006.
17. P. Dittrich, J. Ziegler, and W. Banzhaf. Artificial chemistries—a review. *Artificial Life*, 7(3):225–275, July 2001.
18. B. Drossel. Number of attractors in random Boolean networks. *Physical Review E*, 72(1):016110, July 2005.
19. E. Dubrova, M. Teslenko, and H. Tenhunen. A computational scheme based on random Boolean networks. In: C. Priami, F. Dressler, O. Akan, and A. Ngom, editors, *Transactions on Computational Systems Biology X*, volume 5410 of *Lecture Notes in Computer Science*, pages 41–58. Springer, Berlin, 2008.
20. P. Dwight Kuo, W. Banzhaf, and A. Leier. Network topology and the evolution of dynamics in an artificial genetic regulatory network model created by whole genome duplication and divergence. *BioSystems*, 85(3):177–200, Sept. 2006.
21. P. Eggenberger Hotz. Evolving morphologies of simulated 3D organisms based on differential gene expression. In: P. Husbands and I. Harvey, editors, *Proceedings of the Fourth European Conference on Artificial Life (ECAL 1997)*, pages 205–213. MIT Press, Cambridge, MA, 1997.
22. R. Franke, F. J. Theis, and S. Klamt. From binary to multivalued to continuous models: the lac operon as a case study. *Journal of Integrative Bioinformatics*, 7(1), doi: 10.2390/biecoll-jib-2010-151, 2010.
23. P. S. Freemont and R. I. Kitney. *Synthetic Biology—A Primer*. Imperial College Press, London, UK, 2012.
24. N. J. Fuda, M. B. Ardehali, and J. T. Lis. Defining mechanisms that regulate RNA polymerase II transcription *in vivo*. *Nature*, 461(7261):186–192, Sept. 2009.
25. L. A. Fuente, M. A. Lones, A. P. Turner, L. S. Caves, S. Stepney, and A. M. Tyrrell. Adaptive robotic gait control using coupled artificial signalling networks, Hopf oscillators and inverse kinematics. In: C. A. Coello Coello, et al., editors, *Evolutionary Computation (CEC), 2013 IEEE Congress on*, pages 1435–1442. IEEE Press, Piscataway, NJ, 2013.
26. L. A. Fuente, M. A. Lones, A. P. Turner, S. Stepney, L. S. Caves, and A. M. Tyrrell. Computational models of signalling networks for non-linear control. *BioSystems*, 112(2):122–130, May 2013.
27. C. Gershenson. Introduction to random Boolean networks. In: M. Bedau, P. Husbands, T. Hutton, S. Kumar, and H. Suzuki, editors, *Workshop and Tutorial Proceedings, Ninth International Conference on the Simulation and Synthesis of Living Systems (ALife IX)*, pages 160–173. MIT Press, Cambridge, MA, 2004.
28. T. Glisovic, J. L. Bachorik, J. Yong, and G. Dreyfuss. RNA-binding proteins and post-transcriptional gene regulation. *FEBS Letters*, 582(14):1977–1986, June 2008.
29. T. G. W. Gordon. Exploring models of development for evolutionary circuit design. In: *Evolutionary Computation, 2003. CEC'03. The 2003 Congress on*, volume 3, pages 2050–2057. IEEE Press, Piscataway, NJ, 2003.
30. T. G. W. Gordon and P. J. Bentley. Development brings scalability to hardware evolution. In: J. Lohn, et al., *Evolvable Hardware, 2005. Proceedings. 2005 NASA/DoD Conference on*, pages 272–279. IEEE Press, Piscataway, NJ, 2005.

31. A. Goudarzi, C. Teuscher, N. Gulbahce, and T. Rohlf. Emergent criticality through adaptive information processing in Boolean networks. *Physical Review Letters*, 108(12):128702, Mar. 2012.
32. F. Greil. Dynamics of Boolean networks. PhD thesis, Technical University of Darmstadt, 2009.
33. F. U. Hartl, A. Bracher, and M. Hayer-Hartl. Molecular chaperones in protein folding and proteostasis. *Nature*, 475(7356):324–332, July 2011.
34. S. Huang, I. Ernberg, and S. Kauffman. Cancer attractors: a systems view of tumors from a gene network dynamics and developmental perspective. *Seminars in Cell and Developmental Biology*, 20(7):869–876, Sept. 2009.
35. R. Jaenisch and A. Bird. Epigenetic regulation of gene expression: how the genome integrates intrinsic and environmental signals. *Nature Genetics*, 33:245–254, Mar. 2003.
36. Y. Jin, H. Guo, and Y. Meng. A hierarchical gene regulatory network for adaptive multirobot pattern formation. *IEEE Transactions on Systems, Man, and Cybernetics. Part B: Cybernetics*, 42(3):805–816, June 2012.
37. M. Joachimczak, T. Kowaliw, R. Doursat, and B. Wróbel. Evolving morphologies and controllers for soft-bodied multicellular animats using gene regulatory networks and artificial embryogenesis. In: S. Doncieux, Y. Jin, and J.-B. Mouret, editors, *Evo-Devo-Robo: Evolutionary Robotics and Developmental Robotics at the Fourteenth International Conference on Genetic and Evolutionary Computation*, GECCO Companion'12, pages 357–360. ACM, New York, NY, 2012.
38. M. Joachimczak and B. Wróbel. Evolving gene regulatory networks for real time control of foraging behaviours. In: H. Fellermann, M. Dörr, M. M. Hanczyc, L. L. Laursen, S. Maurer, D. Merkle, P.-A. Monnard, K. Stoy, and S. Rasmussen, editors, *Artificial Life XII: Proceedings of the 12th International Conference on the Simulation and Synthesis of Living Systems*, pages 348–355. MIT Press, Cambridge, MA, 2010.
39. S. A. Kauffman. Metabolic stability and epigenesis in randomly constructed genetic nets. *Journal of Theoretical Biology*, 22(3):437–467, Mar. 1969.
40. S. A. Kauffman. *The Origins of Order: Self-Organization and Selection in Evolution*. Oxford University Press, Oxford, 1st ed, 1993.
41. G. A. Khoury, R. C. Baliban, and C. A. Floudas. Proteome-wide post-translational modification statistics: frequency analysis and curation of the Swiss-Prot database. *Scientific Reports*, 1:90, doi: 10.1038/srep00090, Sept. 2011.
42. M. Kirschner and J. Gerhart. Evolvability. *Proceedings of the National Academy of Sciences of the United States of America*, 95(15):8420–8427, July 1998.
43. A. Koopman and D. Roggen. Evolving genetic regulatory networks for hardware fault tolerance. In: X. Yao, E. Burke, J. Lozano, J. Smith, J. Merelo-Guervós, J. Bullinaria, J. Rowe, P. Tiňo, A. Kabán, and H.-P. Schwefel, editors, *Parallel Problem Solving from Nature—PPSN VIII*, volume 3242 of *Lecture Notes in Computer Science*, pages 561–570. Springer, Berlin, 2004.
44. S. Koos, J. B. Mouret, and S. Doncieux. The transferability approach: crossing the reality gap in evolutionary robotics. *IEEE Transactions on Evolutionary Computation*, 17(1):122–145, Feb. 2013.
45. J. Krohn and D. Gorse. Fractal gene regulatory networks for control of nonlinear systems. In: R. Schaefer, C. Cotta, J. Kołodziej, and G. Rudolph, editors, *Parallel Problem Solving from Nature, PPSN XI*, volume 6239 of *Lecture Notes in Computer Science*, pages 209–218. Springer, Berlin, 2010.

46. S. Kumar. A developmental biology inspired approach to robot control. In: J. Pollack, M. A. Bedau, P. Husbands, R. A. Watson, and T. Ikegami, editors, *Artificial Life IX: Proceedings of the Ninth International Conference on the Simulation and Synthesis of Living Systems*. MIT Press, Cambridge, MA, 2004.
47. H. Kwak and J. T. Lis. Control of transcriptional elongation. *Annual Review of Genetics*, 47(1):483–508, Nov. 2013.
48. W.-P. Lee and T.-H. Yang. Combining GRN modeling and demonstration-based programming for robot control. In: D. Palmer-Brown, C. Draganova, E. Pimenidis, and H. Mouratidis, editors, *Engineering Applications of Neural Networks*, volume 43 of *Communications in Computer and Information Science*, pages 190–199. Springer, Berlin, 2009.
49. R. Lister, M. Pelizzola, Y. S. Kida, R. D. Hawkins, J. R. Nery, G. Hon, J. Antosiewicz-Bourget, R. O'Malley, R. Castanon, S. Klugman, M. Downes, R. Yu, R. Stewart, B. Ren, J. A. Thomson, R. M. Evans, and J. R. Ecker. Hotspots of aberrant epigenomic reprogramming in human induced pluripotent stem cells. *Nature*, 471(7336):68–73, Mar. 2011.
50. Y. Liu, J. Timmis, O. Qadir, G. Tempesti, and A. Tyrrell. A developmental and immune-inspired dynamic task allocation algorithm for microprocessor array systems. In: E. Hart, C. McEwan, J. Timmis, and A. Hone, editors, *Artificial Immune Systems*, volume 6209 of *Lecture Notes in Computer Science*, pages 199–212. Springer, Berlin, 2010.
51. J. T. Lizier, S. Pritam, and M. Prokopenko. Information dynamics in small-world Boolean networks. *Artificial Life*, 17(4):293–314, July 2011.
52. M. Lones, A. Tyrrell, S. Stepney, and L. Caves. Controlling complex dynamics with artificial biochemical networks. In: A. Esparcia-Alcázar, A. Ekárt, S. Silva, S. Dignum, and A. Uyar, editors, *Genetic Programming*, volume 6021 of *Lecture Notes in Computer Science*, pages 159–170. Springer, Berlin, 2010.
53. M. A. Lones, L. A. Fuente, A. P. Turner, L. S. D. Caves, S. Stepney, S. L. Smith, and A. M. Tyrrell. Artificial biochemical networks: evolving dynamical systems to control dynamical systems. *IEEE Transactions on Evolutionary Computation*, 18(2):145–166, Apr. 2014.
54. M. A. Lones, S. L. Smith, A. M. Tyrrell, J. E. Alty, and D. R. S. Jamieson. Characterising neurological time series data using biologically motivated networks of coupled discrete maps. *BioSystems*, 112(2):94–101, May 2013.
55. M. A. Lones, A. P. Turner, L. A. Fuente, S. Stepney, L. S. D. Caves, and A. M. Tyrrell. Biochemical connectionism. *Natural Computing*, 12(4):453–472, Dec. 2013.
56. M. A. Lones, A. M. Tyrrell, S. Stepney, and L. S. D. Caves. Controlling legged robots with coupled artificial biochemical networks. In: T. Lenaerts, M. Giacobini, H. Bersini, P. Bourgine, M. Dorigo, and R. Doursat, editors, *Advances in Artificial Life, European Conference on Artificial Life (ECAL) 2011*. MIT Press, Cambridge, MA, 2011.
57. C. Mattiussi and D. Floreano. Analog genetic encoding for the evolution of circuits and networks. *IEEE Transactions on Evolutionary Computation*, 11(5):596–607, Oct. 2007.
58. B. Mesot and C. Teuscher. Deducing local rules for solving global tasks with random Boolean networks. *Physica D: Nonlinear Phenomena*, 211(1–2):88–106, Nov. 2005.
59. M. Nicolau, M. Schoenauer, and W. Banzhaf. Evolving genes to balance a pole. In: A. Esparcia-Alcázar, A. Ekárt, S. Silva, S. Dignum, and Ş. Uyar, editors, *Genetic Programming*, volume 6021 of *Lecture Notes in Computer Science*, Chapter 17, pages 196–207. Springer, Berlin, 2010.

60. S. Patarnello and P. Carnevali. Learning capabilities of Boolean networks. In: I. Aleksander, editor, *Neural Computing Architectures: the Design of Brain-like Machines*, pages 117–129. MIT Press, Cambridge, MA, 1989.

61. T. Quick, C. Nehaniv, K. Dautenhahn, and G. Roberts. Evolving embodied genetic regulatory network-driven control systems. In: W. Banzhaf, J. Ziegler, T. Christaller, P. Dittrich, and J. Kim, editors, *Advances in Artificial Life*, volume 2801 of *Lecture Notes in Computer Science*, pages 266–277. Springer, Berlin, 2003.

62. S. F. Railsback and V. Grimm. *Agent-Based and Individual-Based Modeling: A Practical Introduction*. Princeton University Press, Princeton, NJ, 2011.

63. T. Reil. Dynamics of gene expression in an artificial genome—implications for biological and artificial ontogeny. In: D. Floreano, D. Nicoud, and F. Mondada, editors, *Advances in Artificial Life, 5th European Conference, ECAL'99*, volume 1674 of *Lecture Notes in Artificial Intelligence*, pages 457–466. Springer, Berlin, 1999.

64. A. Roli, M. Manfroni, C. Pinciroli, and M. Birattari. On the design of Boolean network robots. In: C. Di Chio, S. Cagnoni, C. Cotta, M. Ebner, A. Ekárt, A. Esparcia-Alcázar, J. Merelo, F. Neri, M. Preuss, H. Richter, J. Togelius, and G. Yannakakis, editors, *Applications of Evolutionary Computation*, volume 6624 of *Lecture Notes in Computer Science*, pages 43–52. Springer, Berlin, 2011.

65. S. Sanchez and S. Cussat-Blanc. Gene regulated car driving: using a gene regulatory network to drive a virtual car. *Genetic Programming and Evolvable Machines*, 15(4):477–511, Dec. 2014.

66. M. A. Savageau. Introduction to S-systems and the underlying power-law formalism. *Mathematical and Computer Modelling*, 11:546–551, Jan. 1988.

67. D. J. Taatjes. The human mediator complex: a versatile, genome-wide regulator of transcription. *Trends in Biochemical Sciences*, 35(6):315–322, June 2010.

68. T. Taylor. A genetic regulatory network-inspired real-time controller for a group of underwater robots. In: F. Groen, N. Amato, A. Bonarini, E. Yoshida, and B. Kröse, editors, *Proceedings of the Eighth Conference on Intelligent Autonomous Systems (IAS-8)*, pages 403–412. IOS Press, Amsterdam, 2004.

69. V. B. Teif, Y. Vainshtein, M. Caudron-Herger, J.-P. Mallm, C. Marth, T. Höfer, and K. Rippe. Genome-wide nucleosome positioning during embryonic stem cell development. *Nature Structural and Molecular Biology*, 19(11):1185–1192, Nov. 2012.

70. C. Torres-Sosa, S. Huang, and M. Aldana. Criticality is an emergent property of genetic networks that exhibit evolvability. *PLoS Computational Biology*, 8(9):e1002669, Sept. 2012.

71. M. Trefzer, T. Kuyucu, J. Miller, and A. Tyrrell. Evolution and analysis of a robot controller based on a gene regulatory network. In: G. Tempesti, A. Tyrrell, and J. Miller, editors, *Evolvable Systems: From Biology to Hardware*, volume 6274 of *Lecture Notes in Computer Science*, pages 61–72. Springer, Berlin, 2010.

72. M. A. Trefzer, T. Kuyucu, J. F. Miller, and A. M. Tyrrell. Image compression of natural images using artificial gene regulatory networks. In: M. Pelikan and J. Branke, editors, *Proceedings of the 12th Annual Conference on Genetic and Evolutionary Computation*, GECCO'10, pages 595–602. ACM, New York, NY, 2010.

73. A. P. Turner, M. A. Lones, L. A. Fuente, S. Stepney, L. S. D. Caves, and A. Tyrrell. The artificial epigenetic network. In: A. M. Tyrrell, et al., editors, *Evolvable Systems (ICES), 2013 IEEE International Conference on*, pages 66–72. IEEE Press, Piscataway, NJ, 2013.

74. A. P. Turner, M. A. Lones, L. A. Fuente, S. Stepney, L. S. D. Caves, and A. M. Tyrrell. The incorporation of epigenetics in artificial gene regulatory networks. *BioSystems*, 112(2):56–62, May 2013.

75. J. M. Vaquerizas, S. K. Kummerfeld, S. A. Teichmann, and N. M. Luscombe. A census of human transcription factors: function, expression and evolution. *Nature Reviews Genetics*, 10(4):252–263, Apr. 2009.

76. S. Wolfram. *A New Kind of Science*. Wolfram Media Inc., Champaign, IL, 2002.

77. Y. Yao, K. Marchal, and Y. Van de Peer. Improving the adaptability of simulated evolutionary swarm robots in dynamically changing environments. *PLoS One*, 9(3):e90695, Mar. 2014.

78. P. Zahadat, D. Christensen, U. Schultz, S. Katebi, and K. Stoy. Fractal gene regulatory networks for robust locomotion control of modular robots. In: S. Doncieux, B. Girard, A. Guillot, J. Hallam, J.-A. Meyer, and J.-B. Mouret, editors, *From Animals to Animats 11*, volume 6226 of *Lecture Notes in Computer Science*, pages 544–554. Springer, Berlin, 2010.

79. M. Zanin and A. N. Pisarchik. Boolean networks for cryptography and secure communication. *Nonlinear Science Letters B*, 1(1):25–32, 2011.

80. J. Zañudo, M. Aldana, and G. Martínez-Mekler. Boolean threshold networks: virtues and limitations for biological modeling. In: S. Niiranen and A. Ribeiro, editors, *Information Processing and Biological Systems*, volume 11 of *Intelligent Systems Reference Library*, pages 113–151. Springer, Berlin, 2011.

81. S. Zhan, J. F. Miller, and A. M. Tyrrell. An evolutionary system using development and artificial genetic regulatory networks for electronic circuit design. *BioSystems*, 98(3):176–192, Dec. 2009.

82. N. M. Zivanov, A. Bresticker, D. Krishnaswamy, S. Venkatakrishnan, D. Marculescu, and J. R. Faeder. Emulation of biological networks in reconfigurable hardware. In: R. Grossman, A. Rzhetsky, S. Kim, and W. Wang, editors, *Proceedings of the 2nd ACM Conference on Bioinformatics, Computational Biology and Biomedicine*, BCB'11, pages 536–540. ACM, New York, NY, 2011.

INDEX

Evolutionary Computation in Gene Regulatory Network Research, First Edition.
Edited by Hitoshi Iba and Nasimul Noman.

Wiley Series on

Bioinformatics: Computational Techniques and Engineering

Bioinformatics and computational biology involve the comprehensive application of mathematics, statistics, science, and computer science to the understanding of living systems. Research and development in these areas require cooperation among specialists from the fields of biology, computer science, mathematics, statistics, physics, and related sciences. The objective of this book series is to provide timely treatments of the different aspects of bioinformatics spanning theory, new and established techniques, technologies and tools, and application domains. This series emphasizes algorithmic, mathematical, statistical, and computational methods that are central in bioinformatics and computational biology.

Series Editors: **Professor Yi Pan** and **Professor Albert Y. Zomaya**
pan@cs.gsu.edu albert.zomaya@sydney.edu.au

Knowledge Discovery in Bioinformatics: Techniques, Methods, and Applications
Xiaohua Hu and Yi Pan

Grid Computing for Bioinformatics and Computational Biology
Edited by El-Ghazali Talbi and Albert Y. Zomaya

Bioinformatics Algorithms: Techniques and Applications
Ion Mandiou and Alexander Zelikovsky

Machine Learning in Bioniformatics
Yanqing Zhang and Jagath C. Rajapakse

Biomolecular Networks: Methods and Applications in Systems Biology
Luonan Chen, Rui-Sheng Wang, and Xiang-Sun Zhang

Computational Systems Biology
Huma Lodhi

Analysis of Biological Networks
Edited by Björn H. Junker and Falk Schreiber

Computational Intelligence and Pattern Analysis in Biological Informatics
Edited by Ujjwal Maulik, Sanghamitra Bandyopadhyay, and Jason T. L. Wang

Mathematics of Bioinformatics: Theory, Practice, and Applications
Matthew He and Sergey Petoukhov

Introduction to Protein Structure Prediction: Methods and Algorithms
Huzefa Rangwala and George Karypis

Algorithms in Computational Molecular Biology: Techniques, Approaches and Applications
Edited by Mourad Elloumi and Albert Y. Zomaya

Mathematical and Computational Methods in Biomechanics of Human Skeletal Systems: An Introduction
Jiří Nedoma, Jiří Stehlík, Ivan Hlaváček, Josef Daněk, Taťjana Dostálová, and Petra Přečková

Rough-Fuzzy Pattern Recognition: Applications in Bioinformatics and Medical Imaging
Pradipta Maji and Sankar K. Pal

Data Management of Protein Interaction Networks
Mario Cannataro and Pietro Hiram Guzzi

Algorithmic and Artificial Intelligence Methods for Protein Bioinformatics
Yi Pan, Jianxin Wang, and Min Li

Classification Analysis of DNA Microarrays
Leif E. Petersen

Biological Knowledge Discovery Handbook: Processing, Mining, and Postprocessing of Biological Data
Edited by Mourad Elloumi and Albert Y. Zomaya

Evolutionary Computation in Gene Regulatory Network Research
Edited by Hitoshi Iba and Nasimul Noman